Advances in Science, Technology & Innovation

IEREK Interdisciplinary Series for Sustainable Development

Advances in Science, Technology & Innovation (ASTI) is a series of peer-reviewed books based on important emerging research that redefines the current disciplinary boundaries in science, technology and innovation (STI) in order to develop integrated concepts for sustainable development. It not only discusses the progress made towards securing more resources, allocating smarter solutions, and rebalancing the relationship between nature and people, but also provides in-depth insights from comprehensive research that addresses the **17 sustainable development goals (SDGs)** as set out by the UN for 2030.

The series draws on the best research papers from various IEREK and other international conferences to promote the creation and development of viable solutions for a **sustainable future and a positive societal** transformation with the help of integrated and innovative science-based approaches. Including interdisciplinary contributions, it presents innovative approaches and highlights how they can best support both economic and sustainable development, through better use of data, more effective institutions, and global, local and individual action, for the welfare of all societies.

The series particularly features conceptual and empirical contributions from various interrelated fields of science, technology and innovation, with an emphasis on digital transformation, that focus on providing practical solutions to **ensure food, water and energy security to achieve the SDGs**. It also presents new case studies offering concrete examples of how to resolve sustainable urbanization and environmental issues in different regions of the world.

The series is intended for professionals in research and teaching, consultancies and industry, and government and international organizations. Published in collaboration with IEREK, the Springer ASTI series will acquaint readers with essential new studies in STI for sustainable development.

ASTI series has now been accepted for Scopus (September 2020). All content published in this series will start appearing on the Scopus site in early 2021.

More information about this series at http://www.springer.com/series/15883

Chaham Alalouch · Cristina Piselli ·
Francesco Cappa
Editors

Towards Implementation of Sustainability Concepts in Developing Countries

Springer

Editors
Chaham Alalouch
Department of Civil and Architectural Engineering
Sultan Qaboos University
Muscat, Oman

Cristina Piselli
Department of Engineering
University of Perugia
Perugia, Italy

Francesco Cappa
Luiss Guido Carli University
Rome, Italy

ISSN 2522-8714 ISSN 2522-8722 (electronic)
Advances in Science, Technology & Innovation
IEREK Interdisciplinary Series for Sustainable Development
ISBN 978-3-030-74348-2 ISBN 978-3-030-74349-9 (eBook)
https://doi.org/10.1007/978-3-030-74349-9

Cover photo credit: Chaham Alalouch—the photo shows Oman Ecohouse at Sultan Qaboos University, Muscat, Oman.

This Springer imprint is published by the registered company Springer Nature Switzerland AG
The registered company address is: Gewerbestrasse 11, 6330 Cham, Switzerland

Scientific Committee

Abeer Elshater, Ain Shams University, Cairo, Egypt
Ahmed Hosney Radwan, The American University in Cairo, Egypt
Francesco Cappa, LUISS—Free International University of Social Studies "LUISS Guido Carli", Rome, Italy
Chaham Alalouch, Sultan Qaboos University, Muscat, Oman
Claudia Fabiani, University of Perugia (Università degli Studi di Perugia), Italy
Cristina Piselli, University of Perugia (Università degli Studi di Perugia), Italy
Edoardo Ferrucci, LUISS—Free International University of Social Studies "LUISS Guido Carli", Rome, Italy
Fadzidah Binti Abdullah, International Islamic University Malaysia (IIUM), Selangor, Malaysia
Federica Liberti, Biomass Research Centre—University of Perugia, Perugia, Italy
Federica Rosso, Sapienza University of Rome, Italy
Ilaria Pigliautile, Biomass Research Centre—University of Perugia, Perugia, Italy
Iman Abdel Gawad, Helwan University, Cairo, Egypt
Jens Knispel, Institute of Psychology (RWTH Aachen University), Aachen, Germany
Mahmoud Ghoneem, King Saud University, Riyadh, Saudi Arabia
Marco Seccaroni, University of Perugia (Università degli Studi di Perugia), Italy
Marta Chafer, University of Lleida (Universitat de Lleida—Rectorat), Spain
Mattia Manni, Biomass Research Centre—University of Perugia, Perugia, Italy
Mohammad Saffari, University College Dublin, Ireland
Nabil Mohareb, Beirut Arab University, Lebanon
Pietro Bartocci, Biomass Research Centre—University of Perugia, Perugia, Italy
Riccardo Maiolini, John Cabot University, Rome, Italy
Stefano Franco, LUISS—Free International University of Social Studies "LUISS Guido Carli", Rome, Italy
Viviana D'Angelo, LUISS—Free International University of Social Studies "LUISS Guido Carli", Rome, Italy

Series Editor's Preface

Sustainability is a major concern nowadays—an attempt to reduce carbon emissions, protect environments and keep the delicate ecosystem of our planet in balance. To improve sustainability, the consumption of non-renewable resources has to be decreased to help nature replenish and suffice itself. Moreover, conserving the energy and taking the environment's well-being into consideration are considered to be key roles in improving sustainability.

This book, on Sustainability and its Impact on Developing Countries, contains selected research from the International Conference on Improving Sustainability Concept in Developing Countries (ISCDC)—4th Edition, which was held in Cairo, Egypt. Research from the conference has undergone a rigorous review process thanks to the book editors and scientific committee of reviewers. The book contains several topics regarding sustainability concepts in architecture and urban design, and sustainable design strategies for economics, and environmental and social dimensions. Making this book a great opportunity to help evolve our built environment and also explore new and different approaches with the hope to improve sustainability in different developing communities and countries.

Through this book, readers will be exposed to a number of case studies around the world: from Middle Eastern countries to India, Bangladesh, China and more. It stimulates discussions on planning and design strategies in the development of a sustainable city. It aims to educate readers about the important issues and solutions related to sustainability such as improving the quality of life and built environment in developing countries, poor and rich cities alike.

The book will benefit an array of readers from different backgrounds due to its interdisciplinary nature. It succeeds in bringing together fields such as architecture and urban planning, landscape architecture, environmental preservation and other environmental processes. It is highly recommended to academics, postgraduate students and interested readers.

MAmer

Mourad Amer, Ph.D.
Series Editor, Springer

CEO and Founder, IEREK Cairo Egypt

Preface

Over the years, more countries have started realizing the threat facing our planet—from vanishing species, to catastrophic effects of climate change and depletion of our resources all of which have led to increasing investments in a more sustainable future. According to the Human Development Report by the UN Development programme (2016), more sustainability programmes are being initiated around the Globe. However, as our countries grow more industrialized, the risk of further contribution to an increase in emission levels and environmental degradation remains imminent. Despite recent advancement, there is still more progress to be made and additional room for investment in sustainable practices and strategies, especially in developing countries.

The building and construction industries are among the largest consumers of energy, materials and water, thus there is a great opportunity to achieve sustainability if proper strategies and actions are put in place to make building and construction activities more sustainable. Doing so will give people the chance to live in a healthier environment with improved environmental, social and economic conditions and better quality of life. This book represents an attempt to contribute to the improvement of the sustainability scene of the built environment in developing countries by presenting recent research and case studies from several countries covering the Middle East, Asia, and Africa, and discussing pressing issues related to environmental, social and economic sustainability. The book features 25 chapters structured in three parts: urban sustainability; sustainable building design; and sustainable development management, impact, and education. The first part of the book focuses on strategies for urban sustainability. It features nine chapters representing research and case studies from several developing countries such as Iraq, Egypt, South Africa, Rwanda, Bangladesh, and India. Through these chapters, the reader will learn about the challenges of modern cities and the corresponding state-of-the-art sustainability strategies that could mitigate the negative impact of our cities and improve sustainability practices at the urban level. Issues discussed include urban forms, outdoor thermal environments, urban building blocks, livability, bike routes optimization, TOD, green spaces design, and more.

The second part of the book provides insights into sustainable design strategies at the building level. It includes ten chapters addressing a range of topics that are important to researchers, professionals, and policymakers who have an interest in design strategies for sustainable buildings. The chapters discuss design strategies for several building types such as office buildings, housing, rehabilitation centers, elderly care facilities, hospitals, and more. These design strategies were discussed by the authors from several angles including, but not limited to, materials, building envelop design, pro-environmental behavior, adaptive passive interior design, ventilation, prefabrication and industrialization, and visual comfort.

The last part of the book extends to cover issues related to sustainable development management, impact, and education. It includes six chapters ranging from green project management for sustainable development, global neighborhood sustainability certification and its relevance to the developing countries, Global and regional perspectives on accessibility to cooking fuels, solid waste disposal, CO_2 emissions, and higher education for sustainability.

Through presenting different case studies, this book explores new topics of sustainable development and design in developing countries. The significance of this book lies in its attempts at implementing policies that coincide with sustainable development in areas that are in dire need of urbanization. The authors of the different chapters in this book reunite once more towards one common goal: contributing to new action plans and innovations that tackle the challenges facing our planet while remaining in line with the UN's 2030 Sustainable Development Goals (SDGs).

Muscat, Oman — Chaham Alalouch
Perugia, Italy — Cristina Piselli
Rome, Italy — Francesco Cappa

Acknowledgements

The book Editors would like to thank the authors of the research papers of this book. They would also like to thank the scientific committee of reviewers who contributed with their knowledge and constructive feedback in hopes of ensuring the manuscripts are of the best quality possible. Finally, the Editors would like to express their appreciation to the IEREK team for supporting the publication of this book.

IEREK would like to express its appreciation to all members of the staff and scientific committee for their tremendous efforts and contribution to the growth of this institution and for making our conference on Improving Sustainability Concept in Developing Countries a success. IEREK would like to thank the conference chairperson, Prof. Iman Gawad, Associate Professor of Architecture in the Architecture Department of the Faculty of Fine Arts, Helwan University, who had a hand in making this conference what it is today by providing scientific and logistical support. IEREK takes pride in being an institution that amasses a highly qualified and competent team who restlessly worked for months to make this conference what it is today in hopes of creating a well-rounded society. Last but not least, we cannot neglect the prominent role undertaken by our Editors who made it their duty to help this institution in spreading knowledge to the masses.

Contents

Sustainable Design Strategies at the Urban Level

Toward Resiliency Through Sustainable Urban Formation in Baghdad

Zaynab Radi Abaas

Abstract

This research comes in an attempt to develop the present urban patterns toward resiliency in the developing countries' environment in general and the Baghdad City in particular. This is performed through the looking at the most recent theories designed to extrapolate socio-ecological resilient and sustainable urban formation, viewing their roots and development, by founders and advocates. The aim is to extract the most significant strategies from these perspectives that can serve Baghdad City as a case study. The key finding of this research is the lack of the ideal theory that extrapolated the sustainable urban formation as a structure that can be applied as a prototype of the resilient theory. Each city demonstrates the properties that distinguish it from others, and day after day, applications and indicators are undertaken to serve some cities and not others depending on functionality, stability, and adaptive cycling process, the nature of the city, and the environmental characterization of this city. As a result, this research adopts some integrated strategies as solutions combining the recent theories and includes recent approaches searching for Baghdad urban resiliency, by extrapolating its phases of changing, its flexibility, and its adaptation of the urban fabric according to its master plans. Consequently, some cities might have able to cope with the severe spatiotemporal changes and reach sustainability, and others might fail to achieve urban resiliency.

Keywords

Urban formation • Sustainable theories • Resiliency • Baghdad

Z. R. Abaas (✉)
Department of Architecture, College of Engineering, University of Baghdad, Baghdad, Iraq
e-mail: dr.zaynabr.a@coeng.uobaghdad.edu.iq

1 Introduction

Since the seventies of the twentieth century, sustainability, and resiliency have been the major concern for architects, planners, and specialists. They both have wide controversy. Some of the researchers believe the two terms are identical, or one represents the other, while others see resiliency is adaptable and dynamic, but sustainability is achievable and static (Caputo, 2015). The United Nations in their sustainable development goals is stressed on attaining sustainable, resilient, and flexible cities equally, and used interchangeably (Yamagata & Ayyoob, 2018). Urban sustainability as associated with these terms has lots of applicable theories and seeks to build qualified strategies, indicators, and plans for better futures; however, most of them considered optimizing natural resources use is the solution of urban sustainability (Elmqvist et al., 2019). Resiliency, on the other hand, becomes widely used as an urban development policy, derived from the ability to absorb, reduce, recover, and respond to the disturbance of the socio-ecloogical system (Eraydin & Taşan-Kok, 2013; Yamagata & Ayyoob, 2018), toward the degree of facing, adapting, growing, and working with the challenges in the system or the city keeping its main social, ecological, natural, and technological functions and reorganizing itself (Newell et al., 2016). Additionally, Peres argued the resilient urban system has to have many levels and scales to keep going, starting with the specified level of the ecological and engineering resilience and reaching to the general level, which represents the ability of the system to work with the long-term chronic, slow to emerge stressors of a developing resilience (Peres, 2016). Moreover, resilience and urban form have some theoretical and empirical evidence toward achieving hierarchy and connectivity in multiple city scale (Sharifi, 2019). Still, resiliency insures flexibility and recommended as to be one of the guidelines to achieve solid sustainability, focusing on maintaining the natural resources over a long period of time (Pickett et al., 2019) and (Brand & Jax, 2007).

C. Alalouch et al. (eds.), *Towards Implementation of Sustainability Concepts in Developing Countries*, Advances in Science, Technology & Innovation, https://doi.org/10.1007/978-3-030-74349-9_1

On the other hand, several approaches correlate the structural development of urban form with sustainability and the access of resilience. Jabareen, 2006 identified four types of sustainable urban forms starting with compact and eco-forms and adding neo-traditional and containment development. He provided a matrix in seven sustainable strategies which support the specialists to examine and evaluate the desired form in reaching sustainability (Jabareen, 2006). Forgaci and Timmeren (2014) concentrated on the complexity and adaptability of three terms of resilience (general, spatial, and resilient of urban formation). They asserted on fitness, performance, and sustainability to reach better urban resilient. However, they claimed that the principles of sustainable formation were more important than the structure of formation as long as these forms could adapt some synergies to cope with the climate changes and surroundings (Forgaci & Timmeren, 2014). Sanchez et al. (2017) explored several terms of sustainable resilience concepts and thinking depending on the four principles of sustainability-related to the socio-ecological resilience presented by Sanchez et al. (2016) and detected some short- and long-term actions of urban resiliency (Sanchez et al., 2016, 2017). Farther more, Sharifi and Yamagata (2018) argued that enhancing urban resilience could be achieved through hierarchical urban-scale dynamic framework with multiple attributes (Sharifi & Yamagata, 2018).

As a result, many studies tried to find some solutions and strategies to achieve resiliency by applying several synergies to gain sustainable urban forms. This research is a try to find some solutions accompanied by the recent theories and add new approaches, to explain the transformation nature of Baghdad urban form and how to reach urban resiliency.

2 Methodology

The structure of this paper explains two perspectives, firstly the debate of resilience-sustainable theories and its foundation, and secondly the theories of sustainable urban forms related to these theories. The idea is to look through the urban formation in Baghdad, its recovery, and adaptation abilities in multiple transformation scale and levels. After that, the paper presents the possibility of urban resiliency in Baghdad and some extracted solutions for sustainable urban forms toward resiliency. See Fig. 1.

3 Materials and Methods

The research tries to introduce a conceptual model for urban resiliency in Baghdad following the extrapolating method. The idea consists of an analytical framework to explore the key characteristic of resilience theories gaining the goals of sustainability. It is in an attempt to prove that the spatiotemporal changes in Baghdad City have the ability to reach urban resiliency through its continuation sustainable urban formation. And the diagram below shows the research method in detail. See Fig. 2.

4 Literature Review

4.1 The Foundation of Sustainable—Resilience Theories and Debate

Many theories had been discussing the sustainable-resilience correlation and their impacts on urban sustainability. Lots of them consider the city as a vital dynamic system, blaming the human being and their activities footprint on devastating the natural ecosystem and the cycling process. Holling, Odums, Fath, and Girardets are among the best- and well-known theories. Holling's interpretation theories had captured several meanings since his seminal paper (1973) of socio-ecological systems and conceptualized resilience as the capacity of a system to retain to balance the following disturbance. He produced many viewpoints on the stability and management of resiliency (Holling, 1973). After a while, he developed the adaptive cycle resulting resiliency in multiple dimensions and phases within different scales (exploitation, conservation, release, and reorganization). Later, with his colleagues, he introduced the perspective to clarify the relationship between adaptation and transformation with resiliency. They defined resilience as the ability of a framework to ingest aggravation and rearrange itself while experiencing a change to even now hold basically a similar value and functionality, structure, personality, and response. The most important gaining is to move from the concept of the ultimate goal of gaining sustainability to the functionality of the transformation toward more resilience, adaptive systems, and management (Walker et al., 2004). While Odums Ecological Theory is considered the initial source of the urban sustainability theory that dealt with the city as a living system and tried to link the socioeconomic system with the environment and fit the technological engineering design with the self-ecological system (Odum & Odum, 2003). *The Ecological Theory* was introduced by the scientist H. T. Odum and his brother Eugene P. Odum and described in Odums hypotheses as the thermodynamic orientation or natural tendency that ecosystems follow during succession, supporting the field of ecological engineering and benefit from resiliency, adaptability, mutuality, and facilitation (Mitsch, 2018). Progressively, environmental and planning scientists including Fath et al. viewed ecosystem succession as a series of four growth and development stages: boundary, structural, network, and informational (Fath et al., 2004). Even more, Fath (2017) argued sustainability could be

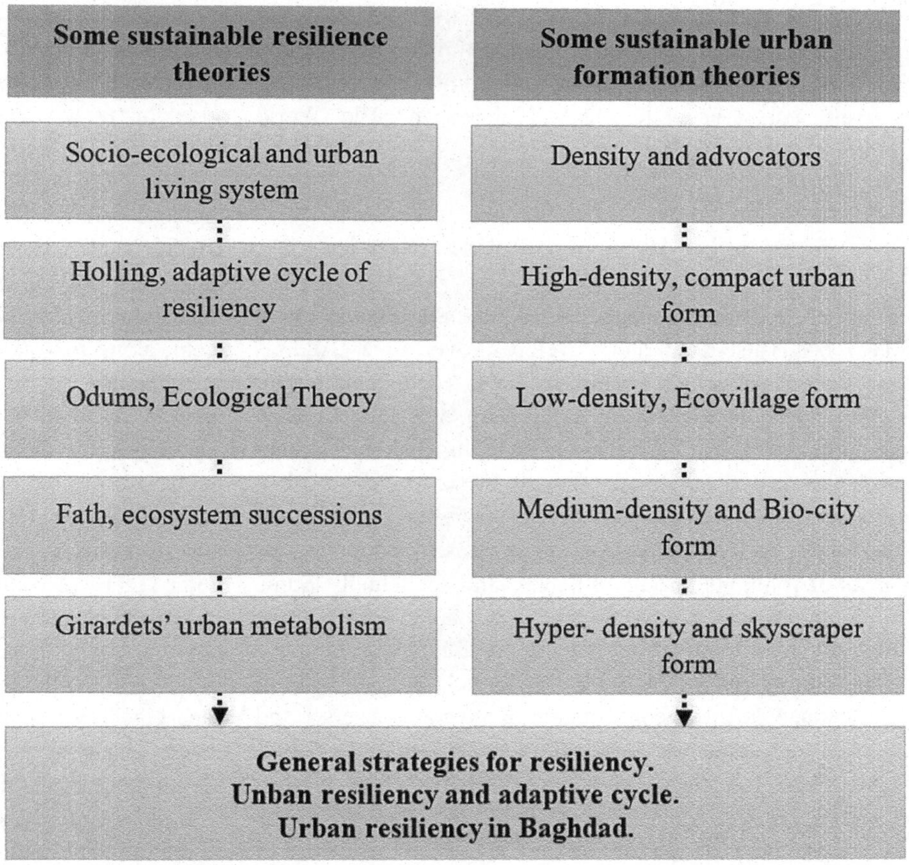

Fig. 1 Graphical framework shows the structure of the paper

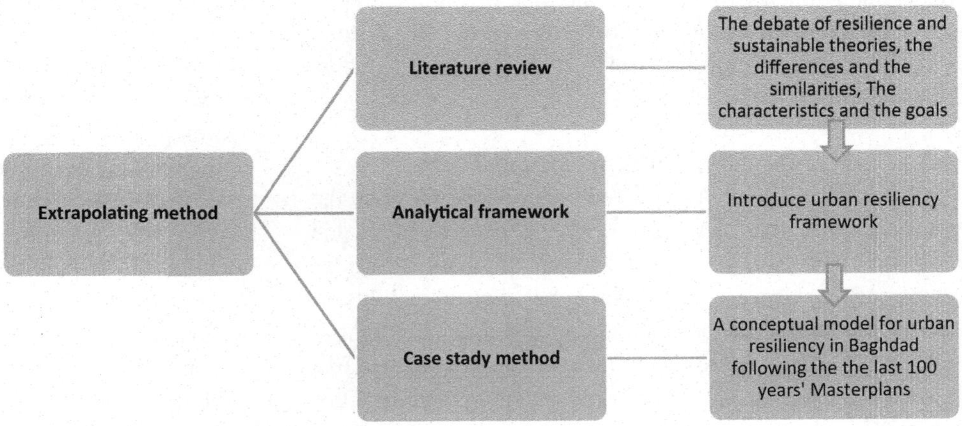

Fig. 2 Method of the research paper

shown by how the ecological system is a complex, adaptational, functional structure and can generate, tolerate, and preserve its dynamic activities (Fath, 2017). On the other hand, many researchers have thought that achieving a sustainable urban system must be linked to the surrounding environment provided by the formative nature of its constituents, as any ecosystem nurtures species structures. These theories started to identify the city as an ecosystem in the sixties of the last century. Its beginning with the ideas of Abel Wolman (1965) whom connected different types of resources called the consumer and the digester or input–output according to a holistic view to the city as a bio-system (Wolman, 1965). Recent methods to analyze urban ecosystem has been presented, biological, spatial, social, and energy flow of urban metabolism and footprint analyses (Piracha & Marcotullio, 2003). After some time, the urban ecosystem scientist Herbert Girardet indicated the possibility of dealing with those energetic inputs as flows entering and exiting the urban environment. This can be observed in the circuit metabolic in nature (*ecosystem metabolism*) (Girardet & Miguel, 2009). The idea of the circular, regenerative metabolism is to reduce the amount of waste through the process of recycling, to enhance the environmental relationship between the ecosystems and the urban structure, and as a result, to promote urban community establishing a self-efficient, regenerative city (Girardet & The World Future Council, 2013).

4.2 Sustainable Urban Formation Theories

From above, the challenges to resilient, sustainable living have been discussed and developed over time. And as mentioned previously, many resilient, ecological, and/or urban metabolism theories, along with others, have been dealing with urban formation as a living dynamic system with the natural inputs and outputs of the energetic urban processes like human activities or environmental forces. Therefore, each of them has taken as a base to adopt various kinds of solutions through different kinds of schools. Some of the urban planners and thinkers today are centered on the idea of urban metabolism and the possibility of taking the compact city and the high-density places. The main concept is to reduce the city's consumption of the goods, energy, and food at high rates, and the environmental pollution. A pivotal debate from the other planning schools' thinkers goes toward the ecological theory with the theme of the eco-village and low-density urban form. The third theory of bio-city can rectify the same energetic consuming and depleting problems by adopting the idea of the integration between the man and his environment as a holistic system. Finally, the hyper-density theory comes solving the urban degradation.

4.2.1 High-Density, Compact Urban Form

Most people studied, saw, lived, or walked through the narrow roads, the Shanshuls, and the courtyards of the compact Arabic city. After hundreds of years, this idea comes again raising a wide controversy about the ability of these cities to reduce or balance the ecological footprint at the urban scale (macro- and micro-scale). By the 1970s, researchers noted that factors advocating higher densities included "nearness to the central area, place of work, transport facilities or an open space" and that within an overall density standard there could represent variations within a development. In the next decade, similar arguments were reinforced, and some were included. Central to the guidance on increased density was the preservation of suitable agricultural land, the prevention of urban sprawl, and the protection of the countryside all of which point to the need for compact development (Jenks & Dempsey, 2005). Rogers (British architect famed for his modernist and functionalist designs) (Astbury, 2019) went toward the concept that the compact or "dense city" model can be adapted to a circular metabolism. That is "where consumption is reduced by implementing efficiencies, where re-use of resources' energy for transport by increasing the viability of public transportation and by reducing distances between facilities, so encouraging walking and cycling is maximized." Rogers (1997) also described the compact city by saying "Urban areas are organized in concentric bands of density, with higher densities around public transport nodes (rail, bus and underground stations)" (Rogers, 1997). According to (Burton, Jenks & Williams, 1996), the compact city could be related to the mixed-land uses and high-density development, along with a convenient public transportation infrastructure, which encourages people to walk and cycle rather than to drive their cars, go some way toward realizing aspirations for the sustainable city. The UK Government published, in 1999, *a strategy for sustainable development* in the UK, and *toward an urban renaissance strategy*. In *toward an urban renaissance,* the report of the Urban Task Force (UTF) very clearly sets out the thinking on the design of sustainable urban form. The current orthodoxy sees the city that approximates to a sustainable form, as a compact and flexible structure in which the parts are connected to each other and the whole, with a clearly articulated public space. The public realm connects the different quarters to each other across the city, while also linking homes to workplaces, schools, social institutions, and places of recreation. The main features of the compact high-density model, outlined in the report of the UTF, *toward an urban renaissance* include (Rogers, 1997):

- High-density mixed-uses development with mixed tenure housing at fifty units per hectare (Mitsch, 2018).
- Development arranged around centers with most homes being within 10–15 min' walking distance from a center.
- These semi-autonomous communities, self-supporting in daily needs, are connected to other centers and to the city main center by public transport routes.
- An important structuring element represents a well-connected public realm of streets, squares, parks, and other open spaces with green areas (Abaas, 2008).
- Additionally, urban managing, designing and regenerating, social well-being, and environmental and ecological balancing were among their targets (Regan, 2000).

Although the name and the properties of the compact cities are extremely close to the Arabic Islamic cities, neither the environmental factors nor the cultural habits allow (what) to build high-rise buildings with more than 3–4 story building. However, the studies demonstrate the ability to reach resiliency through various solutions as synergies of the compact cities. The relationship that the high-density, mixed-land uses of the compact city has on accessibility to local facilities is to provide local facilities that are located within walk distances from one's home reducing the necessity of owning and using private cars prompting the public participation and enhance livability (Jabareen, 2006). It, therefore, follows that the most efficient way in which to make this work is by ensuring that the densities of urban areas are high enough to support services and facilities locally adapted (including public transportation, green infrastructure, passive design, and greening the urban area), and large enough to persuade employers to the area to capitalize on the workforce in the area.

4.2.2 Low-Density, Eco-Village Form

In 1991, the American thinker Robert Gilman and his wife Diana introduced a definition of an eco-village that became widely known in sustainable communities. They defined an eco-village as settlements or cities, in which the community is connected to their natural environment and lives in healthy sustainable development. The aim is to regenerate integrated social and natural environments (Zeybek & Arslan, 2017). In this view, achieving sustainability is not enough; it is vital also to resilient the social and environmental fabric of life, and across all four dimensions of sustainability: social, environmental, economic, and cultural. In 1995, a global initiative called global eco-village network (GEN) supporting the similar ideas. Lastly, GEN concentrates on peaceful life, social justice, and ecological support (Maharani, 2019).

At one time, the term "eco-village" is specifically used to identify low-density settlements in which the traffic routes are widely spaced, and the villas are occupied by wide open spaces consisting of farmland, intensive market gardens. The frontage onto the main roads is a linear strip of buildings and other activities (Moughtin & Shirley, 2005; Maharani, 2019). As a result, eco-village considers an urban development designed to provide an ecological sound and community-oriented alternative to regular individual or impersonal residential housing. Eco-village groups were to respect and restore the native biodiversity of the site. It was decided that only 15% of any site would be built-on while the rest would be landscaped or farmed (Sassi, 2006). However, the advocated planners of the eco-villages cities hope to change the indicator that measures sustainability by the proportion of common land uses and have a new sustainability indicator measuring people's satisfaction with the urban environment. In addition to their intention to have required standards for the quantity and accessibility of urban open spaces, they have a strong ambition to have a target for the proportion of managed urban land in designated greenway strategies. Moreover, they look for more imaginative green space design, habitat creation, river and wetland restoration, and sustainable drainage (Moughtin & Shirley, 2005).

From his studies of the urban forms, Jenks (2005) claimed that the historical pattern results in a net density of 2.8–3.4 units per net acre (6.9–8.4 per hectare), including the alley and a proportional share of the adjacent city streets, but not including other commonly shared public amenities like parks and schools. It can be noted that the low-density standard of 12 or 8 houses per acre in rural areas (20 or 30 houses per hectare), became the norm for the inter-war period and influenced the spread of a uniform suburbia (Mike & Nicola, 2005).

The benefits of the eco-village idea had echoed the earlier pronouncements of Ebenezer Howard and the concept of the "Garden City" and the Garden Cities of Tomorrow, which contains a wealth of ideas for urban development. The eco-village houses followed counteracted health risks ideas by, ensuring healthy indoor air quality by minimizing the use of toxic materials, materials containing organic compounds; ensuring all internal spaces have abundant natural light, and encouraging residents to spend time outside and to enjoy nature (Sassi, 2006). Another benefit from designing and contacting houses with nature is visually and physically reconnects with the outdoors. All windows gain views of the surrounding natural environment and shaded verandas off the living rooms. The site location and design also encourage interaction with nature. By positioning the car parking away from the houses, a few minutes' walking in nature is assured each time one leaves the home. More dynamic interaction is equally possible through the reforestation work, the removal of invasive species, and the general maintenance of the site, as well as the independent gardens (Sassi, 2006). In addition, those who advocate these forms of development pointed out that low densities have many advantages for sustainable development, by installing solar heating for each home, the possibility of extensive vegetable gardens, and recycling of organic' domestic wastes. So, the goal of which represent the harmless integration of human activities into the environment in a way that supports healthy human development in physical, mental, and spiritual ways and is able to continue into the indefinite future (Kasper, 2008).

In summary, it receives a significant role in social transformability toward respecting the natural life and living in sociocultural coherent in human scale settlements (Christian, 2007; Dias et al., 2017). As a result, the new natural and eco-technological integration with the urban environment is best describing the eco-village concept today in multiple urban scales, to get the self-resilient healthy system (Mohan et al., 2017).

4.2.3 Medium-Density, Bio-City Form

After describing the most fundamental properties, advantage, and disadvantage for two famous kinds of the urban forms' adaptation, Moughtin (an architect who authored many books concerning urban form and urban design) suggested a different pattern of sustainable urban form which reduces most of the other forms' defects and combines their benefits. He described his thoughts in the bio-city concept as a combination of compact eco-structures in sustainable urban formation. Adopting medium-density development produces the similar concept of the public transport, smart technologies, integrated green systems, mixed-land uses, and a mix of house types, placing the priorities to the needy for open interconnected spaces in the city (Moughtin & Shirley, 2005).

There is a thought that the roots of the bio-city theory and its name return to the *Baubiologie* movement in Germany in the fifties of the last century which adopted the goals of creation a harmonic relationship between buildings and the environment, depending on the inhabitants' spiritual, physical, and biological needs. Moreover, it is very respectful to the idea of considering the building's system like the ecological system in order to provide the protection, and the insulation to the residents, in aim of insuring indoor air quality and comfortable environment (thermal, humidity, colors, light, and human scale). While the *Baubiologie* movement dealt with the passive, active, and renewable earth energies in a concept of introducing healthy living (Sassi, 2006), the same principles have adopted by the bio-theory which concentrates on the biological urban form and healthy environmental interaction between the building and the surrounding.

In fact, many architects support this idea. Some of them urged that it is challenging task of dealing with market-motivated development without neglecting the traditional aspect of compact built-up areas on a green carpet. Further, planning applications and programs for new developments are permitted more easily as long as the overall green area is unreduced; compensation for ruined green spaces is required through the greening of brownfield sites. This means a radical shift from strict and static land-use planning to more flexible and dynamic management of the city region (Mike & Nicola, 2005). In addition, some planners' opinions are that the polycentric model should be adopted to achieve the virtues of both a high-density compact city and a low-density eco-settlement with enormous tracts of green space. Others said that the convergent images of future megacities might show that a sustainable spatial form requires a regional unity (in the centric form) with a polycentric structure (Mike & Nicola, 2005).

So, the bio-city theory came to combine the monocentric with the polycentric concept in a hybrid urban form. It is seeking to create a resilient livable environment respect the needs of the inhabitant overtime.

4.2.4 Hyper-Density and Skyscraper Form

In 2013, architect and planner Vishaan Chakrabarti wrote about what he termed vertical cities as a key to solving the dense cities environmental and urban degradation, and how to transform them into healthy, efficient environmentally sustainable. He brought up the idea of re-planning the existing cities with a hyper-density green buildings environment with multiple functional, mixed-use high-technology vertical, skyscrapers residential neighborhoods. In which, all will live in affordable housing with improved public transportation and subway services. Residents will share their social and recreational services, which improve the quality of living and increase sociocultural values (Chakrabarti, 2013). It has more than five thousand residents per one square kilometer and overall district density reaching 30,000 residents/km2 or more. This will support livability, connectivity, flexibility, and amenity (Bishop, 2017). Others expressed hyper-dense neighborhoods that consist of 500 home per hectare (Gardiner, 2018). However, some critics focuses on the high-compacted and/or hyper-density, built environment could hinder and resist resiliency during emergency disasters for the possibility of more risk exposures (Sharifi et al., 2017). Honk Kong is an example of hyper-dense city with separated multifunctional podium organization buildings system to provide privacy for the residents' area (Lau & Zhang, 2015).

5 Analytical Framework

5.1 General Strategies for Resiliency

The synergies between various kinds of sustainable urban forms (compact, hyper, low or moderate density) depending on the city nature and population, culture, with the alternative approaches of resiliency (engineering, disaster, socio-ecological, climate change) characteristics might have reform the city toward urban resiliency. See Fig. 3. Hence, urban resiliency is a multidisciplinary, proactive, dynamic approach, addressing, diagnosing, and managing the undesired situation. In which, multiple sustainable synergies with interconnected scales and levels could be applied depending on the urban structure of the city, resulting adaptive, integrative, flexible urban system, ready to respond to the spatial–temporal changes and recover, or sustain or transform its formation structure.

5.2 Urban Resiliency Adaptive Cycle

As Holling defined the affecting forces of his adaptive cycle in three dimensions of connectedness potentialities and resiliency (Holling, 2001), many studies wrote about these disturbing factors that could alter or transform the state of the city (Azevedo, 2016; Galderisi, 2014; Suárez et al., 2016). Resulting from this, and due to the multiple functional urban resilience system natures, each city has its own

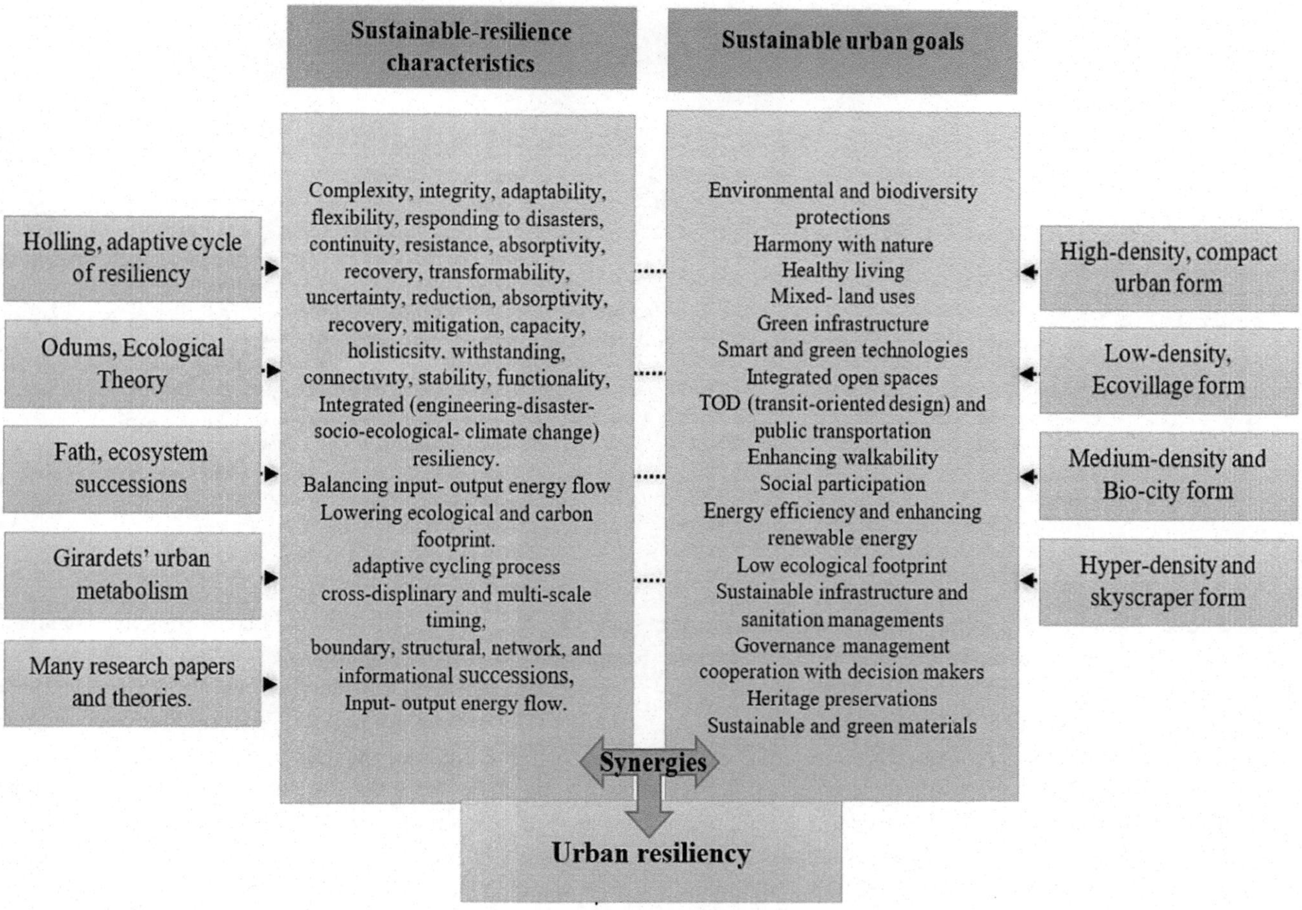

Fig. 3 Graphical framework of the relationship between resilience and sustainable form strategies

internal and external characteristics that force it to take a different form than the others. So, nature's universal re-organization laws might re-cycle these patterns in a hidden matrix according to the homogeneous or heterogeneous internal sociocultural, economic, eco-environmental, political factors, or external natural or unnatural (wars) shocks, disaster, stress factors. And each city could have several bandalas successions goes and back overtime before moving to another phase of transforming to another level of knowledge and formation as seen in Fig. 4a and b. These adaptive or transformative changes could come in a complex integrative pattern and could have a multiple accumulative (might be not rhythm) scale over time, for various unpredictable successions, as seen in Fig. 4.

6 Case Study

6.1 Urban Resiliency in Baghdad

Contrary to the previous time periods of Baghdad City, in which urban changes continued for centuries and the fluctuations were slow and relatively few, the researchers found that during the twentieth century, Baghdad, developments and changes in the urban structure accelerated in a radical and legible manner (Majed & ALslik, 2014). According to Al-Musawi (2017) there are eight basic master plans which were prepared during the last 100 years—(1917–2017), of which only three comprehensive developments were adopted in defining the administrative boundaries of Baghdad City, while the rest were implemented in stages and failed to complete the others due to the lack of flexibility in predicting future changes (Al-Musawi, 2017). The research divided these eight master plans into five crucial stages depending on the urban resilience theories phases and successions, explained in detail in Table 1. Each one these stages took practically twenty year trying to reach its stability, except the fourth one which rises almost forty-five years and still stuck in. That is why the fourth stage called stagnation and took widen phase in Fig. 5. As seen, X path is related to time, and with time, the city should take its stability. Y path is related to the relationships which refer to the connectedness of each phase with others. The multiple changes from highly exploitation growth to the

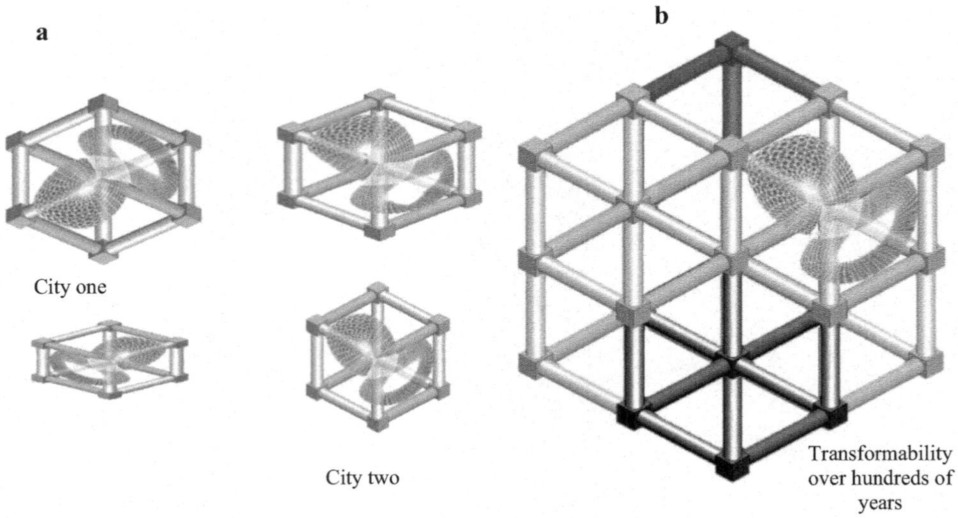

Fig. 4 Urban structure of the city according to the resilience theories. (Author's drawings)

conservation led to the rapid breakdown with a long release in phase four. Z path related to the sources in dynamic goes and back movement. That is why phase five should take enough space because the urban form growth and development in Baghdad experience the same ecosystems' stages, and at this phase, the mistakes should be avoided. Baghdad has its exceptional circumstances that force its adaptive cycle within different successions to take its own formation as seen in Fig. 5. Each phase could possess various names according to the internal and external factors, and these differences came from the complex tribal cultures, believes, and desires which affect its actions. Therefore, the research explains each phase according to its master plans within twenty years.

The First Phase (1917–1935) The Exploration Phase: Beginning with the *boundary growth*, which had come from the basic needs of the British Mandate after World War I. The master plan of the German General Goltz considered the first attempt to construct urban roads. The urban fabric of Baghdad consisted of organic streets and alleys with two-floor courtyard houses. Therefore, it represented the phase of exploration and the start of the transition from the traditional and narrow streets to the expansion and the construction of the first street named Khalil Pasha Avenue (Al-Rashid Street today). In 1920, this master plan is followed by the design of G. M. Willson, who directed Baghdad from its traditional wall and suggested reorganizing the city and creating new residential areas. This stage had a little coup of thinking toward a new modern formation, so it had low exploitation. See Fig. 5 and the related Table 1.

The Second Phase (1936–1955) The Involvement Phase: The "Brecks plan for Baghdad," is moving to the *structural growth stage*. In addition to the developmental process of the physical structures in numbers, quantities, and component volumes, as a result of gaining the primary energy, the ecological system had a positive feedback approach to product subsystems. This move delivered opportunities to develop the main system and earns more energy. The empirical evidence proved that the structural growth in the city had the same power. However, it took a long time to be accepted by the community, so multiple types of the structure appeared at the same place. Resulting in, the start of road construction and the shift toward iron grid planning had presented the essential transform of an urban mix from the traditional structure into the contemporary one. See Fig. 6 and the related Table 1.

The Third Phase (1956–1972) The Integration Phase: The British Miniprio master plan proposal 1956represented the *socio-ecological cycle*, which came with strong steps containing the communication subsystems of the main system through the dynamic transformation activities that lead to increase passes and dynamic recycling. So, this master plan emphasized the necessity of the design flexibility and its ability to change; the radial pattern with a hierarchical system was proposed. It developed into the design of Doxiadis 1959, in which it relied on the theory of geometrical squares named (Ekistics) and network design with separate units and neglected the traditional compact fabric. However, this master plan applied to define the internal division of the city but not the administrative boundaries. This idea of the networked urban system was the start point of the destruction of Baghdad's urban identity (Al-Musawi, 2017; Majed & ALslik, 2014). It was the beginning of the imbalance between the open space and the built structure, neglecting historical characteristics and trying to eliminate heritage. On the other hand, it was an economic flourishing period, many architects went to study in America and Britain, and their return was full of Western ideas, and with expanding the

Table 1 Baghdad urban resilience phases through 100 years (Al Ani, 2011), (Al- Musawi, 2017), and (Al-Shammari et al. 2019)

The administrative boundaries of Baghdad City	Phase name Author description	Description of Baghdad City	Urban form growth—change of Baghdad City	Urban resiliency according to Holling	Map representation
1917–1935 Goltz master plan Willson master plan	(PH. 1) Phase one: Exploration boundary growth	The urban fabric is compacted with organic narrow streets and alleys with two floors. The start of the transition from the traditional streets to the expansion of urban roads and reorganizing the city and creating new residential areas	High-density, compacted Ottoman courtyard type	Start growth—little exploitation	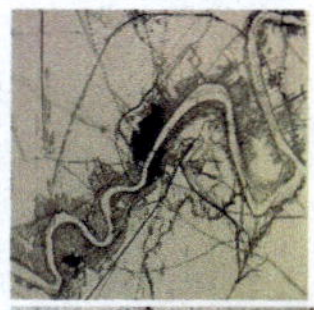 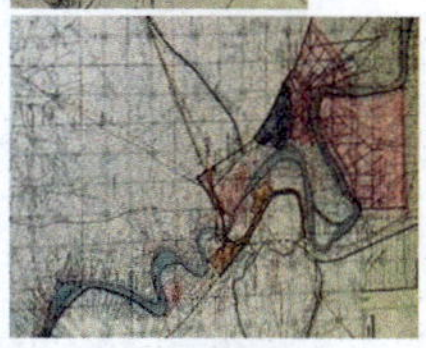
1935–1955 Brecks plan Doxiadis master plan	(PH. 2) Phase two: Involvement structure growth	The start of the streets' network construction and the shift toward iron grid planning represents the main shift of an urban mixed from the traditional structure to the contemporary one	Grid system and some detached urban areas with the courtyard type	Medium exploitation	 
1956–1972 Miniprio master plan	(PH. 3) Phase three: Integration and development structure—networking growth	Start to urban expand, resulting in an imbalance between the open space and the built structure, neglecting historical characteristics and trying to eliminate heritage	Wide streets low-density-detached garden similar to eco-houses	High-exploitation and accumulation, conservation	
1973–2016 Polservice consulting—Comprehensive JCCF consulting firm	(PH. 4) Phase four: Stagnation networking and information then breakdown	Lacked urban details, lacked in defining the main and the secondary centers, the city lost its green areas, neglected mixed-uses, and the most important thing is the increasing of urban sprawl that led to an acceleration of horizontal expansion	Hybrid structure compact of two storys with detached houses and 4–5 mixed-use building within the same street	Highly exploitation, then conservation, and the rapid breakdown and long release	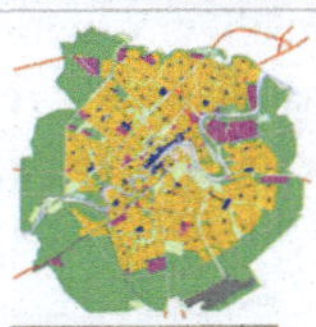 
2017–2030 Khatib & Alami	(PH. 5) Phase five: Rejuvenation regeneration growth	Plans and layouts according to the sustainable development goals of preserving the heritage centers, enhancing the recreational areas, re-growing the agricultural areas, creating jobs within the available capabilities and resources, and investing in a reasonable state budget without infrastructure threatening	Hyper-density and skyscrapers type	Reorganization, and persistency	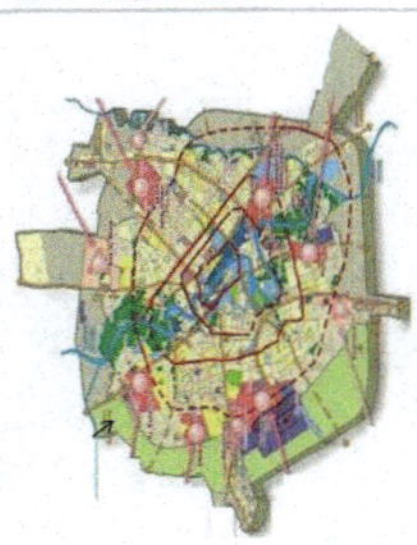

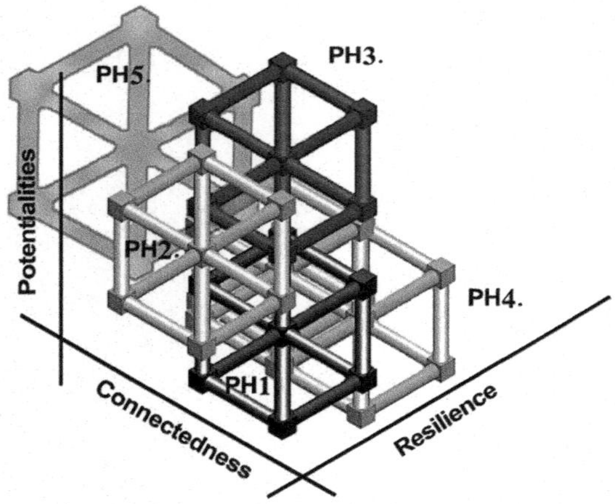

Fig. 5 Urban structure of Baghdad City, according to the resilience theories, shows a pendulum repetition or frequency of the level of changes within the equivalent period. (Author's analysis)

economy in Baghdad, architects started to cooperate with many abroad companies to build the city. Hence, it would be considered as an integration and development stage with high-exploitation, knowledge accumulation, and a little about conservation. See Fig. 6 and the related Table 1.

The Fourth Phase (1973–2016) The Information Growth: This kind is a typical developing in the system behavior from the dissipation and consuming types to the preservative systems more efficient, powerful, integrated, and active. Most of the time, this growth deals with the development of subsystems (other land uses) by increasing their productivity and resiliency with the main system. But in Baghdad, it caused another breakdown and declination point. In the Municipality of Baghdad, specialists realized that the previous master plans did not meet the needs Baghdad, and therefore, the Polish Polservice consulting engineering has been invited in 1965 to work on the comprehensive development plan for Baghdad 2000, and especially after the nationalization of oil in Iraq, new requirements appeared to accelerate growth like studying land uses and absorbing the population density increases. This was produced in 1973, moving toward the vertical construction and adapting the idea of the skyscrapers to accommodate 20% of the population in four administrative circles. Despite the rise in economic prosperity to its peak in the seventies of the last century and the harnessing efforts to provide a comprehensive plan that accommodates all urban changes and the increase in population and housing densities, Polservice master plan lacked in urban details, lacked in defining the main and secondary centers, the city lost green areas, neglected mixed-use amenities, and most importantly the increasing of urban sprawl that led to an acceleration of horizontal expansion at the expense of the scheme (Al-Taif et al., 2014). All these led to the need for preparing a new scheme to avoid past mistakes and reduce urban sprawl. So, the Japanese JCCF consulting firm proposed another plan (the 1990 master plan), which was agreed upon in the eighties, but which was halted in the 1990s due to the Gulf War, that is why it has to be called the stagnation stage that went through multiple changes from highly exploitation growth to the conservation one and rapid breakdown with a long release. See Fig. 6 and the related Table 1.

The Fifth Phase (2017–2030) The Rejuvenation And Regeneration Growth: Khatib and Alami consulting master plan—under study since 2007—supposes to be *the regeneration growth*. It is the recorrecting and rising of tracks. The Municipality of Baghdad contracted with the Khatib and

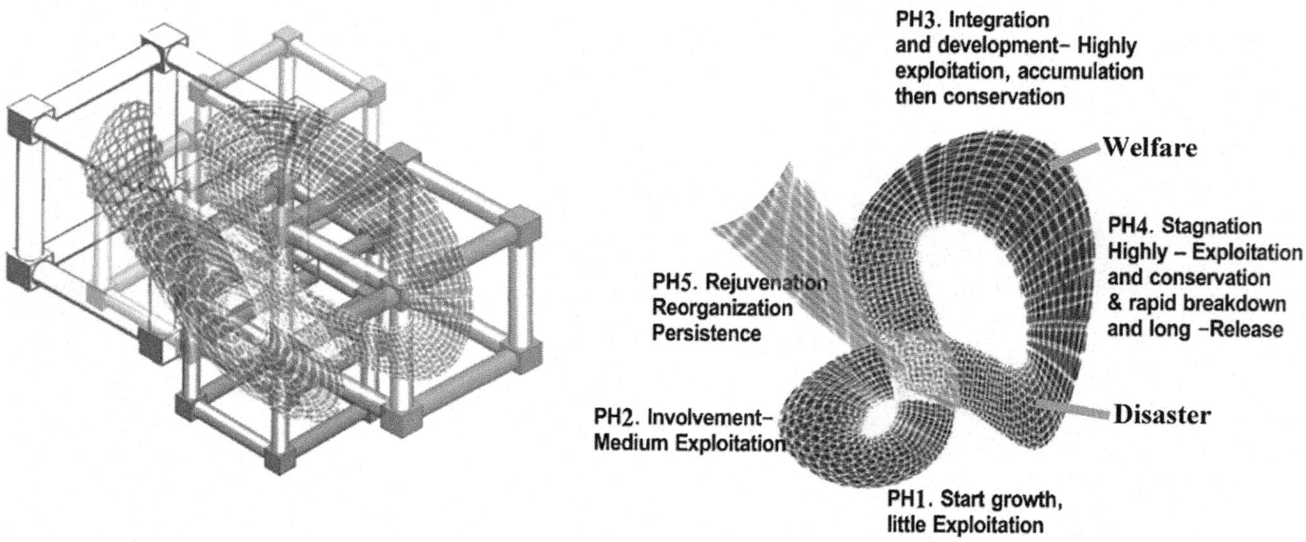

Fig. 6 Urban structure of Baghdad City, according to the last 100 years' master plans. (Author's analysis)

Alami consulting firm as well as the Japanese MEPEX and PCI in preparing the comprehensive development plan for Baghdad City. It is an avoidance of the mistakes that occurred in the previous stages, in which the city of Baghdad is going according to the sustainable development goals in preserving the heritage centers, enhancing the recreational areas, extending the agricultural areas, creating jobs within the available capabilities and resources, and investing in a reasonable state budget without infrastructure threatening. The comprehensive development plan adopted several alternatives and considered the most convenient alternative to the urban sprawl and problems. It consisted of several urban polycentric urban areas and orientated toward the skyscrapers and vertical compacted housing projects. However, all still on papers straggling between the agreements and the reality. See Fig. 6 and the related Table 1.

6.2 Discussion

The comprehensive Polservice plan considered the last approved master plan with clear future vision; what carried on, during the eighties of the twentieth century and after, are proposals. They are delayed and unproven due to the complicated situations and the wars in Iraq. Still, it seems urban changes in Baghdad were interconnected with the phases of resiliency. Each succession has its lines, persistence, precariousness, and panarchy to accept the change and adapt to the new systems. In reversible, each phase of urban growth could have the stages of an adaptive cycle until it reaches stability and moves to the next succession. What happened with the multiple contradictions has been affected the total chain.

After 2003, the conflicts in Iraq negatively affected the urban fabric and its goals to reach sustainability. These conflicts have pendulum changes come and go with political dramatic changes which have adversely affected the integration of urban plans. The result is cumulative, the degradation has become uncontrolled, the over urbanization has spread, and the integration between the traditional area and the whole has lost. Additionally, the specialty of respective large and prestigious construction projects in the early fifties to the eighties has neglected. All these issues led to the missing of connectivity, functionality, and the ability to absorb the multiple disasters faced Iraq. Before 2003, the flexibility is still there, and the city might able to maintain itself; however, many complexities played a significant role to hinder the formation of sustainability toward resiliency.

Socially, hundreds of thousands of people have been displaced from their original neighborhoods to other parts of the city according to their religious faith. With an increased sense of limitations and inequality, it has shattered any sense of social stability. Clear separation and the resulting socioeconomic gaps make restoring urban contact almost impossible, even if public safety is to be restored. Over time, class differences became clear and impacted the changes in the urban social structure.

On the other hand, the inadequate laws regulating building and managing urban planning and the ineffectiveness of the proposed strategic plans led to the deterioration of the built urban environment, the encroachment of land use, the accumulated housing deficits, the informal settlements, the disability of economic development strategy, the insufficient basic services and infrastructure networks, the inadequate road and transportation networks, and the miserly public services and social infrastructure. All these led to the massive loss of the urban balance, the societal harmony, and the healthy living, and drove to the overexploitation and unsustainability (Al Ani, 2011; AL-HASANI, 2012; Al-Saffar, 2017; Banna, 2015).

7 Conclusions

There is a tight connection between the meanings and the goals of sustainable and resilient. And, the study concords that the terms "resilience and sustainability" might have the same main characteristics. However, each city has its different phases and successions and should not use the terms interchangeably. Some cities, like Baghdad, might be able to cope with the changes and resilience over time, but it does not relieved the ability to reach sustainability with its urban formation, and it is inaccurate to describe its situation as an urban resilience. For sure, each city provides the properties that distinguish it from others, and day after day, applications and indicators are undertaken to serve some cities and not others. Therefore, there are no ideal steps to follow, and the forces' factors that extrapolated the urban resilient formation as a sustainable structure are flexible and changeable, which cannot be applied to any city as a typical or prototype resilient theory. Urban resiliency represents a multifunctional, dynamic approach. In which, multiple sustainable synergies with interconnected scales of urban formation could be applied. This result in the adaptive, integrative, flexible urban system is ready to respond to any disturbance changes and recovers itself robustly. However, the research proved some cities, like Baghdad, might be resilience at some levels but fails in others. And, it is misguided to frame the urban formation changes in it with the frame of sustainable synergies. In other words, Baghdad went through multiple adaptive cycles with all the successions of growth in each urban stage to reach resiliency and hybrid system. But this would not sustain with the accumulated pressure. Sustainability means stability over time, and Baghdad in the last one hundred years proved its struggled pendulum resiliency. Whether it is coping with Iraqi identity or not, this demands several studies to identify the potentialities and the challenges to sustain urban resiliency in Baghdad. Today Baghdad includes a hybrid urban formation which consists of many urban types of density. However, future studies need to identify the socioeconomic impacts and how stressful this system is to reach urban resiliency. Furthermore, extended studies are required to identify which urban type is better in adopting the changes and reach urban resiliency. Still, the conceptual model of resiliency needs data and maybe some simulation programs to read the map practically. However, it is very hard to detect the old data. The map might be there, but the information is limited. These simulations could help in identifying exactly the breakdown points or the growth one related to the urban formation.

Eventually, the research provides a comprehensive analytical base for reading resiliency related to cities urban formation over time and diagnosed the points preventing Baghdad City from reaching the resilience level.

References

Abaas, Z. R. (2008). The Green cities: Towards sustainable environment. *The Union of Arabic Universities Magazine*. University of Baghdad.

Al-Musawi, W. A. S. (2017). *Baghdad's Urbanism Atlas (1917–2017)*. (First). Dar AlMada.

Al Ani, M. Q. (2011). *Urban prediction: Towards a new generation to redevelop the Arabic–Islamic city*. University of Baghdad.

Al-hasani, M. K. (2012). Urban space transformation in old city of Baghdad—Integration and management. *CiLT*, (Vol. 7, EK 1, pp. 79–90). Yıldız Technical University, Faculty of Architecture.

Al-Saffar, M. (2017, 27th-29th September). Assessment of the process of urban transformation in Baghdad: City form and function. In: *24th ISUF International conference* (pp. 209–218). City and territory in the Globalization Age Conference proceedings.

Al-Shammari, H., Ismail, N., Hamza, S. (2019). Urban sprawl feature of Arab cities: study Riyadh and Baghdad city case study. *Engineering and Sustainable Development Journal*, 39–56.

Al-Taif, E., Al-Samariy, A., & Ghafar, S. H. (2014). The problems of the built environment in Iraq. *Nineveh Magazine*, *10*(36), 139. https://www.iasj.net/iasj?func=fulltext&aId=86439

Astbury, J. (2019). *Richard Rogers is high-tech's inside-out architect*. Retrieved January 2, 2020, from Dezeen website: https://www.dezeen.com/2019/11/06/richard-rogers-high-tech-architecture/

Azevedo, M, (2016). *The evaluation of the social impacts of culture: culture, arts and development*. Economics and Finance. Université Panthéon-Sorbonne - Paris I, English. ffNNT: 2016PA01E041ff. fftel-01804118v2f.

Banna, S. S. (2015, January 22). *Planning in Baghdad: How years of conflicts have shaped the design of the city*. Retrieved from The Global Urbanist: http://globalurbanist.com/2015/01/22/planning-in-baghdad

Bishop, J. (2017). *Building sustainable cities of the future*. Springer International Publishing.

Brand, F. S., & Jax, K. (2007). *Focusing the meaning(s) of resilience: Resilience as a descriptive concept and a boundary object*. *12*(1).

Burton, E., Jenks, M., & Williams, K. (Eds.). (1996, (1st ed.)). *The compact city: A sustainable urban form?* Routledge.

Caputo, S. (2015). *Urban resilience: a theoretical and empirical investigation*. Coventry University.

Chakrabarti.V. (2013). Building hyperdensity and civic delight. *Places Journal, 2013*. Accessed 06 Jan 2020. https://doi.org/10.22269/13061

Christian, D. L. (2007). *Finding community: How to join an ecovillage or intentional community*. New Society Publishers.

Dias, M. A., Loureiro, C. F. B., Chevitarese, l., & Souza, C. D. M. E. (2017). The meaning and relevance of ecovillages for the construction of sustainable societal alternatives. *Ambiente & Sociedade*, *20* (3). https://doi.org/10.1590/1809-4422asoc0083v2032017

Elmqvist, T., Andersson, E., Frantzeskaki, N., McPhearson, T., Olsson, P., Gaffney, O., … Folke, C. (2019). Sustainability and resilience for transformation in the urban century. *Nature Sustainability*, *2*, 267–273. https://doi.org/10.1038/s41893-019-0250-1

Eraydin, A., & Taşan-Kok, T. (2013). *Resilience thinking in urban planning*. Springer.

Fath, B. D. (2017). Systems ecology, energy networks, and a path to sustainability. *International Journal of Design & Nature and Ecodynamics*, *12*(1), 1–15. https://doi.org/10.2495/DNE-V12-N1-1-15

Fath, B. D., Jørgensen, S. E., Patten, B. C., & Straškraba, M. (2004). Ecosystem growth and development. *Bio Systems, 77*, 213–228. https://doi.org/10.1016/j.biosystems.2004.06.001

Forgaci, C., & van Timmeren, A. (2014). Urban form and fitness: Towards a space-morphological approach to general urban form and fitness: towards a space- morphological approach to general urban. In *20th Annual international sustainable development research conference Norwegian*, (August, pp. 1–11). https://doi.org/10.13140/2.1.4017.3444

Galderisi, A. (2014). Urban resilience: A framework for empowering cities in face of heterogeneous risk factors. *A|Z ITU*, *11*(January 1), 36–58.

Gardiner, J. (2018). *Rapid urbanization and vertical sprawl how can we live like this?* Retrieved January 6, 2020, from WSP firm website: https://www.wsp.com/en-GL/insights/rapid-urbanization-and-vertical-sprawl-how-can-we-live-like-this

Girardet, H., & The World Future Council. (2013). *Towards the regenerative city*.

Girardet, H., & Miguel, M. (2009). *A renewable world: Energy, Ecology, Equality A*. The green books.

Holling, C. S. (1973). Resilience and stability of ecological system. *Annual Review of Ecology and Systematics*, *4*, 1–23. http://www.jstor.org/stable/2096802.

Holling, C. S. (2001). Understanding the Com of economic, ecologica social systems. *Ecosystems*, *4*(5), 390–405. https://doi.org/10.1007/s10021-00

Jabareen, Y. R. (2006). Sustainable urban forms. *Journal of Planning Education and Research, 26*(1), 38–52. https://doi.org/10.1177/0739456x05285119

Kasper, D. V. (2008). Redefining community in the ecovillage. *Human Ecology Review, 15*(1), 12–24.

Lau, S. S. Y., & Zhang, Q. (2015). Genesis of a vertical city in Hong Kong. *International Journal of High Rise Buildings, 4*(2), 117–125.

Maharani, M. (2019). Model of agro-eco-village by using interpretative structural modeling for improving sustainable development. In *The 3rd International symposium on agricultural and biosystem engineering* (pp. 1–8). https://doi.org/10.1088/1755-1315/355/1/012099

Majed, F. A., & ALslik, G. M. R. (2014). Succession of urban structures of the city of Baghdad. *Journal of Engineering*, *20* (November), 1–30. Retrieved from University of Baghdad.

Mike, J., & Nicola, D. (2005). *Future forms and design for sustainable cities (First)*. Architectural Press, An imprint of Elsevier.

Mitsch, W. J. (2018). *Ecology, ecological engineering, and the Odum brothers* (October 2003). https://doi.org/10.1016/j.ecoleng.2003.09.001

Mohan, S. V., Dahiya, S., Velvizhi, G., & Reddy, C. N. (2017). Ecovillages: Resilient approach to sustainable rural development in Indian context. *Journal of Energy and Environmental Sustainability, 2*(January), 55–63.

Moughtin, C., & Shirley, P. (2005). *Urban design: Green dimensions*. Architectural Press.

Newell, J. P., Meerow, S., & Stults, M. (2016). Defining urban resilience: A review. *Landscape and Urban Planning, 147*(March), 38–49. https://doi.org/10.1016/j.landurbplan.2015.11.011

Odum, H. T., & Odum, B. (2003). Concepts and methods of ecological engineering. *Ecological Engineering,* 339–361.

Peres, E. M. (2016). *The translation of ecological resilience theory into urban systems*. University of Pretoria.

Pickett, S. T. A., Cadenasso, M. L., Grove, J. M., Irwin, E. G., Rosi, E. J., & Swan, C. M. (2019). *Science for the sustainable city: Empirical insights from the Baltimore school of urban ecology*.

Piracha, A., & Marcotullio, P. J. (2003). *UNU/IAS report urban ecosystem analysis identifying tools and methods*. https://doi.org/10.13140/RG.2.1.5055.4963.

Regan, S. (2000). Towards an urban renaissance: The final report of the urban task force. *The Political Quarterly, 71*(1), 115–118. https://doi.org/10.1111/1467-923x.00285

Rogers, R. (1997). *Cities for a small plant.* Westview Press, 30–39.

Sanchez, A.X., van der Heijden, J., Osmond, P., & Prasad, D. (2017). Urban sustainable resilience values: Driving resilience policy that endures urban sustainable resilience values: Driving resilience policy that endures. In *CIB World building congress proceedings* (June 2016). Tampere.

Sanchez, A. X., Osmond, P., & van der Heijden, J. (2016). Are some forms of resilience more sustainable than others? *Procedia Engineering, 180*, 881–889. https://doi.org/10.1016/j.proeng.2017.04.249

Sassi, P. (2006). *Strategies for sustainable architecture.* Taylor & Francis Group Inc.

Sharifi, A., Chelleri, L., Fox-lent, C., Grafakos, S., Pathak, M., Olazabal, M., … Yamagata, Y. (2017). Conceptualizing dimensions and characteristics of urban resilience: Insights from a co-design process. *Sustainability*, 1–20. https://doi.org/10.3390/su9061032

Sharifi, A., & Yamagata, Y. (2018). Resilient urban form: A conceptual framework. In *Lecture notes in energy* (pp. 167–179). https://doi.org/10.1007/978-3-319-75798-8

Sharifi, A. (2019). Resilient urban forms: A macro-scale analysis. *Cities, 85*(August 2018), 1–14. https://doi.org/10.1016/j.cities.2018.11.023

Suárez, M., Gómez-Baggethun, E., Benayas, J., & Tilbury, D. (2016). Towards an urban resilience index: A case study in 50 Spanish cities sustainability towards an. *Sustainability, 8*(774), 1–19. https://doi.org/10.3390/su8080774

Walker, B., Holling, C.S., Carpenter, S.R., & Kinzig, A. (2004). Resilience, adaptability and transformability in social–ecological Systems. *Ecology and Society*, *9*(2).. https://www.ecologyandsociety.org/vol9/iss2/art5/

Wolman, A. (1965). The metabolism of cities. *Scientific American, 213* (3), 178–190.

Yamagata, Y., & Ayyoob, S. (2018). *Resilience-oriented urban planning-theoretical and empirical insights*. Springer.

Zeybek, O., & Arslan, M. (2017). Ecovillages: The place where ecotourism turns into educational tourism ecovillages. In *1st International conference on sea and coastal development in the frame of sustainability*, (November, pp. 226–234). Marine Coastal Development Sustainability.

An Exploration of the Effects of Urban Block Design on the Outdoor Thermal Environment in Tropical Savannah Climate: Case Study of Nyamirambo Neighborhood of Kigali

Elyse de la Joie Horimbere, Hong Chen, and Mehdi Makvandi

Abstract

Urbanization is known to alter the microclimate, thus accelerating the effect of climate change. The built environment can have a positive and negative impact on local microclimates and especially at the neighborhood level. Microclimates at this level are created by items such as building form and geometry, street width, surface material types, soil types, trees, and vegetation types which represent urban block design characteristics. At street level, urban block settings have an impact on the overall air temperature, the surface temperature, the wind distribution, and on the solar radiation, and this will influence the outdoor thermal comfort. In this paper, we explore the impact of urban block design on the microclimate and its impact on the outdoor thermal environment by simulating the microclimate using ENVI-met and by assessing urban thermal comfort using UTCI in a carefully selected urban block fabric of the Nyamirambo neighborhood, which is one of the oldest mixed-used neighborhoods in Kigali presenting a particularly dense urban fabric in the fast-growing city of Kigali. While the region has been experiencing extensively high temperatures during the dry season in the recent past years, this paper tends to highlight urban block design strategies that can help to ease the effects of global warming by providing pedestrians with thermally comfortable conditions. We simulated the microclimate at street level of an urban block model in ENVI-met and analyzed its impact on Ta, Va, MRT, and UTCI at the current state and after introducing urban morphological techniques that have proved to enhance thermal conditions outdoors such as adding trees on the roadside, replacing dark and used concrete pavement with a light concrete pavement with a high albedo, and the creation of a small park to offset the positive impact of cool materials. We also analyzed the results based on the impact of the urban block's building geometry represented by SVF and street orientation. Both strategies proved to have satisfactory results when treated separately. The addition of trees alone led to a reduction of Ta by 3.89 °C and MRT reduction by 1.1 °C at noon. Results of simulations based on SVF and street orientation impact on Ta, and MRT did not show a big difference in this particular urban fabric; however, SVF and street orientation proved to play an important role in the distribution of wind velocity. The best results, however, were obtained in the combination of all the mentioned strategies where the UTCI went from the state of "moderate thermal heat stress" to a state of "no thermal heat stress;" the highest MRT which corresponds to the high-angle sun between 12:00 and 13:00 went from 68 °C to 60.35 °C, especially in the area with added trees and regardless of the street orientation and SVF.

Keywords

ENVI-met • Kigali • Microclimate • MRT • Outdoor • Street orientation • SVF • Thermal comfort • Urban block

Nomenclature

H	Height of buildings
SVF	Sky view factor
Ta	Air temperature
Tmrt	Mean radiant temperature
UTCI	Universal thermal climate index
Va	Wind speed
GHG	Greenhouse gases
UHI	Urban heat island

E. de la Joie Horimbere (✉) · H. Chen
School of Architecture and Urban Planning, Huazhong University of Science and Technology, Wuhan, China

M. Makvandi
School of Civil Engineering and Architecture, Wuhan University of Technology, Wuhan, China

C. Alalouch et al. (eds.), *Towards Implementation of Sustainability Concepts in Developing Countries*, Advances in Science, Technology & Innovation, https://doi.org/10.1007/978-3-030-74349-9_2

1 Introduction

Most people live in cities now worldwide which leads to a rather rapid urbanization; rapid urbanization leads to the creation of new cities and expansion of already existing ones which in return leads to an alteration of the microclimate due to the increase of human activities. This forms a very critical environmental situation commonly known as global warming. Higher temperatures in urban areas lead to uncomfortable outdoor areas while at the same time increasing energy use and demand through the use of air conditioning in buildings' interior spaces, creating anthropogenic heat, which, in turn, contributes to the global warming of outdoor air. "We are then faced with a cause and effect loop that needs to stop to improve the conditions of outdoor thermal comfort and so participate in sustainable urban design. As designers, we need to think of an urban design that takes into consideration the impact of our achievements on microclimate" says Achour-Younsi and Kharrat (2016).

According to Wang et al. (2015), "the main factors that contribute to the urban high temperatures are large surface materials (mainly asphalt and concrete) with low albedo and high admittance, reduced vegetation, and permeable surfaces." Mestayer and Anquentin (1994) Oke et al., (1991), add that "The urban microclimate undergoes particular and specific changes such as slowing the wind speed or storage of solar radiation by the construction materials used." "This is a set of parameters, which are mainly due to the building-environment interaction. This complex relationship recently interested researchers and has been studied mainly in the fields of urban climatology. Domain experts agree to recognize the power of regulation of the urban morphology on microclimate parameters;" concluded Achour-Younsi and Kharrat (2016).

That brings us to study the impact of urban morphology at the street level in rapidly expanding cities by taking Kigali city as an example. "Indeed, the street is considered as being the interface between the architectural and urban scale to the extent that it is the common area between the building and the external environment (Ali-Toudert & Mayer, 2006). Therefore, the street is a major element that affects not only the external but also internal microclimate atmosphere," says Achour-Younsi and Kharrat (2016).

At street level, building heights and orientations, spaces between buildings, and plot coverage are variables that can alter solar access, wind speed, and direction. Urban design elements, including vegetation and shading devices, can be used to improve microclimate and comfort conditions (Yahia et al., 2018). Chen et al. (2010) found a linear relationship between SVF and Ta in the warm-humid climate of Hong-Kong, while (Achour-Younsi & Kharrat, 2016) found that the more the H/W ratio decreases, the more comfort decreases.

In the tropical climate of Kigali, climate-responsive urban design has received little attention, although an effort is being made. Strict planning measures, including implementation of the 2013 City Master Plan (Manirakiza, 2014), are in operation.

2 Background or Precedence

2.1 Study Area

This study was conducted during the hot season (long dry season) in the city of Kigali, Rwanda. "Kigali is at 1°57'S, 30° 7'E, 1497 m (4913 ft) and has a tropical wet and dry/savanna climate with a pronounced dry season in the low-sun months, no cold season, and the wet season is in the high-sun months. According to the Holdridge life zone system of bioclimatic classification, Kigali is situated in or near the subtropical dry forest biome. The annual mean temperature is 20.5 °C." (climatemps.com, 2019–2017) During the long dry season, average Ta is about 28.8 °C, average RH is about 58%, while average Va is around 3.4 m/s. In the long and short dry season, temperatures reach 30 °C during the day. Strategic foresight also characterizes Rwanda as a fast warming country, with an increase in the average temperature of between 0.7 °C and 0.9 °C over fifty years. Studies show that the temperature increased with a high frequency of warm days exceeding 30 °C (Strategic Foresight Group, 2013).

Kigali presents various types of urban fabrics. Nyamirambo neighborhood is a sector in Nyarugenge district. It is located in the southwest part of the city of Kigali. The coordinates of the Nyamirambo Sector are: 1°59′37.0``S, 30°02′39.0'E (Latitude: −1.993611; Longitude: 30.044167). It is a mixed residential and commercial sector composed of clustered single-detached buildings (mostly single storied) and a few commercial buildings in the form of mid-rise shape. This is based on observations that revealed that mixed neighborhoods in Kigali are a mixture of planned and informal settlement areas that have progressively merged around a planned site or neighborhoods that resulted from urban transformations, either upgrading or degradation, that affect the initial structure of a planned or an unplanned neighborhood (Baffoe et al., 2020).

Our case study focuses on urban blocks close to the Kigali view hotel in Nyamirambo neighborhood; the area dimensions are 270 × 240 m with an angle of 15°. It is a mixture of single-story buildings with a bunch of mid-rise buildings, little to no vegetation, little to no trees inside properties or on the roadside; a street canopy of 5 m width and a non-built piece of land as portrayed in Fig. 1.

Fig. 1 (Left) Nyamirambo-street map (Google Maps). Square dot (N–S) and Round dot (E–W). (Right) Nyamirambo-street view (Taken on-site 2018)

2.2 Urban Outdoor Thermal Comfort

The outdoor thermal comfort can be classified within fields that began to interest researchers fairly recently. In fact, most scientific research projects regarding thermal comfort are mainly concerned with inside spaces. The issue in understanding the external thermal comfort lies within the multiplicity of factors involved and also the different existing interactions (Achour-Younsi & Kharrat, 2016).

To be effective, the study of thermal comfort is to be conducted by considering its various physical, physiological, and psychological aspects, to put into consideration, the interrelationships between the thermal conditions of the environment, physiological responses, and psychological phenomena (sensation, behavior) (Parsons, 2003). The physical approach of thermal comfort considers the individual as a thermal machine and considers the interactions with the environment in terms of heat exchange. The physiological approach is curious about self-regulatory mechanisms that come into play in a thermal environment, like sweating or shivering. Finally, the psychological approach attempts to see the connection between the physical and physiological variables and sensory results in the individual.

We note that this is a concept that reaches different disciplines. Each specialty has focused on thermal comfort from the precise angle of his discipline. The thermal comfort of an individual depends on the environment first, then her body. Many factors affect the heat flow from the body and therefore the thermal comfort. Researchers have classified these factors into three categories (Auliciems & Szokolay, 1997): climatic factors (air temperature, relative humidity, radiation, and air velocity), personal factors (metabolism, clothing), and contributing factors (acclimation, age, sex…).

The climatic parameters are those that are most considered, since they are the most concrete aspect. Taking into account, the environment requires the quantification of four microclimatic parameters, which are the air temperature (Ta), the mean radiant temperature (Tmrt), the air velocity (Va), and relative humidity (RH). The management of those climate variables simultaneously represents the complexity faced by the designers. They have sought to standardize comfort appreciations by developing thermal comfort indices. Among these, we mention the predicted mean vote (PMV) (Fange, 1970), the physiological equivalent temperature (PET) (Hoeppe, 1999), or the universal thermal climate index (UTCI) (Brode, et al., 2012).

On the other hand, research has also grasped the concept of thermal comfort in trying to understand the influencing factors and might be mastered in urban design (Rizwan et al., 2008). Thus, it was revealed the connection between urban morphology and its direct impact on microclimate variations. Some research has tried to list the most influential morphological indicators on outdoor thermal comfort such as the floor area ratio (FAR), the building coverage ratio (BCR), the H/W ratio, and the SVF (Adolphe, 2001; Ait-Ameur, 2002).

2.3 Urban Climate Change Mitigation

The repercussions of climate change will be felt in various ways throughout both natural and human systems in Sub-Saharan Africa. Climate change projections for this region point to a warming trend, particularly in the inland subtropics; frequent occurrence of extreme heat events; increasing aridity; and changes in rainfall—with a particularly pronounced decline in southern Africa and an increase in East Africa. The region could also experience as much as one meter of sea-level rise by the end of this century under a 4 °C warming scenario. As agricultural livelihoods become more precarious, the rate of rural–urban migration may be expected to grow, adding to the already significant urbanization trend in the region (Salka, 2014).

Emmanuel et al., (2007) argues that the conventional approach to UHI mitigation is primarily differentiated by the scale of intervention. At the local (neighborhoods) and microscale (several buildings to entire streets), mitigation strategies have mainly focused on three aspects: albedo enhancement (Chen et al., 2009; Wang et al., 2015), increased vegetation cover (Chen et al., 2009; Yahia et al., 2018), and 'cool roof' strategies (Chatzinikolaou et al., 2018; Wong et al., 2003). The aim of these efforts being to reduce the negative feedback of urbanization upon air temperature and space cooling of buildings.

Emmanuel et al., (2007) also argued that "the phenomenon of 'cool island' during the day suggests that while the conventional nocturnal UHI exists in tropical cities, the greater 'quality-of-life warming is the uneven daytime warming. Since a typical city has more people using its public spaces during the day than at night, daytime warming diminishes one's ability to enjoy and function effectively in open spaces." (Correa, 1989) demonstrated that this is the case in the tropics where outdoor living is possible for a large part of the year, provided there is no rain.

Emmanuel et al., (2007) finish with "Owing to the high angle of the sun in the tropics, much of the shortwave radiation on building facades is probably reflected and absorbed by the street surface. This makes street-level discomfort to be largely a problem of shortwave radiation from above and long-wave radiation from below. Shading and urban landscape features such as street trees help to reduce this radiation. In this respect, the lack of shading makes low-density developments less suitable for urban climate amelioration in the tropic (Nichol, 1996)."

3 Research Materials and Method

We simulated microscale thermal interactions using ENVI-met 4.4.2 to explore the effects of landscaping and form of urban blocks on long dry season hottest day microclimate of one of the oldest neighborhoods of the city of Kigali. In summary, we will quantitatively analyze the impact of urban block setting to the climate at the street level through simulation and microclimate analysis of carefully selected street canyons in Nyamirambo neighborhood presenting different urban design factors such as vegetation, street width, street orientation, tree type, building height, building orientation, and ground surface cover type.

The results will then be used to calculate thermal comfort through the UTCI.

3.1 Numerical Modeling—ENVI-met

Middel et al., (2014) defines ENVI-met as "a three-dimensional (3D) computational fluid dynamics model that simulates surface–plant–air-interactions in urban environments. Buildings, vegetation, and surfaces in the area of interest are designed on a 3D grid at a typical resolution of 0.5–10 m. The ENVI-met core uses a full 3D prognostic meteorological model to calculate the main wind flow, temperature, humidity, and turbulence. It is coupled with a 1D model that extends up to 2500 m above ground level to simulate processes at the boundary layer. ENVI-met further incorporates a simple 1D soil model (3D for the first grid layer below the surface) that calculates the heat transfer between the surface and the ground, soil temperature, and soil water content up to 2-m depth.

ENVI-met simulations typically cover 24–48 h and result in atmospheric outputs for each grid cell in the 3D raster as well as surface and soil variables in the simulated environment. Also, receptors can be placed in the model domain to record atmospheric conditions at a user-specified location. Selected output variables include long-wave and shortwave radiation; air, surface, and wall temperatures; latent and sensible heat fluxes; and predicted mean vote (PMV), an indicator of thermal comfort, although this model is best suitable for indoor environment" (Middel et al., 2014).

ENVI-met, initially, had been developed for temperate climate zones of Germany. However, in few other studies in different climatic zones, ENVI-met was applied successfully: hot desertic climate (Middel et al., 2014), humid subtropical climate (Tsoka, 2017), hot-summer Mediterranean climate (Achour-Younsi & Kharrat, 2016), warm-humid climate (Yahia et al., 2018), tropical climate (Emmanuel et al., 2007), and hot arid climate (Elwy et al., 2018).

"In the field of UHI mitigation, numerical models have the obvious advantage over field measurements on account of their controllability as well as time and resource frugality" (Emmanuel et al., 2007). A major shortcoming with ENVI-met is that buildings, which are modeled as blocks where width and length are multiples of grid cells, have no thermal mass. Moreover, all buildings have constant indoor temperature and albedo. The thermal transmittance, or U-value, for walls and roofs are the same for all buildings (U-value is defined as the heat flow per unit area under steady conditions from warm to cold side of a building element, per unit temperature difference) (Emmanuel et al., 2007).

3.2 ENVI-Met Model Evaluation

In order to set up the area input files of the ENVI-met evaluation model (Fig. 2), we will digitize buildings and surfaces of the selected urban block using bitmap raster images from google maps. Vegetation was decoded from the field-based plant inventory. From the Google map bitmap raster image file, we produced a 3D area input file in ENVI-met with a grid resolution of dx 3 m, dy 4 m, dz 2 m resulting in an area input file of 90 × 60 × 10 grids.

We ran ENVI-met for a 48 h period (July 23–July 25 2016) starting at 5:00 with constant time steps (2 s) and model output every 60 min using the configuration parameters mentioned above. We discarded the first 24 h of the model run because the ENVI-met model requires spin-up time. This approach doubles the model runtime, but increases the overall performance of the model, especially in the afternoon and evening hours. Daily air temperature data received from the Rwanda meteorological bureau are compared to modeled temperatures in the Nyamirambo neighborhood.

We ran two sets of simulations: using ENVI-met default variables set and using meteorology station provided variables (max temp: 29.4 °C and min temp: 18 °C). Air temperature simulations from both sets of simulations show very good agreement overall for July 24, 2016, but ENVI-met modeled temperatures underestimated observed temperatures by ± 2 °C. Observed maximum and minimum temperatures from meteorology station variables were 28.20 °C and 19.40 °C, respectively, while ENVI-met modeled maximum and minimum temperatures are 26.470 °C and 16.280 °C, respectively. Maximum air temperatures were registered at 16:00 and minimum air temperature at 6:00.

From the above results, we can conclude that it is best to use meteorology stations provided variables or on-site observed variables.

3.3 Method

In this paper, the approach adopted is to simulate different climatic factors such as air and surface temperature and wind speed behavior rising from two common UHI mitigation strategies (albedo enhancement and urban greening) coupled with the building geometry represented by the SVF. We made comparisons based on surface and ambient air temperature and the resulting thermal comfort expressed as UTCI.

Using ENVI-met, we performed numerical simulations from July 24, 2016 (2016 was considered as the hottest year in the past 10 years (Potter et al., 2017)), which was considered as the hottest day in Kigali's long dry season with a clear sky. We ran the simulations for 48 h following an one-hour time step. For the simulations in ENVI-met, the following meteorological parameters were used: low temperatures of 18 °C and high temperatures of 29.4 °C, no precipitation, and no cloud cover, and wind speed of 3 m/s; based on ENVI-met recordings, Ta and RH daily values were obtained from the meteorological station of Kigali city (see Table 1). For roughness length, we used ENVI-met default values. The albedo of the roof used was 0.9, while building walls albedo was set at 0.3 which corresponds to an aluminum single layer (0100AL) which is the roofing material used on the buildings and concrete hollow block which is the default material used for the building walls (0100C3).

Then, we simulated diurnal microclimatic conditions using the parameters determined in the validation process (See 3.2) and conducted a microlevel analysis of urban and landscape design impacts at 1.4 m on the Ta, the Va, and MRT in two steps. The first step was to simulate the microclimate based on the impact of urban block morphology (trees, vegetation, ground surface cover, and street orientation) and based on the urban block geometry (SVF) in the second step.

The four streets were selected based on their orientation (N–S and W–E), presence and absence of trees, and SVF based on building height (Fig. 2):

- Street 1 (S1): N–S oriented without trees
- Street 2 (S2): W–E oriented without trees
- Street 3 (S3): N–S oriented with trees
- Street 4 (S4): W–E oriented with trees.

The area studied was modeled in ENVI-met based on the real-time on-site situation of building height, vegetation,

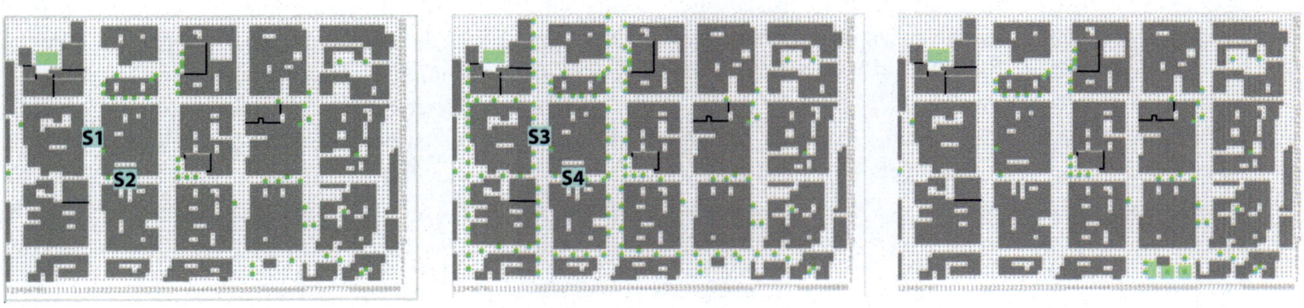

Fig. 2 (Left) Area input file in CS. (Center) Area input file in AT. (Right) Area input file in OS

Table 1 Simulation configuration settings

Configuration data	
Date	24/07/2016
Time	24 h 00
Beginning of the simulation	06 h 00
Ta (low)	18 °C
Ta (high)	29.4 °C
RH	53% (from July 24th, 2015)
Va at 10 m to the ground	3.1 m/s
Wind direction	SE–NW
Albedo (walls)	0.36
Albedo (roofs)	0.9

Table 2 Study area characteristics

Location	1°59′37.0``S
Size	30°02′39.0′E
Maximum building height (m)	270 × 240
Building minimum height (m)	183
Area input cell size	3
Resolution	90 × 60 × 10
Rotation of model from North	3 × 4 × 2
BCR	15°
GCR	51%
Urban form	0.74%
Street type	Low to low rise
N-S street's width (m)	Orthogonal
W-E street's width (m)	9
Tree type	5
Tree crown (m)	Conifer, sparse leaves, small trunk
Tree height (m)	5
	5

building materials, etc. Table 2 contains the physical characteristics of the studied area.

3.4 Simulation Scenarios

To quantify the distinct strategies' impact in ameliorating parameters of local climate, the following scenarios were set up for simulation:

- Based on urban block morphology: current situation (CS), concrete pavements and asphalt street replacement with corresponding cool materials donning high albedo and emissivity (CM), increasing of trees in the street canyons (AT), design a green park within the urban block (OS), a combination of the 4 scenarios above (combination);
- Then, based on the most performing scenarios, we simulated the microclimate in the streets by considering the SVF (case f).

3.5 Microclimate at Street Level

3.5.1 Case A—Current Situation (CS)

Simulations of air temperature showed minimal variations of air temperature on street level; however, very stressful heat conditions are observed on the building level with temperatures reaching as high as 28.52 °C while the maximum air temperature on the street level does not go beyond 25.5 °C (Fig. 3).

Wind speed, on the other hand, is influenced by the street canyon orientation; E-W street canyon registering wind velocity up to 2 m/s, while N–S canyon registers very low wind speed below 1 m/s. This indicates that the buildings form and street orientation play an important role in outdoor thermal ventilation in this specific urban block.

MRT simulation results in this scenario based on street orientation which did not present any particularity, and while wind speed varied according to street orientation, it did not influence much the MRT.

3.5.2 Case B—Additional Trees (AT)

Trees, besides creating a cool canopy and providing shade can enhance or influence negatively the microclimate by slowing down wind flow. We ran simulations of points in the model located in a N–S street canyon with few trees planted on the side and with no trees. The total number of trees in the CS scenario was 50. The trees in the AT scenario were added randomly in the streets surrounding the first two building blocks (see Fig. 2) with a distance of 6 m taking the total number of trees to 130. We chose trees in the family of conifers from ENVI-met database with a small trunk and sparse leaves based on the fact that their height and crown characteristics matched with the trees on-site (Table 2).

Results show that wind speed is less than 1 m/s in both areas with no trees and few trees but lower in the area with trees. This shows how trees can negatively affect the wind speed. In these N–S oriented street canyons, it is best to plant trees and vegetation that will not interfere with the already low wind velocity.

Simulations of points in the model located in an E–W street canyon with few trees and no trees show no particularities of wind, temperature, or relative humidity rather as expected, and the wind speed is higher in the area with no trees than the area with few trees. This shows that in street canyons in the same direction as the wind direction, trees play an important role in the distribution and regulation of wind speed.

Adding more trees on the roadside showcase a cooling effect on the air temperature level by 1 °C in both street canyons; however, little variation is observed on the wind distribution level.

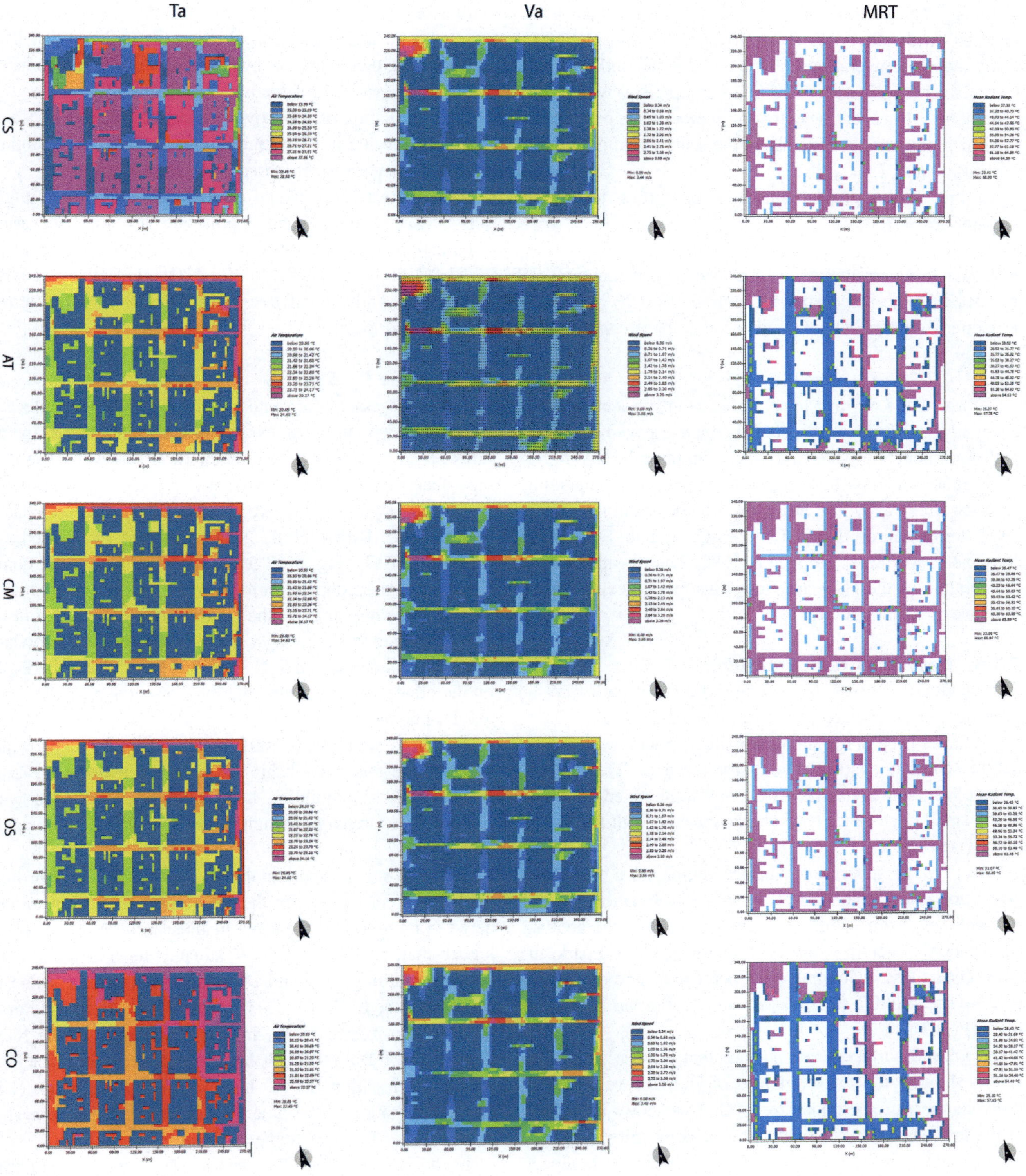

Fig. 3 Ta, Va, and MRT distribution across the five scenarios at noon

3.5.3 Case C—Cool Materials (CM)

Tsoka (2017) defines cool materials as follow: "cool materials are characterized by increased solar reflectivity and high infrared emittance and are considered as a solution toward the urban heat island mitigation. Due to their thermophysical properties, they absorb smaller amounts of solar radiation, resulting in lower heat storage and reduced surface temperatures. As a consequence, long-wave radiation emittance and increase of ambient air temperature through convention can be significantly attenuated." In his study, Wang et al. (2015)

following Taleghani et al. (2014) also showed that "white material (with albedo above 0.9) increased the globe and mean radiant temperature (0.9 °C and 2.9 °C) and produced a cooler air temperature (1.3 °C) in comparison with a dark pavement." However, the extent of microclimate parameters' modification strongly depends on the spatial configuration of the study area (Tsoka, 2017).

In this study, ground surface's properties taken into account are as follow:

- Concrete pavement used/dirty: albedo 0.3 and emissivity 0.9
- Asphalt road: albedo 0.2 and emissivity 0.9
- Concrete pavement light: albedo 0.8 and emissivity 0.9
- Cool gray asphalt

To assess the positive impact of ground surface cover, we changed the current situation used concrete pavement streets with light concrete pavement with a higher albedo.

Simulations show little alterations of surface temperature in N–S street canyons; however, on the building surface level, temperatures dropped by 3 °C, and in E–W street canyons temperatures dropped by 2 °C. The constant surface temperature in the N–S street canyons can be explained by the lack of wind velocity.

3.5.4 Case D—Green Park (OS)

This scenario involves the growing of a small park from an existing non-built piece of land in the urban block.

The trees used in the simulation are 5 m wide with 5 m of height (conifer, sparse leaves, small trunk). Trees, green grass cover, and a pavement walkway were added randomly in the small open space (OS) that is covered with loamy soil otherwise (Fig. 2).

MRT, wind speed, and surface temperature were analyzed to assess the impact of more vegetation on the microclimate from 11 am to 1 pm. The wind speed analysis shows a difference in wind speed distribution; as a park, the OS registers less high wind speed percentage and prevalence of wind speed of 0.7 to 1 m/s probably due to the addition of trees (Fig. 3).

At ground surface level, surface temperature simulation results show significant alterations of ground-level temperatures; temperatures dropped by 7 °C. This shows how parks in urban areas can play a significant role in mitigating heat waves, and it shows how replacing certain high albedo materials with cooler surface materials and vegetation play an important role in cooling the microclimate on a smaller scale.

3.5.5 Case E—Combination

With the three, techniques combined, at 16:00 which is the peak of Ta, a clear distinction can be made between the area with added trees and the area with no added trees. In the area with added trees, the highest Ta is 25.93 °C, while in the area with no added trees, the highest Ta is 27.25 °C. It is worth noting that in the not-built space, we have the same Ta as Ta in the area with added trees.

The simulation of Va shows the lowest value in N–S and the highest value in E–W throughout the day. On the other hand, an increase is registered in N–S.

The highest MRT which corresponds to the high-angle sun between midday and 1 pm with the four scenarios combined shows a distinct decrease in MRT going from 33.91 °C–68 °C in the CS to 30.10 °C–60.35 °C, especially in the area with added trees and regardless of the street orientation (Fig. 5).

3.5.6 Case F—SVF

We also assessed the outdoor microclimate of selected streets by simulating air temperature and surface temperatures based on the sky view factor (SVF) as well. The sky view factor is the fraction of the sky area seen by looking up to the sky (Wang et al., 2015). This parameter evaluates spaciousness (Brown et al., 2001). ENVI-met was used for calculating SVF (Fig. 5). The analysis focuses on the combined strategies coupled with SVF in N–S and E–W.

We, therefore, analyze the results from the simulation of points situated in streets with lower SVF (Fig. 5) than in the rest of the block located in N–S on one hand and in the E–W on the other hand by comparing their respective Ta, Va, and MRT at noon.

Leonardo simulation results of air temperature and wind speed show the relationship between air temperature and wind speed with building height. Slightly lower Ta was registered in areas with high SVF as well as more wind turbulence.

The N–S street's width in the selected block is higher than the E–W streets width; for that reason, the SVF from streets between buildings of the same heights in the N–S streets is higher than SVF in the E–W streets; the SVF in streets between buildings of 5 and 18 m in N–S is of 0.68, while it is of 0.4 in the E–W (Fig. 5).On the other hand, the addition of trees also reduced SVF. In the area situated in the N–S with low SVF from 0.4 to 0.2, a slight decrease in Ta can be shown, but less than 0.5 °C; however, in both scenarios (CS and Combination), the same Ta decrease can be observed between low SVF and higher SVF in N–S.

In the E–W, in the CS scenario, Ta at noon is slightly higher than in N–S. In areas with low and high SVF, in CS scenario, the Ta varies between 24.3 °C and 24.9 °C between high and low SVF, respectively, and in combination Ta varies between 22.71 °C and 23.16 °C which is lower than Ta in the current situation.

Simulation of Va results based on SVF in the combination also shows higher Va between and around buildings with different heights as well as a higher frequency of Va > 1 m/s in N–S and the park.

Simulations of MRT show that the same way MRT in the combination is lower than in the current situation regardless of the street orientation, it is also the case with SVF; however, it is way lower in the area with added trees than the area without.

MRT is closely related to solar radiation; as the sun radiation slowly fades away, MRT also decreases. Additional trees also provide shade, but for these particular trees, the temperature under shade is slightly higher than that of building induced shade; most likely because, at solar noon, trees are more subject to more amount of direct solar radiation than the building facades; which indicates that building induced shading in this situation is recommended over the shade provided by trees.

MRT in the case c (CM) and case d (OS) did not show substantial change compared to case a (CS); which means that ground cover impacts more air temperature than MRT; however, the current situation registers higher MRT than in the rest of scenarios with the scenario with the green park registering lower amount of MRT (Fig. 4).

3.6 Thermal Comfort at Street Level

To assess thermal comfort, we used the UTCI index. Using the data obtained from the numerical simulation, UTCI values in each of the four selected streets were calculated from a typical long dry season day.

The following Fig. 6 presents UTCI value changes on a typical full day of the long dry season based on street orientation on one hand and in each one of the proposed scenarios on the other hand.

Based on street orientation and vegetation, thermal comfort ranged between moderate heat stress and strong heat stress with the highest UTCI value of 33.4 registered around 1 pm in the W–E oriented street with trees. UTCI in the different scenarios showed similar results with 32.8 being the highest value of UTCI registered between 12 and 1 pm corresponding to the time high-angle sun.

4 Discussion

To understand microclimatic dynamics in the tropical climate of Kigali, we systematically analyzed how urban form and landscaping design impact outdoor urban heating and cooling in an urban block composed mainly of single-detached one-story buildings with few 6–7 story buildings.

In this study, the microclimatic effects on a pedestrian are represented by UTCI to assess the amount of thermal stress especially between 12:00 and 1 pm which represents the high-angle sun time.

The results of this study confirm prior efforts that found substantial air temperature cooling, wind distribution enhancement, and surface temperature reduction benefits from the addition of trees and vegetation, introduction of cool road cover, and proper building form and orientation.

According to Wang et al. (2015), “urban trees can reduce the effects of surrounding building mass and help to create a low SVF environment which is cooler in both daytime and nighttime. Deep urban canopy can reduce direct solar radiation during the daytime. Therefore, the thermal environment in open space (high SVF) is worse than that in shaded space (low SVF).”

In this paper, the addition of trees and its cooling effect on the microclimate is seen through the Ta, while Va does not register a big change. In the AT scenario, Ta at noon dropped by 3.89 °C from a Ta max of 28.52 °C to 24.63 °C (see Fig. 3). On the other hand, MRT dropped by 1.1 °C from 68 °C to 66.9 °C (see Fig. 3) which agrees with Wang et al. (2015). The addition of trees along the side of the road did create some shade on the fronts of the buildings and in the areas around the trees. On the thermal comfort level, the AT scenario also proved the above hypothesis although UTCI only reduced by 0.7. This can be explained by the fact that the UTCI was calculated from points from the middle of the street therefore not affected by the effect of shade by the buildings or trees which resulted in the moderate heat stress UTCI range in both CS and AT. On the SVF level, the addition of trees did provide depth to the existing canyon and SVF went from 0.68 (middle of the street) to 0.58 (building front) to 0.31 (immediate tree surrounding) (see Fig. 5). Subsequently, the MRT on the building fronts went from 44.14 °C in areas with no trees to 40.96 °C in areas with trees reducing by 3.18 °C which indicates that the more SVF decreases, UTCI increases creating less comfortable thermal conditions. Here, it is important to note the height and type of trees used in the model. Some trees will provide more shade than others based on their height and LAD. Yahia et al., (2018) recommend that it is needed to plant tall trees with wide and dense canopies along streets, plaza entrances, and setback areas for maximizing pedestrian comfort and to encourage wind flow and activities under the tree crowns, it is recommended to have no branches and leaves up to at least 2 m height.

On the other hand based on the street orientation, while in the CS scenario Ta did not depend on the street orientation, Va on the other hand changed according to street orientation, with Va less than 1 m/s in the N–S oriented streets. After the addition of trees, although the Ta dropped, the Va did not increase in neither N–S and E–W oriented streets. This agrees with Yahia et al., (2018) who stated that trees can negatively affect the microclimate by slowing down Va and

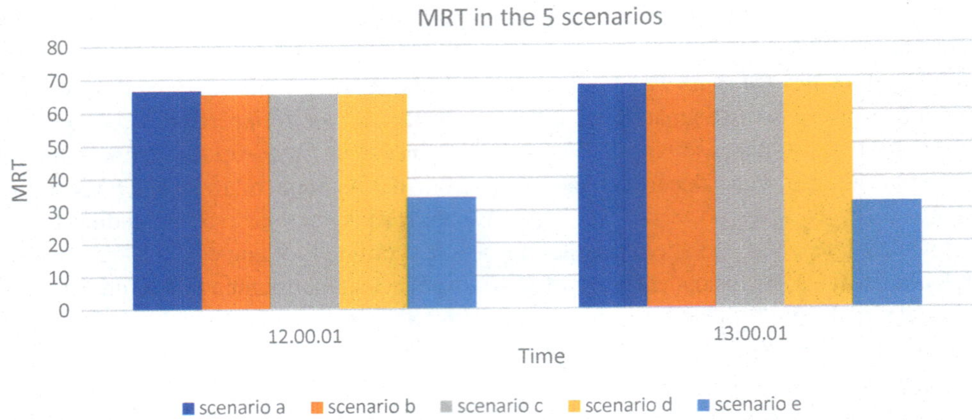

Fig. 4 MRT distribution at solar noon across the five scenarios

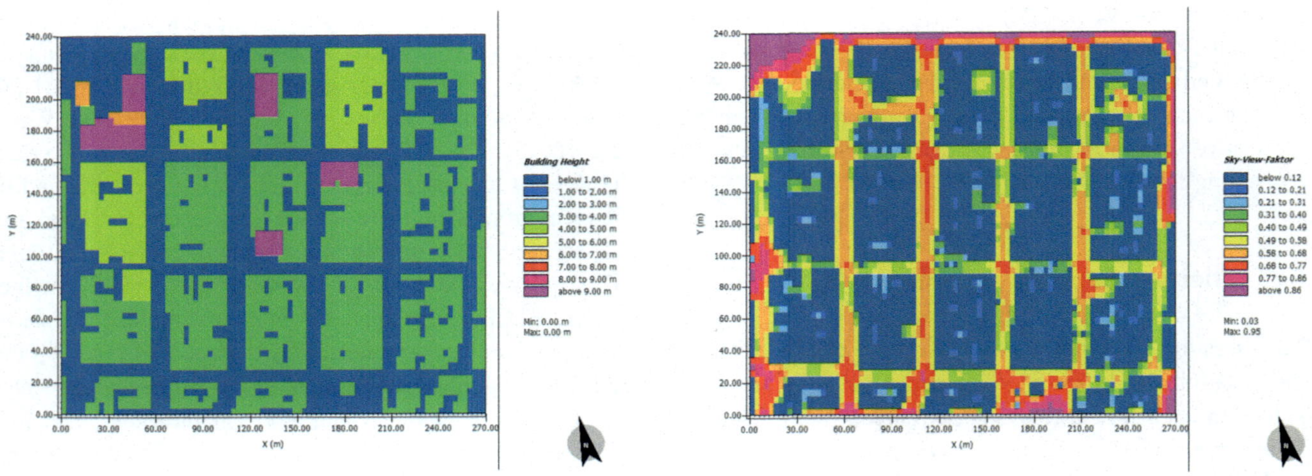

Fig. 5 (Left) Building height. (Right) SVF in the urban block

Fig. 6 UTCI in the six scenarios

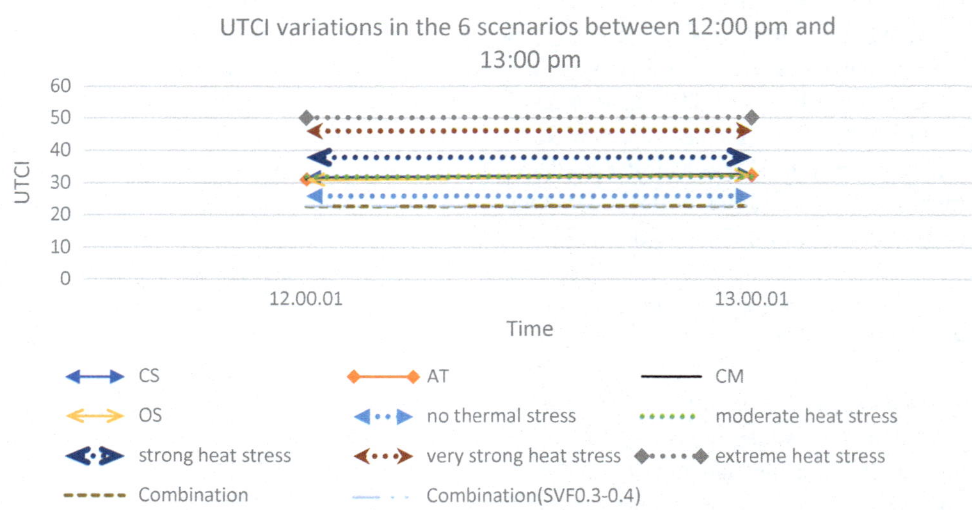

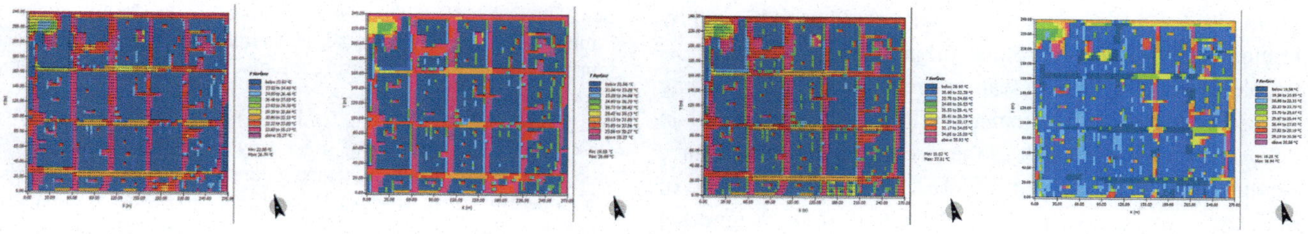

Fig. 7 Tsurf variations in CS, CM, OS, and combination

positively by decreasing Ta. In our case, it is probably best to plant trees and vegetation that do not influence Va at 1.4 m in N-S oriented streets and to carefully choose the type of tree to be used in the E-W oriented streets that can better regulate and distribute Va.

In his study, Wang et al. (2015) stipulates that increasing 10% of urban vegetation could reduce the Ta and MRT all through the day and night-time up to 0.8 °C. In this study, before the addition of trees and the park creation, the urban block had a very negligible vegetation beside the trees here and there. The addition of trees did prove to reduce Ta and MRT, and the created park also proved the theory by registering a drop in Tsurf in the OS of 7 °C on one hand (in the park), and on the other hand, Ta and Tmrt (in the whole block) dropped by 3.89 °C and 1.15 °C, respectively. It should be noted, however, that the daytime MRT is likely to be overestimated by ENVI-met as all of the absorbed radiation on the facades contributes to the warming of the surface, as no energy can be stored (Emmanuel et al., 2007).

The introduction of cool materials did reduce overall min Tsurf; however, the max Tsurf did not change significantly in both the CM and OS scenarios with the OS registering more min Tsurf reduction than the CM although the OS registered highest max Tsurf from 36.74 °C in the CS to 37.81 °C in the OS. The min Tsurf comes from the building surface level dropping by 2 °C (see Fig. 7); this, however, needs further research that takes into consideration the building mass since ENVI-met only considers the façade thermal mass. The same can be observed in the OS scenario where the creation of the park-induced reduction of min Tmax that also coincides with the building surface level dropping by 3.03 °C. This can be explained by the fact that the reflected heat gets in turn trapped and absorbed by the surrounding. On the other hand, Tsurf inside the park dropped of 4.33 °C from 30.86 °C to 26.53 °C (see Fig. 7). On the Ta level, Ta was also reduced in both scenarios. MRT in the whole block in CM, and OS did not drop at street level; this shows that even if some strategies tend toward decreasing ambient Ta, they do not, however, lead to more thermally comfortable conditions. This shows how one technique alone is not enough and that materials albedo, emissivity, and heat capacity are also factors to be considered. It shows how reducing Ta does not necessarily reduce UTCI. While higher albedo reduces the surface temperatures, and consequently, also the air temperature, it increases the amount of reflected shortwave radiation from the environment at the same time. Since the MRT takes into account this increased energy flux, MRT may even increase (Emmanuel et al., 2007). Thus, the insignificant effect on thermal comfort is to be expected.

Lastly, the scenario involving the four techniques combined shows that adding trees on the roadsides together with replacing the used concrete pavement and asphalt with light concrete pavement gives the best results when it comes to reducing outdoor Ta, Tsurf, and Tmrt. The deriving thermal comfort also confirms the latter by falling from the moderate heat stress range of each technique approached individually to the no thermal stress range with the four techniques combined as stated by UTCI (see Fig. 6).

An analysis of the distribution of MRT in the areas showed that the area with single-story buildings had the highest frequency of high MRTs and the lowest frequency of low MRTs. The study illustrates that street canyons with low SVF lead to more stressful urban spaces than streets with high SVF; results showed that streets with high SVF recorded lower value of UTCI than streets with lower SVF (Fig. 6).

5 Conclusion

This study quantitatively sheds light on the contribution of urban design and planning on enhancing or inhibiting outdoor thermal comfort conditions in Nyamirambo neighborhood of Kigali city at a time when Kigali city is implementing its ambitious urbanization through the Kigali city master plan and acts as a contribution to scientific research in new areas by making ideas and knowledge accessible around the world. At the opposite of the vibrant Kigali city center neighborhood showcasing mid- to high-rise buildings, Nyamirambo setting is only composed of low rise buildings; however, it is also one of the most active neighborhoods in the city of Kigali and as such presents residential, commercial and leisure attributes while also being densely populated.

This study shows how the outdoor thermal conditions experienced by pedestrians in the hot and dry season of the selected urban block at its current state could be enhanced by implementing certain cooling urban design strategies as simple as planting trees, creating green parks, and adopting the use of cooler surface materials. It also shows how urban design characteristics such as the street element can enhance air quality by increasing the SVF. UTCI is in the range of moderate heat stress currently; however, if the suggested urban heat mitigation strategies were introduced, the UTCI would fall in the no heat stress range.

Although the orientation of the streets and the SVF did not prove to have a big effect on the microclimate in this particular urban setting, N–S oriented streets are the most thermally comfortable at noon and especially in areas with low SVF (Fig. 6). The study only focused on thermal conditions at solar noon, and it does not consider the thermal characteristics of the building facades in the urban block, but it already allows raising the levers of action to improve the conditions of outdoor thermal comfort. Besides, future studies should take into consideration the temperatures variation during nighttime and in other urban fabrics which present different morphology than Nyamirambo and preferably in areas with fewer activities during the daytime than at nighttime.

References

Achour-Younsi, S., & Kharrat, F. (2016). Outdoor thermal comfort: Impact of the geometry of an urban street canyon in a Mediterranean subtropical climate-Case study Tunis, Tunisia. *Procedia-Social and Behavioral Sciences, 216*, 689–700.

Adolphe, L. (2001). Modeling the link between the built environment and urban climate: Towards simplified indicators of the city environment. In *Building simulations, proceedings of IBPSA (International Building Performance Simulation Association)*.

Ait-Ameur, K. (2002). Characterization of the microclimate in urban spaces through the validation of a "morpho-climatic" indicator system. In *Proceedings of PLEA 2002—The 19th conference on passive and low energy architecture*. Toulouse.

Ali-Toudert, F., & Mayer, H. (2006). Numerical study on the effects of aspect ratio and orientation of an urban street canyon on outdoor thermal comfort in hot and dry climate. *Building and Environment, 41*(2), 94–108.

Auliciems, A., & Szokolay, S. V. (1997). *Thermal comfort*. PLEA with Department of Architecture, University of Queensland.

Baffoe, G., Malonza, J., Manirakiza, V., & Mugabe, L. (2020). Understanding the concept of neighborhood in Kigali city, Rwanda. *Sustainability*.

Brode, P., Fiala, D., Blazejczyk, K., Holmer, I., Jendritzsky, G., Kampmann, B., Tinz, G., & Havenith, G. (2012). Deriving the operational procedure for the universal thermal climate index (UTCI). *International Journal of Biometeorology, 56*(3), 481–494.

Brown, M., Grimmond, S., & Ratti, C. (2001). Comparison of methodologies for computing sky view factor in urban environments. In *International society of environmental hydraulics conference*.

Chatzinikolaou, E., Chalkias, C., & Dimopoulou, E. (2018). Urban microclimate improvement using ENVImet climate model. In *The international archives of the photogrammetry*.

Chen, H., Ooka, R., Huang, H., & Tsuchiya, T. (2009). Study on mitigation measures for outdoor thermal environment on present urban blocks in Tokyo using coupled simulation. *Building and Environment*, 2290–2299.

Chen, L., Ng, E., An, X., Ren, C., Lee, M., Wang, U., He, Z. (2010). Sky view factor analysis of street canyons and its implications for daytime intra-urban air temperature differentials in high-rise, high-density urban areas of Hong Kong: a GIS-based simulation approach. *International Journal of Climatology*, 121–136.

Correa, C. (1989). *The new landscape: Urbanisation in the third world*. Architectural Press.

Elwy, I., Ibrahim, Y., Fahmy, M., Mahdy, M. (2018). Outdoor microclimatic validation for hybrid simulation workflow in hot arid climates against ENVI-met and field measurements, 29–34.

Emmanuel, R., Rosenlund, H., & Johansson, E. (2007). Urban shading—a design option for the tropics? A study in Colombo, Sri Lanka. *International Journal of Climatology*, 1995–2004.

Fange, P. O. (1970). *Thermal comfort: Analysis and applications in environmental engineering*. Danish Technical Press.

Hoeppe, P. (1999). The physiological equivalent temperature—a universal index for the biometeorological assessment of the thermal environment. *International Journal of Biometeorology, 43*, 71–75.

Manirakiza, V. (2014). Promoting inclusive approaches to address urbanization challenges in Kigali. *African Review of Economics and Finance*, 161–180.

Mestayer, P., & Anquentin, S. (1994). *Climatology of cities, diffusion, and transport of pollutants in atmospheric mesoscale flow fields*. Kluwer Academic Publishers.

Middel, A., Hab, K., Brazel, J. A., Martin, A. M., & Guhathakurta, S. (2014). Impact of urban form and design on mid-afternoon microclimate in Phoenix local climate zones. *Landscape and Urban Planning*.

Nichol, J. (1996). Analysis of the urban thermal environment with LANDSAT data. *Environment and Planning B: Urban Analytics and City Science*.

Oke, T., Johnson, G., Steyn, D., & Watson, I. (1991). Simulation of nocturnal surface urban heat islands under 'ideal' conditions: Part 2 Diagnosis of causation. *Boundary layer meteorology, 56*.

Parsons, K. (2003). *Human thermal environment*. Taylor & Francis.

Potter, S., Cabbage, M., & McCarthy, L. (2017, August 7). *National aeronautics and space administration*. Retrieved January 23, 2019, from https://www.nasa.gov/press-release/nasa-noaa-data-show-2016-warmest-year-on-record-globally

Rizwan, A., Leung, Y., & Chumbo, L. (2008). A review on the generation, determination, and mitigation of urban heat island. *Journal of Environmental Sciences, 20*.

Salka, M. (2014). Sustainably localized urbanisms: The role of emerging technology in developing Kigali, Rwanda. *Spring*.

Strategic Foresight Group. (2013). *Blue peace for the Nile*.

Tsoka, S. (2017). Investigating the relationship between urban spaces morphology and local microclimate: A study for Thessaloniki. *Procedia Environmental Sciences*.

Wang, Y., Berardi, U., & Akbari, H. (2015). Comparing the effects of urban heat island mitigation strategies for Toronto, Canada. *Energy and Buildings*.

Wong, N., Chen, Y., Ong, C. L., & Sia, S. (2003). Investigation of thermal benefits of rooftop gardens in the tropical environment. *Building and Environment*, 261–270.

Yahia, M. W., Johansson, E., Thorsson, S., Lindberg, F., & Rasmussen, M. I. (2018). Effect of urban design on microclimate and thermal comfort outdoors in warm-humid Dar es Salaam, Tanzania. *International Journal of Biometeorology, 62*, 373–385.

Where is Wakanda? Who is T'Challa? The Sustainable African City Re-Imagined as a Livable and Lovable City

Amira Osman

Abstract

Where is Wakanda? Is it place? An idea? Who is T'Challa? Who are the African heroes when it comes to the conceptualising of African space and cities? Images of future African cities tend to resemble Wakanda in the movie "Black Panther". These images are dominated by high-rise glass and steel structures—an aspiration of many cities in emerging African economies. While there is a need to imagine alternatives to current trajectories of development of African cities, it is important to consider images that lead to the creation of inclusive spaces, a scale that is accessible and a relationship with the public realm that is meaningful to the majority of city residents. An anecdotal narrative is combined with an analytical text based on a literature review on some of the themes presented. Based on what emerges from this debate, the African City is re-imagined, through the extraction of some lessons and ideas from the analysis of case studies, as a livable and lovable city. These concepts imply that cities are just, attractive, functional and resonate with many communities over time. In this way, cities become sustainable by accommodating for all people and allowing for diversity, choice and change. The future image of these cities thus also strongly relates to their geographical, social, cultural, economic and historical contexts.

Keywords

Wakanda • T'Challa • African cities • Livable-lovable cities

A. Osman (✉)
Tshwane University of Technology, Pretoria, South Africa
e-mail: osmanaos@tut.ac.za

C. Alalouch et al. (eds.), *Towards Implementation of Sustainability Concepts in Developing Countries*, Advances in Science, Technology & Innovation, https://doi.org/10.1007/978-3-030-74349-9_3

1 The Approach: How Ideas Will Be Explored

We need to investigate deeper and look beyond the futuristic imagery of African cities and what it really implies. This links with aspects of rights to space, the city and housing. While issues of small-scale people-centred interventions, appropriate technologies and basic service provision may not be as compelling as the imagery that resembles Wakanda, yet they are crucial. In a quest to explore approaches that achieve a more egalitarian and equitable city, and to give momentum to the realisation of real change in African cities, the power of space and spatial transformation has to be fully grasped. When this is not acknowledged, anger and protest have erupted—not just in South Africa but globally.

To reinforce these ideas, three case studies will be academically "unpacked". I believe that these three case studies have had impact on debates around architecture, cities and space. These are:

(a) Warwick Junction, Durban and a comparison with other international cases of conflict and protest—specifically Takseem Square in Istanbul, Turkey, and the riots around the Khartoum Structure plan in the Sudan which both, similar to the Durban case study, emanated due to perceived lack of access to city space and a restriction on opportunities; the author has offered opinions on both of these matters on various platforms, and this text draws heavily from them (Osman, 2016b, c)
(b) The awarding of a Pritziker to Alejandro Aravana, for his work with low-income residents in cities and how it was negatively received by some in South Africa; I have previously shared many of these thoughts on social media and in a blog (Osman, 2016d)
(c) The death of the "star architect", Zaha Hadid—how her life and work were celebrated by South African architects and the debates generated around the social responsibilities of architects. This text relies heavily on

a blog entry by the author just after the architect's death (Osman 2016a).

The image of an African future (and its manifestations as an African city) needs to be analysed in the light of recent decolonisation debates. It also needs interrogation in terms of how we as architects and built environment professionals are prepared (or not) to work in areas of contestation and with vulnerable groups and fragile economic ecosystems. Problematic entrenched professional attitudes need to be systematically exposed, analysed and dismantled in our educational systems and in our methods of practice. Built environment professions tend to participate in practices that lead to disempowering, humiliating and restricting opportunities for communities. While many decisions are taken by government and private sector actors, these could not be realised without built environment professionals—and these decisions may result in the destruction of livelihoods through disruption of economic networks and damaged ecosystems. This many times results in deepening conflict and reinforcing divisions in cities.

2 Where is Wakanda? Who is T'Challa?

Where is Wakanda? Is it place? An idea? Who is T'Challa? Who are the African heroes when it comes to the conceptualising of African space and cities? Images of future African cities tend to resemble Wakanda in the movie "Black Panther". These images are dominated by high-rise glass and steel structures—an aspiration of many cities in emerging African economies. (Figs. 1, 2, 3 and 4).

And why has a conception of an African city not been produced in the minds of these African heroes? It is a fact that Wakanda is an image of an African city that has been developed, not in Africa, but elsewhere. However, this image, that has captured the imagination of the world, needs closer scrutiny and interrogation. Karam and Kirby-Hirst (2019, p. 12) argue that there have been benefits to the film Black Panther as it has allowed us to engage in a discourse around blackness, identity and afrofuturism.

Sadly, this scrutiny of the imagery presented in Wakanda has not always happened, with some adopting it in an indiscriminate manner:

> Encouraged by the famous blockbuster movie, Marvel's Black Panther, the AU announced Wakanda One, a technologically advanced hub, as a symbol of the new African civilization, envisioned in Agenda 2063. It will be Africa's own Silicon Valley or Shenzhen - the main site for the Fourth Industrial Revolution… Wakanda One seeks to turn Mosi-oa-Tunya (Victoria Falls), on the border of Zimbabwe (2000 hectares), into Africa's technological epicentre, with Zambia (offering 132 hectares) to the project. The irony of the chosen leading country cannot be ignored – Zimbabwe where citizens are on the streets – violently protesting against the deepening economic and political environment (Monyae, 2019).

On the contrary, some of these images are perhaps being taken more seriously than they should be: "… urbanists and city-planning experts agree that some of the design and infrastructure of the fictional place have real-life possibilities". (Malkin, 2018) However, the author does continue to

Fig. 1 Artist view of the Sunut Development in Khartoum (Landsel, 2007)

Fig. 2 Internal view of Corinthia Hotel in Khartoum, on the periphery of the Sunut Development at the meeting point of the Blue and White Niles (photograph by author in 2017)

provide a deeper reflection on what actually constitutes "Wakanda". The streets of Wakanda are seen to be a pedestrian-focused and commerce-filled which is perceived as being desirable in a high-density African city. The level of technology implied may be questioned (such as the high-speed trains) in terms of how feasible this is or how relevant to they are to current African cities.

Wakanda also shows a mix of "futuristic glass-and-steel towers" with African inspired imagery as Malkin's article elaborates: "It looks like regional architecture as opposed to this anywhere-ness that we seem to have in our global architecture these days. I saw architectural expression that was not only organic, but of its place and of its culture." (Fig. 5).

Despite this favourable view of Wakanda, when people think of it, they seem to focus on the shiny towers, perhaps because of where the inspiration for the film set actually came from: "Production designer Beachler turned to the architecture of Zaha Hadid, Buckingham Palace and afro-futurism when creating the fictional world for the new Marvel film" (Yalcinkaya, 2018).

The problem we are faced now, with "Wakanda visions" of African cities, is that these imply a highly exclusive and exclusionary setup and not an environment of equality. (Osman in Smith, 2018) While the film set implies a mix of architectural features, the reality is that, in African cities, there is an ambiguous relationship between governments and informal activities in cities, with local government and city authorities mostly withholding service, maintenance and support to informal settlements, markets and dwellers. (Osman in Smith, 2018) Futuristic visions for African cities

Fig. 3 Incomplete Sunut Development in Khartoum taken from the Corinthia Hotel (photograph by author in 2017)

Fig. 4 Shimmering Kigali skyline in the background, foregrounded by the local mode of transport and a pedestrian (photograph by author in 2018)

Fig. 5 View of Wakanda in the movie Black Panther. *Source* http://africanism.net/the-architecture-of-black-panther/

are many times the result of collaboration between governments and foreign investors in an arrangement that is problematic as it does not consider the need for space and opportunity for the pavement trader as well as the corporate investor.

Many African cities either aspire to have a new city, or part of the city, that has a particular image conceptualised by governments and the private sector capital sometimes (many times?) crossing borders with little regard to the actual needs of the local communities and the climatic context. This has been described as: "…a new form of urban colonialism driven not only by external (Western or multinational) investors but also by native ones, and by local and national governments, who share the same interest as international elites" (Cardoso and Faletto 2008, in Burocco, 2018, p. 62). While conceived in the minds of its creators, it still relates to the aspirations of "big capital/money" and ambitious governments on the African continent who are hungry for power.

Indeed, Burocco (2019, p. 16) interrogates the film in multiple ways, with an understanding that "capital" needs to find "new frontiers accumulation", identified as both "raw resources and subjectivities". She addresses the aspirations reflected and questions the trajectory of "progress" as portrayed in the film which is very much based on a Western conception with little room for what she calls a "rhizomatic" or "multiple" modernity that is different. (Burocco, 2019, p. 15) It is, therefore, not only the spatial and formal features of the film that needs to be addressed, but also the value systems that the film espouses: "The euphoria that accompanied the debut of the film *Black Panther* was not surprising, but it raised questions about the inability to articulate a critical discourse in relation to the film". (Burocco, 2019, p. 1) The same author goes on to interview some people about the film, one of whom states: "I think [the film] puts Africa and the diaspora in a situation of rivalry, competition and humiliation for whites' entertainment. I think it is essentialist, idealistic and bourgeois... I think it disrespects the history of struggle and resistance both in the African diaspora and in Africa." (Ruas, in Burocco, 2019, p. 5).

At its roots, this is a debate around decoloniality and how Africa has been and continues to be portrayed. In South Africa, we are critically aware of the power of the built environment to divide and disadvantage people. There are deep injustices in our society which have forced us to engage with the concept of decolonisation in space, practice and education. Three essays are presented below around how these debates have evolved.

3 Competition in Cities Between Authorities, "Big Capital" and Communities

Warwick Junction is a unique location in Durban. It is a major transport node and provides the setting thousands of informal traders. About half a million users pass through Warwick Station on a typical day, and there are about 8000 informal traders. It is believed that flow of capital within Warwick Junction is similar to that in Durban's city centre and major shopping malls. Despite all the challenges it has faced, this location continues to thrive; many people, families and communities are supported through the co-existence formal and informal economic activities at Warwick (Osman in UIA 2014 Durban 2014).

Warwick Junction has been a site of dispute for many years and has an uncomfortable relationship with the city council (Mdluli, 2010). The site, though functioning, has need for many improvements, interventions, regular maintenance and cleaning that go unsupported. In addition, the traders feel that their livelihoods are under constant threat of being disrupted because of the possible construction of a mall. This has led to numerous protests and anger (IOL Business, 2009). The concern is that the success of the current markets would be disrupted, and their stability perceived as:

> ...an established equilibrium of mixed formal and informal retail providers in the area. Inappropriate development in Warwick, such as the introduction of a major formal retail node, like a mall, in the heart of the market, threatens to undermine its vibrant economic and social fabric... Hundreds of informal traders would be displaced. Customers who rely on the market's affordable and accessible goods would be disrupted, and the livelihoods of a wider network of businesses would be threatened (Dobson, 2013).

Some are concerned that the markets of Warwick survived apartheid and are now under threat in the post-apartheid era (Maharaj, 2010). The situation in Durban can be compared to the Sudanese situation of state-sanctioned mega development and how this leads to displacement of people and disruptions of livelihoods. Protests erupted in Khartoum in 2016 when it became evident that future plans for the capital city may have negative consequences for peoples' daily activities, access to opportunity and public space; there was increased awareness about the political aspects of planning and space (Osman, 2016b). Though the protests were triggered by the rumoured sale to private developers of prime locations along the River Nile, including the sale of the historical site of Khartoum University, it also became a debate about the controversial Khartoum Structure Plan and how the city's unique heritage, natural, social and cultural identity could be disrupted through poorly conceived urban interventions.

Cetin and Has (2014, pp. 52–55) believe that these types of conflicts in cities are the result of neo-liberal urban policies and a reflection of the resilience of local communities in opposing government investments, and government collusion with the private sector:

> ...towards a re-configuring of the spatial economy to the benefit of a specific section of society instead of all urban actors... architecture and urbanism can be considered as spatial dimensions of an ideological war of different interest groups in cities... this struggle manifests itself as the polarisation between corporate sector and public... governments [that] use planning as a means of capitalist control over urban (public) space... space is an instrument of resilience through accommodating multiple identities in a single space on the one hand, and spreading the identity of 'disobedience' in multiple spaces on the other...

The Gezi Park, which is located next to Takseem Square, witnessed resistance to proposed urban development plans, and it was "spatial" resistance at every stage, from the invasion of the park, the construction of structures, the re-transformation by the municipality, the restrictions of movement in space and the defiance of those restrictions (Demir, 2014, p. 56).

> The Gezi Park protests of June 2013 which drew the attention of the world to a very urban conflict in Turkey's most populous city that arose from the government's commodity based decision mechanism for urban land (Demir, 2014, p. 1206).

These conflicts, while triggered by spatial issues, follow in the pattern of the occupy movement and Tahrir square protests; they are also testament to the power of "virtual networks", and they demand that the discipline of architecture responds to evolving sociopolitical circumstances (Goyal & Des, 2014, p. 977). This case study shows us that what is perceived as progress and modernity is a camouflage for much complexity, unhappiness and conflict—something that the polished imagery of Wakanda does not engage with.

4 A Cause for Skepticism Rather Than Celebration: The Aravena Pritziker 2016

Judging by social media, news of Aravena being awarded the 2016 Pritzker generated some contention between South African professionals. Aravena is mostly known for his significant contribution towards re-thinking housing for the poor in well located city sites through a partnership (a "do-tank") known as Elemental (Osman, 2016d). While he has designed many other buildings, his work on housing is the perhaps the most notable.

The South African architect, Thorsten Deckler, questioned the "myth of the creative genius" and celebrated that someone from a developing context was recognised as generally "mainstream architectural media drowns out work like this" (Deckler, 2016). Aravena has been described as "radical" because of its engagement with socially conscious design problems through a: "Robin Hood structure [which] runs throughout Elemental's work, split three ways between social housing, urban planning and more lucrative commercial contracts" (Wainwright, 2016). It is ironic that such an approach would be labelled as "radical" by the proponents of mainstream architecture.

> We invited Aravena as a keynote speaker at the 2005 World Congress on Housing which we hosted at the University of Pretoria. At the time, the Quinta Monroy project (2001-2004) had started to create some ripples in the field – an architect engaging differently with urbanism and poverty, generating debate on these topics in a profession that has generally tended to distance itself from the 'messiness' of informality and affordable housing in cities. Aravena presented on how a hybrid model of funding was used in the project, subsidies, savings and loans. He explained how the subsidy – which needed to include land, infrastructure and architecture – at the time allowed for 30m^2 of built space – which was built with the intention of the users themselves expanding the residential unit to 72m^2 after occupation (Osman, 2016d).

When Aravena delivered this talk in 2004, South Africa's Breaking New Ground policy had just come into effect under Lindiwe Sisulu's leadership of the then newly re-named Department of Human Settlements, implying a re-focus from "house" to "human settlements". Aravena presented high-density urban typologies which offered solutions for urban overcrowding, circulation, light and ventilation while still allowing for future expansion under the control of residents—this resonated with South African concerns in housing and generated great interest from the audience (Osman, 2016d).

Aravena presented buildings that created a good sense of urban presence and spatial definition and a process to negotiate and manage the relationship between the desires of individual families and a larger community through built form. Indeed, the Elemental approach focuses on the high-density, urban "…architecture as an artful endeavour… meeting socio-economic challenges", suggests Franco (2016) and asks the question: "If Chilean architecture is so good, why is social housing so bad" (Franco, 2016).

Since 2005 and the Pretoria Congress, Aravena has been involved in numerous residential projects which replicate the system devised in earlier projects and his practice has delivered around 2500 units (Wainwright, 2016). These projects have similar designs to Quinta Monroy with fixed structures delivered initially and allowing for future additions by the residents. The value of this approach is in its replicability and adaptability to different site conditions. The Elemental website expresses the complexity of the process and negotiations that have to take place with resident organisations, clients and local authorities. Elemental have used the same approach in affordable multi-family and multi-storey buildings as well as higher quality single family homes (Elemental n.d.).

Why then, despite these parallels with the South African condition and despite these achievements, did South African architects not greet this award with the celebration it deserved?

5 May the Day of Your Praise Never Come": The Celebration of Zaha After Her Passing

> Insha Allah yom shukraak maa yaji – a Sudanese saying, the 'day of your praise' meaning the 'day of your death'.

As the General Reporter for the International Union of Architects Congress held in Durban, South Africa in 2014 (UIA 2014 Durban), I have shared my experience regarding the sentiments of South African architects about various high-profile international architects. These sentiments were expressed in various ways and at various platforms. The people involved in the organisation believed that the event was an opportunity to reflect a South African view of professional values, beliefs about education, practice and how the profession viewed life in general (Osman, 2016a). When Zaha Hadid was nominated as a speaker, there was a degree of outrage—yet, when Zaha passed away about two years after the event, there was a massive celebration of her life and work and an outpouring of sadness. While the passing of anyone is a cause for sadness, what puzzled me was at a professional level: where were those voices of criticism that were so evident a few years ago?

Zaha deserved recognition in many ways: at times during her career, in a male-dominated profession, she was the only woman to have achieved such high accolades; she was an Arab working in a hostile environment. Having said that, I still believe that we should critique her work on ethical grounds. Sisson (2016) explains how Dame Zaha Hadid was a much-decorated architect, RIBA royal gold medalist (the first time awarded to a woman) and one of Time's "most influential people" in 2010. Sisson (2016) elaborates:

> It's a bit of a fool's errand to argue who, historically, is the greatest architect, but strictly in terms of awards, how does Hadid stack up? …. A head-to-head between Hadid and Foster gets tougher, making it harder to declare who comes out on top. Both have been given royal recognition (Dame Hadid and Lord Foster) …Hadid has often been the subject of unfair criticism compared to many of her peers… impact and influence go beyond just trophies and medals, both can lay claim to an impressive amount of recognition.

It is something of a paradox, that despite the celebration of her work, Zaha's first building was decommissioned because it failed to fulfil its intended function. Instead of her professional reputation being damaged, the building became a much-celebrated and much-visited museum (it was built as a fire station). As the building became a mecca for architects, her career flourished (Bayley, 2015; Fiederer, 2016).

> She was an architect who pushed the field forward, toward ever more complex, organic shapes that seemed to take their inspiration from the webbed patterns of biological tissue and the globular shapes of cells… She was far more interested in pushing the boundaries of design than of society (Miranda, 2016).

Bayley (2015) states that "The locality Hadid prefers is the back yards of tyrants and dictators", referring to, among others, Zaha's involvement in the Qatar World Cup designs. Amnesty International (2016) names many involved in the World Cup constructions sites in Qatar. Some may argue: why pick out Zaha? It is not only her practice that that has been silent about "safety on sites, exploitation, unethical recruitment, delayed salaries, forced labour, retention of passports and deceptive recruitment"—as listed in the Amnesty International report. Architects are generally seduced by form, perhaps a shortcoming of our educational systems. According to Miranda, Zaha's futuristic designs are so appealing that we have failed to notice the "less seemly" or "unsavoury" aspects of her career (Miranda, 2016). And like many architects tend to do, Zaha denied that there were ethical aspects to her practice: "Hadid's complex legacy reminds us that genius does not operate in a dustless ether of abstractions. It matters where the buildings are built, as well as who builds them." (Nazaryan, 2016).

I have previously stated that "The fact that Zaha has managed to generate such controversy, to capture the imagination with her architecture, to be discussed with such emotion is to be acknowledged. This does not mean that it exonerates her work from being critiqued—or the profession that so highly awarded her from being questioned". (Osman, 2016a) The fact that her work went on to be used as a key informant for Wakanda is something to be contemplated.

6 Decolonisation in Cities, Space, Practice and Education

What these three case studies show us is that there is much to be "read" in imagery. In all of the above cases, architects have participated in the controversial projects that lead to anger and protest. Places inhabited by the poor, and places where the poor carve out livelihood opportunities, continue be highly disputed spaces, and these poor residents continue to have an ambivalent relationship with the city authorities that are never fully supportive of it or its community. While such contexts may be represented in tourist brochures as something to visit, the city also never fully acknowledges its presence by extending the same level of service it would any other part of the city. Hackney (1990) refers to this ambiguous relationship between cities and their residents as "official vandalism".

The awarding of a prize to an architect known for housing for the poor also exposed some concern about the image of what is perceived to be architecture deserving of being awarded—which is usually not aligned with our daily realities in cities of the global south in general, and Africa specifically. The institutions that continue to award the work of star architects far removed from real need also need to be considered—and the lack of focus on the social responsibilities of architects. It is interesting to witness the silence of our professional institutes with regards to issues around practice in contexts of conflict in cities. The profession remains untransformed, as do city structures.

To be able to interrogate and critique the aspirations reflected in the film, the Black Panther, as well as in the imagery of futuristic African cities, we need to "be" and "think" outside of what we have been taught and led to believe as "the truth". This is where theories of decolonisation come into play. There are aspects of decolonisation that are more generic and apply to all disciplines while some aspects are more discipline specific. In the built environment fields, space and power are critical in understanding how alternative approaches may be developed. It is also important to acknowledge that decision-making in the built environment is never neutral. Decisions about the use of resources, the location of projects and the imagery adopted always have deep, multiple and complex meanings.

Many aspects of our lives need to be reinvented by us collectively. (Osman in Smith, 2018) We need to identify our own heroes and value systems on which we design and transform our cities in Africa: "It becomes crucial, from an empowered afro diasporic perspective, within the contemporary economic power relations, to challenge a distorted view of development and modernity. This view seems to lead to a normative assimilation of aesthetic concepts and subjectivities, functional to the new cognitive capitalism and favourable only to its beneficiaries." (Burocco, 2019, p. 20).

7 Future Visions of African Cities and the Architectural Profession

Khartoum's Sunut development has been called "Dubai on the Nile" or "Sudan's Manhattan" (Landsel, 2007). An urban forest, of Sunut trees, was partially destroyed in its (as yet incomplete) development: "They have removed all the trees, built the neighbourhoods, and called them after the trees" as my mother has said (Osman, 2019).

> We certainly need to re-imagine a future for African cities. Yet, the imagery of al Sunut and other similar developments in Africa, implies a scale and interface with the public realm that is highly exclusive and exclusionary. Issues of rights to housing, access of space, appropriate technologies and basic service provision may not be as compelling as the imagery presented, yet they are crucial (Osman, 2019).

By not acknowledging the power of decisions around space in cities, anger and protest many times erupt. I believe that these debates are related to debates around decolonisation:

> If done properly and critically a lot of what we count as great will fall in the process of decolonising knowledge. A lot of formerly unvoiced and unheard ideas will come to light. The process of critical scrutiny is essential to the success of this project – and nobody gets a free pass (Broadbent, 2017).

Al Jazeera's series titled Rebel Architects explains how:

> … the tone of architecture culture has changed in only a few years. In the heady days of the 2000s, architects were in furious competition to produce "iconic" buildings for a global market. Virtuosi such as Frank Gehry, Zaha Hadid and Norman Foster kept the media fed with fabulous images of museums and corporate headquarters, earning the moniker "starchitects". But after the financial crash of 2008, it became clear that the social value of so much of that starchitecture was nil. And there was a correction, to borrow a stock market term, in the architect's image (McGuirk, 2014).

I have previously blogged that:

> "The irony is that the same Euro-centered, male-dominated profession is now entering the arena of poverty, informality and disadvantage with the same kind of 'big attitudes' and 'big thinking' practiced in the rest of the profession. So we now have to deal with 'celebrity' slums (as someone referred to Kibera in Nairobi) and the celebrity architects that have now made those slums their new playgrounds (Osman, 2016a).

In the last few years, working in areas of poverty has tended to generate stardom and adulations similar to those in mainstream architectural practice. A decolonised understanding of cities and space is therefore intrinsically linked to an understanding of ethical practice in the built environment professions. Yet, the architectural profession in particular is not known to take strong ethical positions. Capital and power go hand in hand with big architectural projects: "The projects that get graced by capital 'A' architects are intertwined with the capital required to build them" (Murphy, 2015). Rowan Moore writes:

> Lord Rogers (formerly Richard Rogers), had the ear of ministers to an extent matched by few of his profession… the best in the world it was often said… Rogers, however, acted according to the dictum of the nineteenth-century American architect H. H. Richardson, that the first principle of architecture is to get the job. He argued, charmed and cajoled… (Moore, 2012).

And of Mies:

> For their part architects – some, at least – like to flirt with power. Ludwig Mies van der Rohe lingered longer in Nazi Germany than was decent, apparently in the hope that the regime might adopt his architectural style (ibid).

Architects, their professional and educational institutes promote individuality, image and representation—and these can be used as tools of silencing, sidelining and undermining "other" voices and "other" experiences. A decolonised understanding of cities and space would focus on access to city space, opportunity and small-scale livelihood initiatives and would broaden architectural and built environment discipline to service the wider community and would take into account the multiple role players that shape the built environment. A decolonised understanding of cities and space would recognise that value systems and beliefs play a role in determining our professional attitudes. And finally, a decolonised approach to cities and space would acknowledge the shift the focus from the west to other cultures, other geographical contexts, other religions and other languages.

Spatial justice, politics of space and ethics in the built environment are therefore part and parcel of this conversation and lead one to issues of property rights, access to opportunity and land. The decolonisation of the city is intrinsically linked to the decolonisation of our knowledge and education systems and our professional practice as represented in professional institutions.

The contradictory sentiments in social media with regards to the awarding of the Pritzker to Alejandro Aravena and the outpouring of celebratory rhetoric for Zaha Hadid are a reflection of the confused values of the built environment professionals and what is generally celebrated and what is considered inferior. The lack of in-depth interrogation of the imagery of Wakanda is further evidence of this. This attitude is seen to be further perpetuated by regulatory bodies and voluntary institutions and the directions taken at architectural schools through their academic programmes resulting in perceived exclusion, elitism and further disconnecting students from the contexts where they come from and where they should return to practice.

8 Livability and Lovability as Guiding Principles for the Design of the Future African City

Many African cities aspire to becoming "world class" and "progressive" and "modern". These terms are translated into very specific imagery which is not always innovative or relevant. These aspirations often leave poor communities disadvantage, silenced and hidden. Various measures rate cities based on concepts such as livability or loveability. These are attempts to try and unpack the qualities of cities that appeal to people and to replicate and encourage those qualities in other cities. While these terms can have some benefits, they also fall short in some ways.

The term "livability" has been used to describe the desirable aspects of cities globally. It may be a useful term but also raises intellectual controversies. The ranking of cities based on sets of criteria always seem to rate similar contexts highly and needs further reflection. (O'Sullivan, 2019) Heathcote asks if these polls "eschew diversity?" arguing that it is the juxtaposition of wealth and poverty, the collision between the two worlds, that makes cities work. (Heathcote, 2011).

Many measures and polls exclude issues of the poor in cities leading to the developing world and in emerging economies being labelled as unfavourable. Is it possible that these cities are not desirable at all based on these global measures? What measures, tools and as well as innovative city management strategies have been put in place for these "other" contexts and how have these cities set out to accommodate for degrees of legality/illegality and formal and informal? Perhaps these latter cities could then be labelled lovable? Seeing that they resonate with many people and offer opportunities for many?

Indeed, inspired by Open Building thinking, I believe that the built environment and cities need to be conceptualised and designed with a view to being "lovable" and "open":

> "Open" can mean [freely] accessible, uncluttered, approachable, receptive (amenable, hospitable, responsive), undeveloped, unbuilt, honest, begin… perhaps the term "receptive" and its definitions best explains why some Open Building thinkers refer to Open Buildings as being "lovable" as they have resonance with many users over time, an expression used by Frank Bijdendijk, Dutch developer of THE SOLIDS projects… Open Building has been used both as a noun and a verb to express thinking and practice as well as the products thereof within the context of the built environment professions in general and architecture specifically. Open Building has socio-political, technical, financial and management implications at many levels of the built environment (Osman, 2015).

In this text, a "Solid" refers to a building that is sustainable in the "functional, technical and emotional senses of the word" according to Leupen et al., (2005, p.42). The authors proceed to describe "accommodation capacity" as having the potential to adapt and "preciousness" as the emotional value attributed to a building; the authors then explain that these characteristics extend from the building to the block and city levels as well as at the building level (ibid).

This paper, therefore, adopts these principles and applies them at city level, thus defining cities as "livable" if they are equitable, beautiful, functional and resonate with many people, over many years and "lovable" when they may be defined as being sustainable through their potential to accommodate and embrace diversity, choice, change and by being embedded in their context. (Osman, 2019).

9 Conclusions

This paper has argued that we need to question images of Africa as sold to us through various narrators—many of whom are not from Africa. It also argues that where Africans have bought into these images, there has been a complicit relationship of mutual benefit between governments and companies at the exclusion of the majority residents of African cities. It is therefore argued that "Wakanda" as a place needs further scrutiny and as an idea it still needs to be critically unpacked in terms of what it implies about Africa, Africans and the value systems that it embraces. Three case studies are presented and the principles that could help achieve a re-imagined African City as a livable and lovable. These two terms have been defined as follows:

> Livable = equitable, beautiful, functional and resonating with many people, over many years.
> Lovable = sustainable by accommodating and embracing diversity, choice, change and embedded in context.

The fact that African cities, as are cities of the global south in general, tend to be excluded from such measures is to be challenged; qualities that define these cities need to be better documented, studied, captured and replicated. These may help us, as built environment professions, to completely re-think our approach and to advocate for different, more ethical, practice and more appropriate and relevant decision-making strategies.

References

Amnesty International. (2016, March 31). *The ugly side of the beautiful game: exploitation of migrant workers on a Qatar 2022 World Cup site*. Amnesty International. https://www.amnesty.org/en/documents/mde22/3548/2016/en/

Bayley, S. (2015, August 8). The Heckler: architecture would be better off without Zaha Hadid. *The Spectator*. at https://www.spectator.co.uk/2015/08/the-heckler-architecture-would-be-better-off-without-zaha-hadid/

Broadbent, A. (2017, June 1). It will take critical, thorough scrutiny to truly decolonise knowledge. *The Conversation*. https://theconversation.com/it-will-take-critical-thorough-scrutiny-to-truly-decolonise-knowledge-78477

Burocco, L. (2018). "The darker side of modernity" in an Illuminated Precinct in downtown Johannesburg. In *Reversing urban inequality? Debating spatial transformation in globalizing Johannesburg* (pp. 56–68). Essay, Routledege. http://gentrilogy.com/wp-content/uploads/2018/10/BK-TandF-9781138320444_TEXT_MYAMBO-180617-Chp05LBrevised-.pdf

Burocco, L. (2019). Do not make Africa an object of exploitation again. *Image & Text, 33*. https://doi.org/10.17159/2617-3255/2018/n33a5

Cetin, H., & Has, K. (2014). UIA 2014 Durban architecture OTHERWHERE. In *UIA 2014 Durban congress proceedings, XXV world congress of architecture* (pp. 52–58).

Deckler, T. (2016, January 22). Some 'South African' thoughts on the Pritzker prize and its winner. *LinkedIn*. https://www.linkedin.com/pulse/some-south-african-thoughts-pritzker-prize-its-winner-deckler

Demir, E. (2014). UIA 2014 Durban architecture OTHERWHERE. In *UIA 2014 Durban congress proceedings, XXV World congress of architecture* (pp. 1204–1213).

Dobson, R. (2013, December 9). *Shopping mall for Durban's Warwick junction will be 'step backwards'*. IOLProperty. http://www.iolproperty.co.za/roller/news/entry/shopping_mall_for_durban_s

Elemental. (n.d.) http://www.elementalchile.cl/

Fiederer, L. (2016, April 21). AD Classics: Vitra Fire Station / Zaha Hadid. *ArchDaily*. http://www.archdaily.com/785760/ad-classics-vitra-fire-station-zaha-hadid-weil-am-rhein-germany

Franco, J. T. (2016, January 13). Alejandro Aravena Wins 2016 Pritzker Prize. *ArchDaily*. http://www.archdaily.com/780203/alejandro-aravena-wins-2016-pritzker-prize

Goyal, A., & Des, M. (2014). UIA 2014 Durban Architecture OTHERWHERE. In *UIA 2014 Durban Congress Proceedings, XXV World Congress of Architecture* (pp. 997–983).

Hackney, R. (1990). *The good, the bad and the ugly, Cities in Crisis*. Frederick Muller.

Heathcote, E. (2011, May 6). Liveable v lovable. *Financial Times*. http://ig-legacy.ft.com/content/dd9bba18-769c-11e0-bd5d-00144feabdc0#axzz1LpkdgTGh

IOL Business. (2009). *500 traders protest against new Durban market*. IOL. https://www.iol.co.za/business-report/economy/500-traders-protest-against-new-durban-market-822282

Karam, B., & Kirby-Hirst, M. (2019). Guest editorial for themed section Black Panther and Afrofuturism: theoretical discourse and review. *Image & Text, 33*. https://doi.org/10.17159/2617-3255/2018/n33a1

Landsel, D. (2007, April 24). Outlook: The new Khartoum. *New York Post*. https://nypost.com/2007/04/24/outlook-the-new-khartoum/

Leupen, B., Heijne, R., & van Zwol, J. (Eds.). (2005). *Time-based architecture*. 010 Publishers.

Maharaj, B. (2010). The struggle for the Warwick market in Durban. *ÉchoGéo, 13*. https://doi.org/10.4000/echogeo.12043

Malkin, M. (2018, February 28). The Real-Life Possibilities of Black Panther's Wakanda, According to Urbanists and City Planners. *Architectural Digest*. https://www.architecturaldigest.com/story/the-real-life-possibilities-of-black-panthers-wakanda-according-to-urbanists-and-city-planners

McGuirk, J. (2014). *Activist architects designing social change*. Aljazeera. http://www.aljazeera.com/indepth/opinion/2014/09/activist-architects-designing-2014928103659390595.html

Mdluli, A. (2010, September 16). Market traders sue Warwick Mall. *Pretoria News*. https://www.pressreader.com/south-africa/pretoria-news/20100916/281998963788413.

Miranda, C. A. (2016, April 4). Why we talk about Zaha Hadid's gender and ethnicity even though her architecture transcended both. *Los Angeles Times*. http://www.latimes.com/entertainment/arts/miranda/la-et-cam-zaha-hadid-women-in-architecture-20160331-column.html

Monyae, D. (2019, January 18). *AU must roll up their sleeves to set up Wakanda One*. IOL. https://www.iol.co.za/news/opinion/au-must-roll-up-their-sleeves-to-set-up-wakanda-one-18835565

Moore, R. (2012). *Why We Build*. Picador.

Nazaryan, A. (2016, March 31). Zaha Hadid's complex legacy. *Newsweek*. https://www.newsweek.com/zaha-hadids-complex-legacy-442809

Osman, A. (2015, December 11). *Open Building versus Architecture or Open Building as Architecture?* [web log]. http://uj-unit2.co.za/open-building-versus-architecture-or-open-building-as-architecture/

Osman, A. (2016a, April 20). *The controversy around Zaha* [web log]. http://uj-unit2.co.za/thecontroversy-aroundzaha/

Osman, A. (2016b, August 10). *Khartoum Sudan protest and the KPP5 Stage 1 dissemination* [web log]. http://amiraosman.co.za/2016/08/10/khartoum-sudan-protest-and-the-kpp5-stage-1-dissemination/

Osman, A. (2016c, December 15). *Competing for space, power and authority in the city and on campus* [web log]. http://amiraosman.co.za/2016/12/15/competing-for-space-power-and-authority-in-the-city-and-on-campus/

Osman, A. (2016d, January 30). *Alejandro Aravena, Pritzker 2016 and some reflections from a Residential Open Building perspective* [web log]. http://uj-unit2.co.za/alejandro-aravena-pritzker-2016-and-some-reflections-from-a-residential-open-building-perspective/

Osman, A. (2019, March 19). Resilience thinking for the next generation of designers @ the #cocreate DESIGN FESTIVAL 2019 [web log]. http://amiraosman.co.za/2019/03/19/resilience-thinking-for-the-next-generation-of-designers-the-cocreate-design-festival-2019/

O'Sullivan, F. (2019, June 26). *Death to Livability!* Bloomberg CityLab. https://www.citylab.com/life/2019/06/best-cities-to-live-list-monocle-ranking-zurich-vienna/592492/

Sisson, P. (2016, March 31). *With RIBA Gold Medal, Zaha Hadid Became One of the World's Most Decorated Architects*. Curbed. http://www.curbed.com/2015/9/24/9918080/zaha-hadid-RIBA-gold-medal

Smith, C. (2018, October 8). *There's a problem with 'Wakanda visions' for African cities—academic*. Fin24. https://www.fin24.com/Economy/theres-a-problem-with-wakanda-visions-for-african-cities-academic-20181008

UIA 2014 Durban. (2014). *UIA 2014 International student competition*. http://uia2014durban.org/resources/docs/UIA2014_StudentBrief_FINAL_1906.pdf

Wainwright, O. (2016, January 13). Chilean architect Alejandro Aravena wins 2016 Pritzker prize. *The Guardian*. https://www.theguardian.com/artanddesign/2016/jan/13/chilean-architect-alejandro-aravena-wins-2016-pritzker-prize

Yalcinkaya, G. (2018, March 1). Black Panther's "voluptuous" sets are influenced by Zaha Hadid. *Dezeen*. https://www.dezeen.com/2018/03/01/black-panther-film-designer-zaha-hadid/

The Relation Between Walking and Urban Form: Identifying Gaps in Egyptian Literature

Farah A. Sami and Omar M. Galal

Abstract

Walkability is an indicator of how pleasant a walking area is. It has many benefits that include environmental, health, and economic, and it also affects the quality of life. According to international literature, walkability is mainly affected by two of variables; (a) human-related and (b) urban form-related. Human variables include behavior, age, physical activity, safety perception, socioeconomic level, and others, while urban form variables are related to transportation infrastructure, land-use mix, streets configuration/design, and urban density. Internationally, studies in this area focused on multidisciplinary research methods combining public health, transportation, social behavior, and urban form. However, not many studies have simultaneously linked urban form variables with human behavior-related variables in a quantitative manner. On the other hand, in the Egyptian literature, a clear consensus on the advantages of mixed-use and density can be noticed. Nonetheless, empirical evidence or quantified description for these terms is rarely provided. Furthermore, no significant attention is given to the impact of human-related variables on walking. Therefore, the main aim of this paper is to investigate the international and local methodologies that focused on walkability and its relation with social behavior and urban form. The study at hand classifies and categorizes the main streams of research internationally and locally. Consequently, define gaps in the Egyptian literature and identify possible topics that need to be quantitatively addressed in the local context.

Keywords

Walkability • Urban form • Social behavior • Built environment

F. A. Sami (✉) · O. M. Galal
Arab Academy for Science, Technology and Maritime Transport, Cairo, Egypt

1 Introduction

To a degree, walking is referred to as an easily accessible, safe, connected, functional, and enjoyable mode of transportation. Other references state that walkability is an indicator of "*the urban form quality and availability of footpath infrastructure*" within a specified area (Abley & Turner, 2001). Walking is recognized by the majority of researchers as a means of increasing rates of physical activity. There is growing evidence that walking behavior is affected by the built environment and urban form. The built environment is conceptualized for its walkability as an amalgamation index that incorporates the characteristics of a neighborhood design that can influence pedestrian and travel simplicity (Villanueva et al., 2014).

Lately, many urban and city planners are attracted to the public transportation and walkability concepts (Abedo et al., 2019). However, the term walkability is debatable, according to professionals, researchers, and public officials. Many debates concentrate on environmental amenities and ways to make walkable environments, focusing on traversable areas, compactness, physically appealing, and secure places (Al-Hagla, 2009). Other studies highlight the potential outcome that may result from such environments such as the livability of an area, the development of sustainable transportation solutions, and increasing activity levels (Van Dyck et al., 2009). Many studies refer to the term "walkability" as a tool to enhance the design, either through identifying tangible dimensions or through the provision of comprehensive solutions to urban problems that negatively affect walking. According to Oxford English Dictionary 2013, the term "Walkable" has been used since the eighteenth century (Forsyth, 2015). In contrast, nowadays walkability is rarely defined in dictionaries but commonly used in practice. However, conflicting definitions usually cause problems as they affect how people in practice seek to plan walkable areas, measure environmental livability, and assess the costs and benefits of developing a walkable community.

C. Alalouch et al. (eds.), *Towards Implementation of Sustainability Concepts in Developing Countries*, Advances in Science, Technology & Innovation, https://doi.org/10.1007/978-3-030-74349-9_4

Practitioners and experts may be speaking with great certainty about how to make communities more walkable, but they might as well be suggesting contradictory approaches (Caplan & Nelson, 1973). This vagueness makes it difficult to develop a practical guiding theory to aid municipalities, planners, and architects.

There are many factors which affects walkability that include but are not limited to the presence or lack of and the quality of pedestrian paths and right of ways, roads infrastructure, land-use patterns, neighborhood accessibility and safety (Mid-America Regional Council [MARC], 1998). Moreover, in urban design, more can be done to take into account the many factors beyond physical components that make it possible to walk. These include walking policies and initiatives supporting walkability, demographic characteristics including the residents' preferences, motives, populations, etc. Although these issues are typically considered in the fields of health and transport, they are not always relevant in discussions regarding urban design (Forsyth, 2015).

The United States National Center for "*Chronic Disease Prevention and Health Promotion*" describes walkability as a concept of quantifying walking routes, security, and desirability. Their study has shown how increasing walkability may have a positive effect on a community's well-being such as encouraging social interaction, reducing crime rates, and reinforcing safety perception. With people walking, there are always eyes on the streets of a neighborhood. Another advantage of increased walking is lowering the residents' rates of obesity. Lee and Moudon (2006) stated that up to one-third of people in less walkable environments were more probable to be obese or to have diabetes. Due to concerns regarding the alarming rise in levels of overweight in urban areas, defining and changing the characteristics of the built environment may be an essential policy measure for the prevention of chronic disease (Tawfik, 2017). Furthermore, one of the most significant benefits of walkability is the resulting decrease in the community's vehicle footprint. If more people choose walking rather than driving, carbon emissions can be reduced.

On the other hand, it argued that it is still a matter of subjectivity to determine the role of urban design features at street level (Jacobs, 2014). Streets and urban spaces are considered as a three-dimensional volumetric void. Each street should be identified by some qualities such as unique activity, furniture, or lighting. Those qualities make up the visual characteristics of the space. Urban design qualities associated with walkability include "enclosure, transparency, human scale, complexity and image-ability." Therefore, the implementation of an objective scientific approach to assessing street characteristics will help legislators improve the quality of the pedestrian setting (Abley, & Turner, 2001).

Hence, it becomes clear that an urban space may have some qualities that can be defined as qualitative, and others can be defined as quantitative (Ewing & Handy, 2009; Hansen, 2014). The same goes for walkability. It is internationally acknowledged that new cities are ignoring the pedestrian experience, and streets are losing the human scale and transparency due to barriers and major highways and roads (Forsyth & Southworth, 2008). However, as demonstrated in the international literature, at least there is a deep and diverse, quantitative and qualitative discourse that is rigorously aiming to seek how to define, understand, and implement more walkable urban spaces. However, in the Egyptian literature focused on the topic, the same rigor and diversity were not found until the time of writing this paper. The main aim of the study at hand is to highlight the shortcomings and the blanks within the research conducted on walkability in Egypt in comparison with international studies. The work presented here can contribute to the scientific discussions focused on walking in Egypt.

2 Methodology

In order to investigate research challenges in the context of walkability, a comprehensive literature review is generated to cast light on previous research, using "Egyptian knowledge bank, google scholar and the Egyptian universities libraries consortium" as a research engine with the main keyword "*walkability, walkability and urban form.*" Literature was surveyed to mainly focus on two main scopes; international and national studies in the topic of walkability and urban form. This review process has three main goals: (a) identify urban form parameters that affect walking positively or negatively according to international and national literature, (b) identify gaps in national literature in comparison with international literature, and (c) identify how different research methods and tools can be utilized to examine the impact of urban form on walking. The research sequence is generated accordingly from a general point of view to a more elaborated perspective aiming to investigate relations between walkability, physical environment, and anthropogenic attributes.

Therefore, an overview of both international and national studies was first conducted. This summary provides a holistic understanding of the key findings of these studies with respect to the impact of urban form on walking. Afterward, each of the investigated studies was broken down and classified according to several factors. These include the study's scale (street, neighborhood, city, or regional), targeted group (age, socioeconomic level), urban form (land use, streets network, and density), and the utilized research methods and tools. Hence, through examining each study

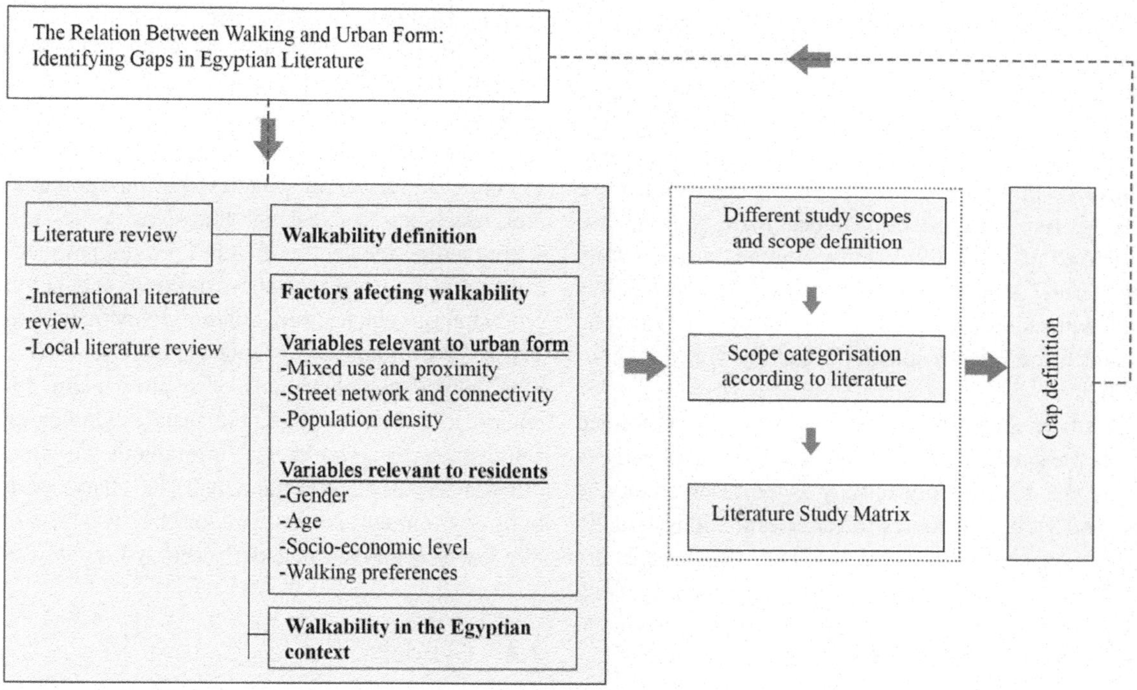

Fig. 1 Research methodology

according to these criteria, it becomes possible to determine the degree of complexity and comprehensiveness of the Egyptian studies in contrast to the international studies (Fig. 1).

3 Factors Affecting Walkability

Walkability is affected by many forces that vary between contexts. These are not limited to but include facilities location, walkway suitability, street design and connectivity, density, esthetic requirements, land use, accessibility, neighborhood activities, and climatic conditions (Jacobs, 2014). However, three main urban characteristics are consistently found to be related to walking which are density of dwellings, streets connectivity, and land use mixing (Owen et al., 2007). On the other hand, different groups within a community tend to walk differently. Previous studies show that the relationship between neighborhood residents socioeconomic status and walkability can be complex (Gullón et al., 2017). Additionally, the walking trends of different age groups differ due to variations in the purpose of the trips and variations in how they react to the built environment. The following part summarizes previous studies focusing on the relation between walking on the one hand and urban form and anthropogenic characteristics on the other.

3.1 Mixed-Use and Proximity

For decades, the combination of residential, family, and multi-family homes encouraged people to walk to several local destinations (Frumkin, 2002). On the other hand, separated land use is argued to cause public health problems associated with less exercising, obesity, and more vehicle carbon emissions (Brown et al., 2009; Booth et al., 2005; Transportation Research Board, 2005). In general, the compact city model is widely recognized as a sustainable urban form in literature (Burton, 2000; Ibrahim, 2011). The two key concepts shaping the overall spatial patterns of cities and building sustainable urban design are high densities and mixed uses. Accordingly, the recent debate on the environment has given significant impetus to develop urban environments that are higher in density and more mixed in uses as it could be an action toward reducing energy consumption and pollution levels (Jones & MacDonald, 2004).

Walking has to compete with other transportation modes and can be at a disadvantage, primarily due to travel distance options (Frank et al., 2004; Owen et al., 2007). Hence, proximity is also a key element when it comes to walking encouragement. Proximity is determined primarily by two main land usage variables: land usage density or compactness; and land usage mix, “the degree of diversity with which functionally specific uses are located in the same space." The more an urban environment is compact and intermixed, the shorter the distances between destinations.

However, the relation between travel and land use has long been debated to be complex by urban and transportation planners (Ewing & Cervero, 2002). For example, some studies have confirmed the correlation between mixed-use and walking for transportation rather than walking for recreation (Cao et al., 2009). Transportation infrastructure, including "*Street lengths, bus riding, and block sizes,*" is associated with some of the land-use factors or with one another, and those criteria affect walkability (Lee & Moudon, 2006).

Choosing between using a motorized or a non-motorized method of transport is also dependable on connectivity (Saelens et al., 2003). Connectivity is represented in the directness and variety of routes toward destinations, usually due to interconnected street patterns. These components are complementary and can be calculated objectively using geographic information system (GIS) tools to assess distances between activities (Aultman-Hall et al. 1997).

3.2 Street Network and Connectivity

Short and direct routes to destinations produced by developing a well-connected street grid provide a pedestrian-friendly neighborhood and enhance walkability. Also, a wide range of destinations attracts a wide option in walking constancy (Brown et al., 2009).The directness of pathways and routes that connect between residences, shops, and work-places based on the street network design is considered a connectivity measure (Saelens et al., 2003). Lack of physical barriers that include freeways, walls, or any physical obstacles and the availability of routes options facilitate direct travel. Furthermore, a regular grid pattern formed by interconnecting streets leads to encouraging residents to walk for transportation. The results from studies concentrate on correlations with traveling to work, shopping, service facilities, and public transportation ensure that the accessibility of destinations along with an integrated road network makes walking more convenient and desirable (Leslie et al., 2007).

3.3 Population Density

Arguments that the density of population is positively correlated with walking leads to decreasing driving demand, overcrowded areas and parking spaces and also create destinations demanding (Brown et al., 2009; Owen et al., 2004;). Studies showed that gasoline consumption increases strongly where community density is less than 12 people per acre. As the lower density has a double effect, in addition to making a longer travel distance, a non-automobile mode is mostly impossible, and also walking to transportation or biking is impractical. Economist suggests models that have extremely important factors in decreasing gasoline consumptions rather than only increase taxes on vehicles and gasoline. A variety of policies that has potentials to save fuels includes: increasing the density of urban form, strengthening the center of a city, providing inner areas land use, enhancing transportation options, and controlling the provision of vehicles infrastructure (Newman & Kenworthy, 1989). Nevertheless, people generally walk more in high-density neighborhoods even after taking into consideration the socio-demographic features. Increased physical activity was nearly identical throughout groups and across different types of environments. The characteristics of the built environment are even more effective for some groups like the unemployed or elderly (Forsyth et al., 2009).

3.4 Gender

Recent studies have shown significant variances in men's and women's mobility patterns, which are evident both at travel distances and in the mode of travel. Traditionally, men use automobile travel mode, while women chose other modes such as walking and public transit, and today women still drive much less than men do and use a more complex mix of modes of transportation. Given of the differences described in the general daily mobility patterns of women and men, it is to be assumed that gender differences will also be evident in walking patterns. Built environmental characteristics associated with fear of crime and pedestrian safety have been proven to affect the walking actions of women more than for men. The limited researches on pedestrian experiences of women point to a gap in walkability research and in understanding of walkability and gender relation (Golan, 2017).

3.5 Age

Studies show that individuals living in more walkable communities, especially for 30 years aged adults and older, which includes higher living densities of mixed land use and interconnected roads, are more likely to walk than people living in less walkable neighborhoods. Although the relationship between walkable neighborhoods and walking is clear, there is a lack of data on how this correlation differs through life stages. Moreover, neighborhood buffer size has possible influences on walking depending on walking preferences. Distance to destinations and buffer size are considered more important to older adults in walking for recreation in particular. In contrast, younger individuals are not significantly affected (Villanueva et al., 2014).

3.6 Socioeconomic Level

The physical activity is generally associated with gender, life stages, and socioeconomic status (SES). Almost all types of physical activity are related to age, but associations with other social and economic-demographic characteristics can differ depending upon physical activity type and intention. Socioeconomic status is an essential aspect to be considered in environmental behavior studies on physical activities (Owen et al., 2007). Neighborhoods' socioeconomic status (SES) and walkability are strongly and inversely associated. In addition, previous research has found that lower SES communities are more walkable. This is to be expected as if an individual does not own a personal vehicle, and he will be obliged to use public transport and consequently walk more. On the other hand, impoverished neighborhoods are lacking aesthetics and safety, which may significantly affect walking behaviors (Gullón et al., 2017).

3.7 Walking Preferences

There are basically three possible aims for the trips that are periodically performed by the residents within any community: work, services, and leisure. Many studies have investigated how walking for a specific trip is affected by the features of the built environment. For example, light rail stops facilitate walking to and from stations, encourage stakeholders who prefer walking or provide environmental benefits through a combination of environmental opportunities and selection influences, allowing residents to respond more easily on their preferences (Brown and Werner, 2009). Researches elsewhere have shown that transit users took 30% more pedometer-measured steps per day than car riders and that people residing in areas with higher subway densities have lower BMIs (Wener & Evans, 2007). However, there has been much recent interest in the ability to exploit this situation in an attempt to increase general physical activity as an important goal for public health (Pucher & Renne, 2003).

One of the problems is the difficulty obtaining accurate data for both "*non-motorized travel activities and the micro-level characteristics*" associated with these travel modes of land use and urban form. "*Grocery stores, restaurants, retailers, colleges, roads, and parks*" are functions that are considered to be positively linked to walking. Conversely, fast food restaurants and large shopping centers are negatively associated with walking. Unlike fitness centers, galleries, museums, and sports facilities can possibly have either direction of relationship with walking (Lee & Moudon, 2006).

Lower weight has also been associated with greater access to parks and other recreational uses. A recent review found that parks are associated with increased physical activity in general. Parks can also be important physical activity venues that might reduce the risk of obesity. Park use and physical activity were correlated with proximity to a park in low-income neighborhoods, particularly in team sports areas (Cohen et al., 2007). Variation in walking frequencies is affected by destinations diversity. Forty–five percent of residents walk to local grocery stores at least once per week, while only 2.2% walk to theaters (Brown et al., 2009).

4 Walkability in the Egyptian Context

There is no doubt that the layout and urban form of the street network are supposed to influence each other over time. Nevertheless, this relationship seems to be unclear in Egypt, especially when it comes to the planning of new cities. New Egyptian cities' planning turns to be unsustainable due to street pattern design that is not relevant to the Egyptian context. It is widely acknowledged that the traditional urban patterns formed over centuries may provide essential future guidelines. Additionally, the current generation of the new urban policy calls for some guidance on urban sustainability based on international standards and, in some cases, on the traditional form (Ibrahim & Alattar, 2017).

Likewise, there are also those who argue against the trend of rising densities on the basis that high population density and mixed-use cities contribute to congestion and overcrowding of traffic. Consequently, it leads to local air pollution escalation, increased crime, noise, and social conflicts (Breheny, 1992). As a way of avoiding the disadvantages specified above, new urban communities plan the new cities to be of relatively lower density. Otherwise, typical Egyptian urban form characterized by a high density compact form, diversity of activities, and mixed uses. Nevertheless, new cities were built not only on a different pattern, but this pattern also has no reference to the contextual background of Egyptian cities (Ibrahim & Shaw, 2009). On the other hand, when coupled with a high mix in uses, density is usually negatively perceived in the Egyptian context either by the residents or by public authorities. Therefore, it is necessary to look for the mixed-use and its impacts on density under the same category of compactness. In this regard, the context of low-income communities culminated in three compact patterns: hard and soft compact patterns, and scattered form. The soft compact is claimed to be the most sustainable model for all future growth and is strongly recommended (Ibrahim, 2011).

Egypt has been pulled into a rapid socioeconomic change over the past few decades that triggers a relative shift in adopted patterns of development. Different types of new residential districts have been transferred into the Egyptian context from a globalized, westernized world. Planners and the authority claimed that solutions to the problems of the old town could be achieved by changing patterns of development from traditional to westernized new communities. On the other hand, numerous researchers have pointed out that conventional "*compact, mixed-use, high-density*" urban types are critical in achieving sustainability goals in terms of ecological, economic, and social benefits. Other studies aimed the sustainability of traditional settlements with modern settlement expansion in the Egyptian context in order to identify lessons learned from traditional urban forms to improve the sustainability of modern developments as a step toward attaining particular sustainability objectives for establishing new settlements in Egypt (Ghonimi, 2017).

5 Discussion

The size of new urban communities is continuously increasing in Egypt. These communities are planned based on policies that lack evidence. However, these new communities do not encourage walking, according to international literature. On the other hand, there is a clear lack of multidisciplinary research on the relation between walkability, urban form, health and human-related factors, in Egypt. The Egyptian studies do not contain ample evidence to verify or justify the applicability of international walkability measures in the Egyptian context. Not much evidence-based guidance is provided to architects, planners, and local authorities to change the current conditions of settlements or plan future communities to be more walkable. As shown in Fig. 2, this makes it hard to accurately and quantitatively estimate the real damages from the current planning trends adopted by local authorities. Consequently, the existing trends remain, and health, environment, and quality of life are negatively affected.

According to the surveyed studies, there is perhaps a lack of local literature in studies that quantitatively investigate the relation between human characteristics' related variables and urban form variables. The investigation of international literature shows that controlling human-related variables is crucial when exploring the impact of urban form on walkability. Most of the investigated local literature in this paper does not include human-related parameters combined with physical environment parameters. Therefore, international recommendations are taken for granted, assuming their suitability to the local context. However, in reality, the impact of urban form can vary with the variation of context either due to variation in social and cultural norms or variations in the natural environment and even climatic conditions.

Studying walkability along with urban form is challenging as researches can have a variety of scales and scopes. Scope definition and categorization are driven from this study's review of the literature. See Table 1. The scale of the focused case studies in the surveyed literature can be classified into the following categories: street level, neighborhood level, city level, and regional level. It was noticed that the scale of the case study affected the utilized methods and tools. For example, most of the studies that focus on neighborhood particular scale use methods that couple quantification of urban form through tools like GIS in addition to quantifying walking through questionnaires. On the other hand, studies that focus on regional and city level use census data collection and depend on statistical analysis

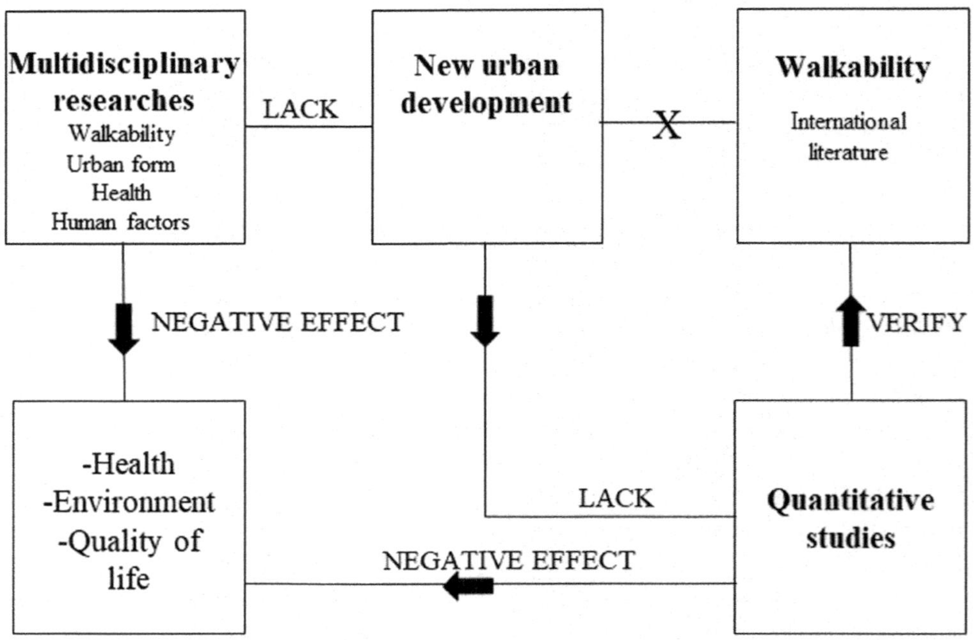

Fig. 2 A diagram that visualizes the broken cycle between walkability research and planning in Egypt

Table 1 (Table 1 is not complete in this version) Part of literature review matrix

Title	year & geo-location	scale	human factor		Urban form variable				methods
			Demographics	health	land use mix	street connectivity& transportation	density	Urban design qualities	
International literature									
(Lee & Moudon, 2006).	2006 Washington	micro 1 km buffer to 3 km							Questionnaire & GIS
(Leslie et al., 2007).	2007 Australia	regional (32 communities)							Survey & GIS
(Peter W. G.Newman , Jeffrey R. Kenworthy, 2007).	2007 United States	regional							Survey
(Barker, 2017)	2012 Seattle	macro							Survey
(S.Hassan Ameli et al .,2015.)	2015 Downtown Salt Lake City, Utah.	micro 32 blocks							observation & GIS"secondary"
(Shamsuddin & Bilyamin.,2012).	2012 Kuala Lumpur	city center							questionnaire & field observation
(Paula et al , 2017)	2017 lisbon	3 neighbourhoods							space syntex survey/observation
(Owen et al., 2007).	2007 Australia	city							Survey
(Jessica Cook, Mallika Bose & Deborah, S.Main., 2014)	2013 Pennsylvania	2 neighbourhoods							questionnaire
(Hansen, 2014)	2014 USA	10 neighbourhoods							Survey
(Villanueva et al., 2014).	2014 Australia	Varriable							Survey & GIS
Local literature									
(Abedo et al.,2019)	2019	2 neighbourhoods"heliopolis& newcairo"							questionnaire &depthmapX
(Tawfik, 2017).	2017	street level							survey
(Abdelkhalek Ibrahim & David Shaw., 2009).	2009	city							literature.review& questionnaire to expert
(Ibrahim ,A.& Alattar, A.,2017).	2017	neighbourhood							GIS
(Ibrahim, 2011).	2011	neighbourhood							site survey & questionnaire
(Ghonimi, 2017).	2017	neighbourhood							survey

Table 2 Study scopes derived from literature analysis

Scale	Street	Neighborhood	City	Regional
Age	School children (6–17 years)	Young adults (18–35 years)	Middle-aged adults (36–55 years)	Older adults (> 55 years)
Socioeconomic level	Low income	Middle class	Upper middle	High
- Method - Software and tools include: GIS Space syntax Depth map X Walk score	Questionnaire	Survey	Literature review, and questionnaire for expert	Survey, and questionnaire
	Data gathering, and analyzing software, and tools	Survey, questionnaire, and software, and tools	Questionnaire, and observation	Urban qualities

through the utilization of tools such as SPSS to investigate the relation between urban form and walking.

Some researches focus on controlling group age while others control the socioeconomic level. Those categories are chosen according to study scope and target. Age is categorized as follows: a) school children, b) young adults, c) middle-aged adults, and d) older adults. Socioeconomic status is categorized as follows: a) low income, b) middle class, c) upper-middle-class, and d) high class. Most of the studies control socioeconomic level either by car ownership or by housing type and location. Furthermore, some studies do not control the socioeconomic level but emphasize and focus on the relation between walkability and socioeconomic status.

Most international studies' methodologies tend to focus on mixing between urban form and human-related variables as done by Lee and Moudon (2006) and as well as per demonstrated in Table 1. They used the 3Ds + R "distance, destination, density & route" as measures using GIS to quantify the built environment on a one-kilometer micro-scale network buffer. Additionally, questionnaires were used to quantify the frequency and duration of walking for individuals per week while simultaneously controlling for age. Owen et al. (2007) used the International Physical Activity Questionnaire Long Form "IPAC" to test the variations in human-related variables that include socioeconomic level, gender, and neighborhood self-selection for 2650 individuals in addition to the GIS database to measure connectedness and proximity on a scale of 32 neighborhoods. In summary, the range of possible scopes covered in walkability studies is shown in Table 2.

Leslie et al. (2007) focused on spatial data analysis for land-use mix, street connectivity, transportation, net retail area, and density using GIS along with survey and data collection for health and physical activity data on a regional scale "32 communities." Cook et al. (2014) have used design quality indicator and questionnaire for two selected neighborhoods for a target age group "youth." While in Egypt, some studies used walk score to measure walkability and focused on street connectivity as a physical parameter. Other used literature review and questionnaire to the expert method for measuring density and mixed-used and its relation to carbon emission. Nonetheless, multidisciplinary methods were rarely found in Egyptian literature.

6 Conclusion

The main aim of the study at hand is to review the Egyptian literature relevant to walkability in light of international studies. Hence, it becomes possible through that review to define gaps within the Egyptian literature and consequently focus on these gaps in future research. After surveying the literature, it was noticed that research on walkability was clustered along with two main focal themes, which are (a) variables that are relevant to urban forms and (b) variables that are related to the residents. Therefore, the part of this study that focuses on international literature was first structured to provide a holistic overview of these two themes. Group (a) was classified into studies that addressed land-use mix, density, and streets network design. Group (b) was classified into gender, age, socioeconomic level, and walking preference. On the other hand, the Egyptian literature was overviewed as well; however, it was not as comprehensive as the international literature to be classified the same way.

The totality of different directions in research related to walkability in terms of the scale of studies, methods, tools, and focused variables were then compiled and summarized for Egyptian and International literature. The study shows that there is a lack in the Egyptian literature, specifically in studies that control for or include residents' related parameters. In many local studies, the recommendations of international literature were used to evaluate local contexts, which may dramatically differ from the contexts of international studies. Future studies in the field of walkability in Egypt may significantly benefit from mixed methods that incorporate qualitative and quantitative research techniques. Coupling questionnaires, interviews, and observations to cover the residents' variables in addition to tools like GIS to cover for urban form variables can provide more impactful results.

References

Abedo, M., Salheen, M., & Elshater, A. (2019). How can urban fabrics enhance walkability? Comparative spatial analyses in Cairo. *Architecture & Urbanism* (pp. 34–43). Researchgate.

Abley, S., & Turner, S. (2001). *Predicting walkability (No. 452).* Transport Agency.

Al-Hagla, K. (2009). Evaluating new urbanism's walkability performance: a comprehensive approach to assessment in Saifi Village, Beirut, Lebanon. *Urban Design International*, 139–151.

Ameli, S. H., Hamidi, S., Garfinkel-Castro, A., & Ewing, R. (2015). Do better urban design qualities lead to more walking in Salt Lake City, Utah? *Journal of Urban Design, 20*(3), 393–410.

Aultman-Hall L., Roorda M., & Baetz, B. W. (1997). Using GIS for evaluation of neighborhood pedestrian accessibility. *Urban Planning Development*, 10–17.

Barker, L. (2017). *Exploring the relationship between walkability and the built environment: A case study of three intersections in Seattle's University District.* University of Washington.

Barros, A.P., Martínez, L.M. & Viegas, J.M. (2017). How urban form promotes walkability? In *20th EURO working group on transportation meeting* (pp. 133–140). Transportation Research Procedia.

Board, T. R. (2005). *Transportation Research Board, Does the built environment influence physical activity? Examining the evidence.* Transportation Research Board.

Booth, K. M., Pinkston, M. M., & Poston, W. S. C. (2005). Obesity and the built environment. *Journal of the American Dietetic Association, 105*(5), 110–117.

Breheny, M. (1992). The contradictions of the compact city. *Sustainable Development and Urban Form*, 138–159.

Brown, B. B., & Werner, C. M. (2009). Before and after a new light rail stop: Resident attitudes, travel behavior, and obesity. *Journal of the American Planning Association, 75*(1), 5–12.

Brown, B. B., Yamada, I., Smith, K. R., Zick, C. D., Kowaleski-Jones, L., & Fan, J. X. (2009). Mixed land use and walkability: Variations in land use measures and relationships with BMI, overweight, and obesity. *Health & Place, 15*(4), 1130–1141.

Burton, E. (2000). The compact city: Just or just compact? A preliminary analysis. *Urban Studies*, 1969–2001.

Cao, X., Mokhtarian, P. L., & Handy, S. L. (2009). Examining the impacts of residential self-selection on travel behaviour: A focus on empirical findings. *Transport Reviews, 29*(3), 359–395.

Caplan, N., & Nelson, S. D. (1973). On being useful: The nature and consequences of psychological research on social problems. *American Psychologist, 28*(3), 199.

Choi, E. (2012). *Walkability as an urban design problem: Understanding the activity of walking in the urban environment.* KTH Royal Institute of Technology, Architecture and the Built Environment.

Cohen, D. A., McKenzie, T. L., Sehgal, A., Williamson, S., Golinelli, D., & Lurie, N. (2007). Contribution of public parks to physical activity. *American Journal of Public Health, 97*(3), 509–514.

Council, M. A. R. (1998). *Creating walkable communities: A guide for local governments.* Planning Commissioners Journal.

Ewing, R., & Cervero, R. (2002). Travel and the built environment: A synthesis. *Transportation Research Record, 1780*(1), 87–114.

Ewing, R., & Handy, S. (2009). Measuring the unmeasurable: Urban design qualities related to walkability. *Journal of Urban Design, 14* (1), 65–84.

Forsyth, A. (2015). What is a walkable place? The walkability debate in urban design. *Urban Design International*, 274–292.

Forsyth, A. & Southworth, M. (2008). Cities afoot—Pedestrians, walkability and urban design. *Journal of Urban Design*, 1–3.

Forsyth, A., Oakes, J. M., Lee, B., & Schmitz, K. H. (2009). The built environment, walking, and physical activity: Is the environment more important to some people than others? *Transportation Research Part D*, 42–49.

Frank, L. D., Andresen, M. A., & Schmid, T. L. (2004). Obesity relationships with community design, physical activity, and time spent in cars. *American Journal of Preventive Medicine, 27*(2), 87–96.

Frumkin, H. (2002). Urban Sprawl and public health. *Public Health Reports*, 201–221.

Ghonimi, I. (2017). Towards sustainable new settlements in Egypt: Lessons learned from a comparison between traditional and modern settlements in Greater Cairo Region—Egypt. *Journal of Sustainable Development.*

Golan, Y. (2017). *Gendered walkability: Building a day time walkability index for women in San Fransisco.* The Faculty of San Francisco State University.

Gullón, P., Bilal, U., Cebrecos, A., Badland, H. M., Galán, I., & Franco, M. (2017). Intersection of neighborhood dynamics and socioeconomic status in small-area walkability: The Heart Healthy Hoods project. *International Journal of Health Geographics, 16*(1), 21.

Hansen, G. (2014). Design for healthy communities: The potential of form-based codes to create walkable urban streets. *Journal of Urban Design*, 151–170.

Ibrahim, A., & Shaw, D. (2009). Egyptian compact-city veracity: The fallacy and credibility of mixed use and high density for a low carbon city. *Compact-City Veracity.* 45th ISOCARP Congress.

Ibrahim, A., & Alattar, A. (2017). Street networks between traditional and new Egyptian developments, problems and learned lessons. Green urbanism (pp. 306–318). International conference: Procedia Environmental Sciences.

Ibrahim, A. (2011). *Veracity of compact urban form for new Egyptian cities: Measuring urban and social sustainability of the low income neighbourhoods.* University of Liverpool.

Johnson, M. (2001). Environmental impacts of urban sprawl: A survey of the literature and proposed research agenda. *Environment and Planning*, 717–735.

Jones, C., & MacDonald, C. (2004). *Sustainable urban form and real estate markets.* Herriot-Watt University.

Lee, C., & Moudon, A. V. (2006). The 3Ds+ R: Quantifying land use and urban form correlates of walking. *Transportation Research Part D: Transport and Environment, 11*(3), 204–215.

Leslie, E., Coffee, N., Frank, L., Owen, N., Bauman, A., & Hugo, G. (2007). Walkability of local communities: Using geographic information systems to objectively assess relevant environmental attributes. *Health & Place, 13*(1), 111–122.

Maria Kockelman, K. (1997). Travel behavior as function of accessibility, land use mixing, and land use balance: Evidence from San Francisco Bay Area. *Transportation Research Record, 1607*(1), 116–125.

Newman, P.W.G., & Kenworthy, J.R. (1989). Gasoline consumption and cities: A comparison of US cities with a global survey. *Journal of the American Planning Association, 55*(1), 24–37.

Owen, N., Cerin, E., Leslie, E., Coffee, N., Frank, L. D., Bauman, A. E., & Sallis, J. F. (2007). Neighborhood walkability and the walking behavior of Australian adults. *American Journal of Preventive Medicine, 33*(5), 387–395.

Owen, N., Humpel, N., Leslie, E., Bauman, A., & Sallis, J. F. (2004). Understanding environmental influences on walking: Review and research agenda. *American Journal of Preventive Medicine, 27*(1), 67–76.

Pucher, J., & Renne, J. (2003). Socioeconomics of urban travel: Evidence from the 2001 NHTS. *Transportation Quarterly, 57*(3), 49–77.

Rafaat, M. H., & Kafafy, N. A. (2014). Approaches and Lessons for enhancing walkability in cities. *International Journal of Education and Research, 2*, 301–322.

Saelens, B. E., Sallis, J. F., & Frank, L. D. (2003). Environmental correlates of walking and cycling: Findings from the transportation, urban design, and planning literatures. *Annals of Behavioral Medicine, 25*(2), 80–91.

Tawfik, M. (2017). Using Urban Design Qualities for Building a New Composite Walkability Index for Cairo Streets. *International Design Journal, 7*(4), 255–266.

Van Dyck, D., Deforche, B., Cardon, G., & De Bourdeaudhuij, I. (2009). Neighbourhood walkability and its particular importance for adults with a preference for passive transport. *Health & Place, 15* (2), 496–504.

Villanueva, K., Knuiman, M., Nathan, A., Giles-Corti, B., Christian, H., Foster, S., & Bull, F. (2014). The impact of neighborhood walkability on walking: Does it differ across adult life stage and does neighborhood buffer size matter? *Health & Place, 25*, 43–46.

Wener, R. E., & Evans, G. W. (2007). A morning stroll: Levels of physical activity in car and mass transit commuting. *Environment and Behavior, 39*(1), 62–74.

Utilization of Neural Network-Based Approach in Bike Routes Optimization for Port Said Urban Road Network

Marwa S. El-Bany

Abstract

Transport department of urban cities always need to keep abreast of sustainable developments such as in Port Said city that is an important urban city in Egypt. Bike routes planning mission is not an easy occupation especially in developing countries. Mixed traffic is a main shape of the transportation system in most of their systems that are far away from the sustainable transport policies. One of the most effective solutions, for a modern sustainable transport, is the increasing of bike users. On the other side, a lot of problems and accidents had been occurred according to bike crossing among other transport vehicles; cars, buses, taxis, and others. This paper aims at studying the introduction bike routes effectiveness. It concludes the driver's reaction to a definite planning scenario and how the objective function is affected by the optimistic impact of the modal modification. The objective function has been supposed by assigning a suitable time saving for bike users using artificial neural network (ANN) approach. It shows the effect of bike speed change on the route using a greedy algorithm of the user's maximization welfare function. The study designated only four streets from seven selected streets to be suitable routes for bike routes introduction. The average bike speed is predicted to increase from 1.5 km per hour to 2.4 km per hour after introducing the bike routes.

Keywords

Bike routes • Neural network • Bike welfare • Sustainable transport systems

1 Introduction

Nowadays, the bike routes are transport techniques that decisions makers tend to use. Many quantitative and qualitative road attributes have to take into account for new policy or introduction of any idea. The bike routes have the following advantages:

- Sustainable, safety, and comfort improvements. It reduces accelerations and decelerations when the driver stops and goes in case of crossing between cars.
- A decrease in waiting and travel time.
- Simple and small form. The only moving parts on a single speed bicycle are the brakes, wheels, and cranks, which makes changes over time minimal.
- Low maintenance cost and small parking space.

The bike is a safe transport mode in case of moving in private lanes but when moving among other vehicles, it becomes dangerous and loss this advantage. So, bike routes are important solutions especially in developing countries with the non-expected behavior of other mode's users.

In major cases, routes have been allocated for only vehicles. Choice of a suitable route by drivers has been expected by a traffic assignment or ANN techniques that had been used to choose the best mode. In this study, ANN technique is used to decide if the route is suitable to be a bike route and combined vehicles or not. There are many previous studies tended to use ANN techniques. Jeong and Rilett (2004) built an ANN model to estimate the arrival time of bus. He used the automatic vehicle location to collect data that was the input of the model. Traffic congestion and every stop data were the inputs. The established model has a great performance when compared with previous models using mean absolute percentage error (MAPE) (Jeong & Rillett, 2004).

Mikluščak et al. (2012) made a congestion prediction and traffic control using ANN and discussed several algorithms

M. S. El-Bany (✉)
Faculty of Engineering, Port Said University, Port Said, Egypt

C. Alalouch et al. (eds.), *Towards Implementation of Sustainability Concepts in Developing Countries*,
Advances in Science, Technology & Innovation, https://doi.org/10.1007/978-3-030-74349-9_5

for a route prediction. (Mikluščak et al., 2012; Gurmu & Fan, 2014) used ANN method to predict the accurate real travel time. Data of real values was based on Global Positioning System (GPS) in order to help people in reducing the waiting time of public transportation under two criteria: prediction accuracy and robustness (Johar et al., 2016). The previous studies try to find the optimum prediction values for times to have a large view of any new transportation policies. They need to reach the maximum usage of the facility which is called "mobility." Nowadays, there is a developing trend of using the effective tools or simulation software, but it needs much cost and software availability. The recent study intends to reach the maximum value of mobility in Port Said city with a new sustainable achievement policy using the ANN techniques and greedy algorithms.

1.1 The Mobility of Port Said's Transport System

Port Said city has a future transport plan that expected the demand of internal trips about 540,000 persons per day; the 86% of them uses the private car, whereas others choose the public transport such as mini-bus, which considers a critical situation for the transport sector in 2019 (Port Said Government Transport Management Administration Plan, 2019). Furthermore, on its inner-city road network, the average travel speed cannot exceed 42 km/h which indicates low mobility. Also, the number of bike accidents per annum within the inner-city area of Port Said is about 227 accidents per year inside the city which considers an alarming data (Port Said Traffic Administration, statistics, 2019). In order to opposite this situation, administrators of transport have recently launched Sustainable Plan of Transport for the next ten years (Elgioshy, 2017), for an additional operative public transport system, which suggests the bike development and give the opportunity to sustainable transport.

1.2 Route Planning in Port Said Transportation Network

The multimodal route planning problem, which seeks journeys joining schedule-based transportation (buses, cars) with unrestricted modes (walking, driving), is even safer, relying on approximate solutions even for the problem inputs and this will be inputs of the greedy algorithm method. Greedy algorithm iterations method is one of the operative methods to determine the optimum routes (Li, 2017). It is an application of mathematical techniques to optimization research. In this study, the main objective is to minimize the total user waiting time (Zheng et al., 2009).

1.3 Bike Routes Model

This study proposes the input for the bike routes model, recording the different transport policy actions for the numerous combinations of the aforementioned criteria. Bike frequency and the width of the road are manually collected and also from Google Map (Port Said, Egypt Google map website, 2019). The policy actions were recognized according to the collected data in working days. Data was collected manually for the seven main streets that are considered the main arterial streets around the city; Mohamed Ali st. (Al-shohada), 23 Jul st., 23 Dec st., el-30th st., el-nasr st., Tarh el-bahr st., and el-Gomhoria st. Each street was classified into homogenous segment with different length as shown in Fig. 1. In Table 1, there are five actions according to policy criteria: have to make bike route, closed to traffic zone, the bike lane have to be weighed, and no action. Techniques for reaching certainty of making a proposed policy (bike routes) are as follow:

- ANN approach for solving many problems in numerous field of sciences.
- The greedy algorithm is a hyper-path minimum algorithm used to illustrate the steps of mobility indication and to determine the optimal set of bike routes by calculating the user's maximization welfare function for each route as described in the following methodology.

2 Methodology

The study commotion used a quantitative methodology. It aims at planning the newest bike routes into the Port Said urban road network. After applying the present lane systems, these have been compared with the model-based. Actually, the study area had been recognized within Port Said with a selection of the roads for its design model concerning the private investment level of service. The study considered parameters such as bike frequency, number of lines, and the road width as a geometric parameter. The trade-off function (T.O.) between the welfares of bike transport users and other significant modes such as car and bus is a multi-component function representing with the following equation:

$$\text{T.O.} = f(dC_{\text{car}}, dT_{\text{Car}}, dC_{\text{Bus}}, dT_{\text{Bus}}, dT_{\text{bike}}) \quad (1)$$

where

dC_{car}: the rise in monetary cost (car users)
dT_{car}: the increase in travel time (car users)
dC_{bus}: the rise in monetary cost for bus users
dT_{bus}: the rise in travel for bus users
dT_{bike}: the reduction in travel time for bike users

Fig. 1 Port Said Google Map with selected streets (Port Said Traffic Administration, statistics, 2019)

Table 1 Considering actions of policy

Criteria		Policy action
Bike frequency, b_f (bike/hr)	Road width at section of the proposed bike routes (excluding parking length), L (m)	
$F_b \geq 40$	$L > 6$	Have to make bike lane
$F_b \geq 40$	$L < 6$	Prohibited traffic zone
$15 \leq F_b < 40$	$L > 6$	The bike lane solution have to be evaluated
$15 \leq F_b < 40$	$L < 6$	No action
$F_b < 15$	Any width	No action

Logically, the behavior of other modes transport users is expected to change when there is a bike lane in the street. The effect of introducing any proposed routes on its travel speed has been assessed using ANN method (Çetiner et al., 2010).

The multi-agent objective function has been used to investigate the optimal bike lane system for Port Said with considering buses and cars movement (the major presence on roads) that have the maximum effect on bike speed if neglect other types of vehicle. That function has five components for each transport mode. These components are computed by using the following equations:

$$dC_{\text{car}} = C_c \cdot dC_{\text{Cb}} \tag{2}$$

$$dT_{\text{car}} = T_c \cdot dT_{\text{cb}} \tag{3}$$

$$dC_{\text{bus}} = C_{\text{B}} \cdot dC_{\text{Bb}} \tag{4}$$

$$dT_{\text{bus}} = T_{\text{B}} \cdot dT_{\text{Bb}} \tag{5}$$

$$dT_{\text{bike}} = T_{\text{b}} \cdot dT_{\text{bb}} \tag{6}$$

As C_{c} and C_{B} are the cost of using car and bus per km, respectively; prospectively, dC_{Cb} and dC_{Bb} are the variation of C_{C} and C_{B} variables when applying bike lanes; T_{c} and T_{B} are the time per kilometer for cars and buses; dT_{cb} and dT_{Bb} are the variation of time when applying bike lanes; and dT_{bb} is the variation of T_{b} variable arising from a present route case that measures the travel time spent by bike. All these values have been estimated according to a sustainable Port Said annual project (Port Said Government Transport Management Administration Plan, 2019). Method of data collection will be illustrated in the following paragraph.

2.1 Data Collection

Data was collected manually for the chosen main streets. Each street was classified into homogenous segment with different lengths. At this instant, lengths are known, travel times by bikes are collected for recent case, and the expected travel times are predicted by making a questionnaire for about 300 bike riders as a sample using the selected routes.

For the private transport demand of Port Said city, the values of dC_{Cb}, dT_{cb}, dC_{Bb}, dT_{Bb} were calculated by constructing an origin–destination matrix. Data was collected in a working day during the peak period except the value of time for bike (dT_{bb}) variable. For determination of such variable, the neural network-based methodology and greedy algorithm have been used. The outputs of the expected average bike speed after bike routes implementation are illustrated in Table 2.

From the table, it can be noticed that the bike speed is increased for all segments of road, but this is not an indication of the mobility without making the greedy algorithm. The greedy algorithm is a hyper-path minimum algorithm used to illustrate the steps of mobility indication and to determine the optimal set of bike routes by calculating the user's maximization welfare function for each route.

2.2 Artificial Neural Network-Based (ANN) Methodology and Greedy Algorithm

The artificial neural network-based ANN methodology is an effective method to solve many problems in science's numerous fields by means of the accessible data (Zhang et al., 2016; Cook et al., 2016). Prediction of speed in Port Said city has been a great concern of this study. Determining the variation of T_b–*hour* variables arising from a present route case, two components want to be evaluated; the bike transport travel speed for links under analysis (namely, the appropriate locations for bike ways) and the number of the bike transport users. This methodology has used a neural network which has multilayer perceptron architecture (Çetiner et al., 2010). For more details, the neural network-based methodology structure is illustrated in Fig. 2.

It contains three layers that reflect the processes of collecting data and functions, input layer, processing layer, and output layer. The input layer contains one of the variables detected for a generic route section which is called a number of neurons. Linear mathematical functions are used to transfer input data into a processing layer. The processing layer is a tool needed to transfer data from input data to output data. A neuron has been hold by output layer for determining the speed of bike before and after implementation of the proposed routes. All data is calculated by a binary stated preference tool. There are many of the qualitative and quantitative road attributes that represent needed for

Table 2 ANN output of the expected bike speed variation

Road name	Segments	Length km	Average bike speed (present case) km/hr	Average bike speed (after making bike routes) km/hr
al-shohadaa	1	0.6	1.5	2.8
	2	0.3	1.6	2.5
	3	1.2	1.2	2.1
	4	0.4	1.9	2.7
23 July st	1	0.6	2.0	2.6
	2	0.5	2.1	2.6
	3	0.6	1.1	1.9
	4	0.8	1.0	2.2
23 December st	1	0.6	1.3	2.4
	2	0.6	1.3	2.8
	3	0.4	1.2	2.3
	4	0.7	1.5	2.1
	5	0.4	1.2	2.3
el-30th st.,*	1	0.5	0.9	2.5
	2	0.7	1.1	2.4
el-nasr st	1	0.4	2.1	2.4
	2	0.3	1.1	2.3
Tarh el-bahr st	1	0.4	1.0	2.1
	2	0.6	1.4	1.8
	3	0.8	1.6	2.5
	4	0.2	1.2	1.9
el-Gomhoria st	1	0.6	1.8	1.9
	2	0.5	2.0	2.5

*This road has mini-buses with many stops and high traffic so it is out of model

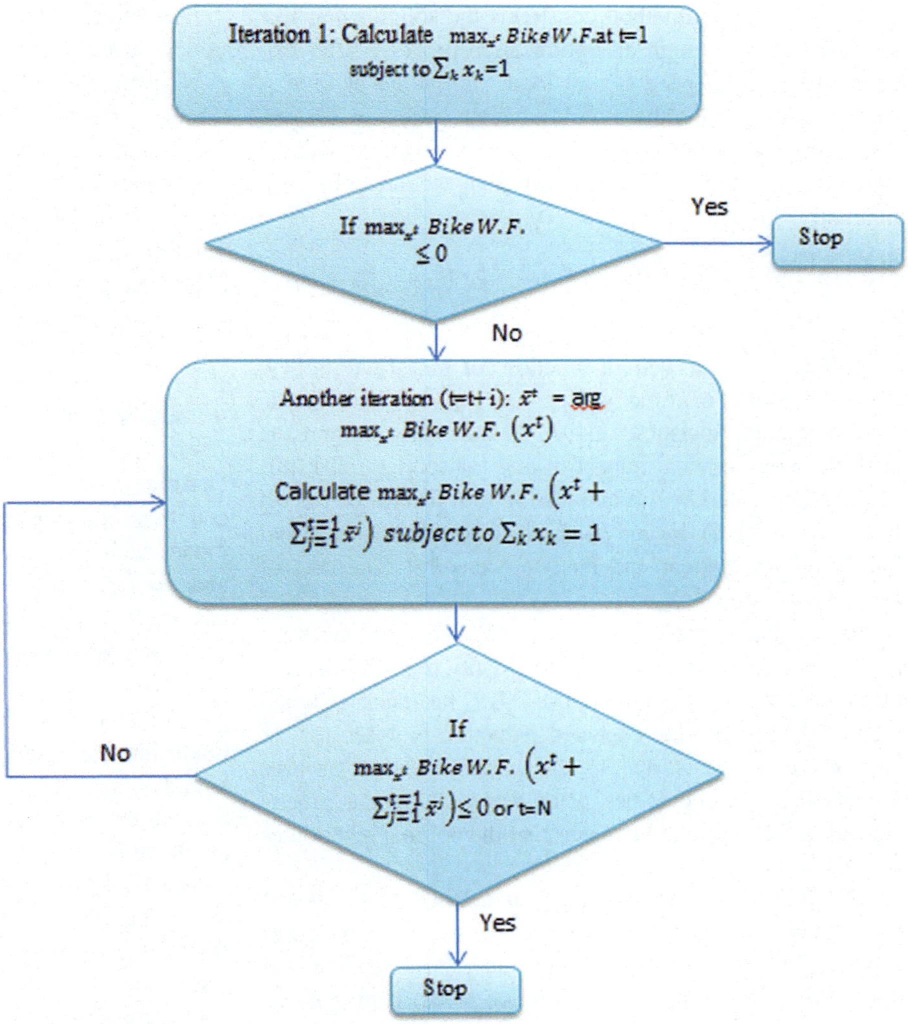

Fig. 2 Welfare greedy algorithm for the proposed policy

processes empirical data which have responds of (Yes/No) such as bike lanes availability, number of pedestrian crossings, parking possibility within the bike lane, misbehavior of the parking, and presence of side shops (La Franca et al., 2004). Some questions are about delay, trip length, travel time, average speed. It looks like a questionnaire.

According to these variables, the analyzed inter-city roads have been subdivided into multi-homogeneous segments. Furthermore, the above-mentioned empirical road data, as comparable and dimensionless inputs, have been normalized to be handled by the neural network. For calculating the number of bike users, manual survey had been implemented in a working day during the peak hour (April, 2019) to assign a corresponding graph using the hyper-path minimum algorithm, with the expected rising of the bike travel speed due to the proposed bike routes.

3 Analysis and Results

From the collected data, some roads were selected to be suitable for reaching the urgent facilities, for example, schools, faculties, government interests, and shopping activities. In this paper, the developed prediction model has been used in any period of time, and then the approach converts the date–time data to a date–time format independent. The data of bike speed has two situations: The first is the present case in which the bikes move among all kind of traffic, and the second is the bike has routes separately that can be estimated as the free speed as the street has minimum traffic flow at non-peak hours. Note that the bike speed obviously varies based on the conditions in which the user riding such as the condition of pavement and traffic.

The experiential method constructs the solution for bike routes problem through an optimization process. After calculating the bike speed for all roads, the value of time is estimated to describe the mobility of Port Said new system. To check the mobility of bike, the bike welfare will be estimated by the following proposed welfare objective function:

$$\text{Bike}_{\text{W.F.}} = a(dT_{\text{bb}} - dT_{\text{cb}} - dT_{\text{Bb}}) - (b_{\text{C}}dC_{\text{Cb}} - b_{\text{B}}dC_{\text{Bb}}) \quad (7)$$

As $\text{Bike}_{\text{W.F.}}$ is a welfare function for bike route policy, a is the value of time (25 l.e/hour) (Port Said Traffic Administration, statistics, 2019). To take into account the positive environmental impact of any transport service, this value is multiplied by 1.3 and also to turn from time to cost (Cook et al., 2016). b_C and b_B are the transport costs (10 and 2 L.E per km for car and bus, respectively). The cost of using bike is equal to zero.

For the determination of dC_c and dT_c, the O-D matrix for the cars transport demand of Port Said city in a working day throughout the peak period (April-2019) has been assigned taking into account the expected reduction in space for car and bus movements due to the implementation of bike routes. The next outcomes were achieved by the greedy algorithm that is applied to determine the optimal bike routes for Port Said city transport system.

The greedy algorithm is used for calculating the user's maximization welfare function when using bike routes (Zheng, 2009). The policy action into road k is estimated after making iterations. If the iteration stops, it means that the selected road is suitable for the policy; if not, it means that it is not suitable (see Table 1 of bike frequency).

where

x: the location pattern vector of bike routes.

x_k: the generic element of x according to those roads with a precarious bike frequency (from 15 to 40). It equals to 1, if the road k has already space for bike routes and 0 otherwise.

After applying the previous algorithm (see Fig. 2), the output helps to choose the suitable route. The outputs are illustrated in Tables 3 and 4.

The values of the welfare function show the impact of the policy on the bike welfare function. It can be noticed that el-Nasr St. has a negative value for all iterations which means it cannot be suitable road for bikes. The optimal solution at every step is written in red font color. The final number refers to the ability to confirm that bike route which can be implemented in the selected road at the previous iteration. So, Al-Shohada, 23 July, 23 December, and Tarh el-Bahr streets are selected to be the bike route policy in the new developing plan.

4 Conclusions

Decision makers of the transport sector in developing countries go toward increasing of sustainable transport modes such as bikes. They tried to reduce the car ownership and air pollutions. An average demand frequency of 40 passengers per hour and an average travel speed of 1.5 km/h are considered the main characteristics of the bike transport. The study predicted that the bike speed will be 2.4 km/h to give more bike mobility in roads when entering the bike routes into the existence transport network. Also, it will have an advantage by reducing the traffic jam and mobility through the existence roads. So, the bike routes seem to be an effective solution to advance the public transport performance at Port Said city, but in general expectations about the urban mobility system, the illustrated planning model excludes bike routes for some roads because of many considerations such as mini-bus stops and road width. It is considered a solution for traffic problems and one of the latest sustainable transport systems.

For describing the cost function of using each of bike or other modes, multi-agent objective function considering buses and cars movement had been formulated and is solved using the artificial neural network-based methodology. Greedy algorithm was used after calculating the user's maximization welfare function to identify the suitable bike routes. It demonstrated that Al-Shohada, 23 July, 23 December, and Tarh el-Bahr streets are suitable links to the bike route policy. This study is useful in taking a precise decision of introducing a green transport mode such as bike.

Table 3 The chosen bike routes location using the greedy algorithm

The cost of impact (L.E) on the welfare function with policy					
No. of greedy steps	Al-shohadaa St	23 July St	23 December St	El-Nasr St	Tarh el-barh St
1st	+25.39	+73.25	+23.16	–4.06	+104.03
2nd	+36.09	+80.34	+44.08	+16.93	–
3rd	+51.78	+94.78	–	+11.03	–
4th	–	–	–	–6.23	–
5th	–	–	–	–10.45	–

Table 4 The model output of selected routes

Road name	Segment	The model output
Al-Shohadaa St	1	*
	2	*
	3	*
	4	*
23 July St	1	*
	2	*
	3[s]	*
	4[s]	*
23 December St	1	*
	2	*
	3	*
	4	*
El-30th St	1[s]	–
	2[s]	–
El-Nasr St	1[s]	–
	2[s]	*
Tarh el-bahr St	1	*
	2	*
	3	*
	4	–

*: means that the road is selected as a bike route, –: means that the road is not suitable to be a bike route, [s]: this road has mini-buses with many stops and high traffic, so it is out of model

The increasing of the bike user's percent is one of the recent sustainable transport solutions. Such qualitative planning approach needs to be extremely attractive for decision makers. The described methodology is an achievement technique but with lack of simulation graphs results.

References

Çetiner, G., Sari, M., & Borat, O. (2010). A neural network based on traffic-flow prediction model. *Mathematical and Computational Applications, 15*(2), 269–278. https://doi.org/10.3390/mca15020269

Cook J. , Kimuyu P., Blum A.G., & Gatua J. (2016). A Simple Stated Preference Tool for Estimating the Value of Travel Time in Rural Africa. published online by Cambridge University Press, 17 June 2016. DOI: https://doi.org/10.1017/bca.2016.1

Elgioshy S. M. (2017). Integration of Egypt's Transportation Plans 2030 with the New Silk Road Project A presentation. https://schillerinstitute.com/media/integration-egypts-transportation-plans-2030-new-silk-road-project/

La Franca, L., Migliore, M., & Salvo, G. (2004). The automatic vehicle monitoring to improve the urban public transport management. *AMAT, Palermo, ITALIA*. ID: 108732627

Gurmu, Z. K., & Fan, W. (2014). Artificial neural network travel time prediction model for buses using only GPS data. *Journal of Public Transportation, 17*, 45–65. https://doi.org/10.5038/2375-0901.17.2.3

Jeong & Rilett L.R. (2004). Bus arrival time prediction using artificial neural network model. In *IEEE intelligent transportation systems conference* (pp. 988–993). Washington D.C., USA. https://ieeexplore.ieee.org/document/1399041

Johar, A., Jain, S. S., & Garg, P. K. (2016). Prediction of bus travel time using ANN: A case study in Delhi. *Transportation Research Procedia, 17*, 263–272. https://doi.org/10.1016/j.trpro.2016.11.091

Li, X. (2017). Solving transportation problems with warehouse locations based on greedy algorithm. In *International conference on advances in materials, machinery, electrical engineering, AMMEE*. Advances in Engineering Research (Vol. 114).

Mikluščak, T., Gregor, M., & Janota, A.)2012). Using neural networks for route and destination prediction in intelligent transport systems. In *Evaluation of microsimulated traffic light optimisation using V2I technology* (pp. 380–387).

Port Said Google Map Website, Egypt (2019)

Port Said Government Transport Management Administration Plan. (2019)

Port Said Traffic Administration Statistics (2019)

Zhang, H.,Wu, H., Sun, W., & Zheng, B. (2016). DEEP TRAVEL: A Neural Network Based Travel Time Estimation Model with Auxiliary Supervision. Published online by Cambridge University Press. https://www.ijcai.org/proceedings/2018/0508.pdf

Zheng, Y., Chuanqing, Xu., & Xue, J. (2009). A simple greedy algorithm for a class of shuttle transportation problem. National Science Foundation. *Optimization Letters, 3*(4), 491–497. https://doi.org/10.1007/s11590-009-0126-9

Measuring the Prospects of Transit-Oriented Development (TOD) to Identify the Potential Zones in Rajshahi City, Bangladesh

Md. Asaduzzaman, Sabbir Ahmed Siddique, and Z. H. M. Monjur Murshed

Abstract

In recent decades, rapid population growth has been a critical issue across the world. Mostly, it is becoming much more alarming in the Asian cities, as the UN (Dept. of economic and social affairs) says by 2050, 90% of the upcoming urban population growth will take place in Asian and African cities. In this situation, transit-oriented development (TOD) offers great opportunities mainly for the Asian cities to face the upcoming challenges. Currently, Bangladesh is the fastest growing country among the Asian cities, and its cities have taken an unparalleled pressure of population with infrastructure growth. So, the concept of TOD can be the ultimate solution for accommodating this ineligible urbanization and motorization happening right now throughout the cities of Bangladesh. This research focuses on evaluating TOD principles in transforming cities like Rajshahi, Bangladesh, where population growth has not yet crossed the red mark and some of its strategic locations which have potentials and opportunities for creating as transit stations.

Keywords

TOD principles • Transforming city • Potential transit station • Sustainable development

1 Introduction

Cities are the manifestation of the social, cultural and economic stimulation which are the outcomes of organized urbanization. Urbanization is undoubtedly the most visible expression of the agrarian-industrial transition. Economies are transferring from agrarian to industrial ones because of productivity growth due to technological progress. Labors from agricultural sectors are moving from rural life to capitalistic society every day for better economic conditions where industrialization is flourishing day by day. Almost 55% of the world's population currently living in the urban environments, which is expected to rise up to 68% by 2050 (United Nations, 2018). More than 60% of that increase will occur in Asian cities where China, India, Pakistan and Bangladesh are in the red mark. Nevertheless, among the South Asian countries, Bangladesh holds the first place as the fastest growing urbanized country with 6% population growth rate since independence while national population growth rate was only around 2.2%. By the data of population census 2011 (BBS), in 1974, the urban population of Bangladesh was 6.27 million which grew to 39 million in 2011. Urbanization rate grew from 8.78 to 27.66% in this period with a 5.18% growth rate and at the same time, rural population growth rate was only 1.36%, where Dhaka was crowned for being the most urbanized district of Bangladesh with the population of 11.20 million at 93% growth rate.

Though in 2016, Rajshahi ranked the top position on becoming a global model for tackling air pollution by cleaning up brick kilns, introducing "Zero soil" project to cover the dust blown up from dry riverbeds, fields and roads, like other divisional cities Rajshahi will experience similar population growth in near future. As an administrative divisional city, Rajshahi City Corporation has already been facing rapid population growth at 8.58% which is increasing day by day.

Additionally, according to the sustainable urbanization policy brief, urban centers are account for both 7.7% of both global energy consumption and greenhouse gas emissions. So, innovative idea for urban infrastructure is very much needed to pace with these issues.

Primarily, we can reduce greenhouse gas emission annually at the rate of 1.5 billion CO_2e by 2030 by making

Md.Asaduzzaman (✉) · S. A. Siddique ·
Z. H. M. Monjur Murshed
Department of Architecture, Rajshahi University of Engineering & Technology, Rajshahi, 6204, Bangladesh

C. Alalouch et al. (eds.), *Towards Implementation of Sustainability Concepts in Developing Countries*,
Advances in Science, Technology & Innovation, https://doi.org/10.1007/978-3-030-74349-9_6

transformative change in transport system like introducing transit-oriented development which can offer Rajshahi city a better quality of life by developing some integrated functions in the highly densified areas, prioritizing walkability and pedestrian facilities with easy accessibility around the transit stations with greater mobility system and comfortable moving. Further research in depth will tell us which kind of TOD is needed for respective locations.

2 Existing Transportation System Profile of Rajshahi City

This research took place in Rajshahi city which is in the central western of Bangladesh. This incorporated area was divided into several wards and four thanas namely: Boalia, Motihar, Rajpara and Shah Makhdum. Spatial structure of Rajshahi city is morphologically monocentric but functionally polycentric.

2.1 Mobility and Modality

Rajshahi city is well-connected with the other divisional cities of Bangladesh by air, bus and train. Thousands and thousands of people are entering and leaving the city per day. Recently, Rajshahi City Corporation carried out the mobility analysis of this city. They also surveyed the purposes of the trips people are making every day and found that people are making about 25% of their trips for going and coming back from their work. Educational purpose 22%, shopping 4%, social 6%, others 3% and back to home 44%. For these trips, they are using various types of public or private transports at different rates, where auto-rickshaw holds the top position. Most of them are unregistered. It's predicted that by 2020, the number of auto-rickshaw will rise to 30,000.

Users of commercial vehicles like truck, bus, utility vehicles, auto-rickshaw and rickshaw is 63% and non-commercial vehicles are 37%. The annual growth rate for both categories together is 6.1%. (Rajshahi City Corporation, 2014).

Moreover, researchers and the transport planners have found that auto-dependent transportation systems are very much high in cost, while public transit-focused transportation systems are more economical. (Hale, 2006) Most people living in Rajshahi city use auto-rickshaw, motorcycle and rickshaw for daily commutes, causing severe traffic congestion, delayed travel time, energy wastage, air and sound pollution during rush hours (see Fig. 1). To accommodate these commuters, it needed the development of transportation facilities with an emphasis on the development of the mass transportation system. (Taki & Maatouk, 2018a).

2.2 Transport Infrastructure

Of all the roads RDA and RCC had built so far, 26% of them are kutcha, 21% semi-pucca and 53% of them are pucca (see Fig. 2). Most of the road surface is peeled off and this makes potholes and open aggregates from the road surface. The dusty aggregates contaminate the air and as a result, the environmental condition of this city is no more in favor. Maximum footpaths of the major node points have already been gone to the control of the illegal authorities. This is increasing traffic congestion and hampering human mobility.

Though city center is known as the core of economic activity, commerce, social and cultural movement, day by day it is becoming quite impossible for the city dwellers of Rajshahi city to maintain their comfortable movement around the city center due to the poor condition of the existing transportation infrastructure facilities. Broken roads, poor walkways and cycling infrastructure and unplanned street networks toward the city center may cause a mountainous chaos throughout Rajshahi city. To avoid these unwanted sufferings, a more efficient transport system is required. That can only be possible by implementing the concept of transit-oriented development (TOD), which presents great opportunities of urban development that can run with this unprecedented urbanization and motorization all together and reduces private vehicles dependency rates simultaneously. (Luh et al., 2015) And that is not possible without the development of infrastructures needed for smooth mobility system. Hendricks says that "if the transit doesn't function in a seamless and speedy manner, the performance and success of the TOD precinct is in question." (Hale, 2006).

2.3 Critical Issues

An integrated transportation system is not only all about accessibility but also depends on the other urban development strategies like compact development, social and cultural movement, more healthier transportation (public transit, cycling and sidewalk) to maintain the population growth and strategic use of existing transportation infrastructures in the urban environments, (Thomas & Bertolini, 2017) which is very much absent in Rajshahi city for the lack of proper maintenance and planning system from both government and private agencies. Poor public transport sector is encouraging people of Rajshahi city to use more private vehicles, and auto-rickshaws day by day are contributing to traffic congestion, noise and air pollution and traffic safety (see Fig. 3).

Additionally, poor drainage system and insufficient knowledge on regulation of parking and road use among the

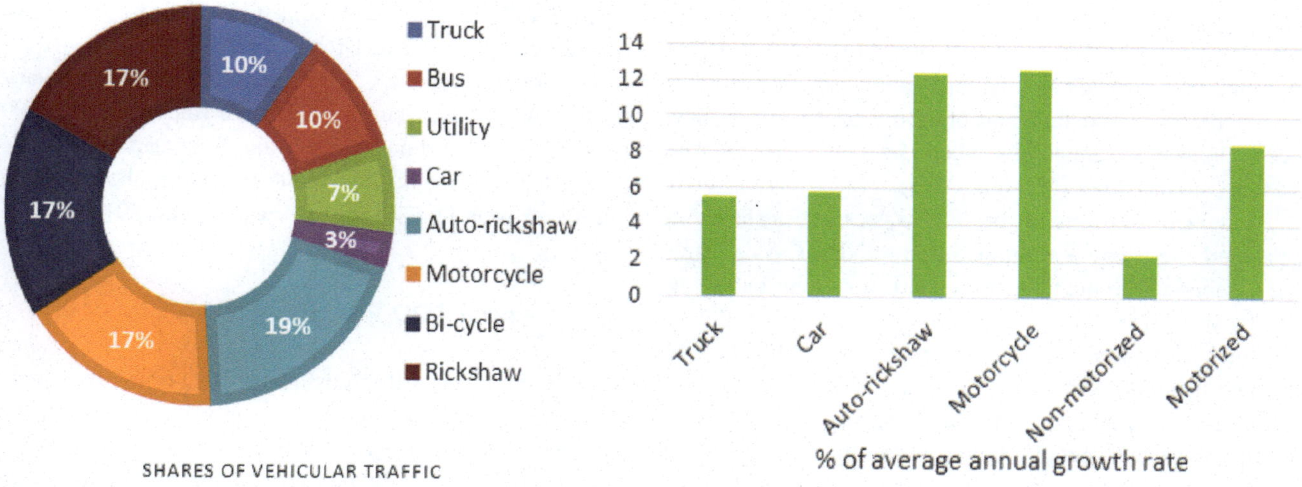

Fig. 1 Mode share of vehicles and growth rate of different vehicles in Rajshahi city. (Rajshahi City Corporation, 2014)

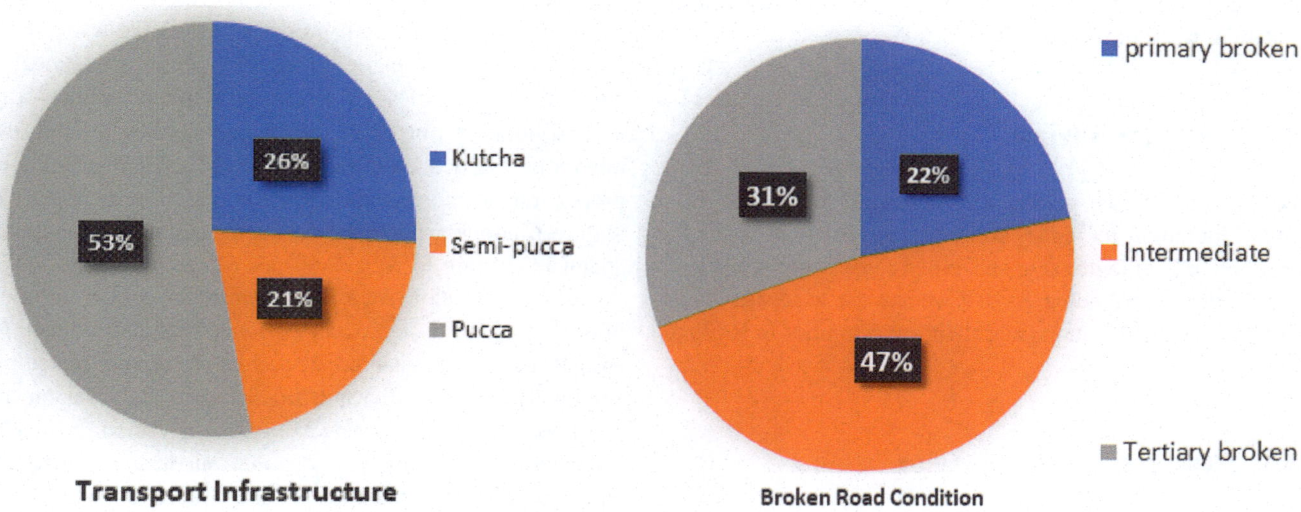

Fig. 2 Road condition and type in Rajshahi city. (Rajshahi City Corporation, 2014)

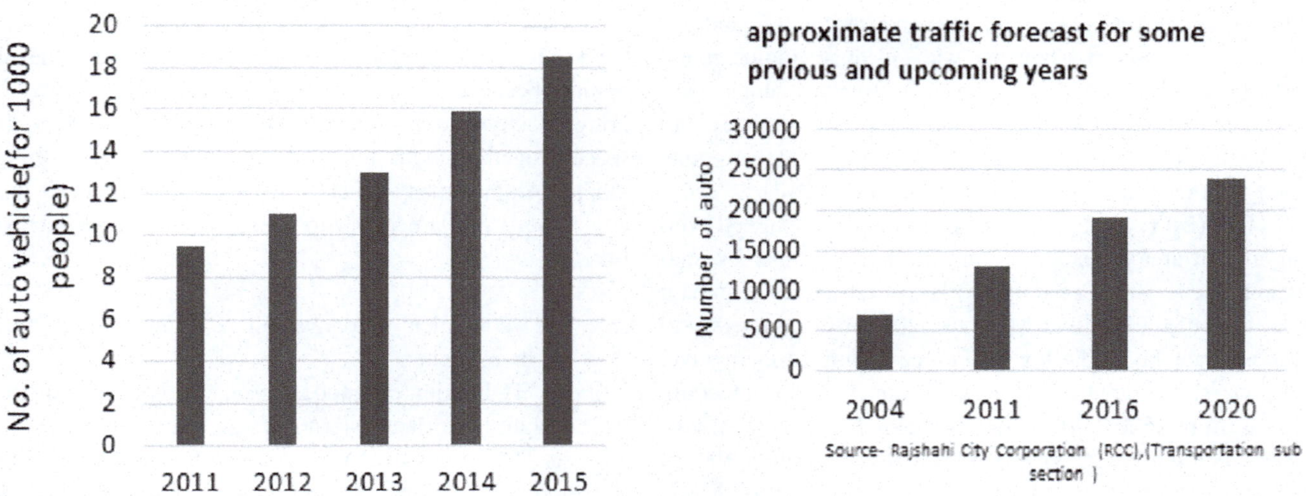

Fig. 3 Increasing rate of auto-rickshaw in Rajshahi city. (Rajshahi City Corporation, 2014)

mass people continuously hampering the mobility of people by occurring frequent road accident throughout the city. A continuous research by city corporation says that there are six intersections and four mid-blocks in Rajshahi city have been found most dangerous where frequency rate of road accident is very high (Mamun, 2015).

Moreover, 140 tons of the total wastes of Rajshahi city produced every day are dumped into the drain. As a result, the numbers of drains filled with solid wastes are increasing day by day. This hampers the drainage water flow; 30% of the total 142.18-km drains are more or less blocked by garbage materials (Rahman & Hasnat, 2018), which is also hampering the mobility of the people during rush hours. So, to attract the mass people, proper transport infrastructure is needed alongside the high quality, cost-effective and timely transport services. We have to keep in mind that "The more that transport provides coverage across the region and connectivity between origins and destinations the greater the potential ridership." (Hendricks, 2005).

3 Literature Review

For this paper, we review a few specific categories from TOD literature to some current transportation issues of Rajshahi city at regional to sub-urban scale which leads us to some criteria and indicators that we can use to identify some potential transit locations and their opportunities in Rajshahi city.

3.1 TOD and Travel Behavior

TOD means a "mixed-use community, that encourages people to live near transit services and to decrease dependence on their driving." (Carmelina et al., 2013) And also "the practice of developing or intensive residential land use near rail stations and housing with adequate public transportation, biking, and walking infrastructure along with complementary public uses, jobs, retail and services, are concentrated in mixed-use developments at strategic points along the regional transit systems." (Barby, 2014).

TOD (BRT, LRT and MRT) stations are designed with significant area coverage around the transit points by analyzing the existing road network, pedestrian environment and land-use. (Taki and Maatouk, 2018b) Commuters will choose their favorable transit type according to their travel behaviors. The area coverage of a transit station generally varies from (500–800)-m radius depending on the transit types. Normally, area coverage of a BRT station is 500 m, while it becomes 800 m for the stations with railway (MRT, LRT) facilities. The station points are roughly within 10–12 min walking distance on both sides of the transit corridor (Ministry of Housing & Urban Affairs, 2017).

Generally, frequency of the buses in BRT stations varies from 2 and 5 min when a BRT route can transport 30,000 passengers in an hour while train frequency is 2–3 min during the peak hours of 7 am to 9 am and about 5–7 min during off peak hours. (Staff Correspondent, 2019).

3.2 General TOD Principles

In this section, we will discuss some TOD or transit-oriented development principles that are only applicable in and around the transit stations. "There are two main categories of consideration to follow in transit supportive assessments: Station Approach and Access and Station Area Development." (Crosby, 2003).

Firstly, Station Approach and Access like creating a walking environment with proper sidewalk, pedestrian and cycling zone which can offer us all to enjoy from local to citywide opportunities and resources in much healthier way. Prioritization of not only quality public transport but also other motorized or non-motorized modes all together with proper management techniques like shared surfaces, driveway restrictions and proper car management by introducing adequate parking facilities will fall under this category.

Secondly, Station Area Development, in which the most important thing is introducing some integrated functions around the transit station areas like designing mixed-use neighborhoods with efficient buildings (compact development) and public gathering spaces. Mixed-use development always promotes vibrant ground floors that encourage community participation among mass people and collective identity.

3.3 TOD's Range of Benefits

Cities are gradually turning toward transit-oriented development because it is becoming the fastest growing way of bringing citywide people, activities, buildings and public spaces together, with the great combination of mobility modes. Among many other benefits of creating TOD in a city, a very few are going to be reviewed in this section. Like:

i. The economical outcome will be around double than the investment.
ii. As TOD-oriented modes promote corridor-oriented design, it will be easier to design infrastructure.
iii. As TOD promotes compact development near stations, unorganized urban sprawl will be minimal. (Newman & Kenworthy, 1999)

iv. People also can maintain affordable housing by reducing household spending on transportation which will help to maintain economic balance to the society.
v. Increased transit ridership (Transit Oriented Development, 2016).

"Transit Oriented Development as an approach to combat traffic congestion and protect the environment has caught on all across the country. The trick for real estate developers has always been identifying the hot transportation system. Today, highways are out; urban transit systems are in."-(Urban Land Institute, 2016).

3.4 Urbanization and Motorization of Rajshahi City

Since the last century, rapid urbanization is happening throughout the world accompanied by massive urban sprawl as well as the cities of Bangladesh (see Fig. 4). The more the growth spread, the more the cost of grocery products will raise. Rapid urbanization is not only giving rise to the unprecedented motorization but also devouring greeneries, threatening estuaries, natural environments and also causing high cost regarding urban infrastructure development. (Cervero, 2014) As there is no transportation system in Rajshahi city which connects regional, urban and sub-urban areas, people from distance are coming to the city center facing lots of transit issues that kill our travel time and at the same time causes massive economical downfall.

Additionally, Rajshahi city is well known for its massive contribution in our national GDP for growing a huge amount of crops (vegetables, fruits or others) throughout the year in the sub-urban areas. For this, farmers from the sub-urban areas have to transfer their goods or products to the city center, from where those goods are distributed throughout the country or internationally. But in the absence of adequate transport system connecting the regional, urban and sub-urban areas, farmers are failing to transfer their goods to the city center straight. On the contrary, adding additional travel time which causes massive price hike is troubling mass people. So, implementation of innovative ideas like TOD, connecting regional to sub-urban areas between Rajshahi city and its surroundings with proper unification can be the utmost effective solution from issues like excessive time consumption on roads, traffic congestion and environmental contamination. (Taki and Maatouk, 2018b). So, this study focuses on to identify some potential locations and their opportunities creating as transit stations at regional, urban and sub-urban scale which can provide Rajshahi city a much healthier life by introducing greater mobility system and comfortable moving.

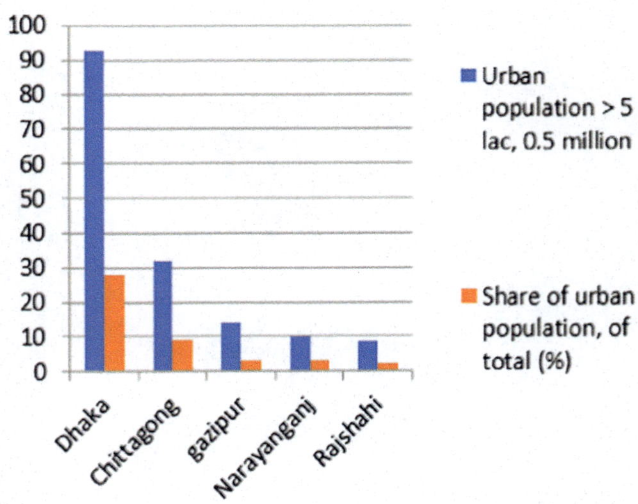

Fig. 4 Urban population growth comparisons with other cities in Bangladesh (Bangladesh, 2013)

4 Spatial Transformation Pattern

"Rajshahi City Corporation (RCC) is planning to expand its area by integrating two municipalities. Apart from Nawhata and Katakhali municipalities, RCC plans to incorporate some portions of horipur union in poba upazila, said the corporation officials. They also said that they are preparing the proposal for expanding its area roughly up to 350 sq.km over three and a half times bigger than the present area of 97 square kilometers." (RCC mulls expanding its horizon | The Daily Star, 2017).

4.1 Major Criteria and Indicators

There are mainly three indicators which will reveal the areas having potentials for future development program which have the capacity to add some economic value to GDP of Rajshahi city and will get overcrowded in the near future. These are:

a. Land-Use Transformation
b. Population Density Pattern
c. Job Density Pattern.

4.2 Land-Use Transformation

The TOD projects present different characteristics and hierarchies on different land-uses. The core land-use pattern and its continuous transformation along with its surroundings can be a key factor while calculating the necessities of transit stations on impactful node points. As Cervero mentioned "tapering of densities with distance from a station could like a wedding cake." (Tong et al., 2018). The development of Rajshahi is mainly three directional. East–west and north-oriented (see Fig. 5). The main reason behind this is the intercity connections. Nature, Chapai Nawabganj and Naogaon are the adjacent cities. The communications between these cities are based on the roads, which are east–west and north-oriented. These roads guided Rajshahi development into three directions.

Major driving forces for predicting future land development pattern:

1. **Distance to sub-city centers**

Distance to sub-city is a vital factor in case of development in Rajshahi. The distance of sub-city or suburban area is the determinant force of the city perimeter and development. It can also guide to sustainable development incorporating the sub-city.

2. **Distance to major roads**

The major arterial roads connect the suburban areas or the sub-city centers. They also can connect the city with the nearby cities. Thus infrastructures and other activities grow surrounding the roads, and roads became the major development line.

3. **Proportion of built-up land**

The location or the position of the land is also a driving force. The expansion of the city or the new infrastructure will be built in adjacent lands. Thus, the availability, position and proportion of land became a driving force in case of development (see Tables 1 and 2).

4. **Distant of government facilities**

The government facilities and its location can guide the development of a city as neighborhood; commercial activities are generated centering them. Thus, it can set a city perimeter and also the development nature.

4.3 Population Density Pattern

"TOD intends to increase land use density at pivotal locations and curb urban sprawl. Higher residential and employment densities in the vicinity of transit station will make residents or workers more likely to use public transit, and support local economic activities. With the distance from transit station increases, development densities will be gradually tapering down and auto modal shares will be increasing accordingly. Higher capacity transit facility requires higher land use density to support it. For example,

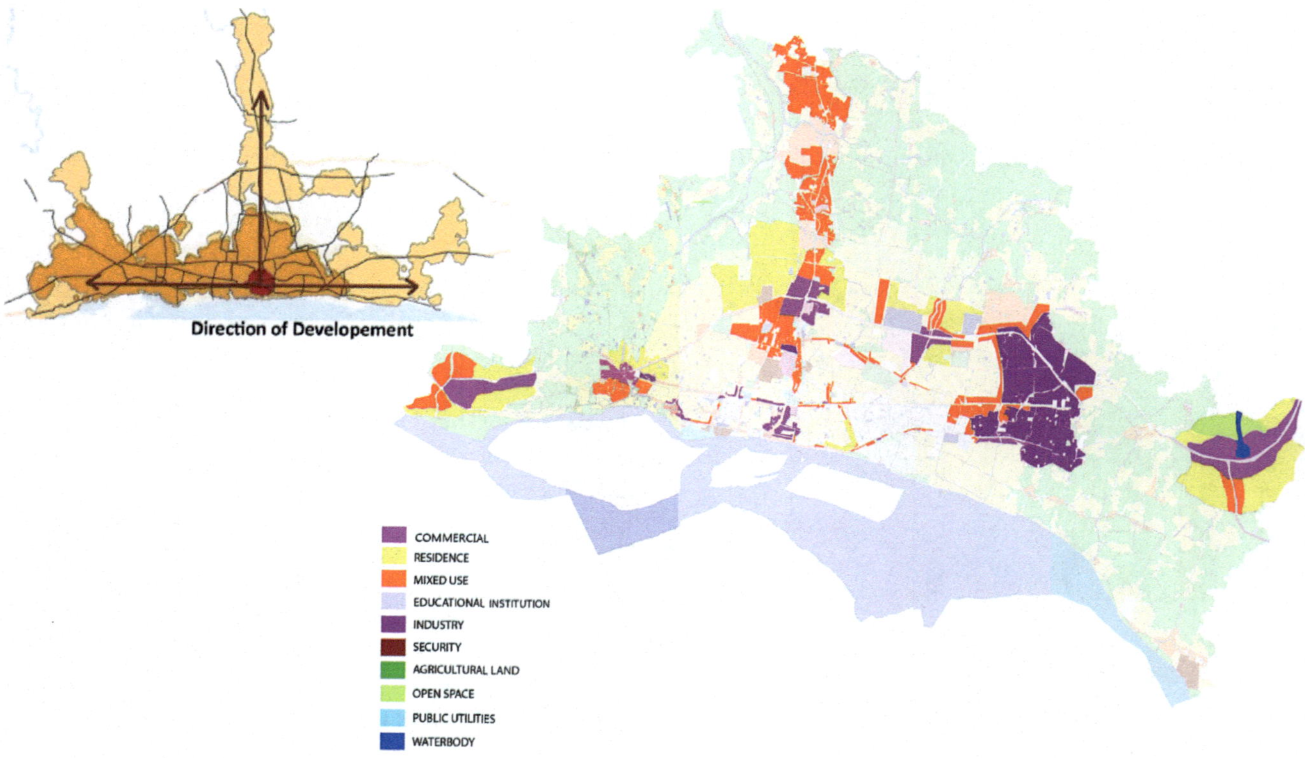

Fig. 5 Morphological growth direction and proposed land use pattern of Rajshahi city (By Author)

Table 1 Analysis of land use distribution pattern with time of Rajshahi city (Rajshahi Development Authority | Rajshahi, 2015)

Study period	Built up change	% Change	Year	Growth rate/year	Average
1990–2000	509.4	88.05	10	8.80	8.58
2000–2010	1161.99	106.80	10	10.68	
2010–2015	845.73	37.58	6	6.26	

Table 2 Analysis of built-up area expansion with time of Rajshahi city (Rajshahi Development Authority | Rajshahi, 2015)

Land Use	1989	1996	2005	2015
Water	11	12	11	11
Bare soil	34	38	24	33
Vegetation	52	46	59	48
Built up	2	4	7	9

in Portland, Oregon, central city TOD had a transit modal share 4 times as high as that of outlying TOD in 1995, reflecting the critical importance of development density in impacting transit ridership." (Chen et al., 2010).

Population is increasing in Rajshahi city rapidly day by Day. In 1991, number of populations was 1,42,12,000, and growth rate was 1.4%. This number moved to 1,63,55,000 in 2001. In the next 10 years, Rajshahi added 10 lakh people more with the growth rate of 1.1% (see Fig. 6).

Population growth:

Change in population size = (Birth + immigration) − (Deaths + Emigration).

Net reproductive rate, r = (births-deaths)/population size or to get in percentage terms, just multiply by 100.

Factors affecting Rajshahi growth pattern:

i. Rural migration
ii. Economic investments and
iii. Urban development.

4.4 Job Density Pattern

As central business district (CBD), areas offer the most employment options and opportunities, Rajshahi will face the similar growth pattern like other divisional cities of Bangladesh because RCC officials said that they have planned to add few more CBD areas around Rajshahi City Corporation which will help to increase their economic condition. There are some criteria (see Fig. 7) which will have to meet firstly for becoming a new CBD area. Like:

i. Home craft industry: Labor training facilities, Silk production industry, Showroom facility
ii. The healthcare industry: Health care facilities, Medical related devices, equipment and supplies, pharmaceuticals and related segments.
iii. Manufacturing industry: Automotive, steel and metal, paper, textile, heavy machinery, automation and plastics.

By following these options, we have some areas which have potential for offering future job opportunities (see Fig. 8).

4.5 Criteria and Indicators

"Measuring TOD-ness is about measuring how oriented an urban area is toward the use of transit. To do so, various indicators need to be measured that define or characterize TOD. The desirable characteristics of these indicators are that they must be quantifiable, fit for forward-looking assessments rather than backward looking evaluations." (Singh, 2015). So, based on the existing analysis of land use and transportation systems of Rajshahi city and the assessment of data availability from above, a few criteria and indicators are established to analyze the potential locations which have opportunities for creating as transit stations in Rajshahi city.

This research mainly focuses on two criteria and indicators: (1) To identify the potential locations developed as transit stations (see Table 3) and (2) to identify the opportunities of those potential locations (see Table 4). (Luh et al., 2015).

After using TOD guidelines over them, we will find the final locations which have opportunities for creating as transit stations.

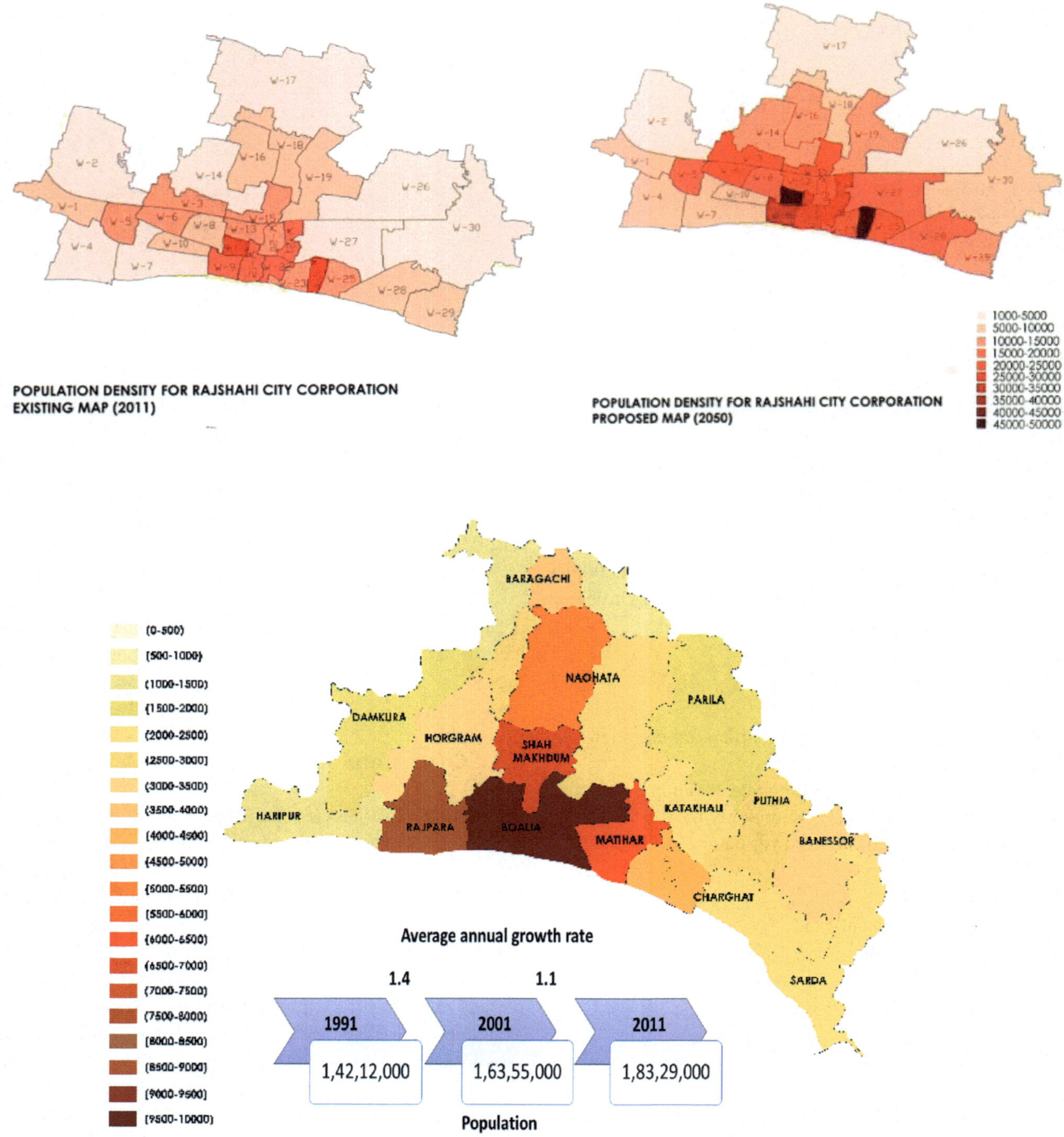

Fig. 6 Population density pattern with time of Rajshahi city. (Rajshahi City Corporation, 2014)

5 Research Methodology

TOD-based development have not done yet in Rajshahi, Bangladesh, due to the lack of collaboration between transportation system plans and spatial transformation patterns. That is why this research focuses on to evaluate TOD principles in Rajshahi city and to find some strategic locations which have potential and opportunities for creating as transit stations at regional, urban and sub-urban scale by considering existing transportation plans and spatial transportation patterns (see Fig. 9).

Firstly, necessary criteria and indicators are needed to be set which is done already, shown in Tables 3 and 4.

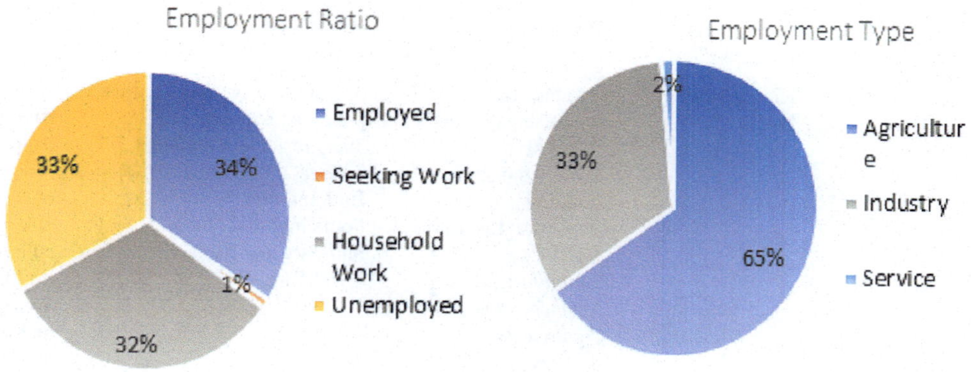

Fig. 7 Employment ratio and type of Rajshahi city (Bangladesh, 2013)

Fig. 8 Potential areas for future job opportunities of Rajshahi city (By Author)

Table 3 Criteria and Indicators No. 1 (To identify the potential locations developed as transit stations) (Luh et al., 2015)

Transit mode	Regional center transit	Urban center transit	Sub-urban center transit
BRT	✓	✓	✓
LRT	✓	✓	✓
ART (Auto Rickshaw)	✓	✓	✓
Frequency of transit	Less than 5 min	Between 5–15 min	Between 5–15 min

Secondly, existing transportation system plans and spatial transformation patterns of Rajshahi city will be examined for identifying major strategic locations in the regional, urban and sub-urban areas and their accessibility, and finally, a contextual framework will be developed by analyzing those strategic locations with the criteria and indicators we have already fixed and their opportunities for creating as future transit stations.

6 Analysis of the Potential Locations of TOD in Rajshahi City

The potential locations and their opportunities for becoming transit stations are going to be analyzed in this section through our criteria and indicators fixed before shown in Tables 3 and 4.

Table 4 Criteria and indicators no. 2 (to identify the opportunities of those potential locations) (Luh et al., 2015)

Criteria and indicators	Regional area	Urban area	Sub-Urban area
Hierarchy of land-use	Around five types of land-use pattern is required 20%- residential building 80%- non-residential building Building type: High-rise, mid-rise, Apartments, condominium	Around four types of land-use pattern is required 30%- residential building 70%- non-residential building Building type: Some high-rise, more mid-rise, low-rise and townhouse	Around three types of land-use pattern is required 30%- residential building 70%- non-residential building Building type: Some high-rise, more mid-rise, low-rise and townhouse
Building coverage (population density)	Very high	Very high–high	High-medium
Shop characteristics (job density)	Regional to local scale	Along with regional to local scale community, scale will be found	Regional, local and community scale

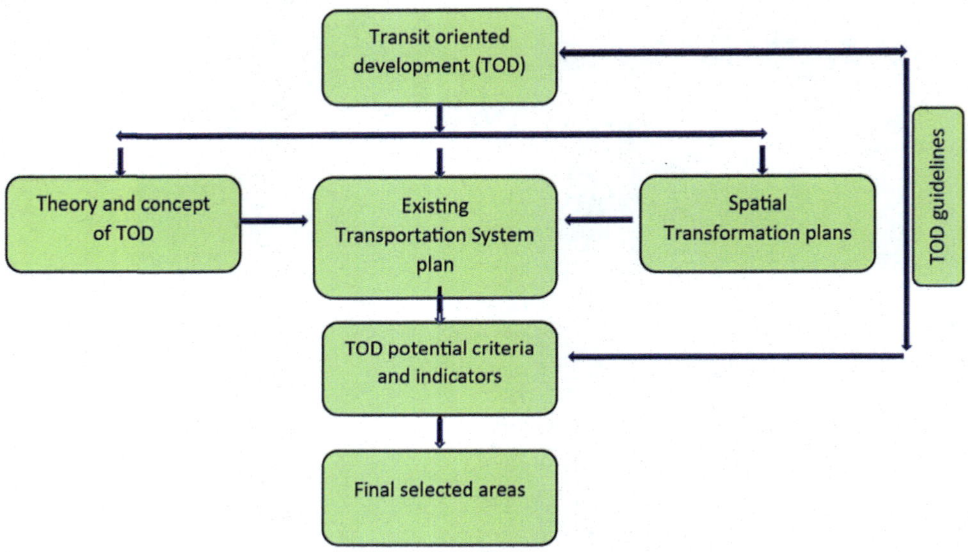

Fig. 9 Flowchart of research methodology (By Author)

Table 5 Prerequisite for transit point selection

Transit mode	Regional center transit	Urban center transit	Sub-urban center transit
BRT	✓	✓	✓
LRT	✓	✓	✓
ART (Auto)	✓	✓	✓

Source By author

6.1 Analysis of the Potential Locations of TOD in Rajshahi City

Before analyzing the final potential locations for TOD, we have to start by identifying transit points according to regional center, urban center and sub-urban center. The prerequisites for selecting transit points in Rajshahi city are given below (see Table 5).

According to the Table 5, following locations are selected which have potential for becoming as regional, urban and sub-urban transit centers are presented in Table 6.

6.2 Analysis of the Potential location's Opportunity of TOD in Rajshahi City

In this section, we will assess the potential location's opportunities based on the second criteria and indicators shown in Table 4 which has been identified as regional,

Table 6 Selected transit locations in Rajshahi city

Regional center	Urban center	Sub-urban center
Katakhali	Baneshwar	Rajabari
Binodpur	Belpukuria	Parila
Talaimari	Kashiadanga	Rajchandrapur
Zero Point	Paba	Dhankura
Laxmipur	Nawhata	Hajuripara
Court	Khorkhori	Modonhati
Vatapara	–	Mohonpur
City bypass	–	–
Naodapara	–	–
Railgate	–	–
Rajshahi Station	–	–
Vodra	–	–
Aam Chattor	–	–

Source By author

Table 7 The potential area's opportunity as regional center transit stations in Rajshahi city. (*Source* by Author)

Criteria	Indicator	Katakhali	Binodpur	Talaimari	Zero point	Laxmipur	Court
Hierarchy of land-use	Around five types of land-use pattern is required	√	√	√	√	√	√
	Building type: high-rise building	×	×	√	√	√	√
	Mid-rise building	√	√	√	√	√	√
	Apartment building	√	√	√	√	√	√
	Condominium houses	×	×	×	×	×	×
Building coverage (population density)	Very high	×	√	√	√	√	√
Shop characteristics (job density)	Regional to local scale	√	√	√	√	√	√
Grade		2	1	1	1	1	1
Radius		400 m	400 m	400 m	400 m	400 m	800 m
Transit type		BRT	BRT	BRT	BRT	BRT	BRT & LRT

Note For calculating the station area radius, we must know that TOD has some common standards regarding this issue starting from 400 m followed by 500, 600 and 800 m standards according to station type (BRT/MRT/LRT). (*Source* Properties & Mucosa, 2015)

Table 8 The potential area's opportunity as Regional Center transit stations in Rajshahi City. (*Source* by Author)

Criteria	Indicator	Vatapara	City bypass	Naodapara	Railgate	Rajshahi Station	Vodra	Aam Chattor
Mixed of land-use	Minimum five types of land-use 20% residential, 80% non-residential Residential type: High-rise, mid-rise, apartments, condominium	√ × √ √ ×	√ √ √ √ ×	√ × √ √ ×	√ √ √ √ ×	√ √ √ √ ×	√ √ √ √ ×	√ √ √ √ ×
Building density (population density)	Very-high Minimum residential FAR 4.0 Minimum office FAR 2.0	√	√	√	√	√	√	×
Retail characteristics (job density)	Regional and local scale	√	√	√	√	√	√	√
Grade		2	1	2	1	1	1	1
Radius		400 m	400 m	400 m	400 m	800 m	400 m	400 m
Transit type		BRT	BRT	BRT	BRT	BRT & LRT	BRT	BRT

Note For calculating the station area radius, we must know that TOD has some common standards regarding this issue starting from 400 m followed by 500 m, 600 m and 800 m standards according to station type (BRT/MRT/LRT). (*Source* Properties & Mucosa, 2015)

Table 9 The potential area's opportunity as urban center transit stations in Rajshahi City. (*Source* by Author)

Criteria	Indicator	Baneshwar	Belpukuria	Kashiadanga	Paba	Nawhata	Khorkhori
Mixed of land-use	Minimum five types of land-use 20% residential, 80% non-residential Residential type: High-rise, mid-rise, apartments, condominium	√ × √ √ ×	√ × × × ×	× × √ √ ×	× × × × ×	√ × √ √ ×	√ × √ × ×
Building density (population density)	Very-high Minimum residential FAR 4.0 Minimum office FAR 2.0	×	×	×	×	√	×
Retail characteristics (job density)	Regional and local scale	√	√	√	√	√	√
Grade		2	3	2	3	1	3
Radius		800 m	800 m	800 m	400 m	800 m	400 m
Transit type		BRT & LRT	BRT & LRT	BRT & LRT	BRT	BRT & LRT	BRT

Note For calculating the station area radius, we must know that TOD has some common standards regarding this issue starting from 400 m followed by 500 m, 600 m and 800 m standards according to station type (BRT/MRT/LRT). (*Source* Properties & Mucosa, 2015)

Table 10 The potential area's opportunity as sub-urban center transit stations in Rajshahi City. (*Source* by Author)

Criteria	Indicator	Rajabari	Parila	Rajchandrapur	Dhankura	Hajuripara	Modonhati	Mohonpur
Mixed of land-use	Minimum five types of land-use 20% residential, 80% non-residential Residential type: High-rise, mid-rise, apartments, condominium	× × × × ×	× × × × ×	× × × × ×	× × √ × ×	× × × × ×	× × × × ×	× × × × ×
Building density (population density)	Very-high Minimum residential FAR 4.0 Minimum office FAR 2.0	×	×	×	×	×	×	×
Retail characteristics (job density)	Regional and local scale	√	√	√	√	√	√	√
Grade		3	3	3	3	3	3	3
Radius		800 m	400 m	800 m	800 m	800 m	400 m	400 m
Transit type		BRT & LRT	BRT	BRT & LRT	BRT & LRT	BRT & LRT	BRT	BRT

Note For calculating the station area radius, we must know that TOD has some common standards regarding this issue starting from 400 m followed by 500, 600 and 800 m standards according to station type (BRT/MRT/LRT). (*Source* Properties & Mucosa, 2015)

urban and sub-urban centers in the following tables (see Tables 7, 8, 9 and 10).

These are the locations which have great potentials for creating as transit locations based on the analysis above. Each of the station areas has their own features that are very much important for having the best outcome for those locations as TOD stations. In this research, the final transit locations are divided into a grading system (Grade 1, 2, 3,) according to their vibrancy.

The final design will take place through three phases (phase 1:2030; phase 2:2040 and phase 3:2050). Each of the phases will contain every type of graded locations (see Figs. 10, 11, 12 and 13).

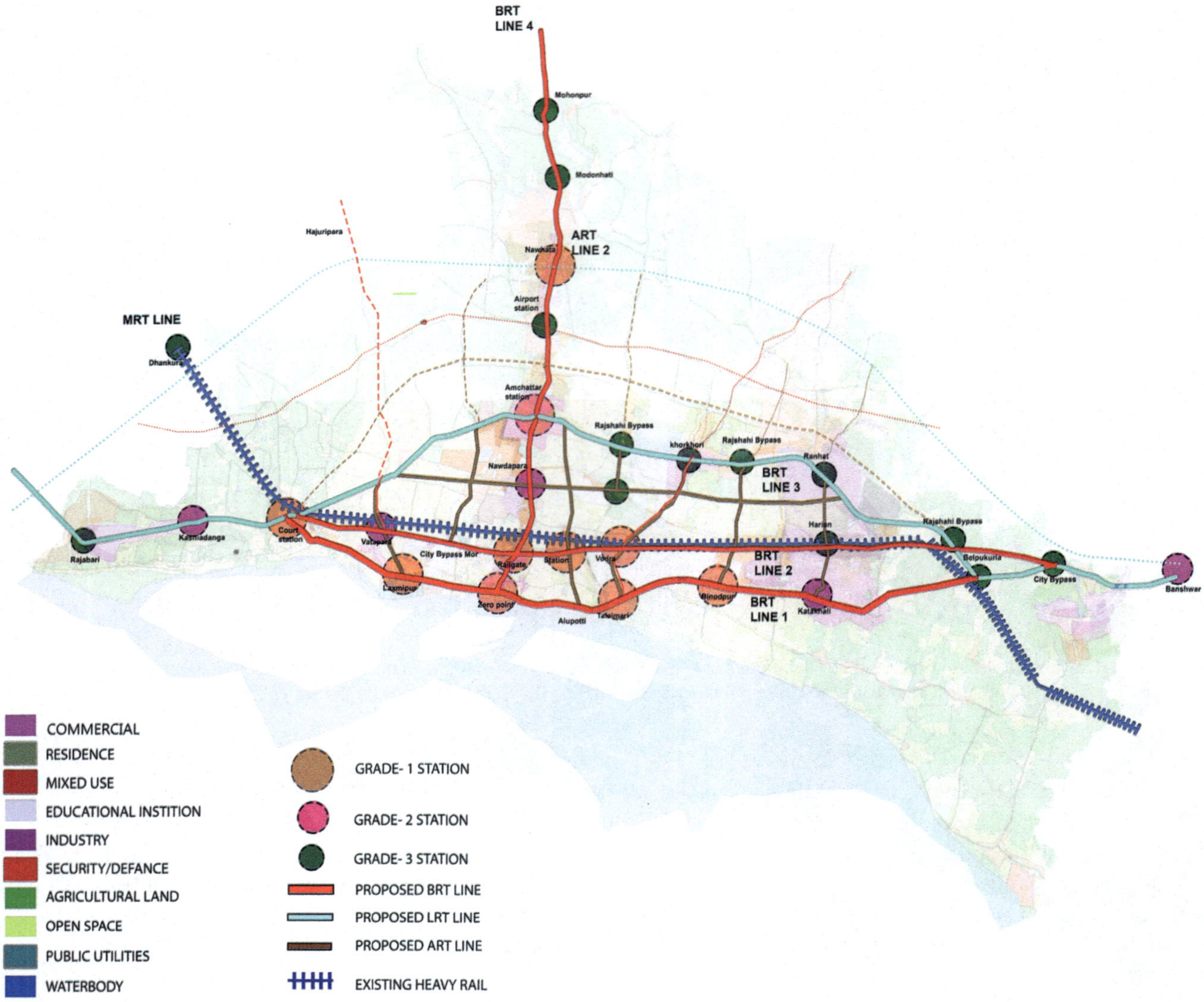

Fig. 10 Potential station areas for transit stations of Rajshahi city, phase01. (By Author)

7 Conclusions

We may conclude from the discussions in the previous sections that designing an area applying TOD principles on spatial structure, urbanization and accessibility complemented with public transit, walkability and cycling facility is a part of designing TOD. However, designing or creating a TOD also includes other things like policymaking and financing. Policymaking turns into quite a big issue while a city-region attempts to implement any kind creative solutions regarding urban growth and healthier transportation systems across the area. Commuter's transit mode choice decision mainly depends on its socio-economic and travel characteristics. (Islam et al., 2019) Implementation of TOD in a third world country like Bangladesh is a complex process, and it needs a concerted effort from various local and international agencies for both economical and field support. The development of TOD cannot be done without the city's studying the existing transportation system plans and spatial transformation plans.

During the TOD implementation process, there will be some sufferers like street vendors. Re-accommodation of these street vendors must be done in a proper way by governmental or private NGO's. After all, TOD has made a revolutionary impact on every country which already has gone through the TOD implementation program. The public welcomed this concept across the world more comfortably than any other as TOD converted their same old environment into the most desired places to live, play and work into. TOD will help the people of Rajshahi to lead a much healthier and comfortable life than ever before. This research

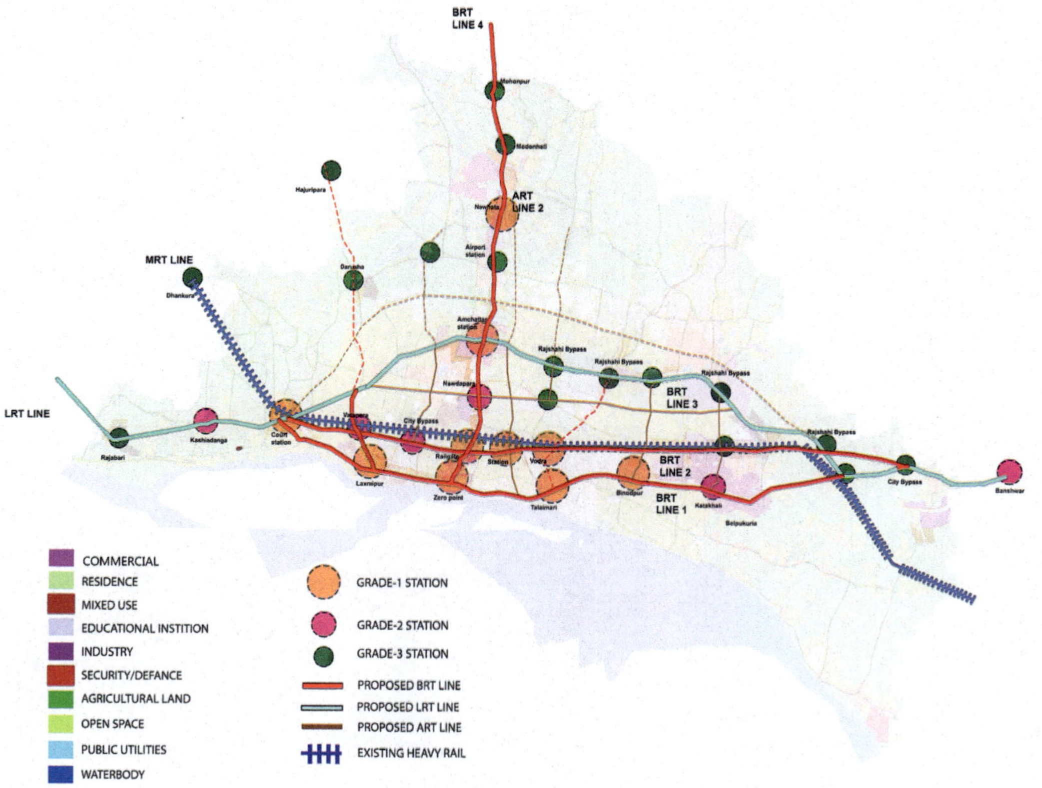

Fig. 11 Potential station areas for transit stations of Rajshahi city, phase 02. (By Author)

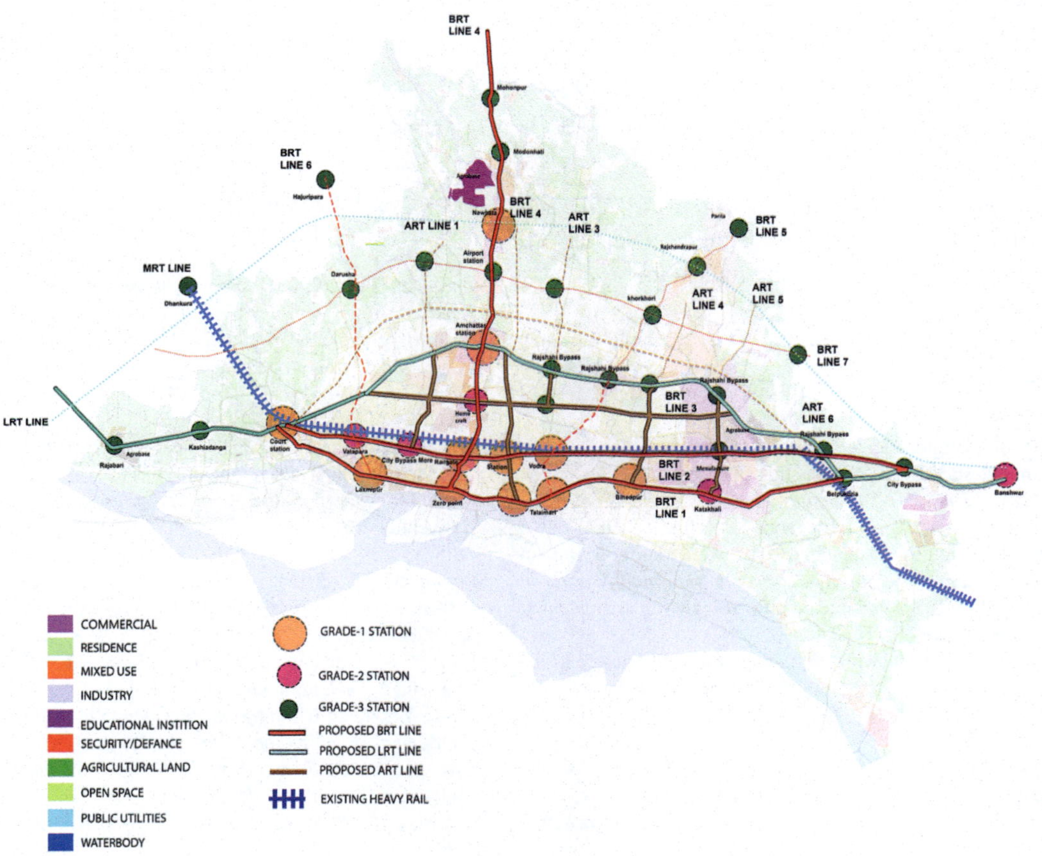

Fig. 12 Potential station areas for transit stations of Rajshahi city, phase 03. (By Author)

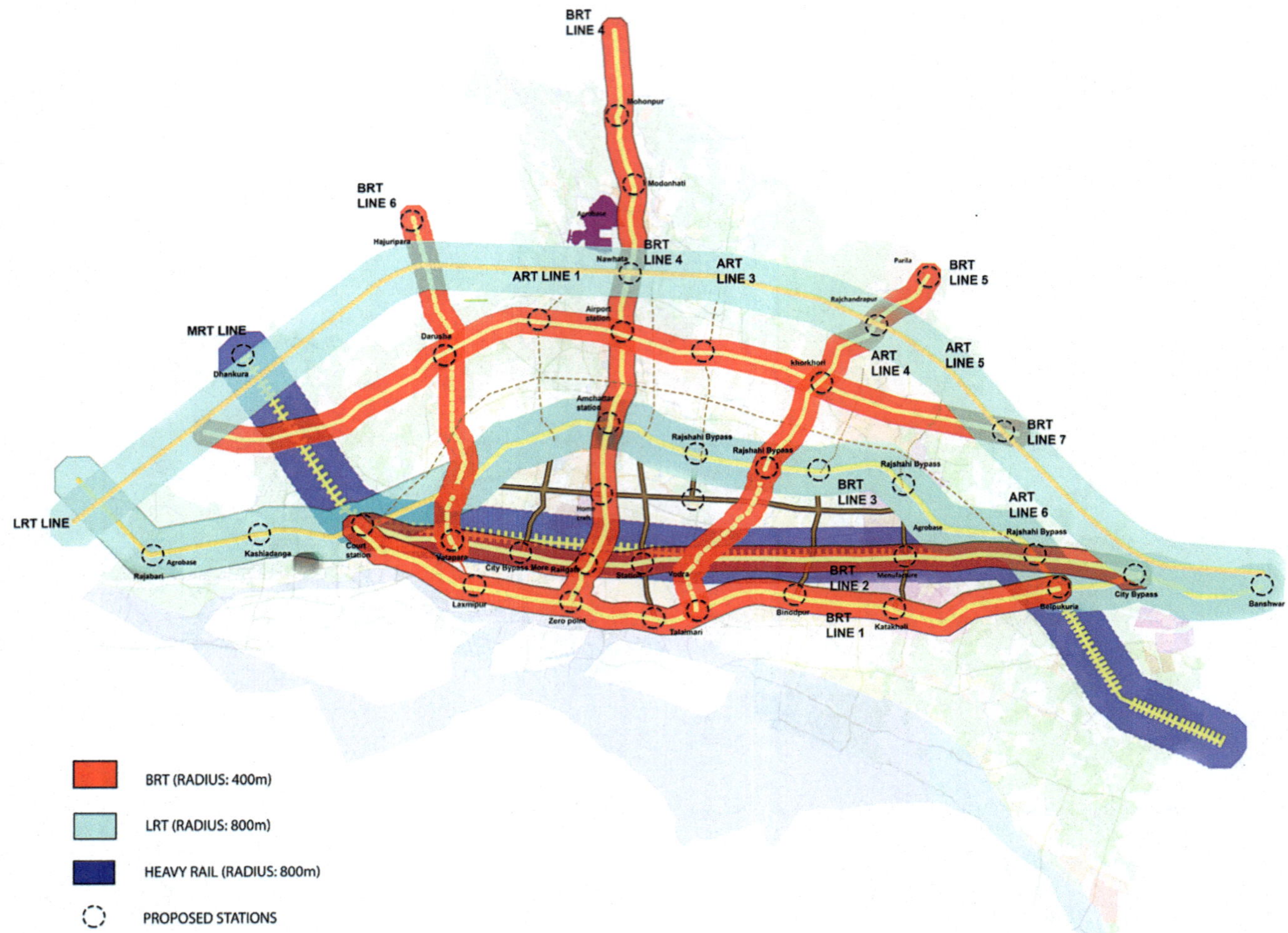

Fig. 13 Final phase covered Areas by transit stations of Rajshahi city. (By Author)

will help to transform cities like Rajshahi with TOD and gives a brief framework suggestion for future potential station areas based on major criteria and indicators.

References

Bangladesh, M. of Planning Government of the People's Republic of (2013) 'Bangladesh Bureau Of Statistics (BBS)'.

Barby, P. V. (2014) *The Next American Metropolis.*

Carmelina, B., Francesco, C., & Luciano, Z. (2013). TOD-Transit oriented development: A sustainable tool towards smart living. In *Proceedings of the 7th conference of international forum on urbanism* (March 2015). Available at: https://www.researchgate.net/publication/273895589%0ATOD

Cervero, R. (2014). Transport infrastructure and the environment in the Global South: Sustainable mobility and urbanism. *Jurnal Perencanaan Wilayah Dan Kota, 25*(3), 174–191. https://doi.org/10.5614/jpwk.2015.25.3.1

Chen, X., Street, W. F., & States, U. (2010). Prospect of the transit-oriented development in China, *2*(1), 83–93.

Crosby (2003). *City of Hartford: Principles for transit-oriented development* (1).

Hale, C. (2006). Making the most of transit oriented development opportunities. In *29th Australasian Transport research forum, ATRF 06* (pp. 1–19).

Hendricks, S. J. (2005). *Impacts of transit oriented development on public transportation ridership* (p. 96, August 2005).

Islam, M., et al. (2019). *Identification of determinant factors influencing modal choice behavior and satisfaction level of commuters: A case of rajshahi city* (pp. 475–482). https://doi.org/10.32474/TCEIA.2019.03.000171.

Luh, N., Widyahari, A., & Indradjati, P. N. (2015). The potential of Transit-Oriented Development (TOD) and its opportunity in Bandung Metropolitan Area. *Procedia Environmental Sciences, 28* (SustaiN 2014), 474–482. https://doi.org/10.1016/j.proenv.2015.07.057

Mamun, M. H. (2015). *Present condition of road traffic accident: A case study of Rajshahi City, Bangladesh* (February). https://doi.org/10.5120/19554-1279

Ministry of Housing & Urban Affairs. (2017). National Transit Oriented Development (TOD) Policy (pp. 1–18). Available at: http://mohua.gov.in/upload/whatsnew/59a4070e85256Transit_Oriented_Developoment_Policy.pdf

Newman, P., & Kenworthy, J. (1999). *Sustainability and cities: Overcoming automobile dependence.*

Properties, M., & Mucosa, H. O. (2015) *Technical Report* (November, pp. 2–3).

Rahman, F. I., & Hasnat, A. (2018). *Case study on transportation system in Rajshahi city, Bangladesh (July).*

Rajshahi City Corporation (2014). Available at: https://erajshahi.portal.gov.bd/. Accessed March 2, 2020.

Rajshahi Development Authority | Rajshahi. (2015). Available at: https://rdaraj.org.bd/. Accessed March 2, 2020.

RCC mulls expanding its horizon | The Daily Star (2017). Available at: https://www.thedailystar.net/city/rcc-mulls-expanding-its-horizon-1415173 (Accessed: 2 March 2020).

Singh, Y. J. (2015) *Measuring transit-oriented development (TOD) at regional and local scales—a planning support.*

Staff Correspondent. (2019). *No Title*. Available at: http://www.theindependentbd.com/printversion/details/185089

Taki, H. M., & Maatouk, M. M. H. (2018a) *Land suitability assessment for the potential location of transit oriented development (TOD).* Springer International Publishing. https://doi.org/10.1007/978-3-319-94180-6

Taki, H. M., & Maatouk, M. M. H. (2018b). Promoting transit oriented development typology in the transportation planning. *Communication in Science and Technology, 3*(2), 64–70. https://doi.org/10.21924/cst.3.2.2018.103

Thomas, R., & Bertolini, L. (2017). Defining critical success factors in TOD implementation using rough set analysis. *Journal of Transport and Land Use, 10*(1), 139–154. https://doi.org/10.5198/jtlu.2015.513

Tong, X. et al. (2018) Correlation between transit-oriented development (TOD), land use catchment areas, and local environmental transformation. *Sustainability (Switzerland), 10*(12). https://doi.org/10.3390/su10124622

Transit Oriented Development (2016). Available at: http://www.tod.org/. Accessed March 2, 2020.

United Nations. (2018). https://www.un.org/development/desa/en/news/population/2018-world-urbanizationprospects.html

Urban Land Institute. (2016). Available at: https://uli.org/. Accessed March 2, 2020.

Heritage Sites: Toward Creative Ambiance in Public Spaces Attached—Impact of Creative Ambiance on Societal Development

Aya Elgobashi and Yasmeen Elsemary

Abstract

Nowadays, a lot of challenges are facing the urban development of spaces in cities, especially the heritage public spaces; these urban spaces provide significant ambiances and senses. Most of the cities have a strong well-acknowledged urban ambiance that is derived from its context, especially if the context of the cities is full of heritage sites that have deep, rich senses of knowledge. These public spaces are rich and full of significant heritage which is important to the cultural heritage of the users. This cultural heritage plays an important role in the perception of current and future generations. The main objective of this paper is to explore the senses of the users toward the historic preservation of those public spaces through adding creative ambiance to the existing atmosphere. Researchers will focus on literature review in analyzing the placemaking as its main tool in making successful heritage public space. Consequently, the analysis achieved through different phases, the main goal of the research, is the role of this attached space in the social approach. A qualitative method based on a theoretical analysis is introduced by theorists and practitioners on designing and understanding the principles of those heritage public spaces as well as integration of ambiance dimensions. This framework will be applied to the selected case study as a new proposal for evaluating the attached space in the heritage site to make it more successful. The results include a thorough evaluation and several questionnaires that are provided to the users as an attempt to shed the light implicitly on guidelines and recommendations to the community in the form of approaches, which in turn, could help in the development of those heritage sites and make them more attractive.

Keywords

Historic preservation • Historical setting • Attached space • Historic urban landscape • Atmosphere • Ambiance • Placemaking • Visitor experience • Users perception

1 Introduction

Nowadays, as steps in the conservation process, a new type of reformation is introduced to help in heritage protection. In the past decade, and to make those heritage public spaces more attractive for users, more functionality has been added to the place, such as educational, social, and cultural activities (Schwartz & Cook, 2002). If the transformation process happened on the basis of critical criteria and deep surveys for the users' needs, it will have siginificant impact on the motivation of the users. Unfortunately, what happened in those heritage sites did not adopt the users' needs, so it became less attractive, which caused those users to lose a lot of heritage, social, and cultural meaning benefits. The idea of a revival of heritage sites was applied worldwide, and it was successful (Poria et al., 2003). Nevertheless, the condition was different in the Middle East, where the preservationist, urban designer, and architect while reviving a heritage site, started to focus on the building itself and totally neglect surrounding space. The heritage site is a piece of the history which this building conveys, so if we remove one of them, a lot of heritage will lose its meaning. Then, researchers are going to focus on the heritage sites defined by UNESCO as historic urban landscapes. The study area will be in the Egyptian context, especially in the Cairene contexts, which is full of heritage sites that have been neglected for many years. The main problem in the Cairene context heritage site is that the architect and preservationist focus on the building and its exterior and interior finishing, while the surrounded space remains green without any attractive ambiance for

A. Elgobashi (✉) · Y. Elsemary
Arab Academy for Science, Technology and Maritime Transport, Cairo, Egypt

C. Alalouch et al. (eds.), *Towards Implementation of Sustainability Concepts in Developing Countries*,
Advances in Science, Technology & Innovation, https://doi.org/10.1007/978-3-030-74349-9_7

users. In this study, the major part focuses on making ideal and creative experiences for the users to be more attracted in addition to its impact on their social quality of life.

Moreover, this paper attempts to investigate the placemaking process principle and add creative ambiance taxonomy as an approach to develop that heritage site or any spaces attached to historical buildings to make it more attractive and useful for the users. Successful historic preservation can be achieved only if those heritage sites become a source of attraction and satisfaction for the users. Many researchers practically found a link between the surrounding atmosphere and the user's sense while buying anything or doing any function. This link should be achieved in the preservation process to make it a successful place. Still, it will not happen even if the designer or researcher adds to existing atmosphere a unique ambiance to the spaces to be more attractive (Bitner & Obermiller, 1984). Recently, the full weight that the preservation process is given to the building's exterior and interior designs and in turn neglects the surrounding urban landscape. In the following discussion, the researcher will focus on the preservation and how placemaking can help develop this heritage public spaces.

2 Methodology (Fig. 1)

The purpose of this analysis is to theorize critical placemaking as a tool for accessing and transforming public places into spaces of dialog, inclusion, and democratic participation. The researcher provided an overview of the significance of public space attached to the heritage site, outlined a definition and brief history of placemaking and how it meets preservation, and then proposed an integrative theoretical framework for critical placemaking and divided it into four parts. The first part is an introduction to historic preservation and a sense of place. The second part focuses on placemaking in meeting preservation by theorizing critical placemaking as a tool for reclaiming public space, revealing contestation, and developing an active, transformative relationship between people and their places, and helping to plant seeds of action. The third part explains the basic principles of designing, studying, and understanding public spaces developed by theorists and practitioners such as W. H. Whyte, J. Gehl, S. Carr, and PPS and how to merge ambiance and placemaking dimensions to get a criterion for the evaluation of those public spaces to make them successful. In the fourth part of the paper, the researcher starts to discuss placemaking in the context of the real world by selecting a case study to make of it a practical one. The researchers start to collect data for the selected case study by conducting a field visit to the heritage site and the whole surrounding area and setting up a questionnaire and semi-structured interviews with the users to learn more about their opinions concerning the main research problem. These findings will serve as guidelines for urban designers and architects to have successful public spaces attached.

3 Literature Review

In societal development, cultural heritage is the backbone of this development, and heritage public spaces involve a significant part of this heritage. Historic preservation is the process that helps in the sustainability of heritage to allow the future generations to benefit from this cultural heritage. Cultural heritage is full of rich resources from history and society culture; it also involves elements from the building environment which include building, objects, streets, urban landscape, and structures (Heintzelman & Altieri, 2013). Preservation is a significant dynamic power with a collective identity, meaning, and sense of place (Adams et al., 2001). In the following section, the researchers will focus on historic preservation and its benefits. An investigation will be conducted to achieve successful preservation in those heritage public spaces to be more attractive, hence, its impact on the user's perception.

3.1 Historic Preservation and Sense of Plac

Farrer is a vital preservation expert who concluded that if historic preservation follows undiscriminating criteria, the nostalgic effect will appear in the place. As a result, original sense of place will be lost in addition to all the cultural heritage meaning it holds (Farrar, 2011). He concludes that many multiple temporalities faced those heritage public spaces, so each period has its value of place and historical background, which should be present for each generation (Farrar, 2011). Farrar agrees with Lynch's argument that those heritage public spaces should be livable and unique activities to be memorable in the city. Consequently, it will affect the cultural and physical life in the city (Lynch, 2005). Preserving is not only to keep heritage as it is; it is a process to make relationships between the cultural heritage and future generation to benefit from heritage (Farrar, 2011). Cities will be smarter and more attractive if historic preservation happens critically. This way should show respect to the users' needs and perceptions (Angelidou et al., 2017). It is essential that we realize the communities' four senses while preserving any heritage site: the sense of place, identity, ownership, and community (Iliyasu, 2014). Nowadays, it is the largest cities that have homogeneous skyscrapers with no sense of the place similar to the heritage sites; consequently, it affects the user's perception (Angelidou et al., 2017). Through the preservation research, in 2005, Feliden observed that there are three main major

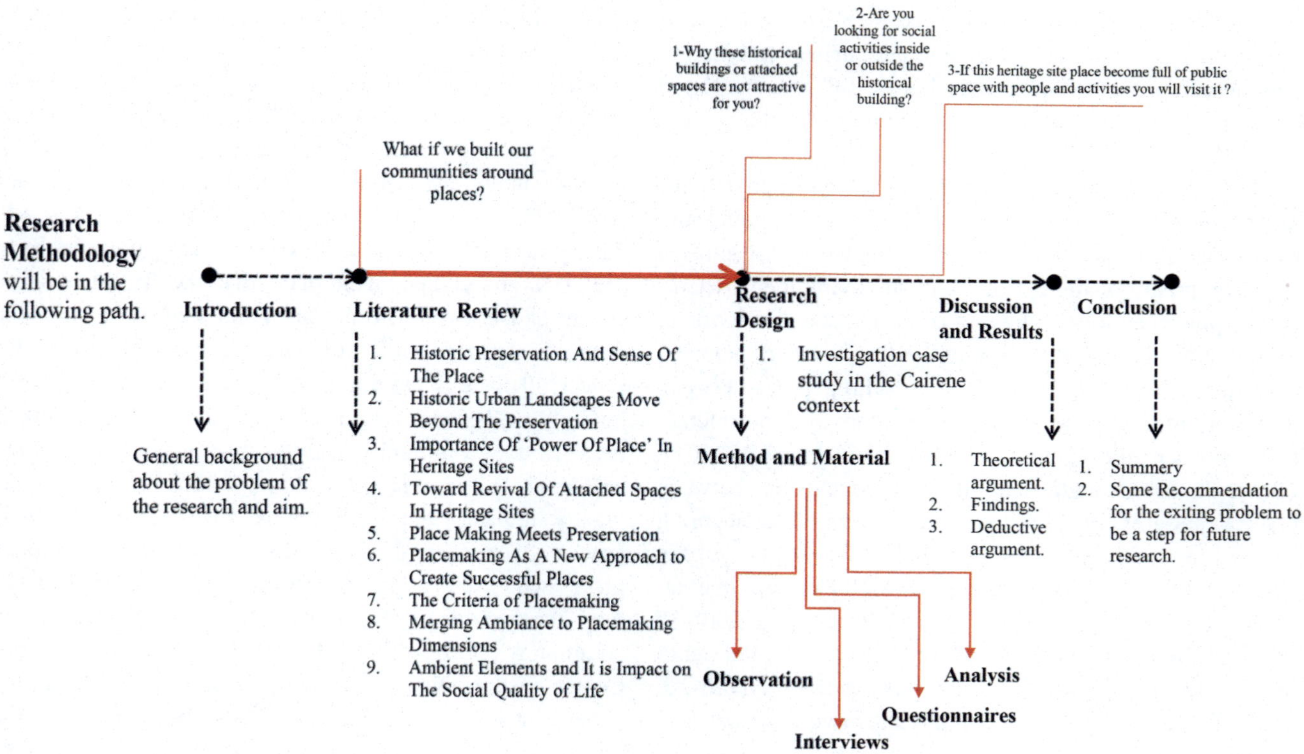

Fig. 1 Research methodology illustrated by the researcher

values. These values are emotional, cultural, useful, and functional values (Feilden, 2005). Those values can be achieved through five steps that need to be considered while preserving any public spaces: identity, sense of the place, investigation for the users, develop, execute and educate (WBDG Historic Preservation Subcommittee, 2019). Before getting involved in preserving the heritage site or element process, preservationists should know how to treat each value (Hyden & Sheckels, 2016).

Rapid development and challenges in life of historic preservation are not only preserving the past alone; it is the process of managing and developing heritage public spaces due to multiple temporalities (Hyden & Sheckels, 2016). Deterioration of the old building or sites without any awareness of this heritage increases the rapid flow of development and challenge in the preserving process (Meeks & Murphy, 2016). In the American communities, due to the significant awareness that happens to preserve heritage, they recognize that they would miss a lot of cultural heritage values if they cut the craft of the built environment (Meeks & Murphy, 2016). Finally, based on the type of preservation, approaches and criteria should be chosen carefully and critically. Therefore, before starting the preservation process, a lot of analyses need to be done for each element. Throughout history, historic preservation had a deep sense of meaning to the community (Meeks & Murphy, 2016).

3.1.1 Toward Preserving Historic Urban Landscapes

Life nowadays is full of rapid expansions with a lot of impacts on the cites. Many cities lose significant historical elements such as heritage building and surrounding areas, streetscape, and local public spaces. To overcome these urban problems, the relationship between urban growth and historical preservation must be re-designed and each element related to the heritage investigated. The preservationists should introduce new design interventions by managing all the dynamic relationship functions in the heritage sites (Zhong, 2012). Through preservations of any historical building, surrounded open space and landscape is neglected, when in fact it is a unique and an integral part of the history building. The heritage public spaces are usually full of rich and significant cultural heritage for the community. Those spaces are a perfect place for social and physical interaction between users (Zhong, 2012). Historical preservation is typically facing many unique urban issues that may occur in society's economic, cultural, and political values, even though the financial crisis is the major issue in preserving (Faris Hmood, 2019).

Many researchers point out that urban issues or challenges can be overcome by overlapping them to achieve efficient historical preservation, but this overlap can not only be accomplished by governance, architects, urban planners, and researchers. A significant part of it relates to the environment. The sense of obligation and understanding the significance of the historical sites that are full of cultural and historical values are affected by community involvement (Faris Hmood, 2019).

Historic urban landscape is an integrated approach for identifying and evaluating, protecting, and maintaining historic sites. It follows a sustainable framework that aims to improve quality of life and the human environment. Therefore, if preservationists follow those sustainable frameworks, it will impact the urban spaces and social and practical diversity (UNESCO, 2011). It is a turning point between past, current, and generations in the future and the interaction between the context and environment (UNESCO, 2011). It is an understanding and conclusion between diverse cultures of tradition and perception concerning national and foreign cultures (UNESCO, 2011). In the "World Heritage and Contemporary Architecture," international conference in Vienna in its 27th session 2005, a proposal for historic urban landscape has been figured out to be a guideline for most of the preservation (UNESCO, 2005). These guidelines are composed of two approaches—tangible and intangible. An image is a tangible asset that defines the historical city with one of its subcategories: urban environment, environment (context/settings), shape, building monuments, descriptions, and natural features (Ginzarly et al., 2019). In contrast, an intangible asset is a photograph of behaviors, phrases, actions, or consumer awareness, and knowledge in ICOMOS Australia Burra Charter declaration in 2013, by cultural significance or cultural heritage. For past, current, and future generations, it means aesthetic, historical, scientific, social, or spiritual meaning. Although it may be embodied in the place, it may differ from person to group in its textile, environment, usage, relationships, interpretations, documents, associated locations, and artifacts (ICOMOS, 2013).

The "historic environments" will be the subject of this study. It plays an important role in improving public policy priorities, education, sustainable economic growth, well-being and health, placemaking, and community cohesion (Parsons, 2010). Heritage public spaces are named in ICOMOS declaration by "Historical setting" as it is full of natural landscape with unique cultural hertiage (Parsons, 2010). One of the most important techniques in the historic urban landscape is "Historical settings" during the 36th conference in Paris, which was powered by UNESCO. The historic urban landscape is becoming a new trend in prompting heritage public spaces to be more attractive. Figure 2 shows the convergence between constructions and environmental land to the concept of the historic urban landscape.

3.1.2 Importance of "Power of Place" in Heritage Sites

At the beginning of the twenty-first century, a new idea for preservation was introduced. It focuses mainly on the historic environment and adds a democratic atmosphere to heritage under the banner of Heritage Protection Review (HPR). Heritage Lottery Fund names this preservation idea the "power of place" and adds to the public policies. The UK government mentioned that the historical atmosphere and environment are an initial step for a promising future for current and future generations (DCMS, 2010). Power of place is the value of heritage and the heritage is not only walls, bricks, paints, and mortar; it is an integration between the community and the heritage site. This integration helps communities do an efficient practice that greatly impacts their perceptions (DCMS, 2010). Public spaces attached to a heritage site or building are significant places where the users can do many activities full of cultural heritage values. Those places add a lot of values to the community (Carr, 1992). In 2013, the World Heritage Center stated that the historic urban landscape moves toward preserving the physical environment and surrounding elements directly affecting the human atmosphere. Atmospheres consist of two aspects, which are tangible or intangible values (World Heritage Center, 2013). Cultural heritage is an interrelated element: participation of the users and surrounding elements (Abdel Tawab, 2014). The community gains a lot of meanings, such as identity, shape, belonging, and local culture. Therefore, the urban environment and culture of the community are also directly connected. "When public spaces are successful, they will increase participation opportunities in communal activity"(Carr, 1992).

3.2 Toward Revival of Heritage Public Spaces

Nowadays, the rennovation of the historic urban landscape or any public spcaces is one of the most critical challenges. Since the right atmosphere and physical setting should embody the past of the place through imaginative atmospheres, such as exhibitions and event areas for users, providing an appropriate experience of each heritage site is required (Boyd, 2002). The space activities should integrate outdoor and indoor to allow users access to huge benefits from the heritage site (Boyd, 2002). Most of the studies in the historical urban field conclude that placemaking is a significant solution. Placemaking creates a unique belonging, ownership, and atmosphere attractive for the users (Lynch, 2005). Sense of the place develops if users are feeling satisfied and comfort.

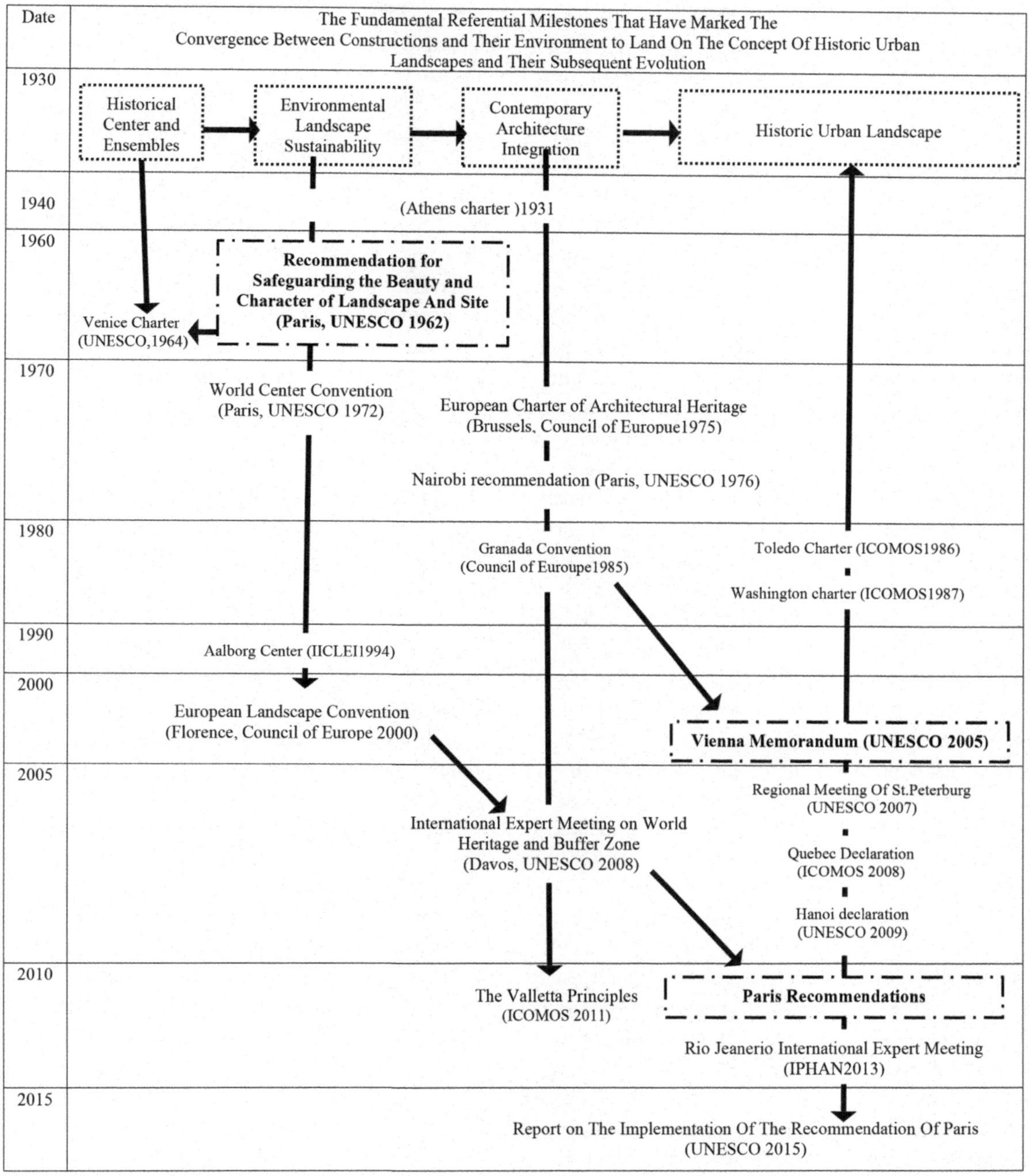

Fig. 2 Shows the convergence between constructions and environmental land to the concept of the historic urban landscape. *Source* Elgobashi and Elsemary (2019)

Consequently, those places should have a unique language to speak to the users for more social aspiration (Lynch, 2005). This integration between users and places can happen in a different city (individual, community) (PPS, 2010). In the following sections, the research focuses on the placemaking and its types and how it meets preservation. Although a narrative study for the impact of adding creative ambiance by placemaking and its impact on the users' perceptions, before studying impacts, researchers will explore a deep study on principles, criteria, and studies of placemaking and its role in heritage public spaces.

3.2.1 Placemaking is the Heart of Historic Preservation

The idea of placemaking started in the mid-1970s; it is an interdisciplinary and multifaceted phenomenon studied by many urbanists and researchers (Andrews, 1974). After ten years of realizing the placemaking concept, William Whyte, mentor of the "Project for Public Spaces" organization, traces and enhances placemaking in urban planning in collaboration with two key researchers interested in urban studies. Consequently, William Whyte wrote his first book, "The Last Landscape," about urban areas or spaces and its development. However, he mentions the bad impact of the rapid urban growth on the user's perception and behaviors (Whyte, 1968). Therefore, he observes the urban setting with the dynamic human life and user behaviors to put a grounded basis for the placemaking process. Those studies were mentioned in City: Rediscovering the Center (Whyte, 1989). "The Death and Life of Great American Cities" a book by Jane Jacobs mentions that any change in the urban setting, even if it is very small in scale, can help improve a transformation or destroy the community. Whyte agrees with Jane Jacobs that any path in cities' urban development will greatly impact the users' social environment and behavior (Whyte, 1989). Placemaking role is not only in the urban setting, it also has spatial and design disciplines in art, social science, education, music, tourism, etc. (Alvarez et al., 2017). Based on all these studies and researches, founder of "Project for Public Spaces" Fred Kent stated that "Placemaking is an act of doing something. It is not planning per say, but it is also doing. That's what is so powerful about it." (Elgobashi & Elsemary, 2019).

Ideal experience plays an important role in creating an ideal atmosphere and physical environment for the user by adding exhibit, activities, gathering, and social attraction. In 2005, in the International versus Domestic Visitors: An Examination of Destination Image Perception, Bonn concluded a relation between the ideal experience and its impact on heritage sites' users (Elgobashi & Yasmeen, 2020). This experience greatly impacts the user's behaviors and perception, contributing to enhancing the heritage site. The project for public spaces, National trust for historic preservation, and National main street center emphasized that ideal experience by a new collaboration for integrating the worth movement of preservation and placemaking as an essential concept in urban planning (PPS, 2010). This collaboration stated some idea after a long term on studies in the same field. They stated the idea of placemaking in the form of questions to be easier for researchers and urbanists: "What if we built our communities around places?" (PPS, 2010). Placemaking inspires people to collectively reimagine and reinvent public spaces to be the community's heart due to the strengthened integration between the users' place and interaction (PPS, 2010). Placemaking is a process for prompting and enchanting urban settings with many different physical and cultural and social values; those values help users define the spaces (PPS, 2010).

Nowadays, most cities face a lot of rapid urban challenges due to rapid urban development. According to this change, the type of placemaking will differ from one city to another (PPS, 2010). This research focuses on the Cairene context where the heritage public spaces suffer from a lack of participant and poor design. In the Cairene context, the transformation will be one form of the three forms of placemaking to find out a focal point focusing on enhancing "successful places" or "quality of space" (PPS, 2010). Standard placemaking is the first form that aims to improve the community as it is suffering from a lack of living public spaces by promoting attractive facilities and activities for the community (Strydom et al., 2018). Strategic placemaking is the second form with a different aim; it focuses mainly on economic and infrastructural development. It targets mainly business projects with attractive green spaces, waterfronts, social gatherings, and entertainment facilities (Wyckoff, 2017). Tactical placemaking is the third form that aims to promote and restore the heritage spaces to the community. It supports the identity, as it is mainly used in the communities with lack of funding the space by adding new policies that adapt to the financial commitments (Bauer et al., 2010). The community's transformation will happen by introducing new creative ambiance and environment such as cultural activities, outdoor display spaces, gathering, and exhibitions (Strydom et al., 2018). Figure 3 presents the three forms of placemaking toward better public spaces (Wyckoff, 2017). In each process of placemaking of the three forms, a comprehensive study and vision should be achieved to help the community reach the quality of spaces (PPS, 2010).

Identifying the place identity context and stakeholders is the first initial step for finding the strengths, weaknesses, and threats. Evaluation of the space is the second vital step to know the current uses of space to be developed quickly. The third step is to select a placemaking form based on the case study context and identity and missing activities; although in this step, it is essential to respect the value of spaces. The last step is a re-evaluation for the improvement in the long term

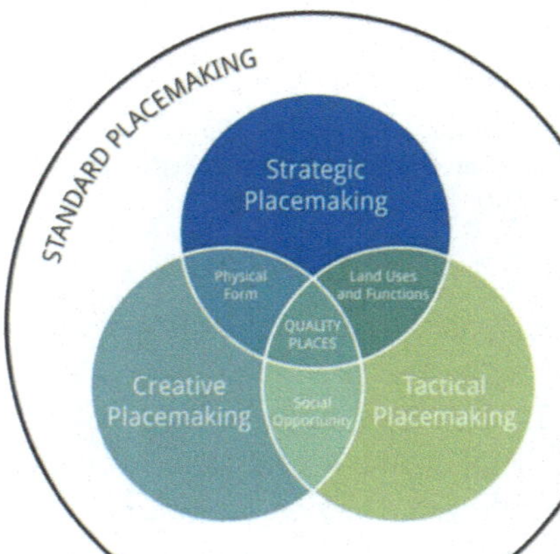

Fig. 3 Shows the relationship between the three specialized types of placemaking (Wyckoff, 2017)

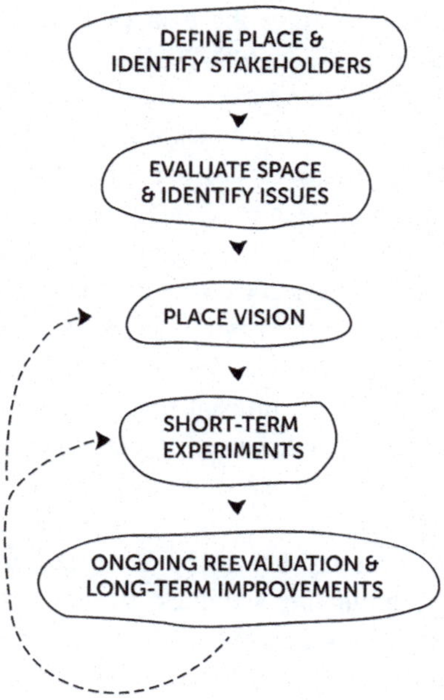

Fig. 4 Illustrates placemaking passes to create Quality in Places (PPS, 2010)

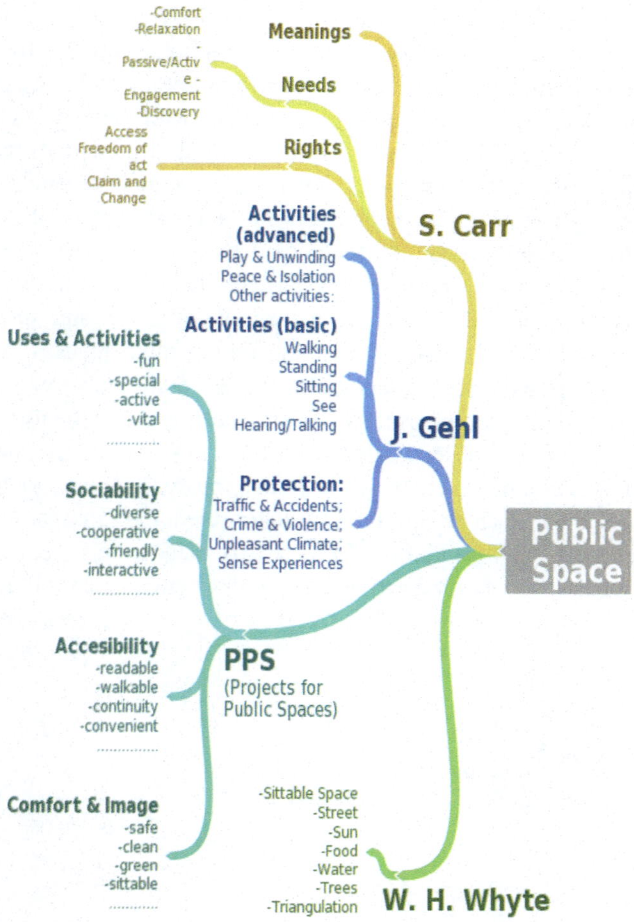

Fig. 5 Successful public space basic criteria. *Source* J. Gehl, S. Carr, PPS, and W. H. Whyte

with management and maintenance plans to make sure that these spaces will keep running without any concerns (PPS, 2010). Figure 4 illustrates placemaking passes to create quality in places.

Placemaking is an observation process; its philosophy mainly depends on observing and interviewing users to know their needs. The needs will be figured out based on the interaction and inspiration of those spaces (PPS, 2010). In the existing or new spaces, placemaking philosophy can be applied with different steps to benefit from each place; so, the user can feel the sense of the place and interact with it (PPS, 2010).

3.2.2 Placemaking as an Approach to Revitalize Neglected Urban

In 1993, in the USA, William Whyte and Carr published a book named "Public Space." A lot of studies were done on human behavior using public spaces in the American community. Moreover, they state the three main pillars of human behavior in public spaces, needs (comfort, relaxation, passive/active engagement, and discovery), rights (access, freedom of act, claim, and change), and meanings (PPS, 2010). Kevin Lynch also states in "A Theory of Good City Form," "A good place is one which, in some way appropriate to the person and her culture, makes her aware of her community"(Lynch, 2005). Finally, atmosphere and the spirit of spaces play an essential role in enhancing spaces successfully. Figure 5 shows the basic criteria to have successful public spaces.

The project for public spaces is a nonprofit organization for planning, design, and education to help people preserve public space to be a strong community. It is the central hub of placemaking in the world as it evaluates thousands of public spaces. This evaluation is based on criteria for great public space. This criterion has four qualities: accessibility, engagement in activities there, comfort and good image of space, and finally, sociability. Figure 5 shows in detail the basic criteria for great public spaces (PPS, 2010). Consequently, researchers will propose these criteria as guidelines for evaluation and integrate the effect of atmosphere and physical environment factors on heritage public spaces.

3.2.3 Placemaking Principle

Placemaking philosophy is based on the fact that every human has the right to access a livable and attrative place. Successful places will greatly impact on their perception and social quality of life (PPS, 2010). The project for public spaces make a development in the basic criteria diagram; they are divided into four sections: "access and linkage, comfort and image, uses and activities, and sociability (PPS, 2010). This new diagram contains many key attributes and new measurements for placemaking, which can be applied to any public spaces, as shown in Fig. 6 (PPS, 2010). In the following section, the researchers will focus on the four sections of the placemaking. First, the access and linkage section focuses on connecting the space with the surrounding physically and visually (PPS, 2010). It also includes the diversity of transportation to this space. The main access and linkage approach is to connect with the urban context of the building by streets, sidewalks, and bicycle, and pedestrian lanes to make those remarkable for the users. All these approaches will affect the perception of users and their lifestyles (PPS, 2010). Second, this section focuses on comfort and image. Suppose, the place is comfortable with a good image for the users it will significantly affect the user's perception. Consequently, it will be more attractive and reachable, but to reach this approach, many facilities should be added, such as sitting area, shading area, the safety of the place, and cleanliness (PPS, 2010). Third, this section focuses on uses and activities: Activities are the central fact of users' attraction to visit and revisit those spaces. The approach here is to make the balance between what users need and activities used (PPS, 2010).

Finally, the fourth section is on social interaction in public spaces. The feeling of comfort with people in those spaces is the approach of this section. This approach is a guideline for the researchers on how users utilized this public space individually or in groups (PPS, 2010). A significant volume of theoretical research on placemaking was done, such as that of Whyte's work. It also allows the urbanist and the preservationist to find a conventional setting which gives people a significant sense of place (PPS, 2010).

A new concept is introduced to placemaking for a better belonging and sense of place in the book of Edward Relph "Place and Placelessness" "directly experienced phenomena of the lived-world and hence is full of meanings, with real objects and with ongoing activities" (Relph, 2008). This new concept was to translate physical and non-physical aspects, which are called ambiance. The following Table 1 shows the concept of ambiance with it in all aspects. These physical

Fig. 6 Shows placemaking criteria required to be created successfully. *Source* PPS (2010)

Table 1 Concept of ambiance is characterized by physical, physiological, sociological, cultural, and psychological aspects (Elgobashi & Elsemary, 2019)

Dimension		Terms	
		Negative	Positive
Physical	Land use (greenery)	Constructed, built, cemented	Green
	Location (urbanization)	Isolated, rural	Suburban, urban, central
	Size	Tiny, small, medium-sized	Panoramic
	Status	Under construction, developing, re-developing	Well-established
Physiological	Touch → temperature	Cold	Warm
	Sight → light	Dark	Shiny, bright
	Sight → colors colorless, gray colorful	Colorless	Gray colorful
	Hearing → sound	Sound silent, quiet	Noisy, loud
Sociological	Cleanliness	Dirty	Clean, immaculate
	Income	Poor, cheap, sustainable, accessible, gentrified	Wealthy, pricey, fancy, rich, luxurious, expensive, exclusive
	Ethnicity	Monocultural, homogeneous	Mixed, diverse, elective, ethnic, exotic, multicultural, international, wordly
	Population	Density deserted, empty	Busy, crowdy, crowded, dense, overcrowded
	Sense of the place	No feeling belonging	Feeling belonging
	Type of activity (night → day)	No nightlife and not safe	Leisure, touristy, sightseeing
	Social activities	Not attractive to people	Popular and different types of socioeconomic target it
Cultural	Education	Vulgar, simple, gritty	Smart, intellectual, sophisticated
	Informality	Formal, decorous, conforming, official	Frivolous, casual, spontaneous, easygoing, familiar, free, unofficial, informal, folksy

and non-physical aspects are highly interdependent and rare (Elgobashi & Elsemary, 2019).

3.2.4 Merging Ambiance to Placemaking Dimensions

Quality of life is a multidisciplinary term with several facets of positive and negative impacts. As mentioned in Table 1, atmosphere and ambiance is one factor that has an important effect on the human environment and quality of life (Redi et al., 2018).Quality of life's layers are divided into objective and subjective aspects. The objective aspects focus on human life and living standards such as infrastructure facilities, and economic issues. Quality of life's subjective element is users experiences and real-life satisfaction (Dajian & Peter, 2006). The theory of quality of life applies to other studies, and it has a significant impact on the years (Elgobashi & Elsemary, 2019).This research is going to focus on the social aspects as its major dimensions in the QOL. The public spaces are hypothetically affected by social activities and content; for without social attraction, it will lose its spatial component (Carmona et al., 2003). One of the advanced aspects of the quality of life is a social quality. Its approach focuses on individual users, well-being, and environmental atmospheres (Wallace & Abbott, 2009).

Depending on the space's atmosphere and ambiance, it can be a supporting or discouraging and motivating or boring one. The place's nature is the intangible, non-material empirical character, and unfocused; this means it is directly affected by the environment. Practically, without imaginative ambiance, this environment does not reflect effectively on the experience of the consumer. Many researchers conclude that satisfaction and attraction give the space a unique spirit and character. In 2010, the project for public spaces joined National trust for historic preservation in a collaborative effort in saving the public spaces by implementing a new vison of prompting heritage public spaces in the American communities in order to be more successful. The National trust's goal is to express the profound resonance of offering consumers a special experience when visiting historic sites with various categories. This principle brings together people to preserve, improve, and enjoy their historical places (PPS, 2010). Also, it contributes to restoring the historical and social functions of building and surrounding space in the historical site (PPS, 2010). The heritage public space that had a positive effect on their communities follows a consistent pattern. PPS presumed that preserving historical buildings with attached open spaces is important since they are critical centers (PPS, 2010). These historical buildings are the backbone of any community, and the building's architecture plays an important role in the users belonging to the space. Consequently, it has a significant effect on the perception of users (PPS, 2010). Every success preservation movement in the space will help users to become involved and learn as much as they can from their heritage (PPS, 2010). Ambiance taxonomy can be defined in three levels: the first level includes a lot factors that have an impact on ambiance (e.g., physiological and psychological); the second and intermediate level consists of dimensions (e.g., senses—"sound" and perceptual characteristics—"joyfulness"); and the third and more specific level consists of terms characterizing each dimension (e.g., "loud" and "cheerful"). Figure 7 shows a color code with the effect of each ambiance dimension.

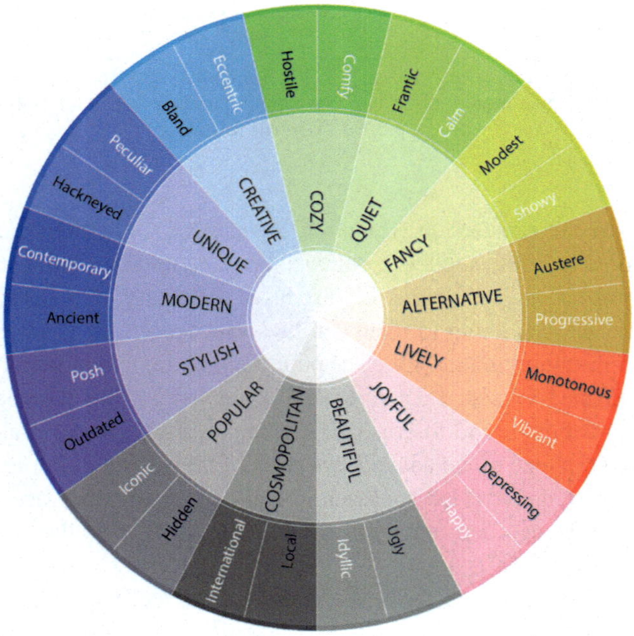

Fig. 7 Ambiance taxonomy wheel. *Source* Redi et al. (2018)

3.3 Ambient Elements and Their Impact on Social Life

While studying the atmosphere and ambiance dimensions, the researcher starts to focus on the users' social behavior as a fundamental dimension. While studying surrounding atmosphere, narrative layers appear in those heritage public spaces.The ambiance is a layer of emotion and feeling for different temporal traces (Elgobashi & Elsemary, 2019).

Ambiant elements connect users emotionally and physically to history or signs in life, rather than being isolated due to many urban challenges (Pallasmaa, 2014). In the following examples, significant merges take place between creative ambiance and placemaking of the heritage public spaces' existing atmosphere. To enhance visitor experience and attraction, this idea was applied to many heritage sites by adding activities, events, or facilities (Figs. 8, 9 and 10).

Finally, the literature review section concluded that the atmospherics and ambiance become a critical issue and merge the placemaking process toward the public space in heritage sites. Researchers will evaluate heritage public spaces based on a proposed structural framework on the previous merging between ambiance and placemaking to reach a creative ambiance in public space in the Cairene context.

4 Research Design

The application of the theoretical proposal concluded in the preceding section can be explored and expanded in various directions. However, for the sake of this research, it is inevitable to limit the application to certain aspects, leaving space for further study and application. The study was performed on an existing case study only. The guideline proposal applies to heritage sites. The principal method of analysis used in this research is a qualitative method for heritage sites. The analysis was achieved through different phases; the main goal of the research is the role of this attached space in the social approach, and this research will apply a qualitative method that is selected on a theoretical basis which is the basic principles of designing, studying, and understanding public spaces developed by theorists and practitioners such as W. H. Whyte, J. Gehl, S. Carr, and PPS. In addition to that, merging ambiance dimensions will

Fig. 8 Renovated Shed 3 historic market in Detroit. *Source* Eastern Market Corporation (2018)

be applied to the selected case study, as a new proposal for evaluating attached space in heritage site to be more successful.

4.1 Data Collection

The application of the theoretical proposal concluded in the past section can be explored and expanded in various directions. However, for the sake of this research, it is inevitable to limit the application to certain aspects, leaving space for further study and application. The study is performed on one case study, which has a lot of public spaces. The guideline proposal is applicable to all heritage sites. The principal method of analysis used in this research is a qualitative method for public spaces. The analysis is achieved through different phases; all supplement the main goal analysis. The researchers start to collect data for the selected case studies by a field visit to the heritage site and the whole surrounding area. There was a chance of seeing, oral communication, and investigation to set up a questionnaire and semi-structured interviews with the users to know their opinions about the main research problem. The researcher set up some questions for the users of those spaces, such as: 1) Why are these historical buildings or attached spaces not attractive to you? 2) Are you looking for social activities inside or outside the historical building? 3) Will you visit those public spaces in the heritage site again? To find out the role of public spaces in the social approach, this research will apply a qualitative method of the potentials of the context, followed by a descriptive analysis for a theoretically based recommendation, and theories will be applied to the case study as a new proposal for public spaces attached to heritage site from a social perspective.

Initially, the questionnaire was conducted on a sample of 150–200 residents representing the local community of the Old Cairo district. The criteria of the questionnaire was based on the placemaking principle. These samples included various categories of community members. The proposed age for participation in the questionnaire is18–40 years old

Fig. 9 The Louvre, Paris. *Source* Louvre Museum, Paris (2019)

Fig. 10 Various activities add to of Nyhavn in Copenhagen to be more attractive. *Source* Smith (2016)

including both males and females. The participants were asked to answer five groups of questions. The first group investigates general information about them, including gender, age, education, and their relation to the district. The second group of questions explored if they do activities rather than religious beliefs. The third group investigated the four indicators of placemaking approach "accessibility, image and comfort, activities and sociability" to look at people's vision and aspirations toward developing this neglected space. Every section was scored in parallel to the ambiance dimensions.

The participants in the questionnaire were asked to answer five groups of questions.

4.1.1 Practical Constraints

For the sake of fair guidelines and receiving precise results for the theoretical proposal, the choice of a historical building for this study was based on the following criteria: ***Geographical Factor***: It should be a landmark location and must be accessible by the surrounding areas and by all levels of the community. ***Historical Significance***: This is related to the historical data, as well as its relation to historical events, significant historical figures as well as owing many historical features which were not utilized. ***Current Existence***: The last factor for choosing the historic building is that the attached space still exists, and the conservation of the building has been done recently with its attached open spaces. ***Public Use***: All the historical buildings in the case study are for public use. It was currently closed for rennovation as its low visitation rate. ***Condition***: in good condition and well maintained.

4.2 Research Case Study: Al Fustat in Cairo, Egypt

The whole area encompassed by what is today Aqueduct (Magr el 'Ayoun), and Old Cairo road (Masr el Qadima) is

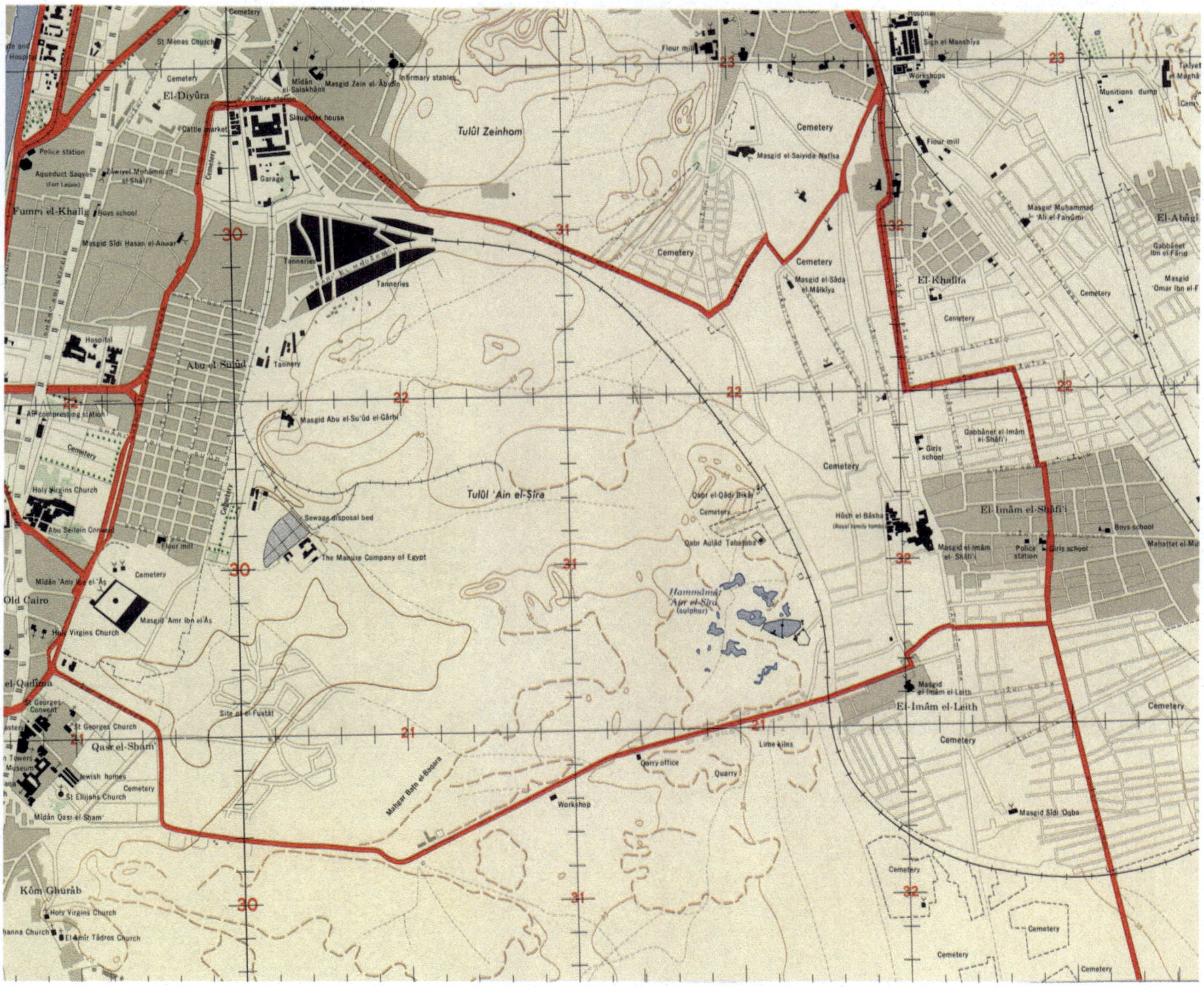

Fig. 11 Fragment of Cairo No. 2, 1:10,000, Edition 4-AMS, Series P971, U.S. Army Map Service (1958)

referred to as North Fustat, while the area from Old Cairo road to the ring road is the Fustat Plateau. In the past, this whole area had a rough topography due to natural erosion and quarrying. Figure 11 shows a rough topography. Figure 12 shows the urban context and the surrounding of the case study.

The researchers select the area of Amr ibn al-As Mosque, the Coptic Church, and Traditional Crafts Center as those spaces are landmarks in Cairo city and very attractive for people all year, but users come for religious beliefs. The location includes several monumental sites, giving it another dimension of potentials. These potentials are illustrated in Fig. 13.

5 Findings

Cairo has many heritage sites; this site is the largest religious complex with many public spaces, yet the user comes for religious beliefs only. A lot of empty outdoor spaces is used for parking lots and non-social activities. Therefore, analysis of the above heritage public space in the above case study proves heritage public space. The evaluation of this case study is based on the criteria of placemaking, which are "Access & Linkage, Comfort & Image", & Uses & activities and Sociability" integrate the ambiance dimensions for a sense of place and better quality of spaces. This integration will have a great impact on the community, especially if

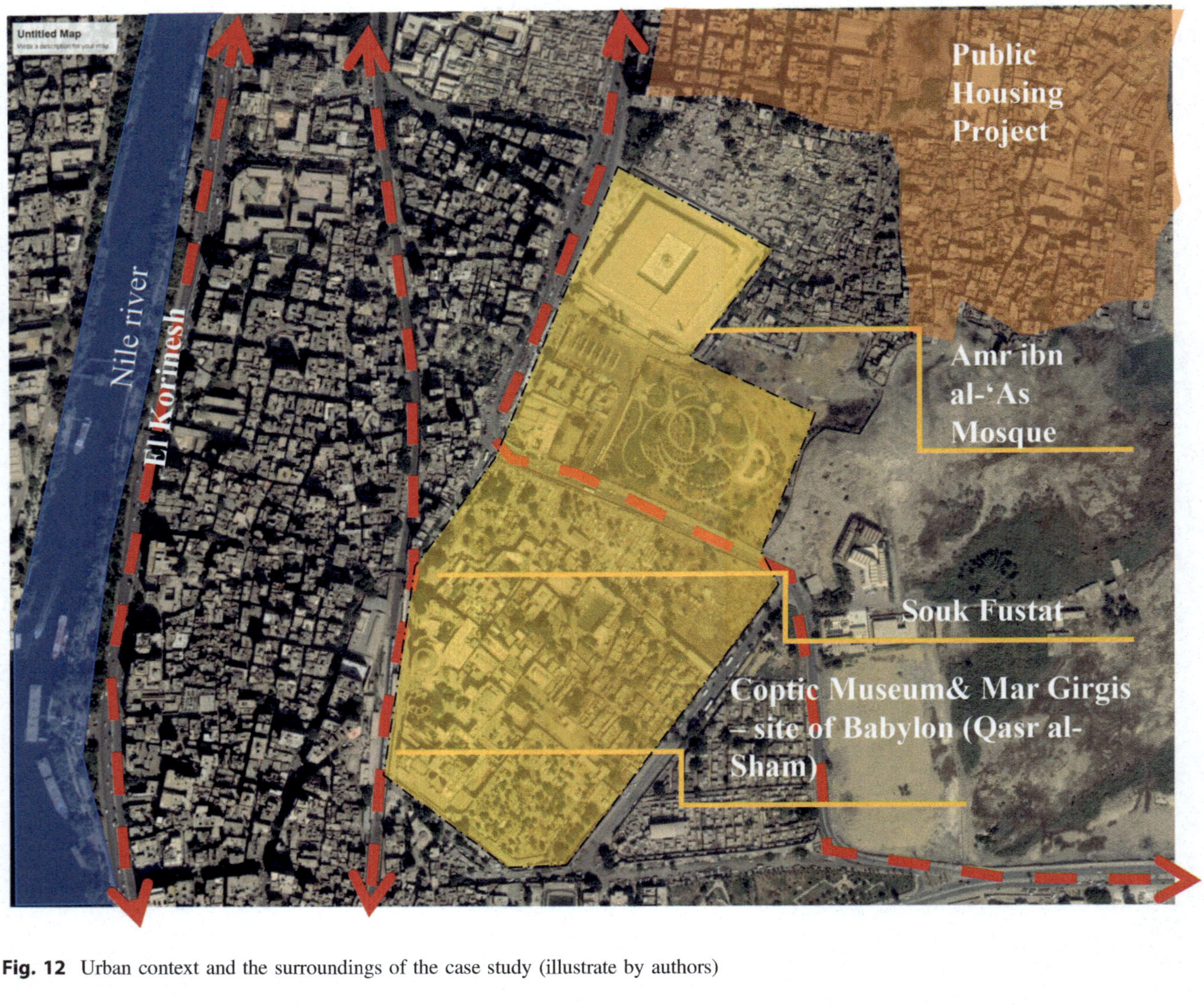

Fig. 12 Urban context and the surroundings of the case study (illustrate by authors)

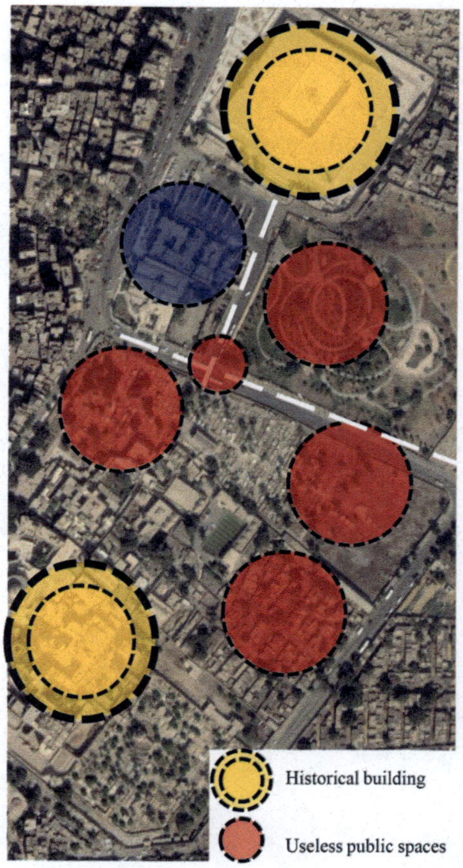

Fig. 13 Potentials of urban context and the surroundings of the case study (illustrate by authors)

those heritage sites have a vital and rich architecture that plays a vital role in community development. This case study encompasses many unique historical buildings, such as Amr ibn el As a mosque, The Hanging Church, Coptic Museum Park, currently demolished, as shown in Fig. 14.

5.1 Observations

After observing and analyzing all the collected data in the case study, this heritage site is unique and in a landmark location. The surrounding spaces are huge enough to help the preservationist to add creative atmosphere and ambiance by adding events, exhibitions' area, and social activities with the same spirit of the building. It is worth saying that the surrounding buildings have unique architecture, so the attached space should help to reflect and define it for the users. This reflection will be done if the attached space supports the buildings by adding engaging activities for achieving the main aim of the research for better social quality of life. An integrated framework was proposed, which gives guidelines to the problem. These guidelines will help to overcome the social problem that users face nowadays in the attached space in heritage sites. This research implicitly provides some recommendations or approaches.

The integrated framework linked together all the theories used—J. Gehl, S. Carr, PPS and W. H. Whyte —and merged ambiance dimensions to create a creative ambiance in those attached spaces. However, this creative ambiance can be achieved only after evaluating the current situation based on the proposed framework by observation and questionnaires for the users. The results of this evaluation have a significant impact and opportunity to make this heritage site with its attached spaces more attractive and full of more activities. These results have been shown in Table 2. As for the four indicators of "comfort and image," "access and linkage," "sociability," and "uses and activities" with introduction to social ambiance, 80% of the respondents expressed their acceptance of the dimensions. Yet, the setting suffers a lot of negative issues based on the proposed criteria. Such non-cleanliness and insecurities leave a bad impression in their minds. About 75% of the respondents mention that they need more commercial, social, recreational, and cultural activities. They suggest that those heritage public spaces need a critical development to improve the sense of belonging and pride of the place. This improvement will significantly impact the user's social behaviors and perception in using the heritage spaces.

Moreover, 90% of the respondents linked their feeling of attachment and sense of place to the process of improving this attached public space. If these places are available for the public with useful attraction, new social bonds and cohesion will develop between the community categories. This bond is one of the essential placemaking credibility outcomes. The respondents need varied activities with a new spirit full of hospitality and attract social network. All respondents expressed their desire to participate in this process by criticizing, developing, and voting on the proposals.

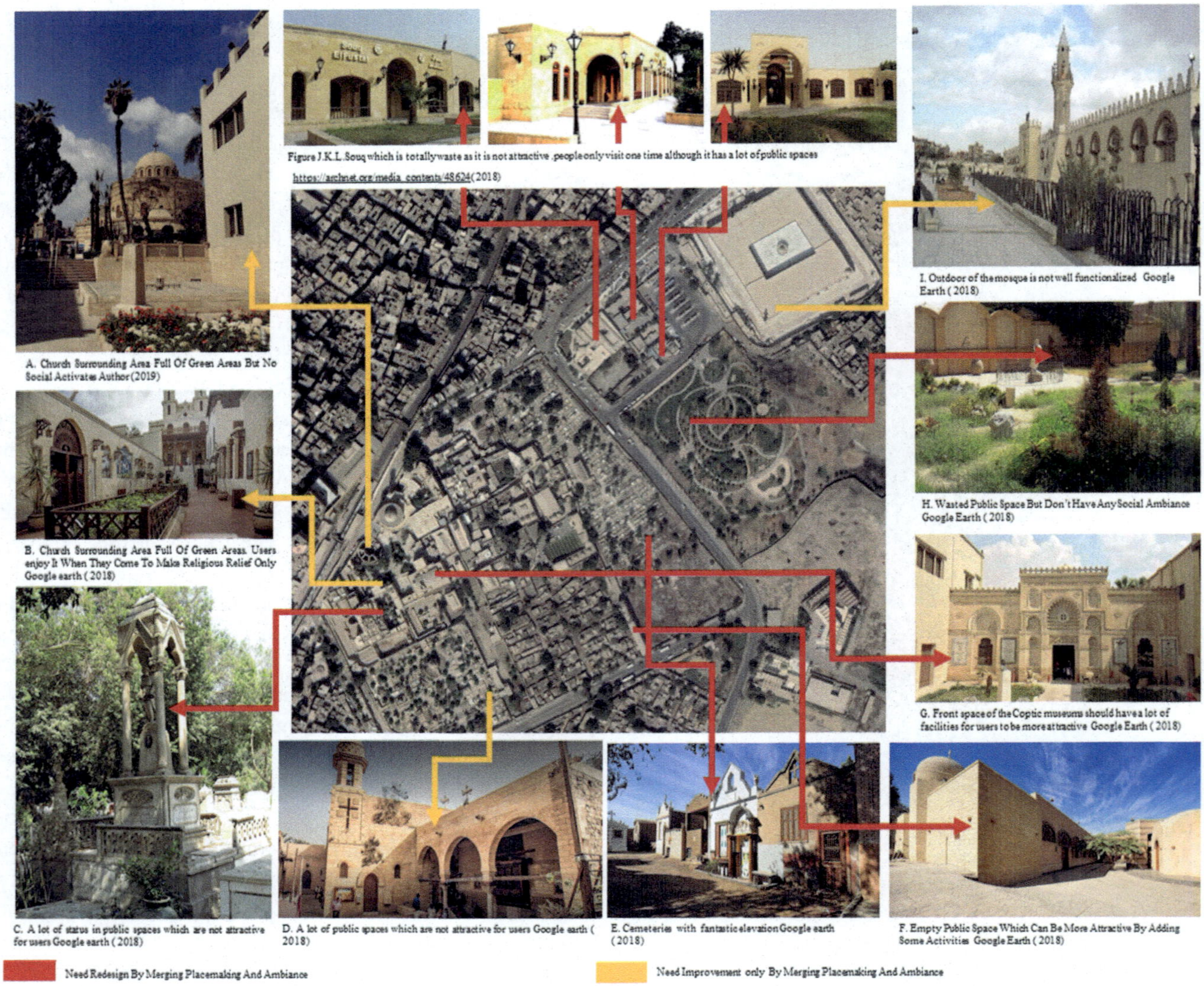

Fig. 14 Urban context and surrounding of case study Google Earth (2018)

Table 2 Evaluation of existing heritage site based on 150–200 questionnaires with the users and the researchers' proposed theoretical framework

Placemaking dimensions			Negative	Positive	Ambiance dimensions		Negative	Positive
J. Gehl	Basic activities	Walking friendly	**X**		Physical	Land use (greenness)	**X**	
		Standing	**X**					
		Sitting	**X**					
		See		X				
		Hearing/talking	**X**					
	Protection	Traffic and accidents		X		Location (urbanization)		X
	Advanced activities	Play and unwinding	**X**			Size		X
		Peace and isolation	**X**					
		Other activities Poor	**X**					
W. H. Whyte		Sittable	**X**			Status	**X**	
		Sun		X				
		Street		X				
		Water		X				
		Food	**X**					
		Trees	**X**					
		Triangulation	**X**					
S. Carr	Needs	Comfort	**X**		Physiological	Touch → temperature		X
		Relaxation	**X**					
		Passive	**X**					
		Active engagement						
		Discovery	**X**					
	Meanings			X		Sight → light	**X**	
	Rights	Access		X		Hearing → sound	**X**	
		Freedom of act		X				
		Claim and change		X				
PPS projects for public spaces	Uses and activities	Fun	**X**		Sociological	Cleanliness	**X**	
		Special	**X**					
		Active	**X**					
		Vital	**X**					
	Sociability	Diverse	**X**			Income	**X**	
		Friendly	**X**					
		Cooperative	**X**					
		Interactive	**X**					
	Accessibility	Readable	**X**			Ethnicity		X
		Walkable		X				
		Continuity		X				
		Convenient		X				
	Comfort and image	Safe	**X**			Population		X
		Clean	**X**			Sense of the place	**X**	
		Green	**X**			Type of activity (night → day)	**X**	

6 Conclusion and Recommendations

Finally, the research tried to validate the hypothesis, which states that if the preservationist and the urban designers put into consideration some guidelines for placemaking and ambiance dimensions, then this would be more effective and would enhance the heritage site and the social quality of life in general. However, placemaking is an interdisciplinary and multifaceted concept that has been widely researched since the mid-1970s (Andrews, 1974). Placemaking may have the ability to create positive social change. This positive social change can include sharing ideas and learning new skills to eventually create an end-product. This sharing and learning of skills to transform an environment suggest placemaking as an enabling tool (PPS, 2010). Despite its importance on a heritage site, there is a limited contribution to placemaking from the theoretical foundation, with special consideration to aspects of function, culture, and society. If all the placemaking and ambiance dimensions were to be applied in heritage-attached spaces, then users will feel the good ambiance, and the heritage sites will be transformed into great public spaces with a cheerful urban atmosphere. Moreover, it is worth to say that the work done was an initial step seeking further studies to be done as a trial for upgrading heritage sites' attached spaces in Greater Cairo with special consideration to creating general criteria for those spaces in order to enhance users' perception and social quality of life.

References

Abdel Tawab, A. G. (2014). The World heritage centre's approaches to the conservation of New Gourna Village, and the assessment of its authenticity and integrity. *Alexandria Engineering Journal, 53*(3), 691–704. https://doi.org/10.1016/j.aej.2014.01.010

Adams, P., Hoelscher, S., & Till, K. (Eds.). (2001). *Texture of place* (1st edn.). University of Minnesota Press.

Alvarez, L., Borsi, K., & Rodrigues, L. (2017). The role of social network analysis on participation and placemaking. *Sustainable Cities and Society, 28*, 118–126. https://doi.org/10.1016/j.scs.2016.06.017

Andrews, G. F. (1974). *Maya cities: Placemaking and urbanization* (1st edn.). University of Oklahoma Press.

Angelidou, M., Karachaliou, E., Angelidou, T., & Stylianidis, E. (2017). Cultural heritage in smart city environments. The *International Archives of the Photogrammetry, Remote Sensing and Spatial Information Sciences, XLII-2/W5*, 27–32. https://doi.org/10.5194/isprs-archives-XLII-2-W5-27-2017

Australia ICOMOS, & International Council on Monuments and Sites. (2013). *The Burra charter: The Australia ICOMOS charter for places of cultural significance 2013*. http://australia.icomos.org/wp-content/uploads/The-Burra-Charter-2013-Adopted-31.10.2013.pdf

Bauer, C., Recreation, S. C., Fletcher, R., Travieso, C., & Woodward, D. (2010). *High performance landscape guidelines: 21st century parks for NYC*. Design Trust for Public Space.

Bitner, M. J., & Obermiller, C. (1984). *The elaboration likelihood model: Limitations and extensions in marketing* (7).

Boyd, S. (2002). Cultural and heritage tourism in Canada: Opportunities, principles and challenges. *Tourism and Hospitality Research, 3*(3), 211–233. https://doi.org/10.1177/146735840200300303

Carmona, M., Heath, T., Oc, T., & Tiesde, S. (2003). *Public places, urban spaces: The dimensions of urban design*. Architectural Press.

Carr, S. (1992). *Public spaces*. Cambridge University Press.

Dajian, Z., & Peter, R. (2006). 2010 World expo and urban life quality in Shanghai in terms of sustainable development. *Chinese Journal of Population Resources and Environment, 4*(1), 15–22. https://doi.org/10.1080/10042857.2006.10677445

DCMS T.D. for D Culture, Media and Sport is a Department of the United Kingdom Government. (2010). *Principles of selection for listed buildings*. The Department for Digital, Culture, Media and Sport is a Department of the United Kingdom Government.

Elgobashi, A., & Elsemary, Y. (2019). *Towards social quality of life through Cairene public parks*. Architecture and Urban Planning Departments.

Elgobashi, A., & Elsemary, Y. (2020). *Reconstructing urban ambiance in smart public places:* (Abusaada, H., Salama, A.M., & Elshater, A. (Eds.)). IGI Global.

Faris Hmood, K. (2019). Introductory chapter: Heritage conservation—rehabilitation of architectural and urban heritage. In Hmood, K. (Ed.), *Urban and architectural heritage conservation within sustainability*. IntechOpen. https://doi.org/10.5772/intechopen.86670

Farrar, M. E. (2011). Amnesia, nostalgia, and the politics of place memory. *Political Research Quarterly, 64*(4), 723–735. https://doi.org/10.1177/1065912910373553

Feilden, B. M. (2005). *Conservation of historic buildings* (3rd ed., repr). Elsevier.

Ginzarly, M., Pereira Roders, A., & Teller, J. (2019). Mapping historic urban landscape values through social media. *Journal of Cultural Heritage, 36*, 1–11. https://doi.org/10.1016/j.culher.2018.10.002

Heintzelman, M. D., & Altieri, J. A. (2013). Historic preservation: Preserving value? *The Journal of Real Estate Finance and Economics, 46*(3), 543–563. https://doi.org/10.1007/s11146-011-9338-8

Hyden, C. T., & Sheckels, T. F. (2016). *Public places: Sites of political communication*. https://books.google.com.eg/books?id=UYpCCwAAQBAJ&pg=PR2&lpg=PR2&dq=Public+Places:+Sites+of+Political+Communication&source=bl&ots=pSVUbwhw86&sig=ACfU3U0GFP96lS-34_3wd0ZvTb5d-q-W8g&hl=en&sa=X&ved=2ahUKEwiP18HU7-TlAhXTTxUIHcfFAyE4ChDoATABegQIBxAB#v=onepage&q=Public%20Places%3A%20Sites%20of%20Political%20Communication&f=false

Iliyasu, I. I. (2014). *Challenges of preservation of cultural landscapes in traditional cities: Case study of Kano ancient city* (13).

Lynch, K. (2005). *The image of the city* (Nachdr.). MIT Press.

Meeks, S., & Murphy, K. (2016). *The past and future city: How historic preservation is reviving America's communities* (2nd ed.). Island Press.

Pallasmaa, J. (2014). *Space, place and atmosphere. Emotion and peripheral perception in architectural experience* (16).

Parsons, R. (2010). *New design in historic setting*.

Poria, Y., Butler, R., & Airey, D. (2003). The core of heritage tourism. *Annals of Tourism Research, 30*(1), 238–254. https://doi.org/10.1016/S0160-7383(02)00064-6

PPS P for Public Spaces. (2010). *Placemaking meets preservation*. https://www.pps.org/article/placemaking-meets-preservation

Redi, M., Aiello, L. M., Schifanella, R., & Quercia, D. (2018). The spirit of the city: Using social media to capture neighborhood ambiance. *Proceedings of the ACM on human-computer interaction, 2*(CSCW), 1–18. https://doi.org/10.1145/3274413

Relph, E. (2008). *Place and placelessness*. Pion Ltd. UK edition.

Schwartz, J. M., & Cook, T. (2002). Archives, records, and power: The making of modern memory. *Archival Science, 2*(1–2), 1–19. https://doi.org/10.1007/BF02435628

Strydom, W., Puren, K., & Drewes, E. (2018). Exploring theoretical trends in placemaking: Towards new perspectives in spatial planning. *Journal of Place Management and Development, 11*(2), 165–180. https://doi.org/10.1108/JPMD-11-2017-0113

The WBDG Historic Preservation Subcommitte. (2019). *Historic preservation*.

UNESCO. (2005). *Vienna memorandum on "World heritage and contemporary architecture—Managing the historic urban landscape"*.

UNESCO. (2011). Proposals concerning the desirability of a standard-setting instrument on historic urban landscapes 36 C/23. *Records of the general conference, 36th Session, Paris, 25 October - 10 November 2011, v. 1: Resolutions*. UNESCO.

World Heritage Center. (2013). *New life for historic cities: The historic urban landscape approach explained*. https://unesdoc.unesco.org/ark:/48223/pf0000220957_chi

Wallace, C., & Abbott, P. (2009). *Social Quailty in Europe* (16).

Whyte, W. (1968). *The Last Landscape*. University of Pennsylvania Press.

Whyte, W. H. (1989). *City: Rediscovering the center*. University of Pennsylvania Press.

Wyckoff, M. A. (2017). *Placemaking as an Economic Development Tool: A Placemaking Guidebook*. 598.

Zhong, J. (2012). *Reconnecting lost place: Comprehensive green space design based on historic preservation principles in compact Changsha City, China* (53).

Understanding Challenges/Barriers and the Motivations of Farmers to Adopt Good Agricultural Practices (GAPs): A Case Study of Sumerpur Tehsil of Rajasthan, India

Aditi Mali

Abstract

State of Rajasthan in India has a semi-arid to the arid climate and is water-stressed with the prevalence of desertification and frequent droughts due to extremely erratic rainfall. This also jeopardizes the sustainability of agriculture and endangers prospects for future generations by affecting the livelihood of the local populace and source of food for livestock. The challenging climate is further negatively augmented with unsustainable agriculture practices such as mono-cropping, extensive use of chemicals, etc. This study builds an understanding of farmers' challenges/barriers and motivations to achieve sustainability in the agricultural sector in this region. This study analyses farmers' perspective of shortlisted ten good agricultural practices (GAPs) for this region which are divided into the following four aspects, (1) cropland management, (2) water management, (3) degraded lands restoration, and (4) soil fertility management. *Results*: There has been as low as 0% of farmers adopting mulching as a GAP and as high as 100% of farmers adopting crop rotation. Koliwara was the only village using advanced micro-irrigation systems; this can be attributed to closeness to the Sumerpur town as well as the engagement of the agriculture department in organising farm field trips to big farms for some farmers. Salodariya has no interaction with the agriculture department as the extension officer rarely visits the farmers hence providing no new technical information to farmers. Chanod has engagement with its extension officer but due to rain-fed agriculture, it does not leave much scope for systems that require irrigation water and technology like mulching, micro-irrigation, etc. *Conclusion*: This paper tells how tasks that might seem simple and easy to do as researchers can be tough for a farmer in this region as they are not well educated and often get neglected by government officials. When research organisations make policies for sustainable land management (SLM), they are theoretically very helpful. But the on-ground application by farmers is only 50–60% due to many reasons which are explored in this paper. Hence, one of the biggest challenges to SLM is its application by farmers.

Keywords

Good agriculture practices (GAPs) • Farmers' perspective • Sustainable agriculture • Challenges of sustainability • Sustainable soil and land management • Motivations and challenges/barriers for adopting GAP • Dry land agriculture • Western Dry Region-XIV • Sumerpur Tehsil • Rajasthan • India

1 Background and Conceptual Issues

> A sustainable agriculture is one which depletes neither the people nor the land.
>
> —Wendell Berry (Illinious Library, 2018).

Sustainable soil and land management (SLM) is very crucial and important for the biodiversity of land and preserving important landforms. Globally and locally different organisations create policy and practice recommendations to tackle the challenges that come with SLM. Theoretically, these policies and recommendations are sound and effective; however, it is critical to understand the challenges a farmer faces to practice them on the ground. This research attempts to build an understanding of farmers' perspective to adopt sustainable practices and what are their challenges and/or motivation to do the same. For this research, the site selected is Sumerpur Tehsil (Tehsil is a sub-district) which is in the Pali District in the State of Rajasthan, India. Sumerpur Tehsil is chosen based on the following factors:

A. Mali (✉)
Royal Agricultural University, Cirencester, UK

C. Alalouch et al. (eds.), *Towards Implementation of Sustainability Concepts in Developing Countries*, Advances in Science, Technology & Innovation, https://doi.org/10.1007/978-3-030-74349-9_8

I. It has a very important and large agriculture produce market in the town of Sumerpur (Solanki & Patel, 2015). This will help analyse the dynamics of proximity to the market.
II. Rajasthan is divided into two agro-climatic regions, the western part is the Western Dry Region-XIV and the eastern part is the Central Plateau and Hills Region-VIII. Both fall under the dry lands zone. This region represents Western Rajasthan and its agro-climatic region is Western Dry Region-XIV (Chancellor, 2007);
III. The author belongs to this region which gives the benefit of knowing the local language, culture and few established resources in the various villages. This will be very helpful to conduct the research;
IV. This research will help in establishing the author's own farm in this area and practice sustainable agriculture by learning from field interaction with farmers and understanding their motivations, challenges/barriers;

Rajasthan State is known to face frequent droughts due to extremely erratic rainfall and high evaporation rate due to extreme heats. Due to lack of proper agriculture practices, Rajasthan has been facing a major collapse of rural employment generation and is facing a migration of workforce towards urban centres (Bhatt, 2009; Dutta et al., 2015; Sharma et al., 2015). Rajasthan is also affected by erosion which is contributing to desertification and an increase in areas of arid and semi-arid regions. In total, 67% of the area is affected due to this. Wind erosion (44.2%) is the maximum contributor followed by water (11.2%), vegetal degradation (6.25%) and salinization (1.07%) (GoR, 2014). This also jeopardizes the sustainability of agriculture and endangers prospects for the future generations by effecting livelihood of local populace and source of food for livestock (Sharma et al., 2015).

Wendell Berry's quote (Illinious Library, 2018) is especially true in case of Rajasthan where due to unsustainable practices the farmer and the farm are in jeopardy. Hence, this paper looks at farmers' motivations and/or barriers that restrict adoptions of few key GAPs which potentially can help them perform sustainable agriculture and protect their farm.

The phrase 'Good Agricultural Practices' has become an umbrella term for defining standards in many aspects of agriculture including production systems, labour conditions, environmental effects, chemical usage, etc. (Poisot et al., 2004). Out of the two kinds of GAP (Food safety and hygiene; food security and sustainability). This paper focuses on food security and sustainability GAPs. This paper also uses GAPs as an umbrella term where aspects of cropland management, water management, degraded lands restoration, and soil fertility management are considered.

Issues like desertification, water scarcity, natural disasters like droughts and floods, erratic rainfall, extreme heat, and unsustainable agriculture practices (like extensive use of chemicals, etc.,) have led to poverty in many parts of the state, and there is an acute need for sustainable and better agriculture practices for the ensured survival of the future generations. There is also a need to look at various aspects of rural development in this region like approachable markets, credit systems, agribusinesses, etc.

This study could help approach of policymakers and agriculture extension services towards farmers by understanding their stance towards practicing sustainable agriculture.

1.1 Research Aims and Objectives

The aim of this study is to capture and understand farmers' perspective towards good agricultural practices (GAPs) in their context and to identify a path towards achieving sustainable agriculture.

To achieve this aim, the following key objectives have been formulated:

I. Identify key GAPs for the agro-climatic region of Sumerpur Tehsil; this will focus the research to targeted practices that are a priority for this region.
II. Find out the rate of adoption of the selected GAPs; this will give an insight on the current practices of farmers in Sumerpur Tehsil.
III. Understand the motivations and/or the challenges and barriers for adopting the selected GAPs; this will help explore the farmer's perceptive and help identify the factors that influence the farmers' decision.

The first section of the paper states the importance of understanding the motivations and/or barriers for adopting good agricultural practices (GAPs) in Sumerpur Tehsil of Rajasthan. The second section explains the research methodology used in this paper. This is followed by the main section of the research where the results are discussed with backing from statistical confidence. The last section concludes the research with further discussion that could be carried out to augment and increase knowledge in this topic.

2 Research Approach

This paper uses mix approaches: qualitative and quantitative approaches. The data collected in each approach are in two categories: primary and secondary. Table 1 shows the tools and resources which are essential to collect data in each category.

Table 1 Methodology approach matrix

	Qualitative	Quantitative
Primary	-Field observations -Farmer interview -Data from agriculture extension	-Farmer interviews -Data from agriculture department
Secondary	-Research papers on GAPs -Research papers on Western Rajasthan -Data from government Web portal on agriculture -Data from other Web sources	-Statistical data from Census of India -Data from government Web portal on agriculture

To achieve the three objectives of this paper, there were three main phases. **Phase 1** tackled the objective I for which the relevant GAPs had to be shortlisted for the agro-climatic zone of Sumerpur Tehsil. The objective I was achieved by studying the local agro-climatic zone of Sumerpur Tehsil and 10 GAPs were considered keeping in mind the local culture and traditional practices from Dr. Jellason's (2017) paper on Environmental Challenges and Linkages to smallholder agriculture in sub-Saharan African drylands: Implications for food security. For this research, key GAPs (Table 2) are identified on their contribution to the four aspects: (1) Cropland management, (2) Water management, (3) Degraded lands restoration, and (4) Soil fertility management. The GAPs that have multiple benefits and are supported by preliminary field observations have been selected. The key GAPs are:

Phase 2 was field work in which farmers from different villages of Sumerpur Tehsil were interviewed and surveyed. As part of the field work for this research, the community engagement was through household surveys and interviews in villages located in different parts of the Sumerpur Tehsil. The selection of these villages is done with the guidance of the Sumerpur tehsil's Agricultural Department. The department has divided the tehsil in three parts, and hence, three villages are selected from each block; Chanod, located further most from the Sumerpur HQ in the north, Koliwara in the east and Salodariya in the west. The farmers are first surveyed followed by the interview. For this, a sample size of 15 farmers is considered in each village which totalled to a sample size of 45. For Objective II, the GAPs listed above in Table 2 are questioned in the interview. It is a close-ended question with options of Yes or No (whether they practice the GAP or not). When this is tabulated, it will give the rate of adoption for the selected GAPs. Each GAP is then further explored as an open-ended question. If the answer is 'Yes', then the motivations for practicing the GAP is explored, and if the answer is 'No', the challenges and barriers are explored. Through this Objective III, understanding the motivations and/or the challenges and barriers for adopting the selected GAPs is achieved. As per research ethics, all the data and information collected is confidential and only used for this research.

Phase 3 was the data analysis of the primary and secondary data collected during the field visit. Forty-five farmers of the three villages were surveyed and interviewed for this research. To merge all the three villages' data, Kruskal Wallis Test is computed to determine the same probability distribution of the data sets. The tests and statistical methods used for analysing the data are based on this data type and are categorised as nominal and parametric. First, the rate of adoption of the selected GAPs is determined by tallying each positive response by the farmer. Post determining the rate of adoption, the factors of influence are analysed through the household survey that is done before the interview which gives an insight into each farmer lifestyle and livelihood. The factors are shortlisted by reviewing the data from the survey, knowledge of the general socio-economic factors, and field observations as to which factors stand out and could have a probable influence. The factors that are chosen are: (1) education, (2) size of farm, and (3) whether the owner is self-employed at the farm. Through these approaches, the rate of adoption and the probable factors that have significance are measured. Second, the motivations and/or the challenges/barriers for each GAP is comprehended. Each farmer reasoned their choice of practising of each GAP which helped in building an understanding of the factors that guided their choice. This gives a detailed insight on how farmers decide on farming systems and which factors dominate this process. After finishing the analysis, the discussion section compares similar studies to this paper leading to the conclusion of the research.

3 Results and Statistical Confidence

3.1 Background of Sumerpur Tehsil

Sumerpur Tehsil is in the Pali District of the State of Rajasthan, India. Sumerpur is situated in the southern part of Rajasthan (Census of India, 2018). The agro-climate zone of India is Western Dry Region-XIV which has arid conditions with hot climate and fragile ecosystem (Chancellor, 2007). The Western Dry Region is further divided into five

Table 2 Key GAPs identified for the research. The benefits of the GAP are green coloured cell. Rows represents the GAP and column state the benefits

	Increase yields	Control erosion	Soil health	Increase fertility	Increase water holding capacity	Conserve soil moisture
Cropland management						
Reduced/zero tillage/no tillage/conservation tillage (light tillage)						
Use of cover crops						
Incorporation of residues						
Use of legumes in crop rotation						
Water management						
Drip and sprinkler irrigation						
Water harvesting: tied ridges or ponds						
Mulching						
Degraded lands restoration						
Re-vegetation plantation of trees (to create wind-breaks)						
Organic nutrient amendments (manures, bio-solids, compost).						
Soil fertility management						
Crop rotation						

agro-climatic regions of Rajasthan. Sumerpur is a part of IIB—Transitional Plain of Luni Basin (Hussain, 2015).

Sumerpur Tehsil has 300–500 mm average rainfall which is higher than the desert parts of the extreme west of Rajasthan. The major Kharif (Monsoon) season are pearl millet, cluster beans, and ground nuts, and the major crops of the Rabi (winter) season are wheat and mustard. The soil type is sierozems with sandy loam to sandy clay soil texture and the colour is yellowish brown (GOR, 2018; Hussain, 2015). Sierozems, also known as grey desert soil, is found in deserts surrounded by mountains. The vegetation is thin cover due to lack of rainfall with low humus soil content. This region has high evapo-transpirative losses which brings the underground salts to the soil surface. The fine texture of the soil and rich plant nutrients makes the soil fertile, hence, suitable for agriculture (PGGC11, 2018).

The coldest time of the year is January to February with temperatures ranging from 3 °C to 35 °C. It is hot in summer from the month of April to June. Sumerpur's highest day temperature in summers is in between 29 °C and 44 °C. It rains from the month of July to September, making it very pleasant and people thoroughly enjoy this period (Triphobo, 2018).

Sumerpur Tehsil has a population of 198,125 as per Census 2011 with 73% population staying in the rural area. The literacy rate is 60.27% in rural area and 74.73% in urban area. There are two municipal towns and 68 villages divided in 29 g-panchayats (Census of India, 2018). Around 25% of the total main workers are in agriculture sector with 15% agriculture cultivators and 10% agricultural labourers (Census of India, 2011). The Sumerpur has a Mandi (wholesale grain market) where one gets wheat, chilli, cotton, oil, etc. It is one of largest markets for mustard seed which is also exported globally (Solanki & Patel, 2015).

For this research, initially nine villages (Balupura, Bankali, Chanod, Dhola Jageer, Dhola Shasan, Kanpura, Khindara Gaon, Koliwara, and Salodariya) from different parts of the Sumerpur tehsil were short-listed with the help of Sumerpur Agriculture Department. Finally, three distinct

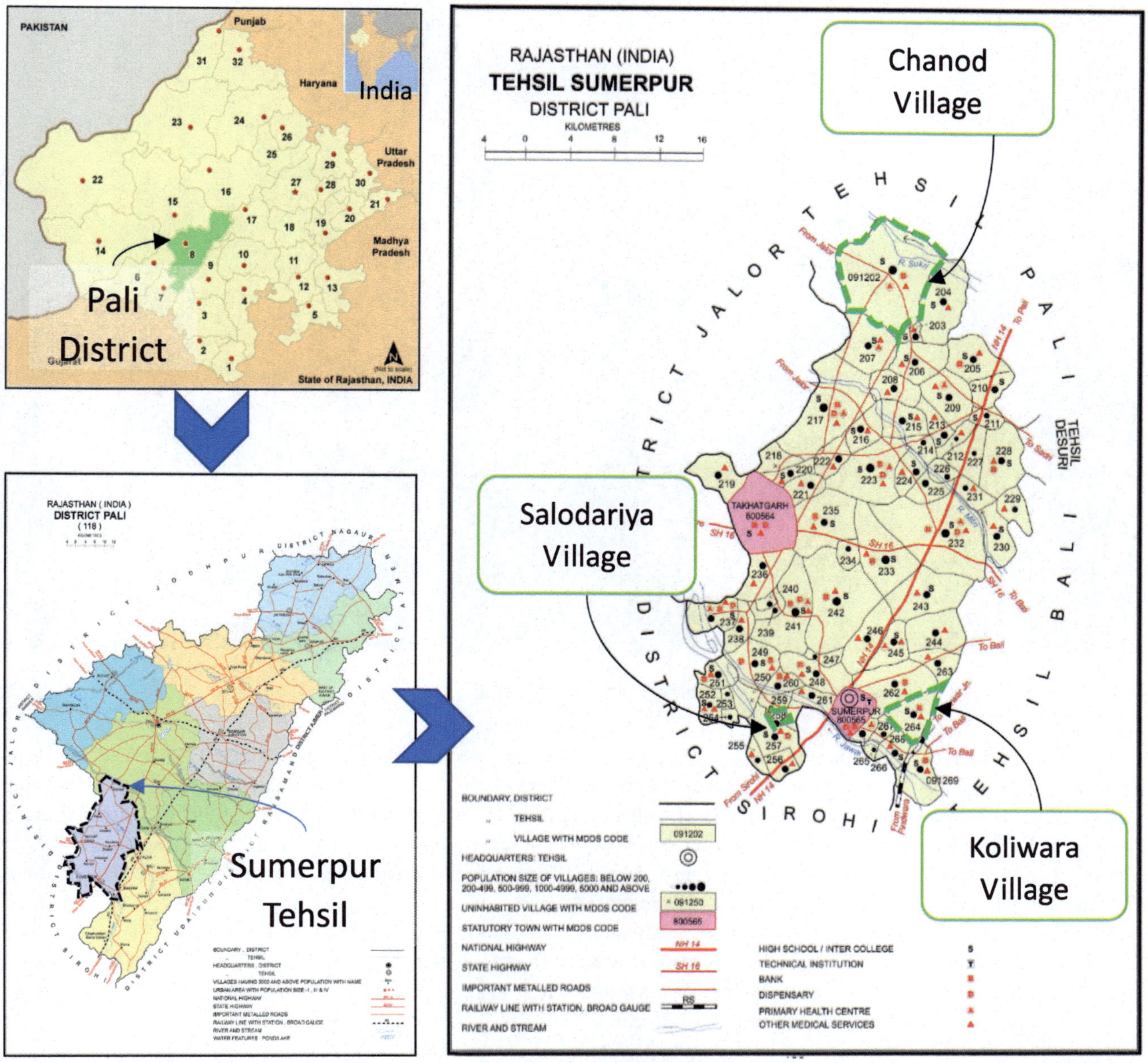

Fig. 1 Regional context of Sumerpur Tehsil and locations of the 3 selected village for this study, (Census of India, 2011)*. *Images have been modified for use

villages are selected: Chanod, Koliwara, and Salodariya (Fig. 1). The agriculture work force for the three villages is Chanod has 36% cultivator and 23% agricultural labourers; Koliwara has 5% cultivator, and 11% agricultural labourers; Salodariya has 32% cultivator and 6% agricultural labourers (Census of India, 2018).

4 Results

For this research, 15 farmers each from Chanod, Koliwara, and Salodariya (a total of 45 farmers) were surveyed and interviewed. Since this paper is an attempt to understanding the motivations and barriers for adopting GAPs in Sumerpur Tehsil, three distinct villages from different demographic are chosen. Chanod (Fig. 2) is selected because the northern region has saline ground water forcing the farmers to depend on rain-fed farming despite this, it has very high numbers of cultivators, which makes it a unique case. Koliwara has advanced agricultural systems in place with comparatively wealthy farmers; it is also a hub for handicraft industry and is the closet village to the Sumerpur town where the wholesale grain market is located. Salodariya is a representative of what most of the villages in this region are like; summer season is not cultivated (unless one has a non-saline irrigation source); animal husbandry is used as a farm diversity especially in times of drought.

Fig. 2 A farm in Chanod Village, Sumerpur Tehsil, Rajasthan, India

The farmers were chosen randomly based on time, willingness, and availability. Despite the randomness, a mixed variety of farmers were interviewed. Categories in which the diversity of farmers interviewed are as follows: type of water source (open well, tube well, canal, dam, and rain-fed), size of farm, type of farm (arable, arable + horticulture, arable + livestock), age of farmer, seasons of cultivation (Kharif season, i.e. cultivating from July to October, Rabi season, i.e. cultivating from October to March, Zaid season, i.e. cultivating in the summers) and education level (primary school, middle school, secondary school, graduate, and postgraduate). Household survey data is used to capture this diversity among the farmers.

4.1 Finding Out the Rate of Adoption of the Selected GAPs

The data collected is from Chanod, Koliwara, and Salodariya in Sumerpur Tehsil. These three villages represent Sumerpur Tehsil for this study, hence to use it as one single data group, Kruskal Wallis test is used to undermine if the three data sets are similar. Table 3 shows the three samples used for the test. Each sample is the tally of positive responses from the farmers for each GAP. The tests and statistical methods used for analysing the data are considered on the basis of this data type and are categorised as nominal and parametric.

The Kruskal Wallis test like any other test has a null hypothesis (H0) and alternative hypothesis (H1) (Mathcracker, 2018). The H_0 = The 3-probability distribution are the same (the three-data set are the same) and H_1 = The 3-probability distribution are not the same. Based on the information provided, the significance level is $\alpha = 0.05$, and the number of degrees of freedom is $df = 3 - 1 = 2$. Therefore, the rejection region for this Chi-Square test is $R = \{\chi 2\colon \chi 2 > 5.991\}$. The formula for Kruskal Wallis test is:

$$H = \frac{12}{N(N+1)}\left(\frac{R_1^2}{n_1} + \frac{R_2^2}{n_2} + \ldots + \frac{R_k^2}{n_k}\right) - 3(N+1)$$

In this formula, N is total number of observations ($N = 30$); $R1$ is sum of ranked values of Chanod village ($R1 = 174.5$); $n1$ is the count of the observed value ($n1 = 10$, this same for $n2$, and $n3$); $R2$ is sum of ranked values of Koliwara village ($R2 = 152.5$); $R3$ is sum of ranked values of Salodariya village ($R3 = 138$).

Hence,

$$= \frac{12}{30(30+1)}\left(\frac{174.5^2}{10} + \frac{152.5^2}{10} + \frac{138^2}{10}\right) - 3(30+1)$$
$$= 0.872$$

After computing the test, the result is $H = 0.872$. Since it is observed that $\chi 2 = 0.872 \leq \chi$ u2 = 5.991, it is then

Table 3 Sum of positive answers, village-wise, for each GAP

GAPs	Chanod	Koliwada	Salodariya
Reduced/zero tillage/no tillage/conservation tillage (light tillage)	0	6	12
Use of cover crops	1	10	4
Incorporation of residues	15	6	8
Use of legumes in crop rotation	15	14	9
Drip and sprinkler irrigation	0	9	0
Water harvesting ponds	15	1	0
Mulching	0	0	0
Re-vegetation plantation of trees (To create wind-breaks)	15	14	15
Organic nutrient amendments (manures, bio-solids, compost)	12	12	9
Crop rotation	15	13	15
Total	88	85	72

concluded that the null hypothesis is not rejected. Using the *P*-value approach: The *p*-value is $p = 0.6467$, and $p = 0.6467 \geq 0.05$, it is concluded that the null hypothesis is not rejected.

It is hence concluded that null hypothesis is not rejected which means that in this study, the three data set of the three villages has a probability distribution which is same. So, all the data collected can be merged into one data group representing Sumerpur Tehsil for this exploratory research.

To find the rate of adoption of the GAPs, all the data collected from the three villages is merged into one data group. Every response of whether the farmer practiced the GAP or not is tallied, and the sum of each respondent is calculated followed by the percentage gives us the rate of adoption of the GAP for each respondent. To illustrate, as we are calculating this for 10 GAPs, 10% indicates that a single GAP is practiced, whereas 40% indicates that four GAPs are practiced by the farmer.

Figure 3 is a breakdown. Example: the light blue slice of the pie says 30%, 1, 2% which means 30% of GAPs are practiced, and 1 Farmer is practicing it which is 2% of the total data group of 45 farmers. The 89% of the farmers have adoption rate of GAPs between 40 and 60%. Half of the farmers adopted 60% of the GAPs, i.e. 6 out of 10 key GAPs. However, not all of them are practicing the same GAPs. There is a single farmer adopting 30 and 80% of GAPs. It would be interesting to compare as to what factors influence his decision. Since the rate of adoption of the GAPs has been determined, let us look at the potential factors that influence this.

The factors of influence can be determined through the household survey that was done before the interview which gives us an insight into each farmer. The factors are determined by looking at the collected data from the survey and field observations as to which factors stand out and could have a probable influence. The factors that are chosen are:

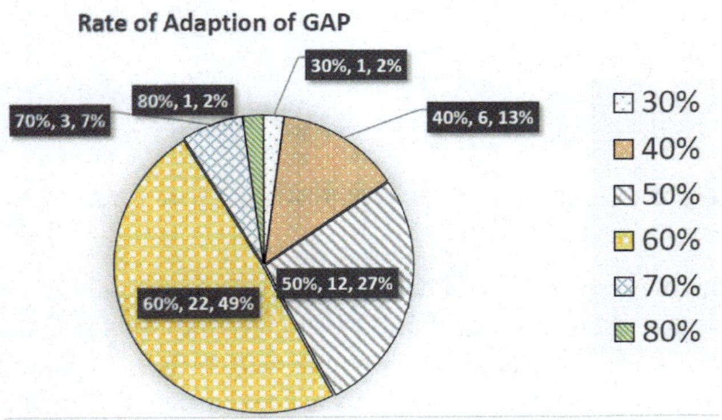

Fig. 3 The rate of adoption of GAPs for every respondent is summed together. Example: 30%, 1, 2% means: 30% GAPs practiced, 1 Farmer, 2% of the total data group of 45 farmers

(1) education, (2) size of farm, and (3) whether the owner is self-employed at the farm.

First factor that could have a probable influence is education level and it is categorised into: no education; primary school (up to 5th standard); middle school (up to 10th standard); secondary school (up to 12th standard); and graduate and postgraduate.

To determine whether the variable rate of adoption of GAPs has a relationship with variable educational level, Chi-squared test (also known as Goodness of fit test is performed) with the significance level of $\alpha = 0.05$. After computing the chi-squared test formula, the $\chi 2$ is 28.08. Since $\chi 2 < 37.65$, the H0 cannot be rejected. Because H0 is accepted, there is a relationship between the variable rate of adoption of GAPs and the variable educational level. Hence, it can be concluded that educational level has an influence on the rate of adoption of the GAPs and that the relation is not an absolute positive or negative. Figure 4 shows that farmers that are graduates and postgraduates have 60% and above rate of adoption of GAPs. Also, the one farmer that has the

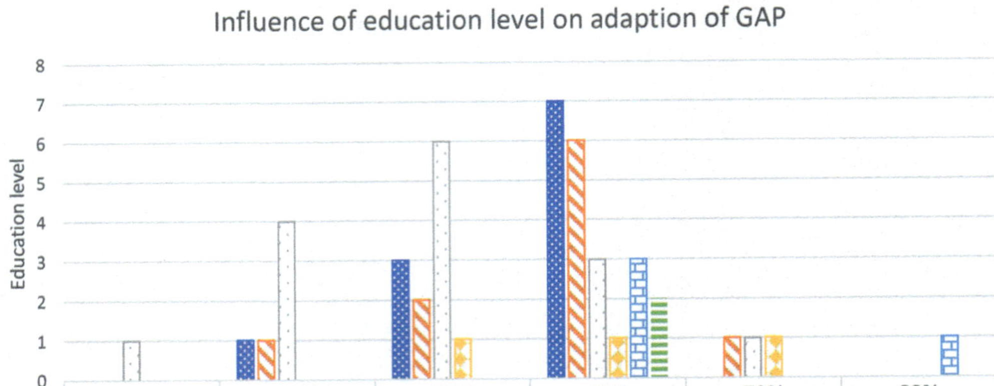

	30%	40%	50%	60%	70%	80%
No education	0	1	3	7	0	0
Primary School	0	1	2	6	1	0
Middle School	1	4	6	3	1	0
Secondary School	0	0	1	1	1	0
Graduate	0	0	0	3	0	1
Post-Graduate	0	0	0	2	0	0

Rate of Adaption of GAP

Fig. 4 Rate of adoption along with educational level in each category

highest rate of 80% of adoption of GAPs is the only farmer with a B.Sc. in Agriculture which gives him the theoretical knowledge as well. But if we look at 70% rate of adoption, the three respondents have only school education. The other factor to observe here is that agriculture is a field where farmers learn through previous generations and not necessarily in formal education. The point here is that the farmer who has formal education in agriculture has the high rate of adoption which can be related to theoretical knowledge gained from the education; also, while interacting with this farmer, his knowledge base was more than many farmers that were interviewed (author's field observation).

Size of Farm is the second factor that could have a probable influence the rate of adoption of the GAPs. The unit used locally to measure land is Bigha; this unit differs in every state. The state of Rajasthan has measure 1 Bigha as 0.2529 Hectares; the data collected for farm size is in Bigha which is converted to Hectares (Ha) for this research (Fig. 5). Majority of the farmers own farmlands which range between 1 and 8 hectares.

Figure 6 is a scatter plot of farm size and rate of adoption of GAPs. The plot does not show negative (falling) or positive (rising) relationship between the two variables. Hence, it is concluded that there is null (non-correlation) relationship between farm size and rate of adoption. The largest farm size also has the 80% rate of adoption of GAPs, but the farms with 70% rate of adoption are drastically smaller and the farm with lowest rate of adoption is larger than the 70% ones. Therefore, it can be statistically concluded that farm size does not influence the rate of adoption of GAP.

The third and the last factor that is tested for influencing rate of adoption of GAPs is whether the owner is self-employed and indulged in the farm work. This is being tested as a factor because the lowest rate of adoption of GAPs is a farm where the owner has employed full time labourers to do the farm and stays in another city, whereas the highest rate of adoption of GAPs is self-employed on the farm as well as stays on the farm (Fig. 7).

To determine whether the variable *rate of adoption of GAPs* has a relationship with variable *working on the farm*, Chi-squared test is performed with significance level $\alpha = 0.05$. After computing the chi-squared test formula, the χ^2 is 13.66. Since $\chi^2 > 11.07$, the H_0 is rejected. Therefore, we accept the H_1 that there is **no relationship** between the variable *rate of adoption of GAPs* and the variable *working on the farm.* The rate of adoption of GAPs is established with 89% of the farmers following 40% to 60% of the GAPs (see Fig. 7 for the calculation).

After testing, the three potential influencing factors (education level, farm size, and working on the farm) it is concluded that only educational has significance level of relationship and that other two assumptions have turned out to be false.

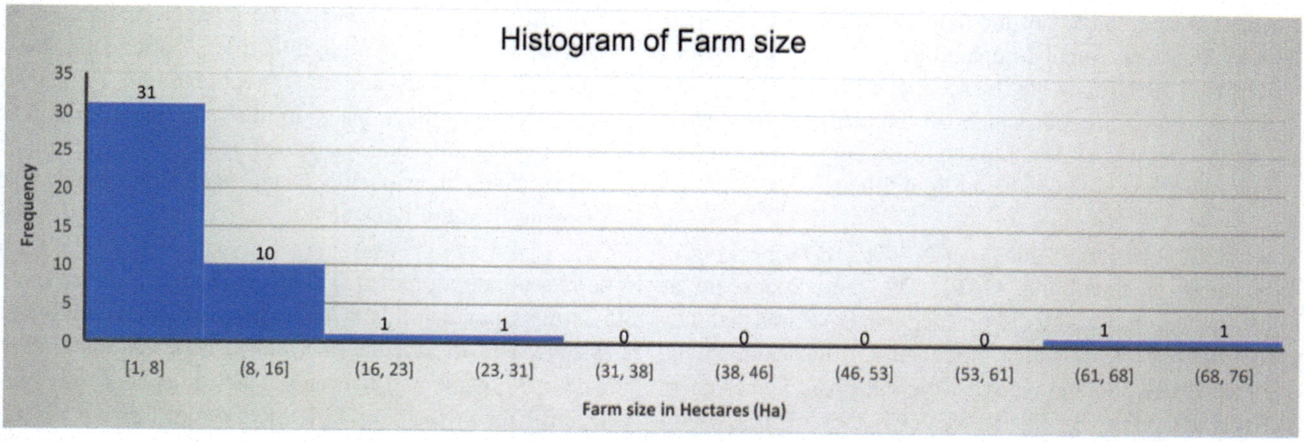

Fig. 5 Histogram for farm size

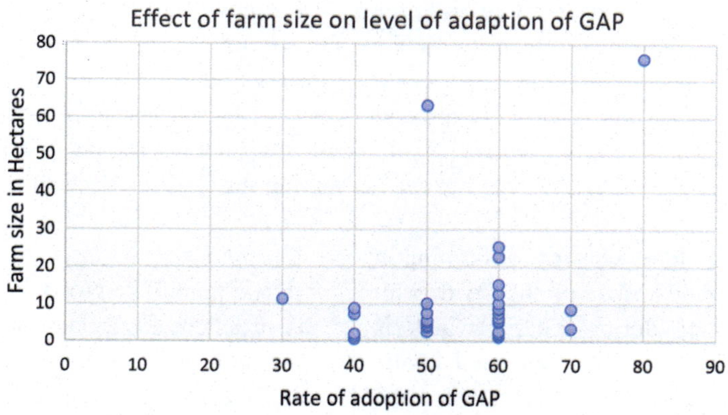

Fig. 6 Scatter plot of farm size with the rate of adoption of GAPs

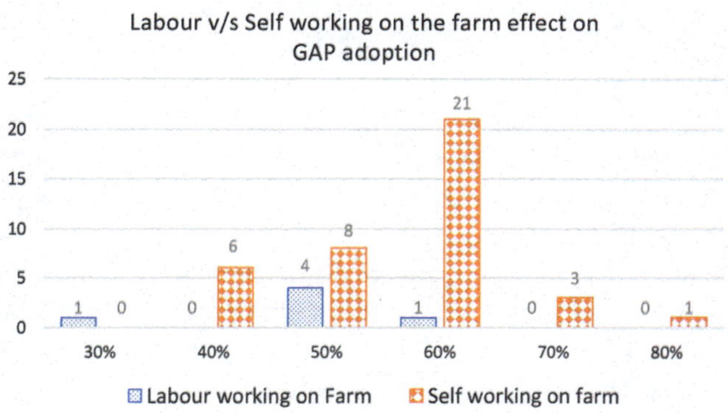

Fig. 7 Labour versus self-work on farm with rate of adoption of GAPs

4.2 Understanding the Motivations and/or the Challenges/Barriers for Adopting the Selected Key GAPs

This section of paper attempts to understand the motivations and/or the challenges/barriers for adopting the key GAPs selected for this research. As part of the interview, every GAP was asked in two parts; first part was a close-ended question of whether they practiced the GAP with option of Yes or No. The second part was about their motivation to practice the GAP (if they answered Yes) or about their challenge/barrier to do so (if they answered No). There are ten key GAPs identified for four purposes: Cropland management (1. Light/reduced tillage, 2. Use of cover crop, 3. Incorporation of residues and 4. Use of legumes in crop rotation), Water management (5. Drip and sprinkler irrigation, 6. Water harvesting, 7. Mulching), Degraded lands restoration (8. Plantation of trees, 9. Organic nutrient amendments) and Soil fertility management (10. Crop rotation). Below are the details for each GAP.

1. *Reduced/zero tillage/no tillage/Conservation Tillage (Light Tillage)*: In the case of Sumerpur Tehsil where the soil is Sierozems which is surrounded by mountains, the soil tends to have rocks in them. Hence, a drilling machine for tillage is not appropriate as the machine could be damaged. Also, these are not the most fertile soils of India like North India where one gets high output; the farmers do not spend high amounts of money in technologies like

these. Also, majority of the farmlands are less than 10 Ha which are very small compared to highly equipped mega farms in the UK and US. Hence, light tillage is considered as up to 6-inch depth which can be achieved by a basic tractor and use of a dish-plough machine. Going deeper than 6-inch is not considered light tillage.

The rate of adoption of this GAP is 40% (18 farmers out of 45 farmers practiced this GAP). The motivations are as follows: the farmers who did practice this GAP was that they used a simple tractor which ploughed 4–6 inches deep and one farmer used a hand plough which dug 1–2 inches deep; in Koliwara, the rocks hit the tractor if they go below 3–4 inches, so they perform light tillage.

The challenges/barriers are as follows: all farmers of Chanod did not do light tillage and the reason for that is the hard land, so they need to perform deep ploughing for completing it faster and loosen the soil to make it soft for crops to grow; in Koliwara and Salodariya farmers with huge lands preferred deep ploughing to make work faster and smoother, and this also killed the weeds, loosened the soil, and brought fresh soil on top.

2. *Use of cover crop*: There are three seasons for farming in India, Kharif, Rabi, and Zaid. Most of the farmers in Sumerpur Tehsil do farming during Kharif, which is rain-fed, and in Rabi, using the water collected in the monsoons. Zaid season is during the summer when the weather is very harsh and can get up to 45 °C which causes high evapotranspiration losses. Use of cover crop can protect the land from losing its moisture.

The rate of adoption of this GAP is 33% (14 farmers out of 45 farmers practice this GAP). The motivations are as follows: 24% of total farmers use this time to grow fodder for their or other farmers' cattle; 7% of the total farmers which are from Koliwara cultivate in summer (Fig. 8) so that their labour does not sit ideal. 2% of the total farmers, which is one farmer from Chanod, are the only ones who cultivates legume in summer when the rain of the given year is good.

The challenges/barriers are as follows: 44% of the farmers do not use cover crop as the underground water source becomes saline by the end of the Rabi season, so they cannot grow anything in summer as that is the only source of water; also, because of evapotranspiration losses if they try and cultivate the surface soil becomes saline. They also leave the soil exposed which is ideal to kill the bacteria from the previous crops. But this also unfortunately kills the good bacteria and soil organisms; 13% of the total farmers have lack of water to produce in summer season; 10% chose to leave their land fallow because they think cover crops reduce the fertility of the soil before the next season.

3. *Incorporation of residues:* The general practice in this region is that people plough their lands after Rabi season or early summer, and it is kept exposed to the sun throughout summer. They do this to prepare the land for the coming monsoons as the exposed loosed soil collects water better than no tilled land, and they directly start sowing then the rains arrive.

The rate of adoption of this GAP is 64% (29 farmers out of 45 farmers practice this GAP). The motivations are as follows: 42% of the total farmers have adopted this GAP to create pro-organic soil nutrients which then acts like a natural fertilizer in the soil; 11% of the total farmers practice this GAP as not only it fertilises their land but also loosens the soil for cultivation; 11% of the total farmers adopt this GAP as they believe that it gives them better production.

The challenges/barriers are as follows: 15% of the total farmers use all the crop residue for cattle fodder; the cattle grazes and unroots all the residue and the cattle dung is good for soil; 19% of the total farmers have lack of water; 2% of the total farmers see chemical fertilizer as the only solution and do not feel the need to use crop residue for soil.

4. *Use of Legumes in the crop rotation:* This region has erratic rainfall so growing crops which need little rainfall is beneficial, especially as the rainfall is not constant in the monsoon but more periodic in nature. Moong gram and cluster beans are a very popular choice along with other pulses in the Kharif season.

The rate of adoption of this GAP is 84% (39 farmers out of 45 farmers practice this GAP). The motivations are as follows: 38% of the total farmers have legume in the rotation for household consumption and sell the excess; moong gram pulse is a key staple food of this region; 33% of the total farmers, all from Chanod, grow legumes as it needs very little water to grow and gives more yield compared to other crops; 9% of the total farmers adopt this GAP for selling as there is demand for moong gram in the market; 4% of the total farmers adopt this GAP as they believe that it benefits the soils and has been a traditional practice of their ancestors.

The challenges/barriers are as follows: 14% of the total farmers belonging to Salodariya do not grow legumes as it is not a priority for them; they have to prepare land for their main season that is Rabi and this is a kharif crop; also, the seeds bursts to spoil the land and create weeds for the next season; 2% of the total farmer do not grow legume crops as there is pig nuisance in this area and these crops attract the pigs.

5. *Drip and Sprinkler Irrigation:* Drip and sprinkler irrigation are micro irrigation methods. They save approximately 40% to 90% of water and hence very useful in

Fig. 8 A farm in Koliwara village where they are cultivating in summer

Sumerpur tehsil as there is lack of water post winter seasons. Government also provides subsidies to install this system.

The rate of adoption of this GAP is 20% (9 farmers out of 45 farmers practice this GAP). The motivations are as follows: all the farmers adopting are from Koliwara village (9 out of 15 have micro irrigation system, mainly drip irrigation), the drip irrigation is mainly for vegetables and fruit orchids; 10% of the total farmers adopted this GAP as it helped them consume less water and save water; 4% of the total farmers installed these systems as it was automatic and hence demanding less labour as well as saving water; 4% of the total farmers adopted this GAP as they had slopes in land, so this was required for pumping water on the higher grounds; 2% of the total farmers adopted this system as it was a part of government scheme where they received electricity connection as a part of it.

The challenges/barriers are as follows: in the whole of Chanod and Salodariya, no farmers interviewed used micro-irrigation system, while in the whole of Salodariya village, only one farmer had adopted this who was not available at the time of interview; 36% of the total farmers which includes all farmers from Chanod village and one farmer from Salodariya (Fig. 9) who did not adopt to this system stated they have flat lands and do not require it; their wells have saline water so they have no source of irrigation. There is a lack of water in this area, and the crops they grow are not premium enough to recover cost if they were to install such systems; 22% of the total farmers who belong to Salodariya have not incorporated micro irrigation as they have enough water and do not feel the need for it; 10% of the total farmers find the process of availing such scheme difficult especially with the reducing subsidy as there is bribing at every stage of approval which they cannot afford. Also, the bankers do not show interest to poor farmers, who also do not get enough help to apply for loans; 8% of the total farmers find it too costly to install such irrigation systems; 2% of the total farmers do not have the requirement for it; 2% of the farmers are planning to install the drip irrigation system.

6. *Water harvesting ponds or tied ridges:* Most of the farm sizes in this area are small. To utilise the farm area, most farmers do not create ponds that consume space on the

Fig. 9 A farm in Salodariya village where due to lack to knowledge and support of extension workers the farmer is practicing conventional flood irrigation from the well

farm even though there are government subsidies available to create water harvesting ponds on the farms. Also, when the summer is harsh, the water collected will not last long in the open.

The rate of adoption of this GAP is 36% (16 farmers out of 45 farmers practice this GAP). The motivations are as follows: out of the 16 farmers who have adopted this GAP, 15 belong to the Chanod village and 1 to the Koliwara village; 34% of the total farmers which are all from Chanod harvest water through Tied ridges; their major source of water is rain which they try and harvest as much as possible by holding the water on the farm specially for the winter crops as that is the only water they have; 2% of the total farmers have a dam on site, and they harvest water using that and distribute to the nearby farms as well.

The challenges/barriers are as follows: 64% of the total farmers who belong to Koliwara, and Salodariya have not adopted this GAP. Open wells are the main reason for this as they collect the rain water, and the smaller opening protects it from evaporation; the well takes up space on the farm so farmers do not want to utilise more of their space for this.

7. *Mulching:* Mulching protects the land from being exposed to sun which, especially in this region, could be helpful as soil moisture is not lost. It also protects weed from growing, and that is seen as a problem in this area. Farmers hire 20–30 labourers even on small lands for a few days in every season just for weeding.

The rate of adoption of this GAP is 0% (0 farmers out of 45 farmers practice this GAP). The challenges/barriers are as follows: farmers from different villages have different take on it; farmers from Chanod have less knowledge of such systems, and they grow less profitable crops which cannot sustain the extra expense of mulching. The farmers from Koliwara have the financial capacity to do it, but since no one does it, there is no trend, hence creating a lack in willingness to set trends. In Salodariya, farmers have a lack of skilful labour to carry out such tasks; also, there is a lack of awareness as well as their main crops do not require such technology.

8. *Re-vegetation through plantation of trees (to create wind breaks):* Wind erosion is a critical problem in western Rajasthan. Trees are a good solution as they create wind breaks and protect the soil. As farmlands are small, planting trees on the periphery serves the purpose.

The rate of adoption of this GAP is 98% (44 farmers out of 45 farmers practice this GAP). The motivations are as follows: all the 44 farmers who adopt this GAP have a similar take on the reasons; trees provide a natural boundary since once delineated, they cannot be moved by neighbours. They also provide shade and block harsh winds; they also give animal fodder and fuel source; and they also provide greenery in the area (Fig. 10).

The challenges/barriers for the one farmer who does not have tree is that he has recently bought the land and planting trees is in process as of now.

Fig. 10 A farm in Salodariya village where the farm boundary has tree plantations

9. *Organic nutrient amendments (manures, biosolids, compost):* The soils in this region has low humus content hence providing organic content will help enrich the soil by restoring the degraded land and curbing further degradation.

The rate of adoption of this GAP is 73% (33 farmers out of 45 farmers practice this GAP). The motivations are as follows: all the farmers using organic matter in the soil use minimal chemicals on their land; 40% of the total farmers put organic matter in the soil as they own cattle and livestock on the farm; 16% of the total farmers adopt this GAP as they have seen the positive impact of this on the production. These farmers also purchase additional farmyard manure; 11% of the total farmers adopt this GAP as it increases productivity and fertility of the soil; 4% of the total farmers state that wheat grown from organic matter are better for health and have a better taste; 2% of the total farmers have an organic farm which makes organic matter vital for the soil.

The challenges/barriers are as follows: 14% of the total farmers do not practice this GAP as they do not have cattle or livestock of any kind, and the organic manure is expensive to buy; 8% of the total farmers feel that chemical is enough for them and they do not see a requirement for organic matter; 5% of the total farmer which belong to the Chanod village face a problem of termites on the land on application of animal manure which damages crops and property.

10. *Crop rotation:* As already established earlier in the paper, India has three seasons of cropping Rabi the monsoon, Kharif the winters, and Zaid the summers. Every agro-climate region has different set of crops that can grow in that region in a season.

The rate of adoption of this GAP is 100% (45 farmers out of 45 farmers practice this GAP). The motivations are as follows: having multiple season, all the farmers naturally have a crop rotation as all crops do not grow in all seasons; 73% of the farmers have legumes in the crop rotation right before their main cropping season which is Kharif; only 27% of the farmers use a cover crop and the rest have fallow period in the Zaid season.

5 Discussion

When policy makers and industry experts recommend sustainable practices or GAPs, they want the farmers to ultimately practice them for improving the farms' productivity and reduce impact on environment and nature. These studies or practices do not always reach the farmers in a way they understand. This paper is an attempt to fulfil this gap in knowledge of farmers' perspective in adopting GAPs. This research paper is about 'Understanding the motivations and barriers for adopting Good Agricultural Practices (GAPs) in

Sumerpur Tehsil' located in state of Rajasthan in India. The potential ten GAPs recommended are divided in these four groups: cropland management, water management, degraded lands restoration, and soil fertility management. These GAPs give solutions to identified problems in this region.

Rajasthan is a very high-risk state for agricultural activities due to irregular and extremely low rainfall and frequent maxims in high temperatures and droughts which are quite frequent in this climate zone (four out of every five years) (Dutta et al., 2015; GoR, 2014). This is an arid area of India, characterized by minimal available land for agriculture, fraught with harsh agro-climatic zones, low water table, water reserves, and fragile ecosystems. The desert in this area has also been growing steadily (Dutta & Chaudhuri, 2015). The low vegetation cover, poor soil quality and texture, and arid climate is further contributing to desertification. The desertification is happening naturally, but the process is being accelerated by human activities (by intensive farming) such as decreasing the soil quality, depleting the soil organic matter and reduction of vegetation by deforestation. The practice of mechanized deep ploughing and increase in the net sown area are increasing the rate of wind erosion which is a primary contributing factor to soil erosion. The gullies and valleys of low-lying Aravalli mountain range contribute to water erosion, and excessive use of canals and wrong drainage has led to increase in salinity and water logging. Water erosion is also being amplified due to incorrect and unsustainable mining practices, increased urban land use and uncontrolled and unplanned deforestation (Sharma et al., 2015).

Consequently, groundwater has become a major resource for irrigation during crop production. Surface water is rare due to high evapo-transpirative losses common in arid and semi-arid regions (Everard, 2015). Recharge of aquifers by capturing rainfall is extremely important to make sure that water security is achieved for agriculture and other fresh water uses. This is made harder due to extremely erratic rainfall during monsoons which is a common climate trend in India. Monsoon arrives in Rajasthan at the end of month of June and can stay until early October. The distribution of rainfall is quite poor and hence cannot be relied upon by farmers for irrigating their lands (Dutta & Chaudhuri, 2015; Everard, 2015). So, water capture and harvesting by lakes on the farms will be vital to these regions. Government of Rajasthan provides subsidies to farmers to create such ponds on the farm but the major barrier to this is the small size of farmlands.

All these challenges make it extremely important for farmers to adopt GAPs for cropland, water, and soil fertility management. Although, not all farmers have means or knowledge to adopt these GAPs. Hence, the focus of this paper lies in exploring the motivation of a famer to adopt GAPs as well as their challenges or barriers to not adopt GAPs. Objective I was achieved through the Phase 1. For objective II and III a field study (Phase 2) was carried out in three distinct villages (Chanod, Koliwara, and Salodariya) which are in different parts of the Sumerpur Tehsil. Fifteen farmers in each village were surveyed and interviewed.

Objective II demonstrated the rate of adoption of GAPs (as shown in Table 3). The maximum rate of adoption of GAPs achieved is 80%, which was achieved by one farmer in the Koliwara village; this same farmer is the only one with formal agriculture education, a B.Sc. in Agriculture. This further strengthens the hypothesis that being more educated, farmers tend to follow more GAPs, making agriculture sustainable. Majority of the farmers have ancestral land, and they have learnt agriculture through generations. In the farmer interviews, the major reason for not adopting new farming technologies and techniques like mulching, micro irrigation, water harvesting, etc. was lack of know-how, skilled-labour, and willingness. Farmers in Koliwara who had the financial capacity to adopt such practices declined to adopt them as it was not done ancestrally, and they refuse to make changes as well. Many farmers also admitted of being lazy to try new things as they did not want to take risks or devote their time in learning new things. Hence, the agriculture department needs to conduct more reach out programs to create awareness and motivate farmers to adopt practices that benefit their farms as well as the environment.

Objective III looks at farmers' perspective, which were each GAP is explored individually. There has been as low as 0% farmers adopting mulching as a GAP and as high as 100% farmers adopting crop rotation. Koliwara village was the only village using advance micro irrigation systems which can be attributed to closeness to the Sumerpur town as well as active engagement of agriculture department in organizing farm field trips to big farms for some farmers. Salodariya has no interaction with the agriculture department as the extension officer rarely visits the farmers hence providing no new technological information to the farmers. Chanod has active engagement with their extension officer, but due to rain-fed agriculture, it does not leave much scope for systems that require irrigation water and technology like mulching, micro irrigation, etc.

The research paper by Tey et al. (2013) on factors influencing the adoption of sustainable practices of Malaysian vegetable farmers has focused on six practices: conservation tillage, intercropping, cover crops/mulches, crop rotation, organic fertilizers/compost, and integrated pest management. The result of this study highlights the major factors that influence the adoption of sustainable practices, like geographical locations, terrain of land, ethnicity of the farmer, financial capital, number of full-time employees, habits, intention, formal education, and more. These influencing factors are similar to the results shown in Sumerpur Tehsil, where the major factors of influence to adopt GAPs are formal education, intentions, habits, ancestral learnings, and others. The paper on factors of influencing farmers adoptions of good management practices and modern rice technologies by Mariano et al. (2012) consider farmers' characteristics like gender, schooling, experience, etc., farm assets and institutional factors like farm size, distance to the nearest markets etc., extension like access to extension workers, training sessions, biophysical conditions like drought-prone area, sufficient water for irrigation etc., farmer behaviours like risk aversion, etc. Many of these same variables are considered in this paper on Sumerpur Tehsil famers and the results show that how many of these variables motivates the farmers to adopt GAPs. Kassie et al. (2013) in their paper focused on rural Tanzania's smallholder farmers' adoption of sustainable agricultural practices. The aspects that influence the Tanzania's farmers are rainfall, effectiveness of extension services, plot size, household assets, and more. Also, the paper by Pongvinyoo et al. (2014) discusses the importance of educational backgrounds of coffee farmers in Thailand, whereby formal education has positive impacts on the adoption of GAPs which is also seen in this study of Sumerpur Tehsil farmers.

Studies on determining the influencing aspects that affect the rate of adoption of sustainable practice or GAPs have similar factors that are common in all the studies like formal education, farm size, support of extension workers, habits, intention, etc. This helps to determine that the results of this study on Sumerpur Tehsil farmers which have some identical results to the other paper is how farmers around different parts of the world have coinciding challenges they face to practice good agricultural practices.

6 Conclusion

This paper tells how tasks that might seem simple and easy to do as researchers can be tough for a farmer in this region as they are not well educated and often get neglected by government officials. When research organisations make policies for sustainable land management (SLM), they are theoretically very helpful. But the on-ground application by farmers is only 50–60% due to many reasons which are explored in this paper above. Hence, one of the biggest challenges to SLM is its application by farmers. With climate change, the weather is becoming more and more erratic and unpredictable, making it a backbreaker for farmers. In 2017, this arid dry region faced major floods for which the farmers were not prepared and had trouble farming during the monsoon season. The soil of this region is very sandy, so the water percolated through it and most of them was lost. If farmers had primitively made small ponds in their farms, they could have saved water for later use. However, the major dams had recharged completely which later benefitted the farmers. In 2019, in this region, a new insect was discovered that ate away the yield before the harvest season leading to major crop loss and livelihood income was lost.

There is very less research done in this region; hence, more research on showing scientific results on the field for these GAPs might help motivate more farmers to practice GAPs. Also, research on the relationship between farmers and agriculture department/extension officers could be done to find ways to strengthen their relationship which will benefit both the groups and better prepare farmer to resist climate change.

Acknowledgements Firstly, I would like to express my sincere gratitude to my supervisor Dr John Conway from Royal Agricultural University, United Kingdom for the continuous support of my research, for his patience, motivation, approachability and immense knowledge. His guidance helped me in all the time of research and writing of this research. I could not have imagined having a better advisor and mentor for my study.
I would like to thank Nathusinghji from Sumerpur Agricultural department for helping me choose the villages for this study. I would like to sincerely thank Lacharamji Mali (Koliwara), Desharamji Mandora (Salodariya), Dineshji Kumar (Chanod) and all the farmers' who participated in the study for their support and time.
Finally, I would like to thank my husband Nandeep Mali for his constant support during this research; from making field trips to helping me edit my papers. Also thank my family, my parents and my in-laws for supporting me spiritually throughout writing this thesis and my life in general.

References

Bhatt, W. (2009). The culture of Indian migration: A theoretical and empirical analysis of Orissa and Rajasthan (1803–2003). *African and Asian Studies, 8*, 147–174. https://doi.org/10.1163/156921009X413199

Census of India. (2011). *District census handbook Nalgonda: Village and town directory: Vol. Series-09* (Issue Part XII-A, pp. 1–1058). DIRECTORATE OF CENSUS OPERATIONS, RAJASTHAN. http://www.censusindia.gov.in/2011census/dchb/2808_PART_A_DCHB_NALGONDA.pdf

Census of India. (2018). *Villages and towns in Sumerpur Tehsil of Pali, Rajasthan—Census India*. https://www.censusindia.co.in/villagestowns/sumerpur-tehsil-pali-rajasthan-604

Chancellor, V. (2007). Final report of the working group agro-climatic zonal planning including agriculture development. Volume I—Main Report. October, I. http://planningcommission.nic.in/aboutus/committee/wrkgrp11/wg11_agrclim.pdf

Dutta, S., & Chaudhuri, G. (2015). Evaluating environmental sensitivity of arid and semiarid regions in Northeastern Rajasthan, India. *Geographical Review, 105*(October), 441–461. http://content.ebscohost.com/ContentServer.asp?T=P&P=AN&K=109307974&S=R&D=eih&EbscoContent=dGJyMNHr7ESeprc4v%2BbwOLCmr0%2Bep7dSr6e4S7CWxWXS&ContentCustomer=dGJyMOzprkmurq9PuePfgeyx43zx

Dutta, D., Kundu, A., Patel, N. R., Saha, S. K., & Siddiqui, A. R. (2015). Assessment of agricultural drought in Rajasthan (India) using remote sensing derived vegetation condition index (VCI) and standardized precipitation index (SPI). *The Egyptian Journal of*

Remote Sensing and Space Sciences, 18, 53–63. https://doi.org/10.1016/j.ejrs.2015.03.006

Everard, M. (2015). Community-based groundwater and ecosystem restoration in semi-arid north Rajasthan (1): Socio-economic progress and lessons for groundwater-dependent areas. *Ecosystem Services, 16*, 125–135. https://doi.org/10.1016/j.ecoser.2015.10.011

GOR. (2018). *Agro-climatic zones: Agriculture portal*. http://www.agriculture.rajasthan.gov.in/content/agriculture/en/Agriculture-Department-portal/departmental-introduction/agro-climatic-zones.html

GoR. (2014). *Rajasthan state action plan on climate change*. http://www.moef.nic.in/sites/default/files/sapcc/Rajasthan.pdf

Hussain, M. M. (2015). Agro-climatic zones and economic development of Rajasthan. *International Journal of Humanities and Social Science Invention ISSN (Online), 4*(2), 2319–7722. http://www.ijhssi.org/papers/v4(2)/Version-2/G0422050057.pdf

Illinious Library. (2018). *Farmer's markets and sustainable agriculture—seasonal guide: Summer—LibGuides at University of Illinois at Urbana-Champaign*. Seasonal Guide: Summer: Farmer's Markets and Sustainable Agriculture. http://guides.library.illinois.edu/c.php?g=651790&p=4571863

Jellason, N. P. (2017). *Environmental Challenges and Linkages to smallholder agriculture in sub-Saharan African drylands: Implications for food security*.

Kassie, M., Jaleta, M., Shiferaw, B., Mmbando, F., & Mekuria, M. (2013). Adoption of interrelated sustainable agricultural practices in smallholder systems: Evidence from rural Tanzania. *Technological Forecasting and Social Change, 80*(3), 525–540. https://doi.org/10.1016/j.techfore.2012.08.007

Mariano, M. J., Fleming, E., & Villano, R. (2012). Factors influencing farmers' adoption of modern rice technologies and good management practices in the Philippines financial inclusion and household welfare: A multi-dimensional perspective view project IndoBeef research program view project author's pers. *Agricultural Systems, 110*, 41–53. https://doi.org/10.1016/j.agsy.2012.03.010

Mathcraker. (2018). *Kruskal-Wallis test calculator—MathCracker.com*. https://mathcracker.com/kruskal-wallis#results

PGGC11. (2018). *Soil Types—1 | PGGC11*. Retrieved March 12, 2018, from http://www.gc11.ac.in/

Poisot, A. S., Speedy, A., & Kueneman, E. (2004). Good agricultural practices—a working concept. In *Background paper for the FAO internal workshop good agricultural practices*. Food and Agriculture Organization, October. http://www.fao.org/prods/gap/Docs/PDF/5-GAPworkingConceptPaperEXTERNAL.pdf

Pongvinyoo, P., Yamao, M., & Hosono, K. (2014). Factors affecting the implementation of good agricultural practices (GAP) among coffee farmers in Chumphon Province, Thailand. *American Journal of Rural Development, 2*(2), 34–39. https://doi.org/10.12691/ajrd-2-2-3

Sharma, H., Burark, S. S., & Meena, G. L. (2015). Land degradation and sustainable agriculture in Rajasthan, India. *Journal of Industrial Pollution Control, 31*(1), 7–11. http://www.scopus.com/inward/record.url?eid=2-s2.0-84938691082&partnerID=40&md5=a0b6bb645855a5156501b5c0918a0eaf

Solanki, A., & Patel, S. (2015). Diversity in urbanism characteristics of medium size twin towns: A case of Sheoganj and Sumerpur. In *XVI N-AERUS Conference*, (vol. 1, pp. 45–63). https://www.researchgate.net/publication/289977156_Diversity_in_urbanism_characteristics_of_medium_sized_twin_towns

Tey, Y. S., Li, E., Bruwer, J., Abdullah, A. M., Brindal, M., Radam, A., Ismail, M. M., & Darmham, S. (2013). The relative importance of factors influencing the adoption of sustainable agricultural practices: A factor approach for Malaysian vegetable farmers. *Sustain Science, 9*, 17–29. https://doi.org/10.1007/s11625-013-0219-3

Triphobo. (2018). *Best time to visit Sumerpur, India | Weather | Season | Months: TripHobo*. https://www.triphobo.com/places/sumerpur-india/best-time-to-visit

Improving Sustainability in Indian Cities Through Expansion of Edible Green Spaces: Exploring Million Plus Cities of Bengaluru, Hyderabad and Ahmedabad

Swati Kothary and Aditi Mali

Abstract

Current food systems for cities are becoming unsustainable, especially with rapidly increasing population to feed. Food for city dwellers travels long distances to reach while there are empty spaces within the city that can be utilised to grow food. Cities generate gallons of wastewater and tonnes of solid waste that pollutes the environment resulting in climate change. Farming has the potential to create nutritious food from this waste. Urban agriculture provides fresh food to the city and generates employment through this process in multiple sectors like farming, food processing, packaging, etc. This study highlights the innovations in idea and practice of urban agriculture and supports sustainable environmental management, preserving local flavour and essence of culture. This study focuses on improving sustainability through the expansion of edible green spaces in the three million plus cities of India—Bengaluru, Hyderabad and Ahmedabad. The role of all the stakeholders in this process like the municipal corporation, urban development authority, horticulture department, markets, rooftop farmers, urban growers, cooperative societies and others are examined to develop a holistic understanding of the existing practice in food systems. *Result*: Bengaluru's urban agriculture movement is people driven; in Hyderabad, it is the horticulture department that has promoted urban agriculture in the city; whereas in Ahmedabad, the urban agriculture movement is mainly through efforts of few individuals and small groups that are engaged in rooftop farming. *Conclusion*: Sustainability through practice of urban agriculture helps increase edible green spaces in a city. This can be achieved through people-driven movement; government incentives and programmes; and individual efforts. There is a need to create similar green movements in other cities to increase the edible green cover.

Keywords

Edible green spaces • Urban agriculture • Sustainability • Green areas • Environmental preservation • Kitchen gardening • Bengaluru • Hyderabad • Ahmedabad

1 Introduction

According to the most recent estimate by the United Nations, there are 7.3 billion people on Earth, and this may reach to 9.7 billion by 2050. The World Bank data suggests that in 2018 the world population living in urban areas crossed 55%. This rapid urbanisation and urban growth along with rising incomes in the developing countries has triggered dietary changes such as eating more protein and meat. This is driving up the global food demand. Food demand is expected to increase from 59 to 98% by 2050.

According to Census of India (2011) data series, India's population rose by 181 million people over the last 10 years to 1.21 billion. The population residing in urban areas in India, according to 1991 census, was 11.4%. There was an increase in this population by 28.53% according to 2001 census, and it crossed 30% by 2011 census, taking the total to 31.16% (Census of India, 2011). The availability of resources is always limited in developing country like India, which is densely populated. This results in the shortage of the most basic resources like food. Rapid urbanisation creates tremendous pressure on cities with increasing demand for better infrastructure, employment, food security, nutrition, health and environmental management. The urban communities hence face challenges in terms of urban growth, changing lifestyles, food preferences and food security. Amongst all of these, providing food security and

S. Kothary · A. Mali (✉)
Institute of Architecture and Planning, Nirma University, Ahmedabad, India

C. Alalouch et al. (eds.), *Towards Implementation of Sustainability Concepts in Developing Countries*,
Advances in Science, Technology & Innovation, https://doi.org/10.1007/978-3-030-74349-9_9

nutritional dietary food in the urban areas is crucial. With ever increasing food demand, the farmers have to 'increase the crop production, either by increasing the amount of agricultural land to grow crops or by boosting productivity in the existing agricultural lands through fertiliser, irrigation and adopting new methods like precision farming. Many other factors, from climate change to urbanisation to lack of investment, will also make it challenging to produce enough food' (Maarten Elferink, 2016). Climate-change-driven obstacles like water scarcity, rising global temperatures, urban heat island effect and extreme weather have severe long-term impact on crop yields and are expected to affect many major agricultural regions. These challenges make sustainable environmental preservation and management a priority.

These days urban greening and urban farming have acquired new significance in tackling food demand. This is done by creating green cover, green buildings, change in construction techniques and increased edible green spaces through urban agriculture with community and rooftop gardens, greenhouses, container gardening, indoor and vertical farms and edible green walls. This paper focuses on improving the sustainability through development of edible green spaces. Formation of edible green spaces through urban agriculture helps in achieving sustainability in a city, and this makes it a pivotal factor in creating a resilient city.

This paper aims to study, explore and understand the development of edible green spaces through urban agriculture in the 3 million plus cities of India, i.e. Bengaluru, Hyderabad and Ahmedabad. These three cities are selected for the different approaches adapted in each to achieve the common goal of increase in urban agriculture.

2 Methodology

Urban agriculture is a new phenomenon gaining momentum in Indian cities. There are no concrete secondary sources through which data and information is available. To collect information and build an understanding on these cities, it was very crucial to conduct surveys. Data was gathered from primary and secondary sources. The analysis uses both quantitative and qualitative information gathered through surveys and interviews. Table 1 shows the resources and various tools that facilitated the conduct of this study conducted from July 2019 to November 2019.

For this study, it was crucial to assess and analyse the existing food system of all the three cities, thus city-level agro-markets were visited and studied. Buyers and sellers of all the important wholesale agri-produce markets in these three cities were surveyed to understand the distance travelled by agri-produce to reach the consumer. The food kilometres of daily perishable goods like different fruits, vegetables, meat, etc., was calculated. The market survey and analysis helped in establishing the need for urban agriculture practice to expand the edible green spaces in the cities. Organic cafes and stores were surveyed to build an understanding on the awareness of food consciousness of the citizens. This analysis supported in understanding the pattern of increasing urban agriculture movements in these cities. It also suggested the increasing awareness of citizens to eat chemical-free and healthy food. Individual urban farmers living in different house typologies across the cities were interviewed and surveyed to understand their motivation and challenges in practicing urban agriculture. The State Horticulture Departments of Bengaluru and Hyderabad play an important role in motivating and promoting urban agriculture by supporting this movement. Officials from these departments were interviewed to understand methods of their support to the citizens. The urban development plans and policies were also studied to check provisions for public spaces allotted for this type of land use. Agribusinesses in these cities were also interviewed to understand their role in providing various products and services that facilitate home growing. This study was conducted as part of 'URBAL' the urban agriculture studio for final year students of Bachelor of Planning, Institute of Architecture and Planning, Nirma University, Ahmedabad, India.

Table 1 Research Methodology approach matrix

	Qualitative	Quantitative
Primary	• Field observations and notes • Urban farmers' interview and focus group discussions (FDGs) • Data from horticulture department Agribusiness interview	• Urban farmers' survey (58) • Market survey Buyer (71) Seller (85) • Organic café (20) • Stores survey (37)
Secondary	Information from government organisations like • Municipal corporation • Urban development authority • Horticulture department	Data from government organisations like • Municipal corporation • Urban development authority • Horticulture department • Agribusiness websites

3 Major Findings

With population of 8.4 million and area of 741 km^2, Bengaluru is the third-largest city in India followed by Hyderabad with population of 6.73 million spreading in an area of 625 km^2. Ahmedabad with population of 5.57 million is spread in an area of 464 km^2 (Census of India, 2011). These cities are chosen for their distinctive character and nature of expansion in edible green spaces through urban agriculture.

Stakeholder network as shown in Fig. 1 displays that each stakeholder has a role in improving and expanding edible green spaces in the city. This section details out the role of these stakeholders in each of the three cities and their contribution in changing the cityscape through urban agriculture and how they play a major role in the expansion of edible greens. The disbursement of power to implement policies related to construction of built form and growing greens is assigned to different departments. Changing priorities and increasing pressure on urban areas with increasing population are resulting in increased food demand. Global climate change is also contributing to the growing awareness of planting edible greens. Though the movements did not start with a focus on reducing food miles, but the result of growing greens in nearby areas is contributing in reducing food miles even though presently the number is miniscule. A systematic method of implementation and increasing awareness can substantiate this effort.

3.1 Bengaluru

Bengaluru is the capital city of the state of Karnataka located in the southern part of India. Always known for its gardens, the city has now become the Silicon Valley of India. Bengaluru is also known for its lakes and pleasant weather (Holidify, 2019). This city is located in the eastern dry agro-climatic zone and gets an annual rainfall ranging from 830 to 970 mm with prominent rainy months as June, July, August and September keeping the climate as tropical wet and seasonally dry. Light showers keep the temperature moderate throughout the year (India Meteorological Department, 2019). Climate of Bengaluru is favourable for growing most of the horticulture crops native to this region making it ideal for rooftop farming in urban areas (Chhabra, 2019).

Major motivation for urban agriculture movement in this city was unhealthy food supply and environment impacts of existing food systems. The air and water quality in Bengaluru started deteriorating in the last decade. A total of 90% of Bengaluru's lakes have been affected by sustained inflow of untreated sewage and industrial effluent. There were

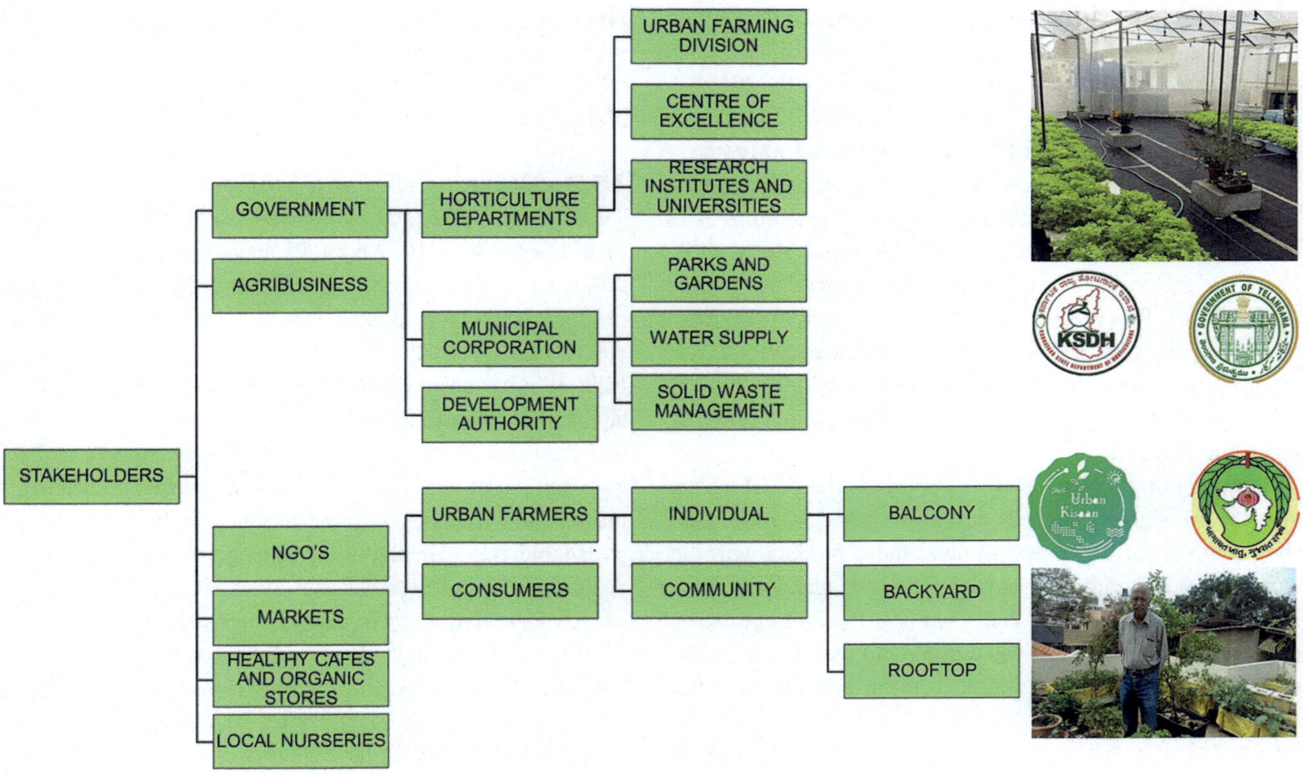

Fig. 1 Stakeholder network diagram

incidences of city's lakes frothing, catching fire thus risking large neighbourhoods. This foam was also reported to be toxic and carcinogenic. Farmers in the surrounding areas of Bengaluru started excessive use of pesticides and chemicals that led to reduced trust in local produce and organic labels. Local media started reporting cases of the toxic environment during 2005, and this increased people's interest to grow their own food. Eating clean, healthy food and helping the environment were the major factors that contributed to the rise of urban agriculture in Bengaluru (Hindustan times, 2017; Kalra, 2018).

Dr. Vishwanath introduced urban agriculture as a concept to the city of Bengaluru in 1995. He formed The Garden City Farmers Trust (GCF) in 2011 to promote urban agriculture in Bengaluru. This NGO is now associated with more than 50,000 rooftop farmers in Bengaluru. These farmers refer themselves as Organic Terrace Gardeners (OTGians). They practice chemical-free farming, and most of them also practice wet waste composting to reduce waste disposal. GCF and OTGians are two main organisations, which have contributed in increasing people's participation in urban agriculture, resulted in increased edible green spaces across the city (Hegde, 2019). Inspired by the Garden City Farmers movement, people of Bengaluru have started growing their own food on the terraces, balconies and backyards of their houses using containers that help in multi-usage of the spaces. Schools have started growing food on their campus premises. Many urban farmers in eastern and southern peri-urban areas are associating with agribusinesses and directly connect with the consumers. (Viswanath, 2019).

3.1.1 Wholesale Agri-Produce Market Analysis

Farmers from the surrounding villages and other parts of Karnataka and India sell their produce to Agriculture Products and Marketing Committee (APMC) and Horticultural Producer's Cooperative Marketing and Processing Society (HOPCOM). This produce is further distributed to two major wholesale markets, KR market and Russell market, located in the city centre. The goods are then auctioned and distributed further to smaller shops and vendors and then finally reach the consumer.

Nine regulated markets were surveyed, out of which two were main markets and seven submarkets. A cooperative society involved in marketing agriculture produce was also surveyed along with the markets. Though, agriculture is the main occupation of India and farming forms a large part of occupation, cities in India see a different trend. In the trading of agriculture produce, farmers form an exceedingly small number of the total people engaged. Others who have the business acumen and are indifferent to the quality of produce carry out most of the business.

Krishna Rajender market (K. R. Market) also known as City Market is operating since 1928 and is considered to be the biggest flower market of Asia. Russell market established before 1947 was created to serve the needs of British convoy living in Bengaluru. This has transformed into a neighbourhood market and sells almost everything. Johnson Market earlier known as Richmond town market was set up in 1929 and named after a former British civil servant Johnson. It is mainly known for selling higher quantity of beef and meat with declining sales in vegetables and fruits.

Food grown in farms around Bengaluru and other areas like Mangalore, Chikmanguluru, Mysore, Ooty and Chennai with distance ranging from 150 to 2500 kms is supplied to the city through Horticultural Producers' Cooperative Marketing and Processing Society (HOPCOM). The markets namely K. R. Market and Electronic Market distribute this produce to malls, super markets, Russel Market, Johnson Market and informal vendors situated within the range of 5–12 kms. The produce then travels a distance of 5–10 kms through consumers for consumption. A network of trucks and tempos usually carries out distribution of the stock (Fig. 2).

Wholesale markets contribute to 60% of the market share followed by retail. Demand-wise local fruits, vegetables and flowers are in demand. Restocking done daily is a common practice in the city. Percentage wise 70% respondents spend an average of INR 500 to 1000, followed by 20% spending more than INR 1000 and 10% spending INR 500 or less on buying vegetables and fruits. Most people in Bengaluru purchase weekly stock and travel up to a maximum of 10 kms to make this purchase. On an average, the food kilometre of vegetables reaching the consumers of Bengaluru is 300 kms (Market Survey, 2019).

3.1.2 Organic Cafes and Stores

Twenty stores were surveyed across Bengaluru with most of them opened in 2010. About 5% have existed since the year 2000 also. Most of the stores surveyed stocked organic products and were located largely in upmarket residential areas. This boosts the footfall and increases their customer base. Regular customers and their specific demands are also attended to. A total of 30% of the cafes and organic stores had their own farms where they practiced organic and permaculture farming. Their products are mostly expensive as the risk in producing organic products is high due to high perishable rate. These outlets manage to make about 30% profits. Each cafe serves an average of 200 customers weekly. The frequency of visits is high and repeat customers visit every 2 to 3 days due to higher perishable rate of the produce. These cafes and organic stores are most popular amongst the age group of 31 to 50 years owing to more awareness about health benefits and higher buying capacity. Corporate sector also forms a large client base for these products as they make a very good gifting option. This indirect marketing has also helped cafes to increase their

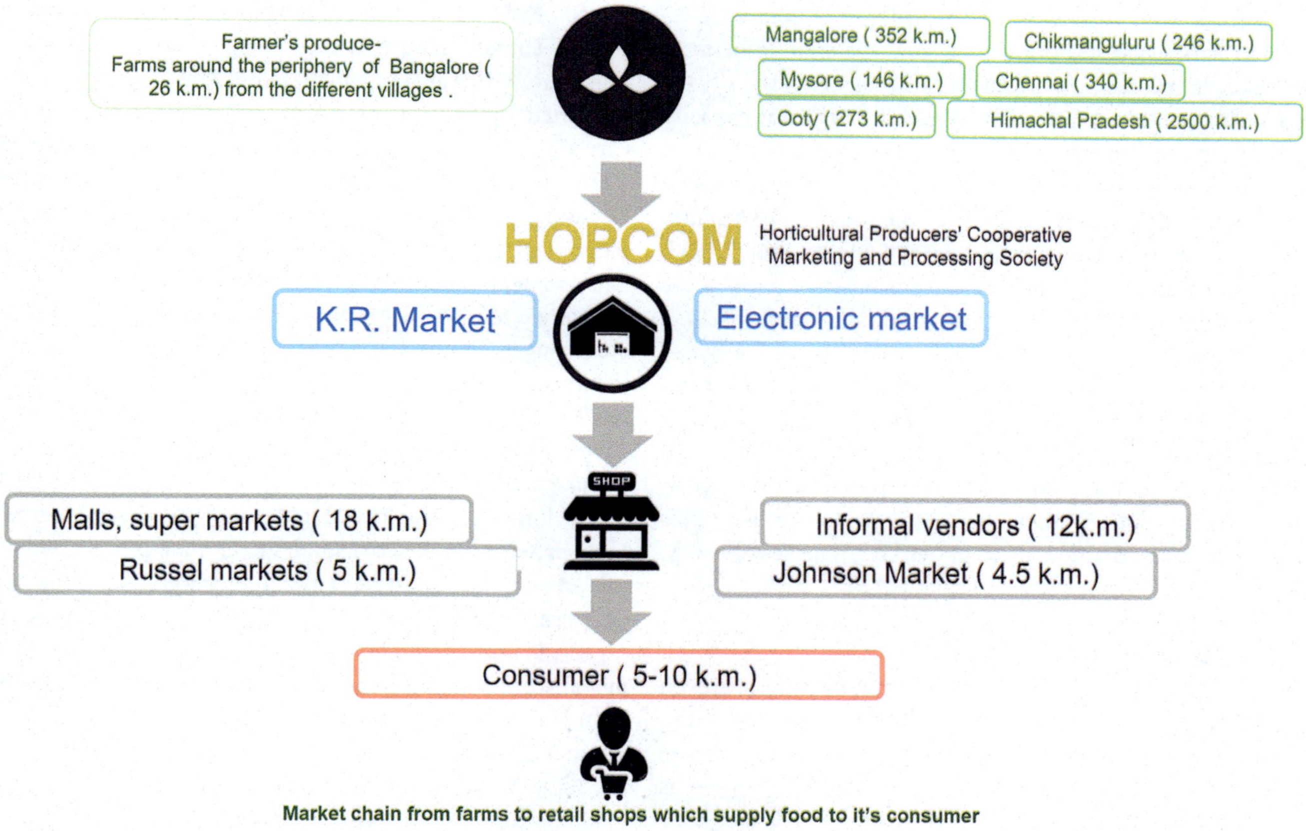

Fig. 2 Market chain from farms to retail shops in Bengaluru

customer base. One of the major observations from the survey was that these outlets do not engage in waste segregation, and only 15% of the cafes were sending their waste for composting. It was suggested to them to take proper care of their waste. Only then the entire chain of activities can be truly considered sustainable.

Only 30% of the stores were found to import organic inputs from various parts of India and about 40 different countries. Such less percentage of stores importing products is an indication that they source products from the city itself. This is also reflected through less transportation cost and higher profit margins. Most common vegetables grown at home are spinach, coriander, mint, basil, tomatoes, capsicum and eggplants. (Cafes & Store Survey, 2019).

3.1.3 Agribusiness and Start-Up

Agribusinesses make system of growing greens stronger by providing services and products to support the urban farmers. Some examples are shared for better understanding. 'My Dream Garden' provides complete solutions to make a garden. They provide potting mixes, magic planter boxes, seed sowing charts, pest managements, grow bags and tools. They complement these products with services like workshops on gardening, gardeners' team, research and maintenance and designing (My Dream Garden, 2019). Woolly Farms are building farming techniques like hydroponics, vertical farming, mobile technology and data sciences to help people become self-sustained in growing their vegetables and herbs. They use vacant terraces and space to set up small green houses and assist people in growing their own produce and sharing the excess quantity. They are connecting people through Woolly digital cloud platform. Multiple rooftops and farms where they grow different crops within a community including peri-urban areas are connected, and their product collected and redistributed to provide all kinds of vegetables and herbs to people on the cloud (Wolly Farms, 2017). Farmizen is an app-based agribusiness where a person can rent a mini farm of 600 sq. ft. and grow seasonal veggies of their choice in the nearest community farm. This agribusiness gives an opportunity to people who do not have space to grow or time to grow their own veggies but wish to still grow. This way a farmer in peri-urban and urban area who is cultivating in the community farm gets employment and an assured income to retain the land for agriculture instead of leaving it barren or selling it (Farmizen, 2017).

Agribusinesses have been growing over the years. Maximum numbers were added between the years 2013 and 2016. A total of 41% agribusinesses have their own farms to grow agri-produce and 15% of them bring the organic vegetables from farmers in nearby villages. A total of 22% of the agribusinesses bring their raw material from local forests, horticulture department and other parts of the country. A total of 39% of agribusiness organise capacity building programmes for their staff, whereas 22% do not find it necessary. A total of 39% agribusiness owners were clueless about concept of capacity building. In Bengaluru, 18% firms were creating awareness and adding skill through training programmes. Most of the customer base for these agribusinesses is online marketing forming 48% of their business followed by household customers and organic stores. Out of 12 agribusiness surveyed, 58% provided finished products as food along with personal hygiene products, 25% provided seeds and helped in setting up farms and supported in its maintenance and 16% provided seeds, grow bags, home kits, potting and composting bins. Two agribusinesses provided all solutions (Agribusiness Survey, 2019).

3.1.4 Urban Farmers

A total of 53 rooftop farmers of Bengaluru were surveyed for their experience on growing edible greens. Most common produce through rooftop farming is vegetables (55%) followed by fruits (25%) and herbs. Rooftop farming is commonly practiced in an area of 50 m^2. There are some rooftop farmers who are practicing in 51–150 m^2 area also. Grow bags are most common for rooftop farming with 53% of the survey respondents using them; only 5% are using hydroponics or aquaponics. Other practices include grow bags, earthen and cement pots. People are using Jeevamrut made from cow dung and jaggery as pesticide and compost as fertiliser. Only 6% were found to be using chemical fertiliser or pesticide. The survey also suggests that the people who produce 1 and 5 kg yield are using only Jeevamrut, whereas others are using compost and neem cake. The people who are producing 4, 5 and 10 kg yield are using grow bags, and this shows the popularity of container gardening in urban area. This also reflects that a combination of grow bags and Jeevamrut produces best result in urban farming. Looking at the trend of growing urban greens on rooftop, it was observed that earlier it was only vegetables that were grown and gradually people moved to growing herbs. Some have also added fruits and beans depending on the area of roof available to them. This can be correlated to the increasing awareness to health, wellness and changing food preferences. Interestingly, in contrast to cafes and stores, 86% of the rooftop practitioners use their waste for composting. These urban farmers save 300 food kilometres by producing their own food. The large amount of organic cafes and stores directly reflects in the high awareness in consuming chemical-free food in Bengaluru. (Rooftop Farmers' Survey, 2019).

3.1.5 Role of Stakeholder in Developing Edible Urban Green Areas

Bengaluru's urban agriculture movement is people-driven supported by non-governmental organisations like Garden City Farmers Trust (GCF) and Organic Terrace Gardeners (OTGians). The other major stakeholders are Horticulture Department of Karnataka and many agribusinesses like Farmizen, My Dream Garden, Woolly Farms, Oncrop Agro Sciences and more.

Horticulture department provides basic beginner kit consisting of grow bags, bio-mix and tool kit along with monthly training programmes. Garden City Farmers Trust (GCF) organises quarterly event called Oota from your Thota (OFYT). It is a knowledge-sharing platform where urban growers come together to share seeds, raw materials and techniques. An informal interaction with rooftop terrace farmers helped in understanding the farming process and the role of seeds, compost, design and crops growth. They also organise workshops on composting and organic gardening especially for school children and freshers to orient them to this field. Expert lectures along with exhibition stalls by farmers, agribusiness, homemakers, etc., help in awareness. They also proposed progressive and successful urban farmers to take leadership roles. Social media is playing an important role in promoting urban agriculture in Bengaluru. Urban farmers are constantly in news through Facebook and WhatsApp groups. There are area-wise WhatsApp group where the farmers share knowledge and provide solutions and services. They use this medium to barter the raw materials, machinery and techniques reducing their monetary cost of operations. Agribusinesses further strengthen their system by providing services and products, thus making it easy for all kinds of consumers; the ones who have space and want to set up their own farm; the citizens who have empty space that can be let out to grow food; and those growers who do not have space and through app-based solutions grow in peri-urban areas. All these stakeholders together have created an empowering environment for urban farmers in Bengaluru, which is improving sustainability and expanding edible green spaces through urban agriculture.

3.2 Hyderabad

Hyderabad is the capital of the Telangana State located in the southern part of India. Earlier known as the city of pearls, city of Nawabs and for world famous biryanis, it is now more popular for its IT industry, hosting Google and Microsoft headquarters. Hyderabad falls under the southern Telangana agro-climatic zone which is a tropical wet and dry climate, with maximum temperature of 40 °C and minimum temperature of 13 °C. The average annual rainfall ranges from 850 to 900 mm, occurring mostly in the months of July, August, September and October (India Meteorological Department, 2019). The microclimate of this city is very suitable for cultivating both tropical and temperate vegetables (Horticulture Department TS, 2018).

Urban agriculture became popular in Hyderabad mainly due to availability of open spaces for growing vegetables and increasing pollution from traffic and industry. Most buildings had terraces that generally were not used, and they received sufficient quantity of sunlight, ideal to sustain plant life and propagate its growth (Madhusudan, 2019). People in the city initially practiced urban agriculture using polluted and later made adaptations in the farming by changing crop choices and irrigation methods. Urban farmers cultivating along the Musi River of Hyderabad showed a good example, when the river water used for irrigation became polluted by city waste. Within a short period of time, these farmers shifted from vegetables and rice growing to fodder and tree crops. This change brought awareness in people about the polluted food they were eating resulting in the shift to growing their own chemical-free food on terraces and backyards.

In the year 2012–2013, The Department of Horticulture, Government of Telangana State introduced 'Vegetable Growing in Urban Area' scheme under the RKVY scheme (Rashtriya Krishi Vikas Yojana which means National Farmers Welfare Programme) (Horticulture Department TS, 2018). Under this scheme they formed an 'Urban Farming Division' as part of the Department of Horticulture, Government of Telangana located right in the centre of Hyderabad city. This is a unique, one of its kind division in the whole country that works on urban farming exclusively. They create awareness through wide publicity in newspapers, local television channel, YouTube and social networking sites like Facebook and WhatsApp. The idea is to bring awareness about fresh home-grown vegetable health benefits, delicious taste and safety from chemicals. They also promote kitchen gardening as an outstanding way to spend quality time with one's family and how it is a beneficial fun activity for children. The department continuously organises awareness and training programme and provides low-cost grow bags and seeds. To further convince the citizens to adopt urban farming in their homes; the department grows vegetables in many innovative ways on its premises. There are over 8000 farmers practicing urban agriculture in the city since this scheme was introduced in Hyderabad.

3.2.1 Wholesale Agri-Produce Market Analysis

Farmers' from the surrounding villages of Hyderabad and nearby districts in the range of 100–200 Kms sell their farm produce directly to wholesale markets of Hyderabad called the Rythu markets which are located in different parts of the city. Farmers also sell their produce to village commission agents who resell these goods to collector/transport agent. Produce is again sold to wholesalers at Madannapet Market that is one the largest wholesale market in Hyderabad. Smaller informal vendors and store owner buy these from the wholesale markets like Maddapet or farmers market like Kothapet Rhythu, Eraganda Rthythu or other similar markets.

In the city of Hyderabad, eight regulated markets were surveyed to understand the food supply chain and business. Along with them, informal markets of Shapur vegetable market, Madannapet Vegetable and Fish Market were also surveyed. Lingampally, a commercial place for Nawab's and fruits garden in ancient times, has many commercial shops, restaurants, fruit juice centres and vegetables market. Rythu market or bazaar is a type of farmers' market in Andhra Pradesh and Telangana State. The Governments of Andhra Pradesh and Telangana run it for farmers with small landholdings. Shahpur, a neighbourhood market in Shahpur Nagar, houses wholesale retails for vegetables. Monda Market established more than 100 years ago to cater to resident British army units is a vegetable market located at Secunderabad. This market located approximately at 500 m from Secunderabad Railway Station is one of the largest wholesale and retail markets in Hyderabad.

In the city of Hyderabad, selling of agriculture produce mostly happens through wholesale markets followed by retail and informal. Vegetables are grown and traded in the city. Fish and meat are also common commodities marketed through wholesale markets. Tempos are the most preferred mode for transportation for goods. Rickshaws also carry items from one place to another around the city. A total of 90% shopkeepers do daily restocking of their goods in the city. Average amount spent by buyers in the city is INR 500 to 1000 (70%), 20% buyers spend less than INR 500 on their purchase and very few about 10% spend more than INR 1000. Buyers spend weekly and only travel up to 5 kms for their daily purchase (Fig. 3). On an average, the food kilometre of vegetables arriving to the city consumers is about 300 kms (Market Survey, 2019).

3.2.2 Organic Cafes and Stores

From the 38 cafes and stores surveyed, more than half of them have come up in the last five years, though 27% opened between 2010 and 2015. Almost 60% of the cafes and stores surveyed were organic in nature. They are mainly located in upmarket residential areas owing to repeat demand, affordability and increasing frequency of customers. A total of 30% of the cafes have their own farms where they practice organic and permaculture farming. These outlets easily make up to 30% profit despite high perishable rate. Very few responded about the number of customers they serve weekly. An approximate number suggested by few is about 225 people per week. Due to less

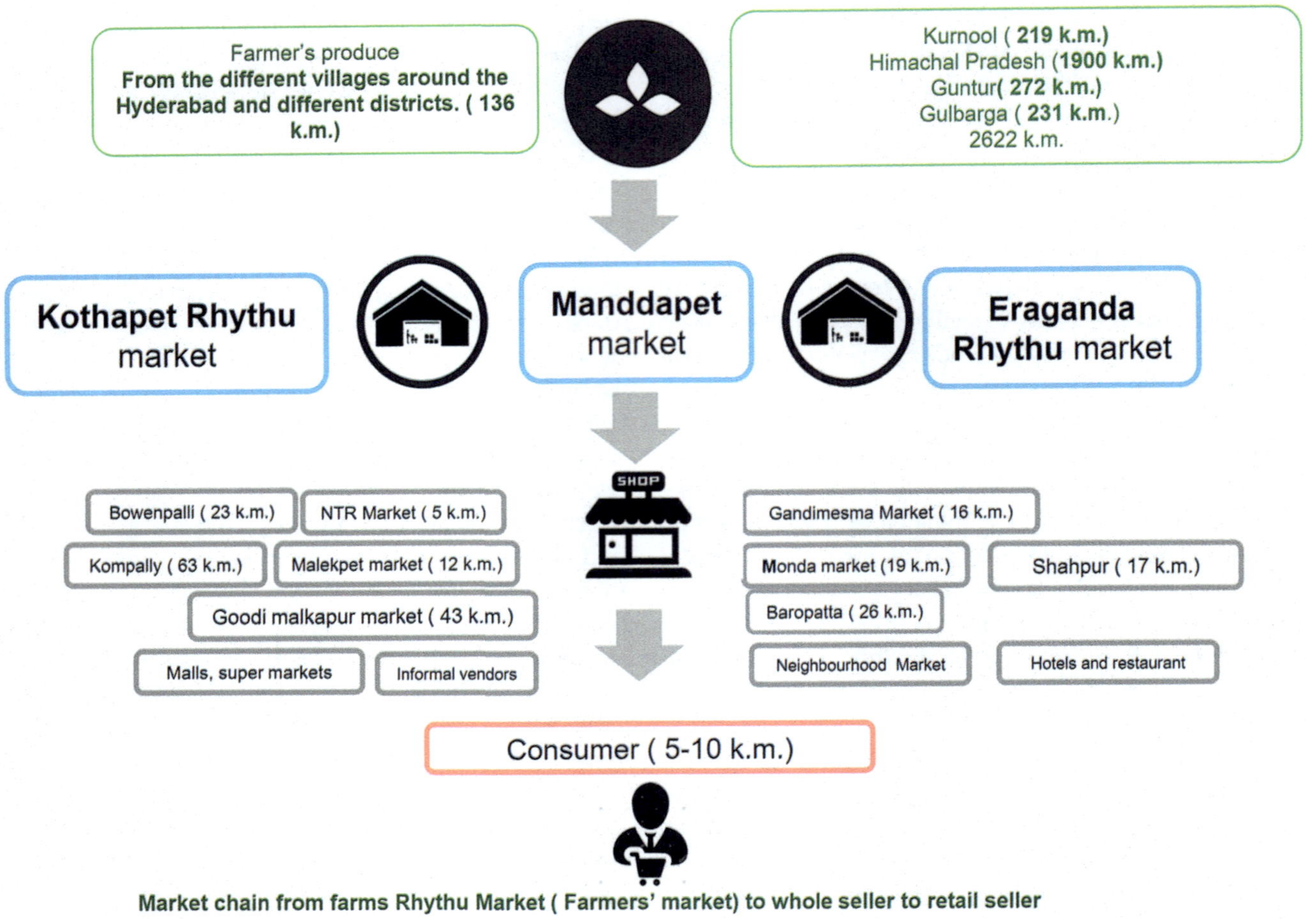

Fig. 3 Market chain from farms to retail shops in Hyderabad

shelf life, the frequency of customer visits is three times in a week with the age group largely between 31 and 50 years. Unfortunately, most of the cafes are not practicing composting which is easily possible from the type of waste generated.

Most common produce grown is vegetable like spinach, coriander, mint, basil, tomatoes, capsicum; eggplant can also be easily grown at homes. Increase in this activity will reflect in minimising of carbon footprint that is generated due to travelling of food. A total of 45% cafes and stores in Hyderabad are sourcing the produce from farms within the periphery of the city and not importing organic inputs. Some products are sourced from Delhi, Northeast, Rajasthan, Chennai, Assam, Spain (Eastern Logistics Hub), Australia, New Zealand and South Africa. (Cafes & Store Survey, 2019).

3.2.3 Agribusiness and Start-Up

Agribusiness supports in promoting urban agriculture in Hyderabad by providing services and useful products. 'Home Crop', for example, provides complete solutions to kitchen gardening ranging from grow bags, kits and maintenance services to gift bags that have a starter crop growing kit as a gifting option. Their other product is the smart grow bag with handles made from light special textile material. Available in many shapes and sizes, they are quite easy to carry and move around. They also conduct School Crop Programme and create awareness through workshops with school children and staff promoting healthy eating and growing habits (Homecrop, 2019). The other such agribusiness is Urban Kisaan that sells home kits as well as vegetables from their rooftop automated greenhouse farm on the seventh floor located in prime location of Hyderabad. Through this vertical farm, they grow leafy greens by hydroponics systems. They also sell small units of hydrponics systems commercially (Urban Kisaan, 2019). Farmizen also operates in Hyderabad in similar capacity as in Bengaluru (Farmizen, 2017).

Though some initiatives were taken between the years 2004 and 2007, maximum agribusinesses were established between 2013 and 2016. A total of 41% agribusinesses have their own farms to source raw material and 15% of the agribusiness source the organic vegetables from farmers in nearby villages. A total of 22% of the agribusinesses source their raw material from horticulture department, local forests and sometimes from other parts of India. A total of 50% of agribusiness houses believe that training helps in increased and improved production, but only 39% organise capacity building for their staff. A total of 22% did not find it necessary, and 39% were also clueless about initiatives like these. Most of the customers for these agribusinesses are households, forming 41% of their business followed by online markets and organic stores. In Hyderabad, 25% agribusiness surveyed, provided home kits, grow bags, potting and composting bins along with help in setting up farms. They also continued supporting further by providing fertilisers, pesticides and regular maintenance. (Agribusiness Survey, 2019).

3.2.4 Urban Farmers

Traditionally, vegetables are the most frequent produce from rooftops irrespective building age. Most frequent practice of growing urban greens takes place in flats where area available for farming is up to 50m^2. A total of 75% of the respondents were growing edible greens in area of this range. Grow bags, tarolines and earthen pots are commonly used techniques to grow greens. Use of Jeevamrut improves quality and compost helps to increased quantities of the produce. Most of the growers practice composting as it helps in multiple ways from saving manure cost to disposal issues. These urban growers save 300 food kilometres by producing their own food. The growing number of organic cafes and stores reflects the growing of awareness of consuming chemical-free food in Hyderabad (Rooftop Farmers' Survey, 2019).

3.2.5 Role of Stakeholder in Developing Edible Urban Green Areas

Hyderabad's urban agriculture movement expanded due to good governance in the state. The government as mentioned above has a unique urban farming division through which they initiated the urban agriculture movement. The city development authority and agribusiness in Hyderabad promoted urban agriculture further with support from the horticulture department as the main driving force.

Horticulture department provides a complete beginners kit with grow bags, seed packets with 12 varieties of vegetable seeds, soil mix, neem cake (25 kg), 1 curry leaf plant, neem oil, tool kit and poly feed. This kit is available with a subsidy of 50% provided by the government. The department officials mentioned that they can provide free kit, but the idea is to inculcate responsible practice of doing kitchen gardening in the people. They conduct six days' workshop on composting, water management, pest and disease control management. During these six days, they train the user and equip them with the above-mentioned kit to successfully practice urban agriculture in their homes/offices. The workshops are conducted in the main building of the department where they are practising urban agriculture in containers that are upcycled from waste products like industrial drums. They also provide training in soil-less farming through techniques like deep-water culture, hydroponics and controlled environment and fertigation with automation unit. One farmer can avail the kit provision only once in a lifetime but can come back for training unlimited number of times. The staff is very friendly and also available on social media platforms like WhatsApp or over a phone call. The department staff also goes for a time-to-time inspection to a farmers' houses to ensure 100% use of the kit which the farmer has been provided. They have also formed area-wise WhatsApp groups to connect the people living in same localities to create support for each other by sharing resources like knowledge, seeds and excess goods, to name a few. Successful farmers were included in providing training as they also inspire others in setting up an urban farm (Begum, 2019; Chandrashekar, 2019; Horticulture Department TS, 2018).

Good governance is reflected in the Horticulture Department and the Hyderabad Metropolitan Development Authority (HMDA) whose joint efforts have made urban agriculture vision a reality. HMDA Metropolitan Master Plan 2031 for the Hyderabad City has conservation zones within the city and the peri-urban areas. These conservation zones are green belt buffer zone where urban forestry and agriculture are permissible. Many areas located in this zone practice urban agriculture and strengthen the resources for edible green spaces in the city.

3.3 Ahmedabad

Ahmedabad is one of the major cities of Gujarat situated in the western part of India. The city is known for its old textile mills, prominent educational institutes, and is an UNESCO world heritage site (Ahmedabad Municipal Corporation, 2017). It is a part of the North Gujarat agro-climatic zone with arid to semi-arid climate keeping it hot and dry. Rainfall received is around 735 mm per annum. (India Meteorological Department, 2019).

Urban agriculture is still a very new concept in Ahmedabad. There are few collective efforts done by groups like the Gujarat Vidyapeeth, Tree Walks and individuals like Jagruti Bhatt, Bhavna Shah, Anuj Patel and Siddharth Rana, to name a few. They practice and preach methods to do urban agriculture in the city. They all are practicing kitchen gardening in their homes, at balcony, rooftop and backyards. They started doing urban agriculture to adopt a more sustainable lifestyle and challenged their gardening hobbies by converting the greens into edible greens. There is a strong culture of gardening in the city, and the efforts to convert greens into edible greens is slowly picking up.

3.3.1 Wholesale Agri-Produce Market Analysis

Agricultural Produce Market Committee (APMC) has the biggest wholesale markets in Ahmedabad. There are many APMC markets in Ahmedabad that markets variety of crop produce. Farmers from villages and other states directly or through agents sell their produce at APMCs. From the wholesale markets, the produce is resold to other major private wholesale markets around the city like Kalupur and Jamalpur markets that have special allocation for fruits, vegetables and flowers. Retailers from Ahmedabad and other surrounding areas come to these markets daily to collect stocks, which are further sold to informal and formal shops of the region. It is a common sight to find large informal markets on the streets at neighbourhood level all over the city in the evening. Vegetable vendors also sell their produce on hand driving cart/lorries directly at the neighbourhood level.

A total of 19 regulated markets were surveyed in Ahmedabad. Major informal markets surveyed were Kalupur fruit market, Kalupur vegetable market, Manek Chowk vegetable market, APMC market (Jamalpur, Juhapura) and Naroda fruit market. There are about 200 rooftop farmers actively growing greens. Naroda fruit market is functioning since 1998 under APMC and distributes fruits received from the farmers. Jamalpur flower market is an early morning flower market selling different types of flower sourced from nearby places. Kalupur wholesale market, a former financial centre of Ahmedabad, now stocks vegetables and fruits. Manek Chowk, a retail and wholesale market, stocks vegetables and fruits with other utilities (Fig. 4).

The food market is mostly wholesale, and some informal markets are also popular. Very few retail outlets operate in the trade of food produce. Most common item traded is vegetables and fruits to some extent. Unlike Bengaluru and Hyderabad, rickshaw is the most preferred mode of transportation of the food produce. Tempos are also used sometimes. Restocking of produce takes place daily. Average amount spent per visit by people in Ahmedabad is about INR 500 or less (68% respondents shared this information). This is lower as compared to the expenditure made by people in Bengaluru and Hyderabad. Very few, about 28%, spend between INR 500 to 1000. People fill supplies daily, and the distance travelled to fill these supplies is also high with more than 10 kms. The average food kilometres of the vegetables reaching Ahmedabad city is 250 kms. (Market Survey, 2019).

3.3.2 Organic Cafes and Stores

Six cafes and stores were surveyed and most of these have come up between the year 2000 and 2010. The city has no organic cafes, but some stores do sell organic products. A total of 33% cafes and stores surveyed reported that they are not importing organic inputs as against 17% that are importing their inputs. Imported inputs are from various regions of Gujarat, Bengaluru, Rajasthan, Delhi, Kolhapur and Kerala. In Ahmedabad, majority of products are imported from areas around India. The survey also suggested that food travelling to the city travels shortest food kilometres as compared to Bengaluru and Hyderabad. (Cafes & Store Survey, 2019).

3.3.3 Agribusiness and Start-Up

Agribusinesses have been running since 2016 in Ahmedabad. Ones like 'Greenman' sells all types of products from kitchen gardening like seeds, soil mixes, fertilisers, containers and more. To prompt its customers to grow vegetable and herbs, it has created its own rooftop farm as a display site with different innovative growing methods. Such efforts promote farmers to shift from growing ornamental greens to edible greens. 'Urban Khet' is an upcoming agribusiness that is providing innovative solutions to practice urban agriculture in Ahmedabad. Other plant nurseries also provide plants, seeds, soil mix and containers to the customers.

These businesses do not believe in the concept of training for people who are engaged in these activities. This reflects in the lack of awareness and initiatives for capacity building. Approximately 67% buyers are households and 33% belong to the online portals. These agribusinesses support mainly through fertilisers, pesticide and finished products. (Agribusiness Survey, 2019).

3.3.4 Urban Farmers

There are isolated cases of rooftop farming in the city. Gujarat Vidyapith, founded by Mahatma Gandhi in 1920, and some citizens use rooftops for farming. Techniques commonly practiced here are hydroponics, aquaponics and drip irrigation using compost. Paints container and raise beds made of bricks are also common in practice. Mostly fenugreek, okra, spinach, capsicum, tomatoes and all kinds of lettuce are grown. These urban growers save 250 food kilometres by producing their own food. Less number of organic cafes and stores reflect lack of awareness on consuming chemical-free food in Ahmedabad. (Rooftop Farmers' Survey, 2019).

3.3.5 Role of Stakeholder in Developing Edible Urban Green Areas

Ahmedabad's urban agriculture movement is mainly through individual efforts and some small groups engaged in raising awareness. The other stakeholders are Gujarat Horticulture Department, Parks and Garden Department of Ahmedabad Municipal Corporation, Ahmedabad Urban Development Authority (AUDA), agribusiness and local plant nurseries.

The Horticulture Department of Gujarat with its headquarters located 20 km from Ahmedabad at Gandhinagar provides seeds and manure at subsidised rates to the people

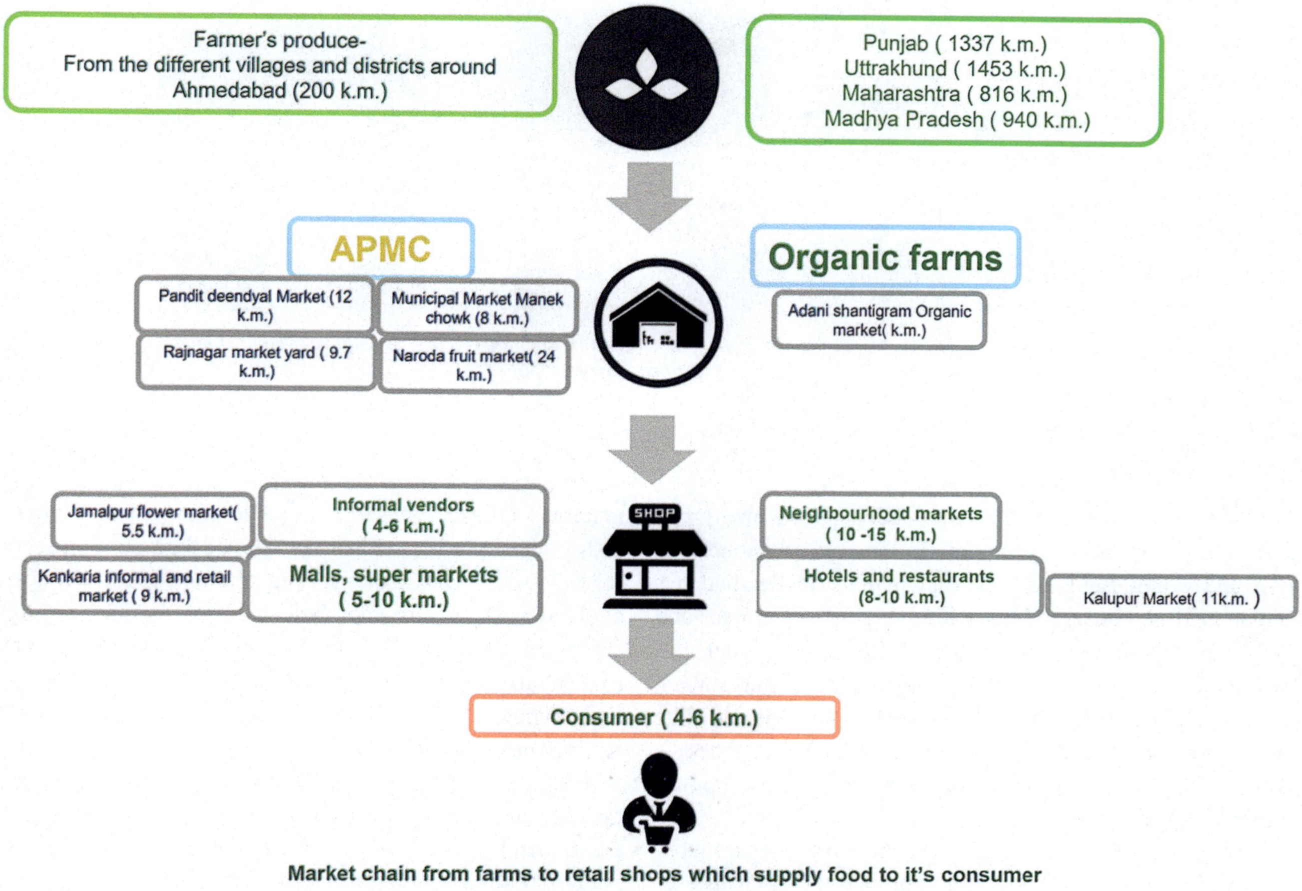

Fig. 4 Market chain from farms to retail shops in Ahmedabad

who are interested in growing their own food. The distribution is carried out by their office in Ahmedabad. As part of the Canning and Kitchen Gardening Centre, they provide all local vegetable seeds at INR 1 for each packet. This packet contains a handful of seeds and neem-based manure. There is no initiative or scheme directed towards the promotion of urban agriculture in the state. Some people who want to start growing their own food do not have access to resources or knowledge about various techniques and technologies available. This is preventing the growth of urban agriculture in the city.

Parks and Garden Department of Ahmedabad Municipal Corporation is engaged with the citizens in providing public land parcels and support systems to create urban forest lands in recent times. They also conduct monthly basic gardening workshops for citizens in the major city gardens. This activity has a potential to promote urban agriculture. Ahmedabad Urban Development Authority (AUDA) as part of their city master plan has provided dedicated agriculture zones as a buffer to city's development in the peri-urban areas. This has helped in controlling the development and promotes agriculture in the peri-urban areas. There are small initiatives and attempts towards increasing the edible green spaces in Ahmedabad. However, collective efforts are needed to make it a pioneering movement in the city.

4 Discussion

Innovations have helped all the three cities initiate growing of edible greens in some way or another at building level. Bengaluru sees expansion of green spaces with inspiration from Garden City Farmers movement. People grow food on the terraces, balconies and backyards of their houses. The Telangana Horticulture Department has supported people in growing vegetables, fruits, herbs and exotic greens on their terraces, balconies and backyards. There are some households that have started practicing rooftop farming in Ahmedabad but Urban Agriculture as a common practice as not picked up on a large scale due to lack of awareness about the techniques of growing their own food. Figure 5 shows some methods of farming in all the three cities.

Use of vacant plots in urban and peri-urban areas to grow edible greens in these cities is introduced through various methods. Bengaluru sets the example by providing BM English School Garden and Lalbaugh for growing greens.

HOME - Heaven On My Earth, Bangalore

Rooftop of Telangana Horticulture Department, Hyderabad

A rooftop in Ahmedabad

Fig. 5 Different methods of growing edible greens in the three cities

There are urban farms which are associated with many private enterprises such as Farmizen in the peri-urban areas of East and South Bengaluru. Master plan of Hyderabad has demarcated a conserved agricultural zone in urban and peri-urban areas. Currently, some of the land is under cultivation, and the rest is under development. At Ahmedabad, institutions like Gujarat Vidhyapeeth and schools like Shreyas Foundation are practising urban agriculture. There is also a demarcation of agricultural zone in the proposed Master Plan 2021 (Fig. 6).

Bengaluru has an advantage of conducive climate, technologically savy people and large number of agribusinesses that provide easy, accessible and user-friendly solutions. They also connect the peri-urban farmers to the citizens which in turn strengthens rural economy. Continuous communication through social media platforms like WhatsApp and Facebook amongst the urban farmers is also beneficial for farming. Conscious healthy and clean eating habits and open knowledge-sharing platforms have supported the movement.

Bengaluru operates mostly through wholesale (68.8%) and retail (31.3%) markets, whereas Hyderabad has about 15.8% informal markets along with wholesale and retail markets. Informal markets (33.3%) play a major role in distribution of edible greens in Ahmedabad coming second to only wholesale markets. The retail markets are not as evolved as Bengaluru and Hyderabad for organic and edible greens in Ahmedabad (Fig. 7). Government plays a major role in distribution of supplies in Bengaluru and Ahmedabad, whereas private distributors take up the responsibility in Hyderabad (Fig. 8).

Presence of research cell and training institute for urban farming and strong initiative from Telangana Horticulture Department and Hyderabad Metropolitan Development Authority for Urban Agriculture supports the movement in Hyderabad. Low-rise housing typology is also conducive to promote urban farming. Different housing typologies promote urban agriculture. Organised supply chain in market catering to the demand and supply of food supports future commercialisation of the practice.

Agribusiness and their start-ups have played an important role in building awareness amongst citizens and increasing their participation in expansion of edible greens in all three cities. The movement started early in Bengaluru and has the advantage of being the IT capital of India (Fig. 9). Gradually, their involvement has increased in the last 15 year, and

Lalbaugh Horticulture Department, Bengaluru

Hyderabad Peri-urban Area

Rooftop farming at Gujarat Vidyapeeth in Ahmedabad

Fig. 6 Edible green spaces in the three cities

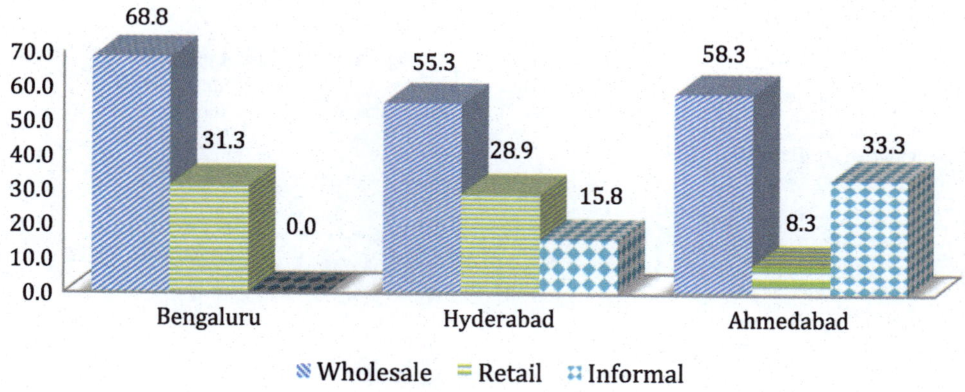

Fig. 7 Types of markets operational in three cities

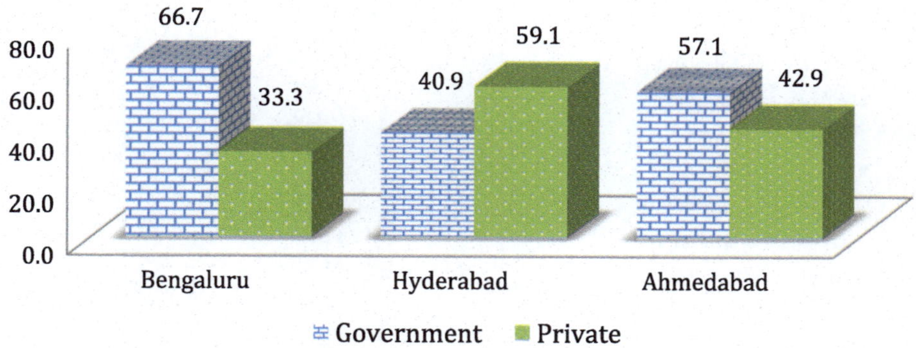

Fig. 8 Role of government and private sector in product distribution

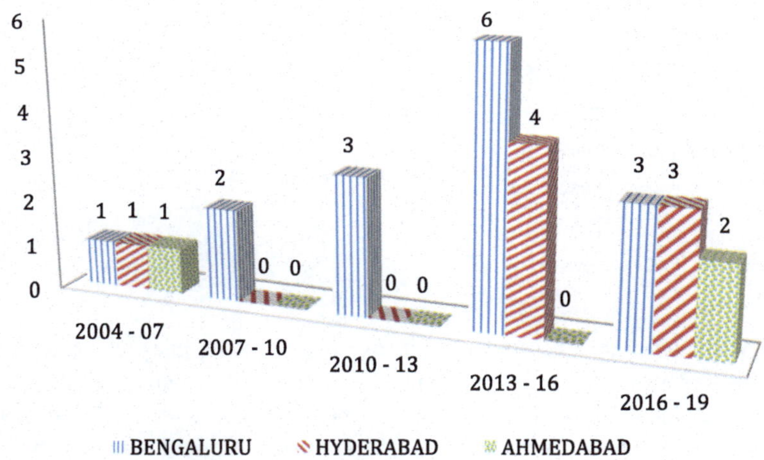

Fig. 9 Timeline of agribusiness start-ups

now these start-ups engage in a range of activities from providing farming kits, grow bags to sharing technical innovations and capacity building through frequent training programmes (Table 2). (Cafes & Store Survey, 2019; Market Survey, 2019).

With the involvement of horticulture department, the development of edible greens can reach the next level in Bengaluru. Expensive raw materials and tools restrict people to practice urban agriculture. Hyderabad is not able to protect land use conversion, and this may hamper the growth of edible greens in the city. Ahmedabad faces climatic, environmental and social challenges in expansion of green spaces. Lack of initiative from horticulture department and conversion of agriculture land use to non-agriculture due to urbanisation pressure is also retarding the process of urban greening.

5 Conclusion and Way Forward

Growth of urban agriculture in cities of Bengaluru and Hyderabad has different driving forces contributing to the awareness and success at household, institutional and commercial levels. This reflects that both the approaches, be it people-driven, or government-driven, can create awareness and movement resulting in improved edible green cover.

Table 2 Services provided by agribusinesses

City	Services Agribusiness	Home Kits	Grow Bags	Setting up Farms & its Maintenance	Potting and Compost (mix & bins)	Seeds	Fertiliser & Pesticides	Finished Products (food & personal hygiene)
Bengaluru	Back to basics							✓
	Seed paper					✓		✓
	Herbal strategi							✓
	On crop	✓	✓		✓	✓	✓	
	Agroscience							✓
	Parampara							✓
	My Dream Garden	✓	✓		✓	✓	✓	
	Millet amma							✓
	Happy Hens							✓
	Woolly Farms			✓				
	Hosachiguru			✓				
	Farmizen			✓				
Hyderabad	Homecrops	✓	✓	✓	✓			
	Urban Kissan	✓	✓		✓			
	Neem Consultancy						✓	
	Bokashi Decoders							
	Kbs Pvt. Ltd				✓			
Ahmedabad	Harvesto			✓				
	Greenman	✓	✓		✓	✓	✓	
	Domes Ltd							✓

Source: Primary Survey, Site Visit

Ahmedabad is witnessing a rise in awareness, and people are coming forward to experiment. Beginner kits, training, proper handholding and technical support will improve and accelerate the movement towards urban agriculture. Awareness about benefits of local crop varieties, ease of their growth and quality of product will also facilitate the entire movement.

Best practices can be derived and applied from this study of three cities to increase their edible green spaces. Formation of a dedicated urban farming division can act as accelerator for a city trying to begin its journey to promote urban agriculture and kitchen gardening. This division can be a part of the horticulture department in the region, as this department is already quipped with the technical and practical knowledge in this sector. Urban farming division should carry out activities like formulating awareness programmes, creating beginner kits, conducting training workshop, selling subside input materials and more. Engaging the community in this process has proved to be beneficial and encouraging to newcomers. Making community level leaders and support groups can engage people in adopting kitchen gardening. This can also be a platform to share knowledge and barter input materials along with experience. Social media platforms like Facebook and WhatsApp where people are already connected can be utilised for this purpose.

Including urban agriculture as an aspect in building by-laws can further ensure increase in edible green spaces. Similar to laws on rainwater harvesting, a new chapter can be introduced wherein building can provide dedicated space and infrastructure at individual unit level and community level to practice urban farming. This can promote people's well-being by engaging gardening activities and bringing the community together.

Edible green spaces should be provided at city and neighbourhood level by utilising the vacant plots for urban agriculture. Urban voids like land along the railway lines, under the bridges, buffer zones, empty plots, commercial compounds, empty rooftops and more can be explored. Farming in these spaces can create employment as well as provide local source of vegetables especially for economically weaker section of the society creating an example of inclusive planning.

Promoting agribusiness that provide goods and services in this sector can further support the process, providing solutions like renting spaces for people who do not have space but wish to do urban farming, leasing unused private spaces to farm, management of gardens/farms, farming kits, agri-inputs, etc. Such solutions can further encourage people to involve in farming activities in an urban set up.

The ultimate question that arises in process to increase the edible green spaces is how to meet the infrastructure requirement for water, soil and space. The recycled water should be used to water the crops. This helps in reduction of wastewater quantity. Wet waste can be composted and used as nutrient for edible greens. This is an exceptionally good medium for growing when mixed with coco peat and soil. Soil-less farming which is lightweight as compared to conventional soil-based farming can reduce dead weight on building structure making urban farming more practical. Most of the indoor farming is done in container making it easy to move around and creating multipurpose space to carry out other occasional activities as well. All the above practice are proven to increase edible green spaces in the three cities making it a successful model to follow.

The vision to take this movement to the next level shall benefit greatly with inclusion of information on options to grow greens at various levels from building unit, to neighbourhood and master plan. Alterations in building by-laws to accommodate provisions for edible green spaces shall help in putting the process on fast track. Bringing together various stakeholders like departments of agriculture, horticulture, environment, forest and climate change, urban development and urban housing along with orientation of real estate developers to improve built form with provisions for edible green best suited for the given climatic and weather conditions will bring a revolution that would benefit all.

Acknowledgements The authors acknowledge the Institute of Architecture and Planning, Nirma University for giving them an opportunity to conduct a unique studio on urban agriculture. Sincere gratitude to Prof. Utpal Sharma, Riya Thomas and students of fourth year, Bachelor of Planning who opted for a unique subject like this and giving it their one hundred per cent. The authors remain grateful to all the organisations, institutions, citizens, owners of cafes and stores and government officials who shared their journey and insights on growing edible green in the cities of Bengaluru, Hyderabad and Ahmedabad.

References

Agribusiness Survey. (2019, August). *Agribusiness Primary Survey in Bengaluru, Hyderabad and Ahmedabad* (Urbal Studio, B Plan VII Semester Students, Institute of Architecture and Planning, Nirma University, India, Interviewer).

Ahmedabad Municipal Corporation. (2017). *Home Page*. https://heritage.ahmedabadcity.gov.in/

Begum, M. N. (2019, August). *Successful terrace garden farmer—Asif Nagar, Hyderabad* (Urbal Studio, B Plan VII Semester Students, Institute of Architecture and Planning, Nirma University, India, Interviewer).

Cafes and Store Survey. (2019, August). *Cafes and Stores Primary Survey in Bengaluru, Hyderabad and Ahmedabad* (Urbal Studio, B Plan VII Semester Students, Institute of Architecture and Planning, Nirma University, India, Interviewer).

Census of India. (2011). *District Census Handbook*. https://censusindia.gov.in/2011census/dchb/DCHB.html

Chandrashekar, Y. (2019, August). *Hortliculture Department of Karnataka—DDH Lalbaugh, Bengaluru* (Urbal Studio, B Plan VII Semester Students, Institute of Architecture and Planning, Nirma University, India, Interviewer).

Chhabra, M. A. (2019, August). *Organic Terrance Gardner—Bengaluru* (Urbal Studio, B Plan VII Semester Students, Institute of Architecture and Planning, Nirma University, India, Interviewer).

Farmizen. (2017). *Frequently asked questions*. https://www.farmizen.com/faq/

Hegde, R. (2019, August). *Garden city farmers—Representative & Organiser of OFYT Oota from your Thota, Bengaluru* (Urbal Studio, B Plan VII Semester Students, Institute of Architecture and Planning, Nirma University, India, Interviewer).

Hindustan times. (2017, August). *Bengaluru lake froth on streets: Here's what causes the toxic foam and how it is harmful to people*. https://www.hindustantimes.com/india-news/bengaluru-lake-froth-on-streets-here-s-what-causes-the-toxic-foam-and-how-it-is-harmful-to-people/story-yuAhx2f4wIYlJYPReVxz7O.html

Holidify. (2019). *Bangalore*. https://www.holidify.com/places/bangalore/

Homecrop. (2019). *Home Page*. https://homecrop.in/

Horticulture Department TS. (2018). *Urban Farming*. http://horticulture.tg.nic.in/UFarming/UFarming.html

India Meteorological Department. (2019). *State Composite AAS Bulletin (Archive) | India Meteorological Department*. http://www.imdagrimet.gov.in/node/18071

Kalra, S. (2018, June). *An urban farming group in Bengaluru is teaching people home agriculture to fight pollution*. https://www.indiatimes.com/news/india/an-urban-farming-group-in-bengaluru-is-teaching-people-home-agriculture-to-fight-pollution-322845.html

Maarten Elferink, F. S. (2016, April 07). Global demand for food is rising. Can we meet it? In *How to meet global food security without increasing pressure on land* (p. 7). Retrieved December 13, 2019, from https://hbr.org/2016/04/global-demand-for-food-is-rising-can-we-meet-it

Madhusudan, S. G. (2019, August). Horticulture Department of Telangana—Asst. Director of Horticulture, Urban Farming Section (Urbal Studio, B Plan VII Semester Students, Institute of Architecture and Planning, Nirma University, India, Interviewer).

Market Survey. (2019, August). *Wholesale agri-produce market's buyers and sellers survey in Bengaluru, Hyderabad and Ahmedabad* (Urbal Studio, B Plan VII Semester Students, Institute of Architecture and Planning, Nirma University, India, Interviewer).

My Dream Garden. (2019). *About Us*. Retrieved from mydreamgarden.in: https://www.mydreamgarden.in/about-mdg.html

Rooftop Farmers' Survey. (2019, August). *Roofttop farmers and Urban Growers Survey in Bengaluru, Hyderabad and Ahmedabad* (Urbal Studio, B Plan VII Semester Students, Institute of Architecture and Planning, Nirma University, India, Interviewer).

Urban Kisaan. (2019). *Home Kits*. https://www.urbankisaan.com/homekits

Viswanath, D. B. (2019, August). *Garden city farmers organisation* (Urbal Studio, B Plan VII Semester Students, Institute of Architecture and Planning, Nirma University, India, Interviewer)

Wolly Farms. (2017). *About wolly*. http://woolly.io/who-we-are/

Sustainable Design Strategies at the Building Level

Potentials of Plant's Strategies for an Adaptive Building Envelope

N. Nour ElDin and A. Abdou

Abstract

Architects nowadays use many design approaches to reach a sustainable architecture; one of those rapidly growing approaches in the past few years is the "biomimetic approach" (Benyus, Biomimicry, William Morrow, 1997). This approach seeks solutions inspired by nature; however, some architects could encounter some difficulties in finding the relevant; appropriate solutions that fit their architectural challenge they are trying to solve. Therefore, they need to collaborate with other disciplines like biologists for assistance, consequently consuming more time and effort. There are several methodologies developed to facilitate the translation process from the biological domain to the architectural application. However, the difference in this paper is to generate a methodology tailored specifically for architects. This methodology seeks solutions for building envelope to be adaptive, for the environmental challenges, using solutions and strategies developed by plants. It will offer alternatives for specific environmental challenges at a certain adaptive level, according to the architect preferences and building typology. Consequently, overcoming the need for other disciplines in this translation process: consuming less time. Accelerating the design process is where all the data will be gathered in one platform and categorized according to the aforementioned criteria.

Keywords

Biomimicry • Adaptation • Plants strategies

1 Introduction

The building envelope contributes to the building's overall energy consumption to reach the indoor thermal comfort; thus, more attention should be taken to the façade's design. In this paper, the focus would be on solutions corresponding to environmental challenges that were classified into four categories: heat, light, water, and air. This paper is part of an ongoing Ph.D. thesis entitled "Towards generating a methodology for an adaptive building envelope—A biomimetic approach in hot dry regions" by the researcher. The paper seeks to integrate the plant's adaptive strategies to the building envelope. The motivation was to tackle solutions for the building envelope environmental problems mainly for heat, water, air, and light challenges. The "biomimicry" approach follows two methodologies: either "problem-based" or "solution-based." In this paper, the "problem-based" approach would be followed. It begins by identifying the key environmental challenges that face the building envelope, an initial selection of 200 research publications corresponding to 40 plant strategies. Approximately, 30 plants were chosen for this paper and analyzed. The equivalent function and possible architecture application in the building envelope are then identified according to the hot dry context through the proposed methodology.

Several methodologies and other categorization systems like the one proposed in this paper are available; however, this paper focuses on the building envelope and the environmental challenges it faces. The examined plant's strategies below were classified into four categories, corresponding to environmental challenges: heat, light, water, and air and onto three levels: behavioral, morphological, and physiological, and some plants have more than one strategy and on different adaptive levels.

N. Nour ElDin (✉) · A. Abdou
Architecture Department, Faculty of Fine Arts, Helwan University, Helwan, Egypt

C. Alalouch et al. (eds.), *Towards Implementation of Sustainability Concepts in Developing Countries*, Advances in Science, Technology & Innovation, https://doi.org/10.1007/978-3-030-74349-9_10

2 Plant Adaptive Strategies

The following section would present some plant's adaptive strategies for the four selected environmental challenges (heat, light, water, and air), on different adaptive levels (morphological, behavioral, and physiological) adaptations. The presented plant examples do not represent all the available plants, but what the researcher has reviewed up to this date (Fig. 1).

Heat regulations utilize multiple methods like prevention, dissipation, retention, or gaining. Some plants evolved heat, behavioral, and regulatory strategies related to orientation, such as "acacia" (Batanouny, 2001). Narrow leaves, less projected to the sun" or, "Bending," like in the "Chinese silk tree" (Vogel, 2005a, b) that direct leaves down for not freezing avoid radiative cooling. The "Rhododendron" (Barrett & Barrett, 2016) avoids freezing by creating a microclimate in "yellow crocus" [a] and kinetic reaction of plant structure in response to changes in temperature.

Moreover, plants developed "morphological" adaptations to regulate "heat." For example, the fan-shaped leaf of "palm trees" allows air to pass through. The "tree bark's" (Baumeister, 2007) overall morphology and "round cross-section" dissipates heat and light, thus stays cool. The ribs and folds in "cactus" (López et al, 2017) provide shade to be casted on portions of its surface, which improves its ability to combat heat radiation and therefore cool. The "density" in "cushion-shaped tree" (Cavieres et al., 2007) creates a microclimate. The relationship in "fungus" provides thermal protection: hot springs panic grass providing insulation. Also, the "density, porous," and "compacted leaves" surface in "grass tree" form efficient heat insulation by keeping air. "Gynadriris setifolia" uses its "surface texture (hairs)" to reflect sunlight from their surface (López et al, 2017).

Thus, heat prevention could be fulfilled through minimum radiation exposure through the envelope's morphology or orientation or other "physiological" adaptations by minimizing the heat load utilizing changing colors, the surface-to-volume ratio, and reflectance. Heat retention could be a result of increasing the insulation within the structure for example or through heat exchange.

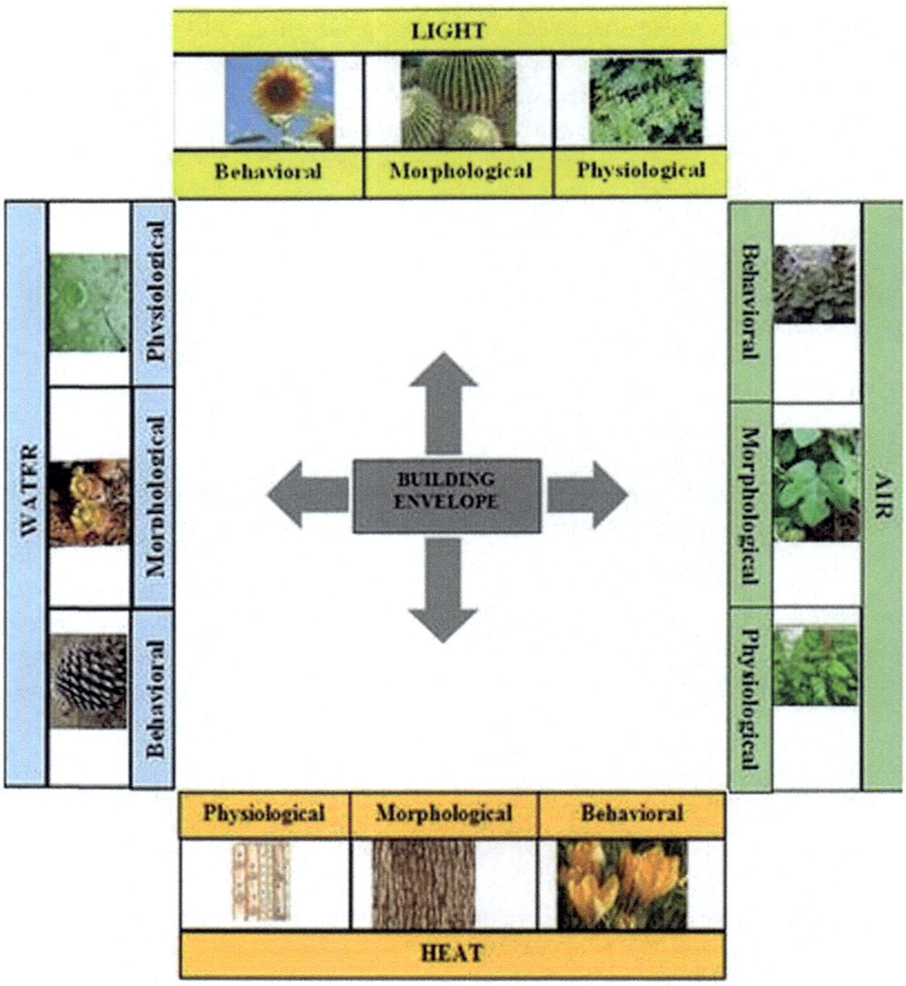

Fig. 1 Example of the relationship between plant strategies and building envelope requirements on different adaptive levels (The researcher)

Several plant species are also capable of solar tracking; they move and adapt their position according to the sun's direction to capture more light or to avoid it. An example of "light behavioral adaptations" is "solar tracking" of plants like "sunflower" (Badarnah, 2017), and sunflower tracks sun path throughout the day by bending toward light and maintaining radiation perpendicular to the surface. The "unidirectional movement" of "Lilium casablanca" (Schleicher et al., 2015) where it is and is firmly packed and consists of six petals, three inner and three outer, that is prominently curved. "UV protection" and moderate microclimates forming a compact, dense group in "alpine plants" (b) unite into dense pads and retain moisture and temperature by creating a collective temperate microclimate. Some plants "transmit light selectively"; climbing plants move to places with better light conditions like "goose grass" or "folding" as in the "Aldrovanda vesiculosa (water wheel plant)" (Bischoff et al., 2017).

The "surface texture, peeling, and rough surface" of "tree bark" resulted from the growing process of the tree. The xylem tissues push the circumference outward which causes the outer layer of the bark to tear and die. Some barks form peeling crusts, while others react by forming deep cracks, which dissipate light. The outer skin of the stem of the barrel cactus (Pohl, 2011) is covered by a thick waxy coating to reflect the solar radiation. Also, cactus's "spines" are densely spaced to create loose semi-transparent skin which helps shade the skin of a cactus and reflect and diffuse the direct incident solar radiation. "Quiver tree" (c) also provides shading through its branches protected from the sun. The "overall morphology" of "window plants" and crystals draws sunlight into the plant, transmitting maximum light. (Vogel, 2005a, b).

Other "light physiological" adaptations like in "Selaginella ferns," (d) photonic structure and leaves change colors under different lighting. "Spinach" (e) performs "energy conversion" where its leaves convert photons to energy. "Ivy plant" (f) generates energy, producing renewable energy while also helping to provide shade. Those concepts could inspire climate adaptive building skins. Adjusting the amount of light, the spatial arrangement, shape, orientation/inclination, and dynamic movement abilities of plants makes them an ideal source of inspiration for shading applications. The aim is to reduce or control the amount of light incident upon a surface, thus, improving the comfort of the space for users and the building's functionality.

Additionally, heat regulation could be by using evaporative cooling to cool the envelope. "Pinecone," for example, "opens and closes" in response to moisture (Krieg, 2004). The "water harvesting" in "desert shrubs" through roots can extract hard to remove water from soil using negative pressure. "Leaf curling" in "Poikilohydric" plants is to avoid dehydration and maintain/control specific moisture contents.

The "hydrophobic surface" of the "lotus leaf" (Mauseth, 2003) due to the micro and nanostructure of the leaves can run water off its surface. The "waxy outer layer" of "cactus" produces extra barriers during the hottest time of the day to prevent water from leaving the stem in addition to its "material thickness" which is a thick and dense outer layer to store water. This helps the cacti delay any water evaporation. The long needle shape of the "spines" also helps the cactus catch the rainwater and directs the water toward the shallow roots of the cactus. These spines are not only able to collect the rainwater but also able to collect any water vapor in the air.

The "mescal cactus" also can adapt to seasonal water availability via dehydration-induced "shrinking" below the desert floor and hydration-induced "swelling" to reemerge after rainfall. The "round shape" of "pebble plants" (g) reduces water loss, evapotranspiration.

Water conservation in hot dry regions could be through reducing the exposure of the surface by self-shading or reflection, reorientation, folding, and flexibility. Another method to achieve water conservation is through reducing the evaporation process by adjusting the temperature, SV ratio, or permeability. Water transport could assist in water regulation; this transportation could be through the effect of gravity or the effect of capillary action. Finally, water loss could be fulfilled through evaporation.

Plants also adapted strategies for air regulation; for example, on the morphological scale, the "vines" and "lobes" in "grape leaf" allow air to pass between them. The "phragmites" (Van der Valk, 2006) offers "permeability" where the stems move air, generate/use air pressure difference, or temperature for ventilation. On the behavioral scale, "black mangrove" (h) performs "gas exchange," where pressure makes air move and generates permeability to ventilate. "Vascular plants" allow "air exchange."

Hence, the building envelope could perform air regulation through air exchange, diffusion, or air movement. This regulation could be fulfilled as a result of natural convection like temperature gradients or pressure differences. The pressure difference could be due to a variation in volume or velocity gradient.

3 Methodology Generation

Remarkably, the study of every plant analyzed offered interesting solutions and adaptive options for building adaptation. Consequently, we could conclude the following methodology that shares similar factors between the three pillars: plants, building envelope adaptive levels, and environmental requirements. This methodology derives its main components from the above variables where each plant strategy is to perform a type of regulation based on the envelope challenge (Fig. 2).

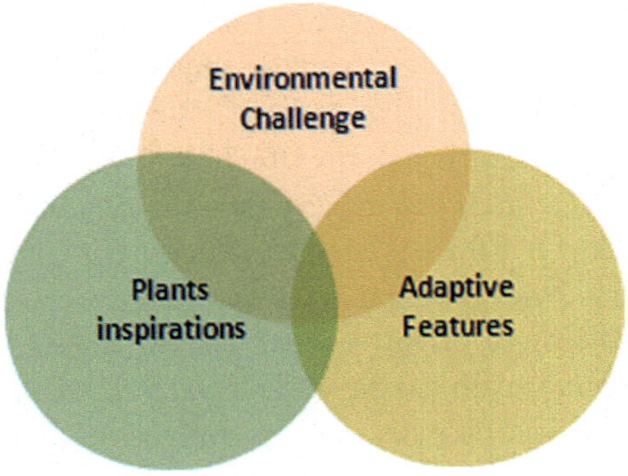

Fig. 2 Methodology main three pillars (The researcher)

3.1 Methodology Motivation

The abundance of information demanded a specific methodology and a classification system consistent with the objectives of the paper. Therefore, a preliminary database was developed; this database follows the proposed methodology in analyzing and categorizing the plant's adaptive solutions. They are analyzed according to the environmental factor they perform and their adaptive level in performing the strategy. Relevant architectural applications inspired by plants would be also presented to provide the designer with a methodological database. As a result, a methodology was generated in the form of a 3D matrix developed by the researcher (Nour El Din & Abdou, 2020) where plant's strategies are collected and analyzed with their possible architectural applications. The following figure shows the general outline of the proposed methodology. One plant could offer other solutions and other strategies if another environmental challenge or other adaptive level was selected (Fig. 3).

As could be observed from the previous matrix, out of approximate of 40 plant's solutions, the search results were filtered to 11 when light regulation was chosen. However, if a further adaptive level was chosen, for example, "physiological," the solutions would be narrowed down to 2 solutions out of 40. Consequently, the matrix accelerates the decision-making process offering only ideas related to the designer's preferences.

3.2 Possible Architectural Applications

The following table shows another layer of the previous matrix (sub-matrix) where some of the possible architectural applications that could be solved by some of the previously presented plants are summarized (Table 1).

3.3 Methodology Development

This methodology was further developed to facilitate navigation within the previous matrix (Figs. 4, 5), as well as more user-friendly, visually appealing, and less time-consuming.

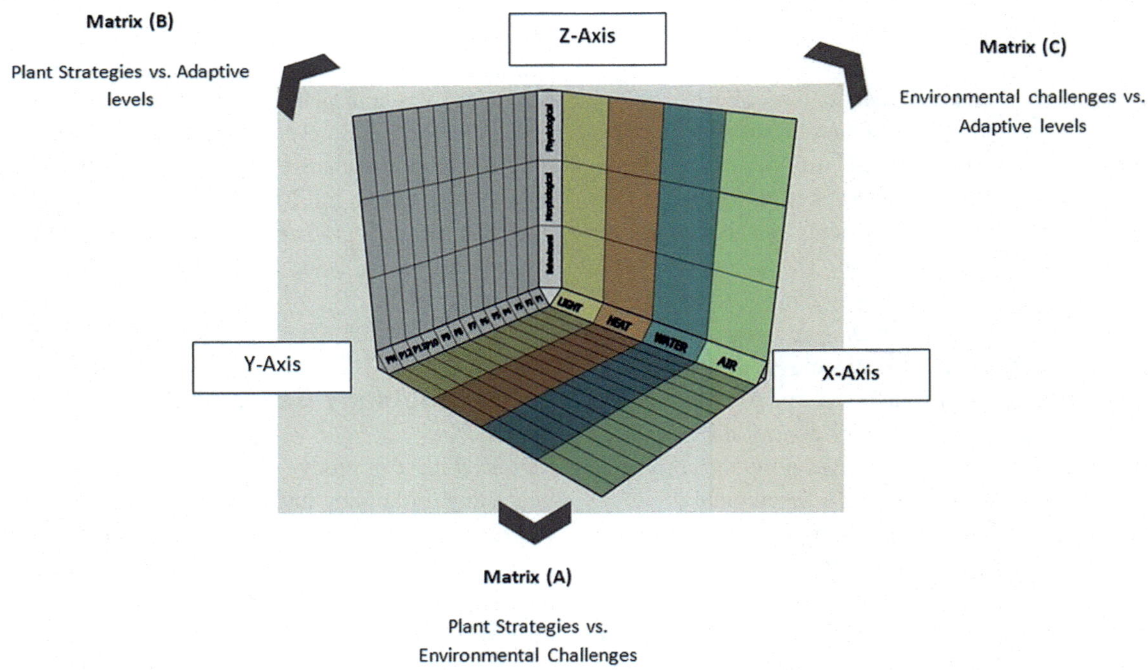

Fig. 3 Methodology motivation and relationship between plant strategies and building envelope requirements on different adaptive levels (The researcher)

Table 1 Example of some possible architectural applications, for an adaptive building envelope, inspired by plants

Plant name	Environmental challenge					Level of adaptation		Possible architecture application
	Light	Heat	Water	Air	Morphological	Behavioral	Physiological	
Sunflower [9]	•					•		Creating a shading system, rotatable dynamic responsive structure, controlling "light" transfer through louvers, and paneling systems with embedded sensors and actuators mechanisms
Pinecone			•			•		Hygroscopic materials like "wood". For example, "hygro-skin" project. The architectural skin developed through this process can close or open depending on the climatic changes without any need for a power supply or other sensors
Cactus [7]	•	•				•		Using an increased reflective value of the material and folded structure of corrugated skin
Black mangrove				•			•	Breathable structure porous surface membranes
Palm tree		•			•			(a) Multilayer and lightweight, instead of masses to deal with thermal performance of building envelope pattern (roof)
Grape leaves				•			•	Perforations allow air to pass
Lotus leaf [20]			•				•	Super-hydrophobic surfaces (e.g., Lotusian paint)
Tree bark [6]	•	•			•		•	Surface texture (rough/peeling/cracks) creates shades; applying thermal insulation materials reduces the amount of energy consumption for both cooling and heating loads. Proper installation for the insulation will define the maximum thermal insulation or R-value or buildings overall morphology. Also, the pipe system could be implemented within the envelope
Acacia [2]	•				•	•		Microstructures that enhance light scattering
Lilium casablanca [10]	•					•		Controlling "light" through the envelope material using "thermotropics," "piezoelectric" as sensors for closing mechanisms
Gynadriris setifolia [8]		•			•			Reflective surfaces
Phragmites [22]				•		•		Passive ventilation systems for buildings generate permeability to air
Alpine plants [11]	•					•		UV protection and moderate microclimates by forming dense groups
Euphorbia			•		•			Keeping water at the external envelope to act as an insulator with a high thermal capacity
Vascular plants				•		•		Breathable structures
Pebble plant [21]			•		•			In geometric adaptation, water could move throughout the façade through evaporation and use the wastewater as a constant nutrient source to the living building envelope. Water penetration within the envelope could be controlled through drainage systems and series barriers and rain screens that prevent the water to reach the indoor spaces

(continued)

Table 1 (continued)

Plant name	Environmental challenge					Level of adaptation		Possible architecture application
	Light	Heat	Water	Air	Morphological	Behavioral	Physiological	
Spinach [17]	•						•	Using solar radiation for energy generation
Ivy Plant [18]	•						•	Photovoltaic leaves that can be attached to building facades and other vertical spaces. Solar ivy system has a modular design (a Brooklyn-based firm—SMIT: Sustainably Minded Interactive Technology)
Yellow crocus [5]		•				•		The active component in the air flow(er) device is a custom manufactured shape memory alloy (SMA) wire. It is commonly regarded as a smart material because of its unique response to changes in temperature. When an SMA is in its lower temperature form, it can be easily deformed into a new shape
Aldrovanda vesiculosa (water wheel [12])	•					•		The bio-inspired pliable shading system is ideal for architectural applications with double-curved, free-form geometric. (Schleicher S. bio-inspired compliant mechanisms for architectural design [Ph.D. thesis]. Stuttgart: ITKE—Universität Stuttgart Forschungsberichte; Jan. 2016)

•Main improvement

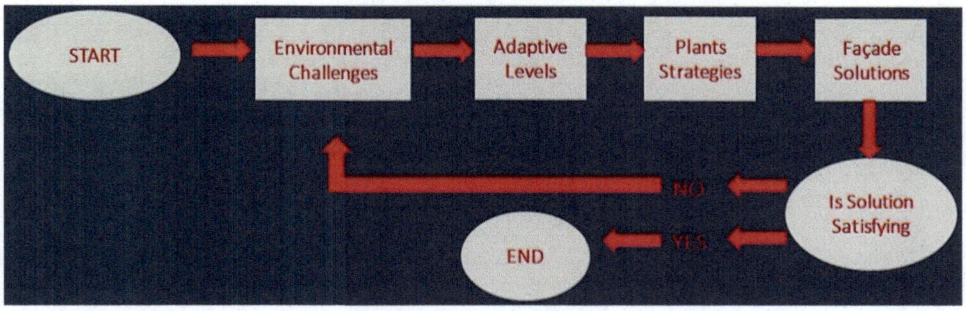

Fig. 4 General outline to the methodology (The researcher)

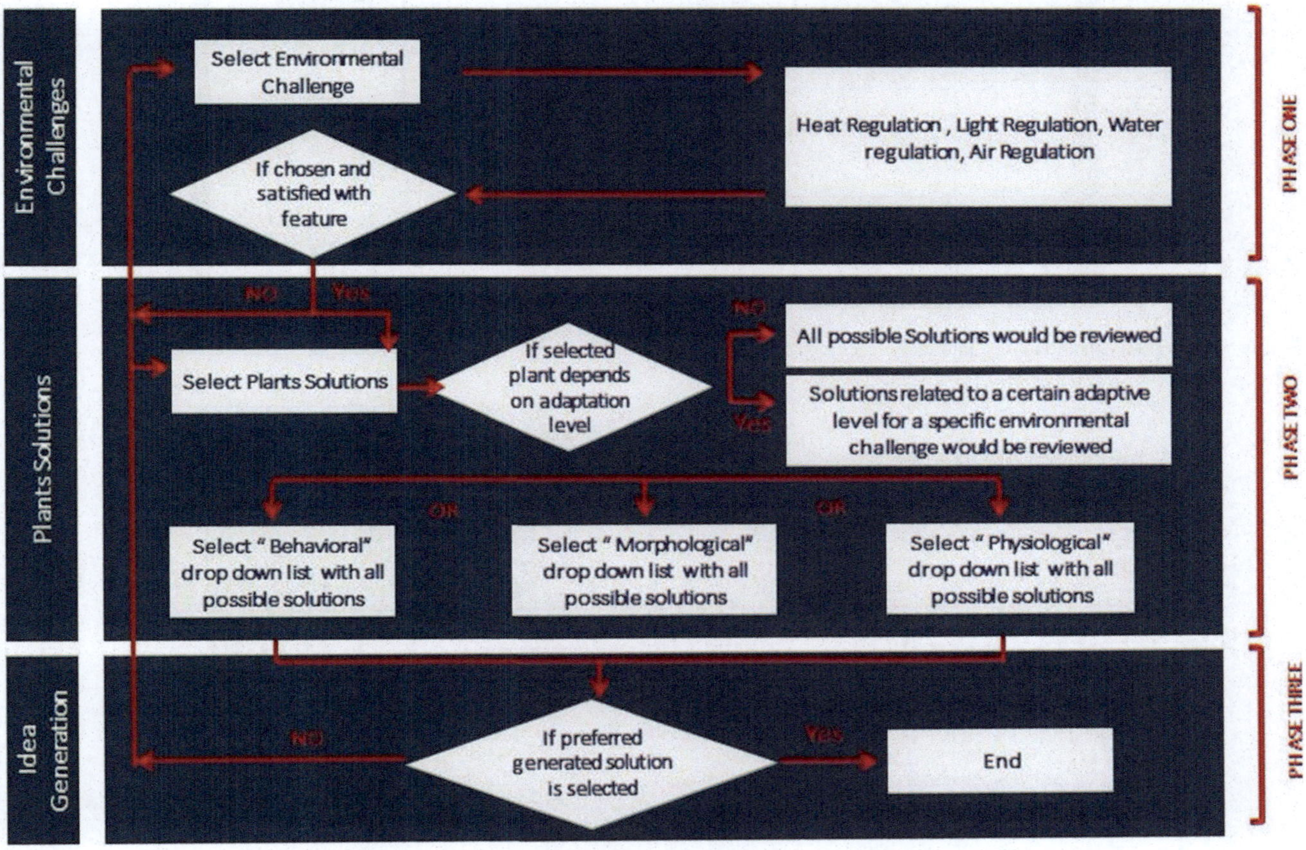

Fig. 5 Methodology development showing the translation steps from a plant domain to an architectural application idea (The researcher)

4 Methodology Application

The following figure represents an image of the plant-developed database based on the previous methodology. Each sheet refers to one plant, with its general biological information, strategies, and environmental challenge they perform as well as the level of adaptation they perform to fulfill this challenge. It also attaches some sketches with their references for further search. It also adds some personal notes with other possible architectural applications that could be useful (Fig. 6).

The following is an example of navigation within this tool. "Cactus," for example, has more than one strategy; thus, if the designer's preference was to design on the "morphological" level to regulate "light," then a description of strategies developed by cactus would be generated. The generated datasheet also attaches an example to a project that used a similar adaptive strategy for the same purpose (Fig. 7).

However, if the same adaptive level was selected, but by changing the input to "heat" or "water" as in figures (Figs. 8 and 9), the output and offered solutions projects would be different.

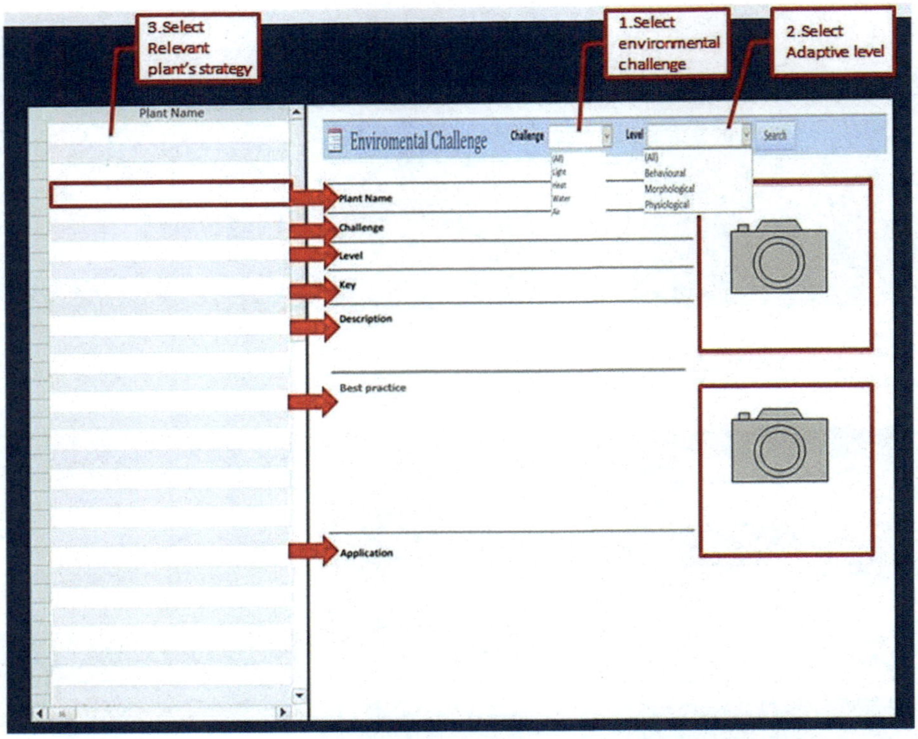

Fig. 6 User-friendly methodology interface developed by the researcher

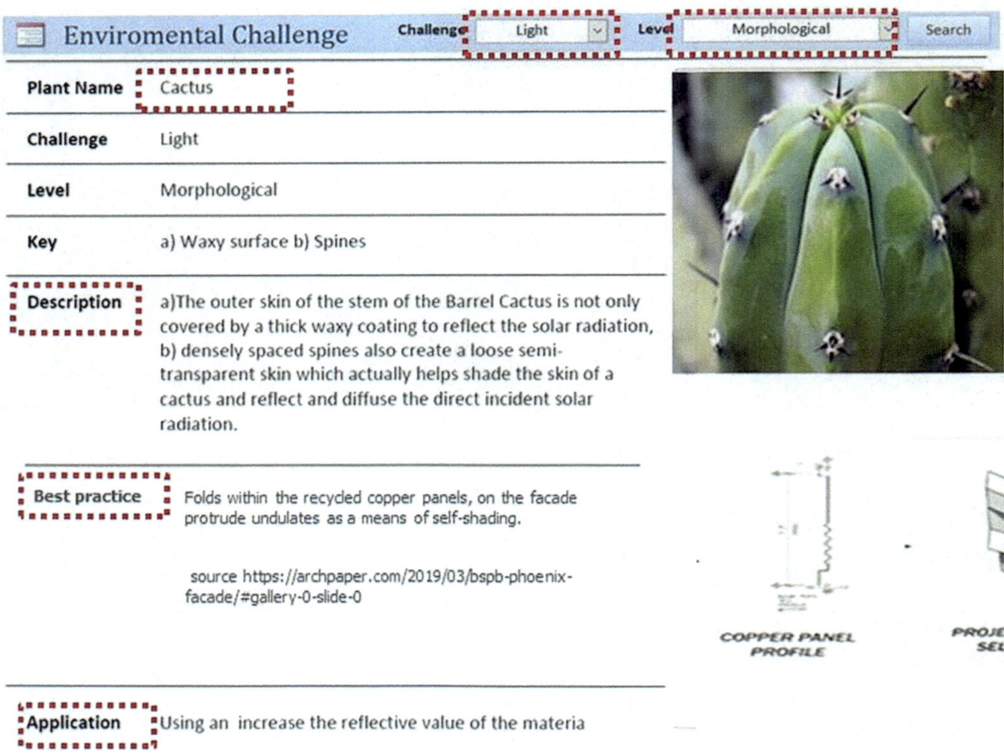

Fig. 7 Cactus adaptive strategies for "light/morphological" requirements

Enviromental Challenge — Challenge: Heat — Level: Morphological — Search

Plant Name	Cactus
Challenge	Heat
Level	Morphological
Key	Ribs and folds
Description	Ribs and folds provideshade to be cast on portions of its surface which improves the ability of the Barrel Cactus to combat heat radiation and therefore cool down the skin. (Jens Pohl, Building Science: Concepts and Application, (Chichester, Wiely-Blackwell 2011), 52.)
Best practice	Aluminum tile with octahedral surface articulations, testing penumbra shade, concave and convex surface topography. Source :(https://www.witpress.com/Secure/elibrary/papers/DN08/DN08025FU1.pdf)
Application	folded structure or corrugated skin

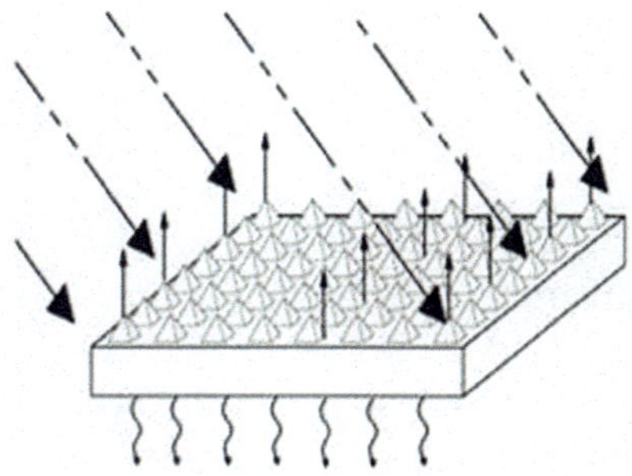

Fig. 8 Cactus adaptive strategies for "heat /morphological" requirements

Enviromental Challenge — Challenge: Water — Level: Morphological — Search

Plant Name	Cactus
Challenge	Water
Level	Morphological
Key	a)Waxy outer layer b) Material thickness c) Spines
Description	a)Cacti produce extra waxy barriers during the hottest time of the day to prevent water leaving the stem. B)thick and dense outer layer to store water. This helps the cacti delay any water evaporation. C)l ong needle shape of the spines also helps the cactus catch the rainwater and direct the water toward the shallow roots of the cactus. These spines are not only able to collect the rainwater, but are also able to collect any water vapor in the air
Best practice	The water holding capabilities and the regulated flow of air, a dome shape was used to visualize the idea, using it to cool and add moisture to the air at the peak hours of the day but be closed the rest of time. exc clay as material. source: (https://spatialexperiments.wordpress.com/2016/09/18/the-architecture-of-desert-leaves/)
Application	Keeping water at the external envelope to act as an insulator with a high thermal capacity

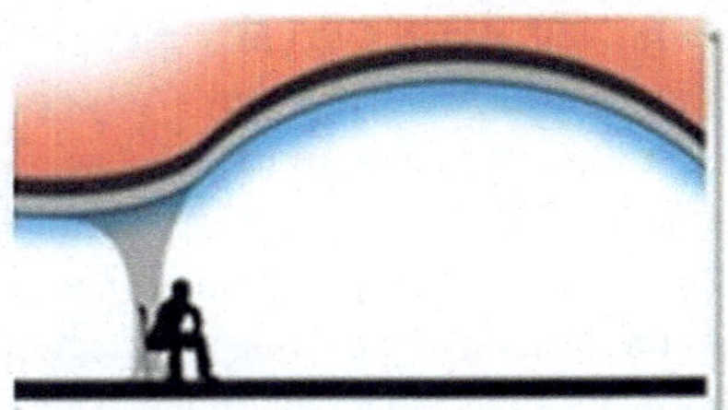

Fig. 9 Cactus adaptive strategies for "water/morphological" requirements

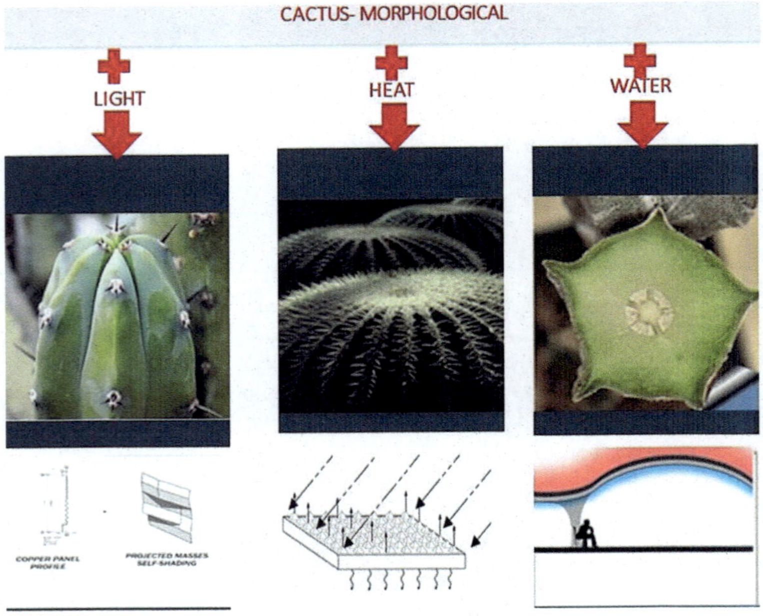

Fig. 10 Methodology findings, an example of cactus morphological adaptations. (The researcher)

5 Findings

As could be observed in the following figure, one plant could have more than one possible architectural application. Those results differ depending on the designer's preferences and building typology. Therefore, the methodology emphasizes the following process as follows: Selecting the design problem (environmental problem), the level of application within the building's envelope, accordingly selecting the most suitable plant that has previously solved this problem efficiently (Fig. 10).

6 Conclusions

This paper has investigated simplified plant's problems and challenges to assist understanding and underlying their main concepts. Those concepts could be translated to corresponding architectural applications depending on the building's typology. The methodology was further developed to a user-friendly interface that could be further developed to a tool that facilitates navigation within the comprehensive database offered by plants. The plant's ideas presented in this paper are not all those available plant's strategies in the plant's kingdom. However, they only present what the researcher has investigated up to that date. The offered plant solutions and database set just an example to show the methodology application possibilities and the ideas behind each plant translated to a possible architectural application. It was additionally noted that each plant could have more than one biomimetic strategy; thus, the database was necessary to facilitate and simplify the selection process. The database also aids in deciding which strategy and which plant could be further explored.

The proposed interface represents the development of the proposed methodology and could be open-source. Further plant strategies with related projects could be added following the same methodology. Consequently, assist the "idea generation" process to a bio-inspired building envelope. The designer could also combine more than one strategy or more than one challenge.

References

Asknature. Retrieved August 2019 from https://asknature.org/

Asknature. Retrieved January 2020, from https://asknature.org/strategy/elevated-leaves-reduce-evaporation/

Asknature. Retrieved January 2020, from https://asknature.org/?s=Pebble+Plants#.XIWPvnduIq0

Asknature. Retrieved January 2020, from https://asknature.org/strategy/pressure-makes-air-move/

Asknature. Retrieved January 2020, from https://asknature.org/?s=Selaginella%20Ferns&page=0&is_v=1

Asknature. Retrieved January 2020, from https://asknature.org/strategy/density-moderates-microclimate/

Badarnah, L. (2017). Form follows environment: Biomimetic approaches to building envelope design for environmental adaptation. *Buildings, 7*(2), 40.

Barrett, R. M., & Barrett, R. P. (2016, September). Thermally adaptive building coverings inspired by botanical thermotropism. In *Smart materials, adaptive structures and intelligent systems* (Vol. 50497, p. V002T06A008). American Society of Mechanical Engineers.

Batanouny, K. H. (2001). *Plants in the deserts of the Middle-East, Series "Adaptations of desert organisms."* Springer.

Baumeister, D. (2007). *Biomimicry presentation at the University of Washington College of Architecture.*

Benyus, J. (1997). *Biomimicry.* William Morrow.

Bischoff, M., Sachse, R., Körner, A., Westermeier, A., Born, L., Poppinga, S., ... & Knippers, J. (2017, September). Modeling and analysis of the trapping mechanism of Aldrovanda vesiculosa as biomimetic inspiration for façade elements. In *Proceedings of IASS annual symposia* (Vol. 2017, No. 8, pp. 1–9). International Association for Shell and Spatial Structures (IASS).

Cavieres, L. A., Badano, E. I., Sierra-Almeida, A., & Molina-Montenegro, M. A. (2007). Microclimatic modifications of cushion plants and their consequences for seedling survival of native and non-native herbaceous species in the high Andes of central Chile. *Arctic, Antarctic, and Alpine Research, 39*(2), 229–236.

Krieg, O. (2004). *HygroSkin—Meteorosensitive pavilion*. University of Stuttgart.

López, M., Rubio, R., Martín, S., & Croxford, B. (2017). How plants inspire façades. From plants to architecture: Biomimetic principles for the development of adaptive architectural envelopes. *Renewable and Sustainable Energy Reviews, 67*, 692–703.

Mauseth, J. D. (2014). *Botany: An introduction to plant biology* (2nd ed., pp. 214–231). Jones & Bartlett Publishers.

Nour El Din, N., & Abdou, A. (2020). An approach to adaptive sustainable facades inspired by plants. In *International conference on improving sustainability concept in developing countries (ISCDC)* (4th edn.).

Pohl, J. (2011). *Building science: Concepts and application* (p. 52). Wiley-Blackwell.

Schleicher, S., Lienhard, J., Poppinga, S., Speck, T., & Knippers, J. (2015). A methodology for transferring principles of plant movements to elastic systems in architecture. *Computer-Aided Design, 60*, 105–117.

SMIT SOLAR IVY. Ask Nature. http://www.asknature.org/product/256b9f497821497773d9f0c442ab367a

van der Valk, A. (2006). *The biology of freshwater wetlands* (p. 173). Oxford University Press.

Vogel, S. (2005a). Living in a physical world IV. Moving heat around. *Journal of Biosciences, 30*(4), 449–460.

Vogel, S. (2005b). Living in a physical world V. Maintaining temperature. *Journal of Biosciences*, 581–590.

Pro-Environmental Behaviors in LEED and Non-LEED-Certified Workplaces: A Comparative Study

Mona A. Mohamed, Manal A. S. Abou El-Ela, and Hala B. El Naggar

Abstract

LEED, a green building rating system, provides a framework for sustainable building and assesses a building's energy and resource efficiency. LEED prerequisites and credits are achieved by employing sustainable strategies. The efficiency and success of some of these strategies are dependent on future occupant behaviors. As occupant behaviors shape their building's energy-use, resource-use, and waste production, this paper aims to understand whether the achievement of LEED credits shapes occupant pro-environmental behaviors (energy-saving, resource-saving, and sustainable waste management). People spend the majority of their time at work; therefore, this paper takes the work environment as its setting. Further, it explores occupant behaviors in one LEED-certified (CAE) and one non-LEED-certified (ABBE) office building. Occupant surveys extract self-reported energy-saving, resource-saving, and sustainable waste management behaviors of building occupants in both settings. Results reveal a possible impact of LEED credit (SSc4.1) on commute behaviors and (M + R prerequisite) on sustainable waste management behaviors in LEED-certified building occupants. Results showed insignificant differences in both populations in terms of resource-saving behaviors. Further, a negatively significant difference was found in energy-saving behaviors of LEED-certified building occupants which were possibly influenced by limiting building policy.

M. A. Mohamed (✉)
Department of Architecture and Urban Design, GUC, New Cairo, Egypt

M. A. S. A. El-Ela
Department of Architecture, Faculty of Engineering, Banha University, Shoubra, Egypt

H. B. E. Naggar
Department of Architectural Engineering and Environmental Design, AASTMT, Giza, Egypt

Keywords

Pro-environmental behavior • LEED • Sustainable design • Workplace • Employees

1 Introduction

Built environments consume 40% of all primary energy worldwide and significantly contribute to greenhouse gas (GHG) emissions, 78% of which is caused by human activity. Emissions caused by human activity have the greatest observed impact on climate change and other global issues that pose a threat to the environment we live in. Traditional buildings deplete resources, generate waste, and raise emissions in the air and are costly to run. High operation costs and global environmental crises have led to a rise in sustainability practices.

Sustainable design has taken significant strides forward in the last 30 years. Sustainable building involves the design, construction, and resource-efficient operation practices, minimize negative environmental impacts, cut life-cycle costs while simultaneously enhancing occupant well-being and productivity. However, to make the necessary cuts in energy and resource-use, buildings must incorporate sustainable strategies that require occupant participation. Occupants must engage with their buildings and make conscious decisions about how energy and resources are used. Governments around the world attempt to create legislation to decrease our carbon footprint and enhance sustainable human behavior in all settings (i.e., households, workplaces, educational institutions, etc.) (Blok et al., 2015). Accordingly, green building rating systems such as Leadership in Energy and Environmental Design (LEED) were created.

LEED provides the necessary framework for sustainable building and evaluates a building's energy and resource efficiency. Buildings influence the productivity, well-being,

C. Alalouch et al. (eds.), *Towards Implementation of Sustainability Concepts in Developing Countries*,
Advances in Science, Technology & Innovation, https://doi.org/10.1007/978-3-030-74349-9_11

health of occupants, and, ultimately, the planet. Given that people spend over 90% of their time indoors, and at work, a balance must be achieved between the conservation and excessive consumption of energy and resources to efficiently achieve overall occupant comfort. Occupant activities directly impact building energy-use, resource-use, and waste production rates. However, it is argued that human activities reflect the existing constraints in everyday contexts. Design can make environmentally friendly behaviors easier to do. Thus, existing settings of day-to-day activities may impact the sustainable behaviors of building occupants (Lockton, 2013).

1.1 Research Problem

All human activity takes place in one environment or another. Any activity that occurs within these settings has significant consequences for both the people and their natural and/or built environments (Gifford, 2013). Therefore, it is important to understand how contextual factors of the built environment have the potential to enable sustainable occupant behavior. Human behavior plays a major role in the conditioning of an environment (Steg & Vlek, 2009). Engaging building occupants in supportive, pro-environmental human behavior is now an essential issue in green building (Morgan, 2012). As people spend so much of their time indoors, activities in these settings should be given more research attention. Little documentation describes the extent to which sustainable work environments promote pro-environmental behavior (Kirk, 2010). There is an insufficient gap in the literature explaining PEBs exclusively in the workplace. Therefore, a more acute perspective of workplace settings and their influence on employee PEBs behaviors is necessary.

1.2 Research Objective

Various settings and elements of the built environment which have the capability of influencing the way people use and experience the space (Fischer, et al., 2012; Kirk, 2010; Kopec, 2006; Lee, 2007; Lockton, 2013; Manning, 2009; Rapoport, 1994). In the context of Cairo, this is a relatively new topic. The objective of this exploratory study is to investigate the pro-environmental behaviors of occupants in sustainable work environments.

– The aim is to understand the extent to which PEBs are promoted or enhanced in a sustainable work environment.

1.3 Research Question

The main question tackled in this study is:

– To what extent does the sustainable environment affect and enhance the pro-environmental behaviors of its occupants?

2 Literature Review

This research will expand upon the existing body of knowledge concerning the workplace and PEBs. References from various fields such as social and environmental psychology are necessary and will be drawn accordingly.

2.1 Leadership in Energy and Environmental Design (LEED)

Green building rating systems have been developed to assess the environmental impact and resource efficiency of buildings (Pretlove & Clarke, 2010). Leadership in energy and environmental design (LEED) was created in 1998 and is now the most widely used green building rating system in the world. LEED is a global symbol for the achievement of sustainability (USGBC, LEED for New Construction & Major Renovations, 2005). LEED offers certifications for all project types, with pre-determined credit requirements tailored to the project type. These requirements guide the design process and suggest methods to design and construct high-performance green buildings. LEED awards buildings with certified, silver, gold, or platinum certifications using a credit point system. This is based on the project's evaluation according to a fixed criterion (Prakash, 2005).

2.1.1 LEED BD + C: New Construction (LEED-NC)

LEED BD + C: New construction (LEED-NC) category is one of the six types of certificates offered. LEED-NC is designed to evaluate commercial, institutional, educational, and retail buildings, among others, with a focus on office buildings (Prakash, 2005; USGBC, 2005). According to LEED v4.1, the most recent version of LEED-NC, there are eight major credit categories: Location and Transportation, Sustainable Sites, Water Efficiency, Energy and Atmosphere, Materials and Resources, Indoor Environmental Quality, Innovation, and Regional Priority (USGBC, U.S Green Building Council, 2019).

The LEED-NC system has 110 obtainable credits. Each credit amounts to a number of points; thus, the maximum number of attainable points is 139 (USGBC, U.S. Green Building Council, 2019). The rating system categories are as follows:

- Certified: 40–49 points
- Silver: 50–59 points
- Gold: 60–79 points
- Platinum: 80+ points

Kirk, (Sustainable Environments and Pro-Environmental Behavior, 2010), identified credits associated with the LEED-NC certification that may have the potential to influence occupant pro-environmental behaviors. The study analyzed all prerequisites and credits (v2.1) for their potential to influence behaviors of building occupants and identified the following as applicable (Table 1).

2.1.2 Pro-Environmental Behaviors (PEBs)

To first define the behaviors considered in the study, it is necessary to understand the specific LEED initiatives and campaigns applied by the LEED-certified building. According to the literature, LEED-NC certification requires credits from eight different categories. Among these are the Materials and Resources prerequisite (MRp1) which provides occupants with storage for the collection of recyclables; Sustainable Sites credits (SSc4.1, SSc4.2) which provide access to public transportation, bicycle storage, changing rooms; and Environmental Quality credits (EQc6.1, EQc6.2) which provides occupant temperature and lighting controls, as shown in Table 1. The selected credits, shown in Fig. 1, are all dependent on occupant behaviors and are only as beneficial as the number of occupants engaging with them.

This suggests the availability of the necessary environmental infrastructure to facilitate the following behaviors:

- energy-saving behaviors
  The efficient use of plug loads, temperature and lighting controls.
- resource-saving behaviors
  Minimal consumption of materials and resources; driving and printing less.
- sustainable waste management behaviors
  Recycling behaviors

3 Methodology

3.1 Scope of Research

According to the literature, LEED-NC certification incorporates strategies that require occupant participation to lower negative environmental influence and operational costs. The implementation of these strategies has the potential to enable employee PEBs. Therefore, for this research, LEED-NC standards will be adopted as a benchmark for sustainability.

Table 1 LEED-NC credits requirements that provide room for occupant involvement (USGBC, 2019)

Credit Category	Credit	Requirement	Intent
Sustainable Sites (SS)	SSc4.1	Provision of access to alternative transport methods such as railway stations, bus stops, and other rideshare options.	Decrease potential pollution and impacts on land development. This builds the necessary environmental infrastructure that could alter employee car commutes and minimize car use
	SSc4.2	Provision of bicycle racks, bicycle storage, shower/changing facilities in/near the building for a percentage of the occupants	
Materials + Resources (M + R)	M + R prerequisite	Provision of areas dedicated to collection/storage of recyclables throughout the building	Waste reduction by occupants
Indoor Environmental Quality (IEQ)	EQc6.1	Provision of occupant lighting controls for 90% of occupants	Controllability of lighting, temperature settings to foster productivity, comfort, and occupant well-being
	EQc6.2	Provision of occupant temperature controls for 50% of occupants; provision of operable windows for occupants that work 6 m inside and 3 m to either side of an operable window	

Fig. 1 LEED-NC credits related to occupant pro-environmental behaviors

3.2 Data Collection

To tackle the research objective, the strategies identified in Table 2 were designed. Strategies employ the use of online occupant surveys, informal interviews, and observations. Interviews provide the needed statements utilized in survey questions. Survey questions were mostly multiple-choice questions based on results from interviews and offer a larger sample of data collection. Surveys were designed based on previous studies and collected self-reported employee behaviors in two workplace settings (Banwo & Du, 2019; Cudmore, 2015; Fishbein & Ajzen, 1975). Finally, observations complement the obtained results with added notes.

To measure PEBs, occupants are asked to indicate the frequency in which they engage in specific energy-saving, resource-saving, and sustainable waste management behaviors on a Likert scale of 1 (Never) to 5 (Always). A low score indicates a low engagement in a specific behavior.

3.3 Case Study Selection

People spend the majority of their time indoors and at work. Thus, the rational selection for the setting is the workplace and more specifically the office setting. This research explores employee PEBs in two office settings: one LEED-certified and one non-LEED-certified office building. This allows us to compare the impact of the different office environments on occupant behaviors and reduces variables that could affect or influence behaviors. Additionally, buildings must also be located near one other.

Finally, research shows that for a newly learned behavior to become a natural habit, it will take around 18–254 days. The length of this period depends on several factors, such as the frequency and the length of time spent practicing the behavior (Mulvey, 2011). Naturally, there is no set answer for how long it takes for a behavior to become a habit. It depends on the nature of the behavior you are trying to change and that varies from person-to-person. This was taken into consideration in the range of the building's operational time. For the aforementioned reasons, we formed our case study criteria. The following are the criterion for inclusion:

- Located nearby
- Standalone buildings
- Operational: over 12 months.

New Cairo's business and commercial district is home to a large number of head offices in Egypt; therefore, a prime location as a base for this study. As of September 2019, there are 18 LEED-certified buildings in Egypt, 9 of which are LEED-NC certified and have been operational for at least 12 months (Table 3). However, the Crédit Agricole Egypt (CAE) headquarters is the only LEED-NC certified building located in New Cairo. Accordingly, the CAE was selected as the sustainable building for this study. While several New Cairo office buildings satisfied the criteria, the building was selected based on convenience and the ease of connection. In this case, Al Baraka Bank Egypt (ABBE) was selected as the traditionally designed building in this research.

4 Case Studies

4.1 Credit Agricole Egypt

The Crédit Agricole Egypt (CAE) headquarters is situated in New Cairo and was built by the Engineering Consultant Group (ECG) in 2014 (Khaled, 2017) (Fig. 2). It was awarded the LEED-NC 2009 platinum certification in March 2016 and was placed in the top 6% of all LEED-NC 2009 activities (GBIG, n.d.).

Table 2 Research strategies, data collection, and methods of analysis to tackle research questions

Research objectives	Research question	Strategies	Data collection	Analysis of results
Recognize the extent to which an environment can shape occupant PEBs	To what extent does the sustainable environment influence PEBs?	Investigate occupant PEBs in LEED-certified and non-LEED-certified buildings / comparative analysis between occupant behaviors	Survey: self-reported behaviors	Statistical analysis: two-sample t-test assuming unequal variances

Table 3 LEED-NC certified buildings in Cairo

Building Typology	Company	Rating	Credits
Communication & IT	Outsourcing Service Building - Mb4	*silver*	50/110
Industrial	L'oreal Pyramids	*silver*	54/110
Manufacturing	Mars Egypt Site Development	*gold*	61/110
Investor/Bank	HSBC Bank Egypt Cairo Processing Centre	*gold*	44/69
Investor/Bank	**Credit Agricole Egypt Headquarters**	***platinum***	**81/110**
Oil & Gas Company	EMEC Headquarters	*gold*	66/110
Warehouse	Aramex Mashreq Warehouse—Cairo	*silver*	51/110
Distribution	Info Fort Warehouse Project	*silver*	53/110
Corporate office	Dar Al-Handasah New Premises In Cairo	*gold*	62/110

Fig. 2 Crédit Agricole Egypt headquarters

CAE prides itself on its sustainability governance. Some of the CAE sustainability initiatives include (CAE, 2019):

1. **Efficient Consumption of Resources**
   - "*Go digital*" initiative focused on paper reduction in the office by digitalizing processes in the bank. Staff awareness campaigns were also created, and e-statements were then launched to allow customers to partake in this initiative.
2. **Paper recycling**
   - A CAE initiative that collects and delivers old paper (from their branches) to a charity organization that turns them into handmade products. These products are then sold, and all profits finance the underprivileged makers of these products.
3. **Renewable Energy-Use**
   - A 200KWp-capacity solar panel station was built on the rooftop of the CAE head office, reducing the energy consumption rate by 7%.
   - Pursuing an increase of renewable energy usage in the rest of its branches.
4. **Clean Energy-Use**
   - Provide employees with buses to reduce gas emissions.
   - Motivate employees to cycle with on-site bicycles (Khaled, 2017).

4.2 Al Baraka Bank Egypt

Al Baraka Bank Egypt (ABBE), shown in Fig. 3, is a branch of the Al Baraka Banking Group (ABG). ABG headquarters are situated in Bahrain, with Al Baraka Bank Egypt taking its head office in South 90 street, New Cairo. The building was constructed and completed in 2018 by Projacs, an Egis Group company. It is a corporate and institutional building that has a built-up area of $8500m^2$ (Projacs, 2018).

Fig. 3 Al Baraka Bank Egypt head office

The Al Baraka head office consists of 7-floor levels. The ground floor is open to the public and makes up the bank's branch, along with the first floor. The remaining 5 floors are occupied with offices. Open-plan shared offices are the most common throughout the building with very minimal small enclosed office spaces.

5 Analysis of Results

5.1 Socio-economic Data Summary

General demographics of a sample of employees from two banks were collected through occupant surveys. A total of 61 full-time employees contributed to this study. Participants were selected based on convenience sampling. A convenience sample of the ground and third-floor employees in Al Baraka Bank received access to printed surveys to fill out. 37 Al Baraka Bank employees manually returned completed surveys. Soft copies of a similar survey were sent out to Credit Agricole employees and 24 employees responded. The average age of the respondents ranges from 30 to 50 years old. 64% of respondents were male and 36% were female. Table 4 presents the socio-economic status of the sample selection from both sites.

5.2 Inferential Statistics

LEED-NC certified buildings incorporate methods of occupant involvement, thus may have the potential to influence pro-environmental behaviors. To examine the relationship between sustainable design and pro-environmental occupant behaviors, behaviors linked to LEED-NC credits were studied. Surveys collected data regarding three types of behaviors: energy-saving, resource-saving, and sustainable waste management behaviors. Data were extracted to determine the frequency of participation in pro-environmental behaviors. Inferential statistics were employed to test the significance of the difference between the two population means for each survey question. Descriptive and inferential statistics were computed using the Excel add-in data analysis software. The summary of the results is organized in Tables 5, 6, 7, 8, 9, and 10.

A t-Test: two-sample assuming unequal variances (Keller & Warrack, 2003) was conducted to test the significant difference between the mean population responses for Credit Agricole Egypt (CAE) and Al Baraka Bank Egypt (ABBE). Ultimately, the test was used to determine the effect of different environments on occupant pro-environmental behaviors in two office buildings.

The following hypotheses were tested:

$$H_0 : (\mu_1 - \mu_2) \leq 0, H_1 : (\mu_1 - \mu_2) > 0$$

where μ_1 *is the CAE mean population response and* μ_2 *is the ABBE mean population response.*

H_0 states that μ_1 is greater than or equal to μ_2. This would suggest a greater frequency of pro-environmental behaviors in CAE than that of ABBE. Alternatively, H_1 states that the μ_2 is greater than μ_1. Thus, these hypotheses can identify the environment in which occupants are more inclined to engage in environmentally friendly behaviors.

Table 4 Socio-economic status of participants

Socio-economic oeconomic		Credit Agricole		AlBaraka Bank	
Category		Number	%	Number	%
Gender	Female	9	37 5	13	35
	Male	15	62 5	24	65
Age	Under 30	2	8	7	20
	30–50	22	92	22	65
	50*	0	0	5	15
Education	High School	0	0	3	8 5
	Bachelor's Degree	24	100	29	83
	Master's Degree	0	0	3	8 5
Type of Work	Administration	15	62	2	6
	Supervisory	5	21	12	35
	Customer Service	0	0	1	3
	Accountant	0	0	IS	44
	Support Services	0	0	3	9
	Credit Bank Officer	0	0	1	3
	Technical	4	17	0	0
Duration at Current Workplace	Under 6 months	0	0	2	6 5
	6-12 month$	0	0	2	6 5
	Over 1 year	24	100	26	87

5.2.1 Energy-Saving Behaviors

LEED Environmental Quality credits (EQc6.1 + EQc6.2) provide the controllability of lighting and temperature to a majority of building occupants. This helped define the use of temperature and lighting controls as the main energy-use behaviors to be studied in this research. Additionally, other energy-using behaviors relevant to the office setting, including computer and elevator use, were also studied. Occupants were asked about the frequency in which they engaged in several energy-saving behavior statements. The statements described simple energy-saving actions in which they can engage in on a day-to-day basis. Occupants scored statements according to their frequency of participation, from 1 (Never) to 5 (Always). A low score indicates a low involvement in a specific behavior.

It is important to note that building policies put into effect are not always aligned with LEED credits. This was confirmed by the Credit Agricole Head of Property, who acknowledged the use of automated lighting and HVAC systems in shared office spaces in the building. Such a policy aims to limit unnecessary HVAC and lighting use and, thus, limits occupant control in the space. Limiting occupant lighting and temperature controls have a direct impact on the results of this portion of the study.

Table 5 summarizes the energy-saving behavior results in both environments using descriptive (mean, variance, standard deviation) and inferential statistics (t stat, p-value, and df). Results show significantly higher scores in overall energy-saving behaviors in Al Baraka Bank ($M = 2.748$, SD = 1.44) than in Credit Agricole ($M = 2.435$, SD = 1.01); conditions; ($t(372) = -.469$), $p = 0.007$. This indicates a greater frequency of participation in energy-saving behaviors in Al Baraka Bank than in Credit Agricole, thus, rejecting the null hypothesis and accepting the alternative hypothesis.

Table 5 Total energy-saving behavior—Data summary

Total energy-saving behaviors		
	Credit Agricole (CAE)	Al Baraka Bank (ABBE)
Mean	2.435	2.748
Variance	1.03	2.08
SD	1.01	1.44
t Stat	-2.469	
t Critical	1.649	
$P(T <= t)$	0.007	
df	372	

Table 6 further breaks down the scores of each behavior statement allowing easy comparison between different behaviors in Credit Agricole and Al Baraka Bank. Sustainable plug load behaviors (*computer and elevator use*) show higher mean scores and, thus, higher participation than HVAC and lighting use behaviors in occupants of both buildings. Results indicate statistically significant differences in the two populations in the following statements:

Table 6 Breakdown of energy-saving behavior statements—Data summary

Energy saving behaviors												
	"I switch the lights oft during the day"		*"I rely on blinds/curtains to help regulate the temperature indoors"*		*'I rarely use the AC"*		*"I use the fan instead of the AC"*		*'I turn my computer off when not in use"*		*"I take the stairs instead of the elevator"*	
	CAE	ABBE	CAE	ABBE	CAE	ABBE	CAE	ABBE	CAE	ABBE	CAE	ABBE
Mean	1.929	3.265	2.333	1.970	1.958	2.853	1.900	1.647	3.174	3.938	3.250	2.742
Variance	0.379	2.261	0.667	1.593	0.650	1.463	0.832	1.387	1.150	1.544	0.457	1.531
SO	0.616	1.504	0.816	1.262	0.806	1.209	0.912	1.176	1.072	1.243	0.676	1.237
t Stax	−4368		1.319		−3.379		0.881		−2.436		1.942	
t critical	1.679		1.674		1.673		1.677		1.675		1.677	
p-value	0.000035		0.096		0.001		0.191		0.009		0.029	
df	46		54		56		48		51		48	

"I switch off the lights during the day"

Results show a significant difference in scores regarding lighting use during the day in Credit Agricole ($M = 1.928$, SD = 0.6157) and in Al Baraka Bank ($M = 3.3$, SD = 1.503) conditions; $t(46) = |-4.368| > t$ critical = 1.678, $p < 0.01$. Results indicate that Al Baraka Bank occupants switch off unnecessary lights more frequently than Credit Agricole occupants that may be the result of the lack of lighting controls for Credit Agricole occupants.

"I rarely use the AC."

Results show a significant difference between average population responses in Credit Agricole ($M = 1.958$, SD = 0.806) and in Al Baraka Bank ($M = 2.853$, SD = 1.209 conditions; ($t(56) = |-3.379| > t$ critical = 1.673, $p < 0.01$). Although both Al Baraka Bank and Credit Agricole occupants exhibited low levels of participation in said action, Al Baraka Bank occupants showed stronger levels of involvement. A lack of temperature controls for Credit Agricole occupants may be the cause.

"I take the stairs instead of the elevator"

Results for this statement indicate a higher score in occupant behaviors in Credit Agricole (M = 3.25, SD = 0.676) than in Al Baraka Bank (M = 2.742, SD = 1.237; conditions; (t 48) = 1.942, > t critical = 1.677, $p < 0.05$). This suggests a greater inclination to taking the stairs in Credit Agricole occupants rather than those in Al Baraka Bank. This difference may be attributed to the number of floors in each building; Credit Agricole consists of 3-floor levels while Al Baraka Bank consists of 7.

"I turn my computer off when not in use"

Computer use behaviors significantly differed in Credit Agricole ($M = 3.174$, SD = 1.072) and in Al Baraka Bank ($M = 3.938$, SD = 1.243) conditions; $t(51) = 2.436 > t$ critical = 1.675, $p < 0.01$. Credit Agricole occupants showed a lower frequency of engaging in this specific behavior.

5.2.2 Resource-Saving Behaviors

LEED Sustainable Sites credits (SSc4.1 + SSc4.2) provide alternative transport methods, rideshare options, and the necessary infrastructure to cycle on-site. This helped defined sustainable commutes as the main resource-saving behaviors to be studied in this research. Other resource-saving behaviors to be studied are all day-to-day actions relevant to the office setting and based on Credit Agricole sustainability Initiatives. Credit Agricole initiatives including the "efficient consumption of resources," and "paper recycling" both aim to reduce the amount of paper and materials used in the office. Thus, paper and resource-use behaviors are also studied.

Table 7 summarizes the resource-saving behavior results in both environments using descriptive (mean, variance, standard deviation) and inferential statistics (t stat, p-value, and df). Results show significantly higher scores in overall resource-saving behaviors in Credit Agricole ($M = 2.755$,

Table 7 Resource-saving behaviors—Data summary

Total resource-saving behaviors		
	Credit Agricole (CAE)	*Al Baraka Bank (ABBE)*
Mean	2.755	2.461
Variance	1.90	2.01
SD	1.38	1.42
t Stat	2.218	
t Critical	1.648	
$P(T <= t)$	0.014	
df	444	

Table 8 Commute behaviors—Data summary

Commute behaviors								
	"I walk.."	*"I ride a bicycle.."*			*I take public transportation*		*"I carpool.."*	
	CAE	ABBE	CAE	ABBE	CAE	ABBE	CAE	ABBE
Mean	2.737	1.607	1.563	1.231	3.136	2.172	3.417	2.710
Variance	1.760	1.581	0.529	0.745	1.647	1.362	1.471	2.146
SD	1.327	1.257	0.727	0.863	1.283	1.167	1.213	1.465
t Stat	2.926			1.335		2.762	1.957	
t critical	1.687			1.688		1.681	1.674	
$P(T <= t)$	0.003			0.095		**0.004**	**0.028**	
df	37			36		**43**	**53**	

SD = 1.38) than in Al Baraka Bank (M = 2.461, SD = 1.42); conditions; *(t (444) = 2.218), p < 0.05*. This indicates significantly higher scores in resource-saving behaviors in Credit Agricole than in Al Baraka Bank, thus, accepting the null hypothesis and rejecting the alternative hypothesis.

Table 8 breaks down the results of the commute behaviors. Results indicate higher engagement in "*public transportation*" and "*carpool*" commutes than in "*walking*" or "*biking.*" Results show significant differences in the following statements: Results are further explored according to the following statements:

"I walk".

Results indicate a highly significant difference in terms of walking behaviors of occupants in Credit Agricole (M = 2.737, SD = 1.327) and Al Baraka Bank (M = 1.607, SD = 1.257; conditions; (t(37) = 2.926) > t critical = 1.687, p = 0.001). This shows that Credit Agricole occupants walk more frequently than Al Baraka Bank occupants.

"I take public transportation."

Results indicate that the provision of public transport methods may have had a positive impact on the occupant commute choices in Credit Agricole *(M = 3.136, SD = 1.283)* than those of Al Baraka Bank *(M = 2.172, SD = 1.167; conditions; (t(43) = 2.762) > t critical = 1.681, p = 0.004).* This suggests that taking public transportation is more common in Credit Agricole occupants.

"I carpool."

The provision of rideshare options may have promoted carpooling in Credit Agricole (M = 3.417, SD = 1.213) than in Al Baraka Bank (M = 2.71, SD = 1.465; conditions; (*t* 53) = 1.957) > *t* ritical = 1.674, *p < 0.05). Results show a significantly higher response in Credit Agricole occupants.*

"I ride a bicycle."

Results indicate that the provision of the necessary infrastructure to allow for cycling on-site has not had a major effect on Credit Agricole occupants *(M = 1.563, SD = 0.727)* than Al Baraka Bank occupants without the proper infrastructure *Al Baraka Bank (M = 1.231, SD = 0.863; conditions; (t(36) = 1.335) > t critical = 1.688, p = 0.095).* Although Credit Agricole occupants performed better than Al Baraka Bank occupants, the difference was insignificant. This indicates that the environment did not successfully promote cycling behaviors in its occupants. However, this may be due to several unexplored reasons such as societal factors and safety issues in the car dominant city of New Cairo.

Table 9 breaks down the results of office resource-saving behaviors in Credit Agricole and Al Baraka Bank occupants. Sustainability initiative: "efficient consumption of resources" aims to digitalize bank processes to limit the consumption of resources in the workplace.

Mean population scores suggest that occupants occasionally engaged in paper reduction behaviors over other forms of resource-saving behaviors. Results reveal an insignificant difference in the majority of office resource-saving behaviors in the two populations. This suggests that "paper recycling" and "efficient consumption of resources" initiatives were not so successful in behaviors such as "*using scrap paper for memos,*" "*reusing envelopes,*" "*using reusable bottles,*" and "*using reusable rather than disposable cups.*" However, results indicate a statistically significant difference in the following statement:

"Printing double-sided."

Sustainability initiatives may have promoted paper reduction in printing behaviors of Credit Agricole occupants (M = 3.5, SD = 0.722) than in Al Baraka Bank occupants (M = 2.5, SD = 0.916; conditions; (t(54) = 4.567) > t critical = 1.674, p < 0.01). Results indicate significantly higher scores in Credit Agricole occupants than in Al Baraka Bank occupants.

Table 9 Office resource-saving behavior—Data summary

Resource-saving behaviors										
	"Printing double-sided"		*"Use scrap paper for memos"*		*"Reuse envelopes"*		*"Use reusable bottles"*		*"Use reusable rather than disposable cups"*	
	CAE	ABBE	CAE	ABBE	CAE	ABBE	CAE	ABBE	CAE	ABBE
Mean	3.500	2.500	3.250	3.000	3.208	3.118	2.208	2.000	3.125	3.457
Variance	0.522	0.839	0.717	1.882	0.868	1.865	1.824	1.688	1.071	1.491
SD	0.722	0.916	0.847	1.372	0.932	1.365	1.351	1.299	1.035	1.221
t Stat	4.567		0.864		0.301		0.584		-1.125	
t critical	1.674		1.673		1.673		1.677		1.674	
P(T <-t)	0.000015		0.196		0.382		0.281		0.133	
df	54		56		56		49		54	

5.2.3 Sustainable Waste Management Behaviors

LEED Materials + Resources prerequisite (M + R) helped defined sustainable waste management behavior to be studied in this research. This credit aims to reduce occupant waste by providing areas for the collection and storage of recyclables. Thus, occupant recycling behaviors were examined.

Table 10 presents the descriptive statistics (*mean, median, mode, and variance*) and the inferential statistics (*t stat*) of the results relating to the sustainable waste management behaviors of Credit Agricole and Al Baraka Bank employees.

"Recycle materials (plastic, cardboard, paper, glass, etc.)"

The provision of the proper infrastructure for the collection/storage of recyclables may have significantly impacted occupant recycling behaviors in Credit Agricole ($M = 2.833$, SD = 0.917) than those without the necessary infrastructure, Al Baraka Bank ($M = 1.485$, SD = 0.795; conditions; (t (45) = 5.793) > t critical = 1.679, $p < 0.01$). Results show that the M + R prerequisite was successful in promoting sustainable waste management behaviors in Credit Agricole occupants in comparison with Al Baraka Bank occupants.

Table 10 Waste management behaviors—Data summary

Waste management behaviors		
	"Recycle materials (plastic, cardboard, paper, glass, etc.)"	
	Credit Agricole (CAE)	Al Baraka Bank (ABBE)
Mean	2.833	1.485
Variance	0.841	0.633
SD	0.917	0.795
t Stat	5.793	
t critical	1.679	
P(T <= t)	0.00000063	
df	45	

6 Conclusions and Recommendations

Exploiting design settings in a way that promotes sustainable behaviors would be an effective way to combat negative environmental impacts caused by human activities. This would substantially contribute to those dedicated to minimizing our ecological footprint.

This paper explores the influence of sustainable design on pro-environmental occupant behaviors. The LEED-NC certification was used as a standard measure for sustainable building. Several LEED credits and prerequisites incorporate sustainable strategies whose efficiency and success are determined by future occupant behaviors. Accordingly, LEED prerequisites and credits defined energy-saving, resource-saving, and sustainable waste management as the main themes of the behaviors studied in this research.

Results imply that LEED credits and prerequisites may have positively impacted the sustainable waste management (M + R prerequisite) and resource-saving commute (SSc4.1) behaviors of LEED-certified building occupants. Insignificant differences were found between "cycling" (SSc4.2) behaviors responses of both populations, regardless of the available infrastructure. Insignificant differences were also found in office resource-saving behaviors in the two populations. However, mean population scores reveal that occupants most frequently engaged in office resource-saving behaviors than in any other behavior. Finally, results also showed significantly lower energy-saving lighting (EQc6.1) and temperature (EQc6.2) control behaviors in LEED-certified building occupants than in non-LEED-certified buildings occupants. This may be the result of the automated systems which caused a lack of occupant temperature and lighting controls in the sustainable building. Therefore, this should be taken into consideration. Further analyses are required under different settings to determine whether energy-saving behaviors can be significantly impacted in LEED-certified environments.

The topics tackled in this paper merely shed a light on the potential sustainable design settings have to shape pro-environmental occupant behaviors. Further studies should be applied in various settings to better describe the relationship between design settings and occupant behaviors.

References

Banwo, A. O., & Du, J. (2019). Workplace pro-environmental behaviors in small and medium-sized enterprises: An employee level analysis. *Journal of Global Entrepreneurship Research, 9*(34), 1–20.

Blok, V., Wesselink, R., Studynka, O., & RonKemp. (2015). Encouraging sustainability in the workplace: A survey on the pro-environmental behaviour of university employees. *Journal of Cleaner Production, 106*(1), 55–67.

CAE. (2019). *Crédit Agricole Egypt*. Retrieved September 25, 2019, from https://www.ca-egypt.com/en/corporate-social-responsibility/respect-the-environment/?bank_segment=personal-banking

Cudmore, T. (2015). *Delivering workplace pro-environmental behaviour change using evidence-based practice*. Institute of Energy and Sustainable Development.

Fischer, J., Dyball, R., Fazey, I., Gross, C., Dovers, S., Ehrlich, P. R., Brulle, R. J., Christensen, C., & Borden, R. J. (2012). Human behavior and sustainability. *Fronteirs in Ecology and the Environment*, 153–160.

Fishbein, M. A., & Ajzen, I. (1975). *Belief, attitude, intention, and behaviour: An introduction to theory and research*. Addison-Wesley.

GBIG. (n.d.). *GBIG*. Retrieved October 30, 2019, from Green Building Information Gateway: http://www.gbig.org/activities/leed-1000005034

Gifford, R. (2013). Environmental psychology matters. *Annual Review of Psychology, 65*(1), 541–579.

Keller, G., & Warrack, B. (2003). *Statistics for management and economics* (6th ed.). Brooks/Cole.

Khaled, F. (2017). *Invest Gate*. Retrieved April 21, 2019, from https://invest-gate.me/features/egypts-young-architects-seek-to-turn-construction-industry-green/

Kirk, J. L. (2010). *Sustainable environments and Pro-environmental behavior*. Interior Design Program.

Kopec, D. (2006). *Environmental psychology for design*.

Lee, S.-A. (2007). Environmental psychology research in practice: Thoughts on sustainable architecture and urban design. In *Environmental psychology: Putting research into practice* (Chapter 2). Cambridge Scholars Publishing.

Lockton, D. (2013, March 20). *The Guardian*. Retrieved October 9, 2019, From Designing For Sustainability: Making Green Behaviour Easy: https://www.theguardian.com/sustainable-business/design-sustainability-green-behaviour

Manning, C. (2009). The psychology of sustainable behavior. *Minnesota Pollution Control Agency*.

Morgan, E. (2012). *Does design affect behavior? A case study of Pomona and Sontag Halls*.

Mulvey, J. (2011, August 31). *Redbird*. Retrieved September 25, 2019, from https://www.redbirdonline.com/blog/how-long-does-it-take-adopt-new-behaviour

Prakash, P. (2005). *Effect of indoor environmental quality on occupant's perception of performance: A comparative study*.

Pretlove, S., & Clarke, J. (2010). *Sustainable buildings that encourage sustainable behaviour*.

Projacs. (2018). *Projacs*. Retrieved November 27, 2019, from http://www.projacs.com/Home/Projects/Our-Projects/Project-Details?selectCountry=&ProjectID=1abb03ae-4ca3-4836-9db8-4695cc3fd927

Rapoport, A. (1994). On 'the invisible in architecture': An environmental-behaviour studies perspective. In O. Bouman, & R. Toorn, *The invisible in architecture* (pp. 66–73). Academy Editions.

Steg, L., & Vlek, C. (2009). Encouraging pro-environmental behaviour: An integrative review and research agenda. *Journal of Environmental Psychology*, 309–317.

USGBC. (2005). *LEED for new construction & major renovations*.

USGBC. (2019, April 9). *U.S. Green Building Council*. Retrieved April 12, 2019, from https://new.usgbc.org/leed-v41#bdc

Evaluation of the Sustainable Building Materials for Economic Housing in Egypt

Bishoy Magdy Tawfeeq, Hossam Eldin Hassan Othman Elborombaly, and Ali Kamal Ali Altwanasy

Abstract

Buildings are considered the national valuable assets need to preserve and maintain, because of their functional, civilizational or historical value. The buildings are suffering from design problems, such as architectural drawings and taking decisions that take many periods of time or implementation and maintenance by the implementation of the structure of buildings and finishes … etc. Sustainable housing development has been become an important issue that needs to be taken into subject to provide adequate housing for families, but methods of work are still inadequate for the needs of users. The importance of this subject is to find suitable solutions to provide adequate housing for low-income families through the availability of appropriate methods and materials that achieve the appropriate cost, quality and sustainability as well. In Egypt, people are suffering from providing suitable housing for the social and making the buildings environmentally friendly and administratively successful. Egypt is characterized by a continental climate with strategic locations distinct from any other location and has the most important natural resources that can be used but quickly can be used badly and the buildings are incompatible with nature and lack of respect for the site and this negatively affects the community. The problem is the failure to take advantage of local materials that help to reduce costs, time and the completion of the implementation period. The research aims to identify the housing projects in Egypt and the extent to the criteria of sustainability are met or not available. Consequently, the focus of this study is not only to look at these projects in problems but also to focus on developing future visions and how to implement sustainability elements, especially the idea of sustainable construction of the construction industry as well as on individuals and construction processes. This paper presents studying and analysis of economic housing projects in Egypt and the evaluation of the elements of sustainability achieved or vice versa, such as using of local building materials and how to apply them to achieve the sustainability of those projects and the requirements of users. The research finds that the lack of use and utilization of local materials leads to high costs of housing units and increasing problems, which need to be maintained periodically because of the inability of those materials of weather conditions.

Keywords

Purchasing power • New cities • Local materials • Economic housing

1 Introduction

The way of adequate housing to obtain them is a real problem for many sectors of households in Egypt, especially those with low incomes. All families are facing many of difficulties in providing the required physical housing. The principle of facilitating access to housing includes the provision of multiple housing patterns, different levels of costs that can all meet the needs of multiple segments of households and provide housing to low-income families (Simms & Kamal, 2008). The mechanism should be including programs to develop the use of housing project sites, adjust building densities and requirements to achieve higher a measure of the economic side of the adoption of these programs for flexible policies for the development and maintenance of existing residential buildings (Qutb, 2013).

The problem is also the lack of use of local materials that help to reduce costs, time and completion of the

B. M. Tawfeeq (✉) · H. E. H. O. Elborombaly
Faculty of Engineering, Ain Shams University, Cairo, Egypt
e-mail: 9804@eng.asu.edu.eg; eng.Bishoy-89@hotmail.com

A. K. A. Altwanasy
Department of Architecture, Faculty of Engineering, October 6 University, Cairo, Egypt

C. Alalouch et al. (eds.), *Towards Implementation of Sustainability Concepts in Developing Countries*, Advances in Science, Technology & Innovation, https://doi.org/10.1007/978-3-030-74349-9_12

implementation period that leads to people resort to slowness and encroachment on the land because of their inability to bear the costs without the existence of mechanisms to reduce them and the occurrence of thefts from energy sources, such as: electricity, water, lack of knowledge and awareness of the role that sustainable design strategies play in minimizing these impacts as shown Fig. 1.

In recent years, urban sectors are also have lack of non-renewable sources of energy as the result of increased daily consumption to provide for user needs and appear in the pressing environmental issues. The states are seeking to develop strategies to control environmental and energy problems to meet user needs and leave part for future generations. Technology and rapid development have also emerged in many sectors to provide for the needs of users including: (agricultural, industrial, commercial … etc., which have led to environmental problems and increased pollutions rates, leading health, social, cultural … etc. (Mayson, 2014). All of which need to develop strategies, foundations, standards and how to apply sustainable development housing (UN Habitat Housing, 2019) as shown Fig. 2.

The research is based on the following questions:

i. What is the reflection on the dimensions of sustainable development in housing projects?
ii. What are the tasks of the role of the state in reviewing the housing problem and amending the legislation in housing projects?

2 Methodology

A. Theoretical Approach

It includes a comprehensive theoretical study of housing problems, relationship of low-income housing problem with income source and sustainable housing.

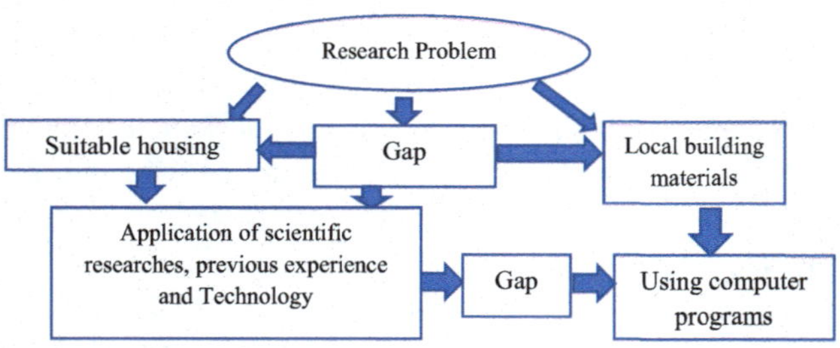

Fig. 1 Problem of providing suitable housing and using local materials

B. Comparative Analytical Approach

The study includes a comparative and analytical framework for analysis study of economic housing projects in new cities in Egypt and make of comparison and evaluation of the local examples in Egypt that suffer in achieving the housing projects for sustainable design and the availability of sustainability criteria from its components and services.

3 Housing Problems

Young people are the future of the state to achieve their needs, requirements, psychological comfort to help them to work and production significantly (Ahmed et al., 2012). The problem of housing is the increasing of population, internal and external migration, which led to the inability of the state to control of this problem of this phenomenon and the delay of legal procedures; thus, the absence of a future plan is to achieve adequate housing units for young people.

Housing projects were established in Egypt but these projects were failed to meet these basic needs of citizens so these problems are:

1. Internal space area in houses is small, within each unit of housing that does not suit most people and their needs to enlarge the space to make easy movement and doing their lives better.
2. The weakness of using building materials and finishing materials, which expose these establishments to environmental factors and therefore the occurrence of erosion and the appearance of cracks on the façades, which cause danger these facilities and the lives of citizens.
3. Facades are exposed to weather conditions, such as high temperatures inside spaces and direct lighting, which makes users feel discomfort inside spaces and thus the use and consumption of electric power, such as air conditioners for cooling spaces.
4. Lack of suitable open areas, such as public parks, services and long distance from their homes and places of work.

The process of building design and implementation is important to have direct and indirect effects on the environment, where the life cycle of buildings is affected by humans and the environment that need to provide principles and strategies to develop the process of sustainable architectural design (El-Attar et.al, 2018) as shown Fig. 3.

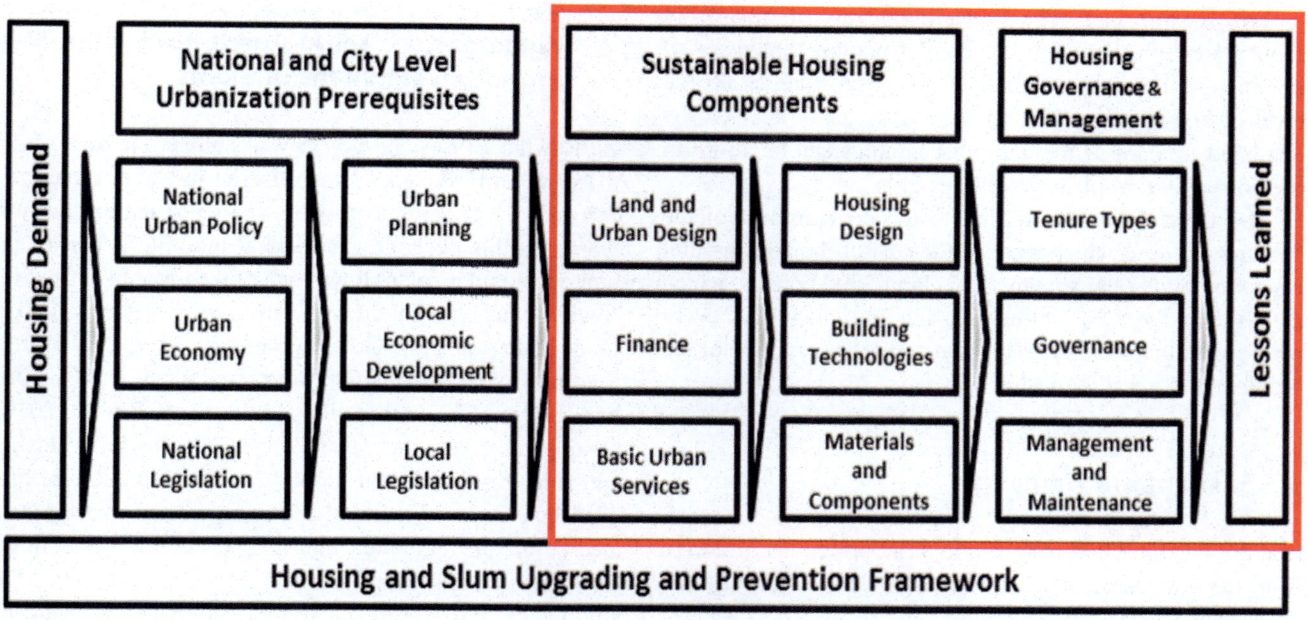

Fig. 2 Housing and slum upgrading and prevention framework for sustainable housing components

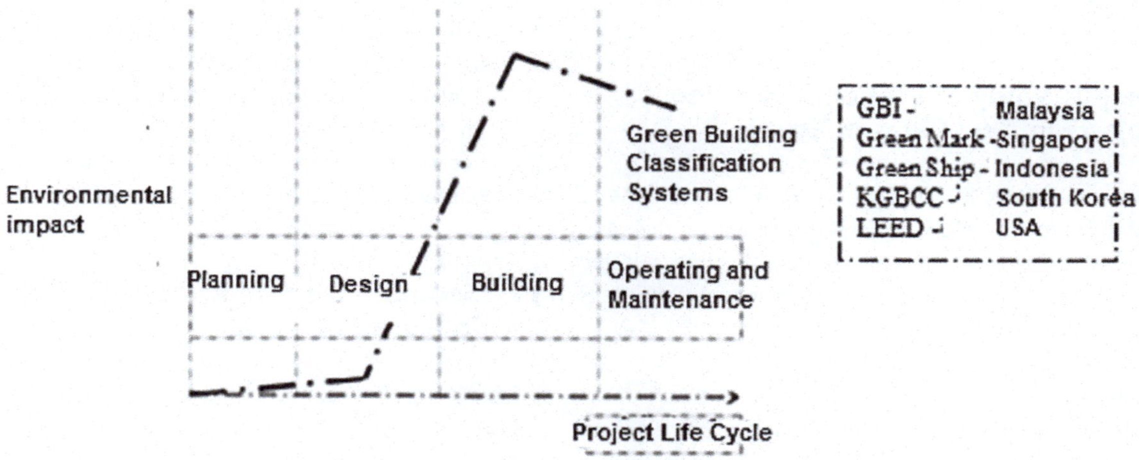

Fig. 3 Impact of buildings on the environment during the life cycle of building

4 Relationship of Low-Income Housing Problem with Income Source

Economic studies in low-income Egypt are generally defined as groups whose income level does not match their living requirements and have no savings or possibilities to access adequate housing through the formal housing market mechanisms. Previous research have indicated that the low-income population may reach 50% of the total households, while about 48.5% of Egyptian households see the source of income as insufficient for their purchasing power and monthly needs (Shawkat, 2014).

The Central Agency for Public Mobilization and Statistics (CAPMAS) has pointed out through income and expenditure research indicators for 2013, ranging from 400 to 1500 LE per month, representing 40% of the total population (Musa et al., 2014).

The problem of housing has increased of low-income people, which prevents their ability and access to adequate housing that meets the minimum basic requirements, which led to the resort to informal areas; the problem is shown by the following facts:

The first fact: There is a tight crisis in the housing of this category despite the large supply in the real estate market and manifestations of this crisis emerged in random gatherings.

The second fact: The prevailing levels of income in society are representing a social and economic dimension of low-income housing and lack of jobs opportunities on the level of housing that the family budget cannot bear on the one hand to increase the gap what families can be afford to own or value rent of the dwelling.

The third fact: It is related to the development of the housing unit cost which reached the cost of the housing unit from the popular housing in 1975 about 2800 pounds, which means that the cost of the housing unit increases every period with the stability of income and the magnitude of the problem and get the housing unit.

5 Sustainable Housing

It is the dwelling that follows the basic principles of design from the compatibility and observance of the natural environment with all elements: natural and industrial to achieve functional and environmental efficiency by providing comfort to users and reduce the negative impact on the environment and public health (Samir, 2018) as shown Table 1.

6 Analysis Study of Economic Housing Projects in New Cities in Egypt

The research includes the application of the foundations of sustainability criteria, which is a measurement on the samples of existing and new Egyptian cities that are selected according to the criteria contained in the research study as follows: (Jomaa & Mohammed, 2017).

1. The city should be able to accommodate as much of the population with the provision of suitable areas for them and also take a reserve to provide spaces of increasing the population to accommodate the future population.
2. The city should be existing with the requirement of different housing types (economic, medium and high) with no focus only on a particular housing type.
3. The target population should not be less than 500,000.
4. The project should be implementing at a rate of 60–70% of the city plan within a certain period of time.
5. The occupancy rate should be 60–70% of the city plan.
6. 60–70% of the facilities and infrastructure of the city plan are implementing.
7. The services in the plan should be of suitable areas to meet the needs of the population.

6.1 First Project: Al-Sayeda Zeinab Kindergarten (Tal Al-Aqarib Project): A Step in the Development of Slums

The Rawdat El Sayeda project was built in the area of Tal Al-Aqarib in El Sayeda Zeinab district in Cairo, on an area of 7 and a half acres to save the nest population that was erected on this place after removing the hill, as part of the slum development plan. It was one of the unsafe slums in the Sayyida Zeinab neighborhood. The gravity was classified as "inadequate housing" and is completely erected on a limestone hill, with a population of approximately 3500, representing about 500–600 families (Al Asmarat Project, 2019). It was planned that the citizen will receive his fully finished unit, fully furnished, "on the key," with a simple rent provided that the inhabitants of the same area, who were confined before the temporary transfer procedures. The building consists of a ground floor and 5 recurring floors. The buildings were designed according to three models that cater to the needs of the citizens living in the region as shown Fig. 4.

The quality of the finishing is done by luxury finishing system compared to social housing, adding that the facades of residential blocks are finished with Hashemite stone. Mashrabiyat was established in the outlets to take the Islamic form in architecture, pointing out that this architectural character is not designed in another project as shown Fig. 5 (El Mahrousa Project, 2019).

6.2 Second Project: Al Asmarat Project in Mokattam

In May 2016, President Abdel Fattah El Sisi was inaugurated the first and second phases of the project "Long live Egypt" for social housing in Asmaraat district in Mokattam. The Asmarat project is considered one of the most important national projects implemented in Cairo governorate (Grape Gheit Project, 2019). The project "Asmarat" is implemented by the Cairo governorate in cooperation with the fund (Long live Egypt) and the engineering authority in Mokattam to move the population of dangerous slums and is the hope of many eagerly waiting for completion to transfer to their new homes. It was built on an area of 65 acres. There are 124 buildings in the project. The number of housing units is 7304 housing units and 176 shops. The number of floors in each building in one building (ground + 9 typical floors) is shown Fig. 6 (Hilal and Mahdi, 2014).

Table 1 Basic principles of sustainable housing

Bask principles of sustainable housing	Energy efficiency	Efficient material handling	Efficient water handling	Sustainable architectural design	Functional efficiency—environmental quality
Site	– Accurate location selection – Take advantage of the natural potential of the site	– Use of natural materials – Use local materials to reduce transportation distances and cost	– Choose local plants with low water consumption – Take advantage of rain and flood water	– Respect for the characteristics of the site (terrain - soil - climate) – Respect for ecosystems – Design to resist ambient conditions	– Responsibility to site markers in design – Distance from dangerous and unhealthy areas
The climate	– Control the building envelope to reduce thermal loads – Employ climatic characteristics in energy production	– Use of climate-friendly materials – Good exploitation of materials and raise its thermal performance	– Use of plants that tolerate local climatic conditions	– Design compatible with climate characteristics	– Utilization of natural resources and good exploitation in the design (solar radiation - lighting - ventilation)
Technology	– Efficiency of operating systems technology – Using renewable energy technology	– Use of economic technology (recycling–reuse) – Accurate execution and operation	– Improve the efficiency of health systems – The use of economic irrigation systems in the work of site landscape	– Use simple local technology. - Introduction of local arts and crafts in design	– The use of clean technology and away from materials that harm the environment, which produces emissions and pollutants
Community culture	– Use and develop community technologies in energy production to improve efficiency	– The use of local labor trained to deal with materials – Reuse materials for old buildings or waste	– Raise awareness of water efficiency – Improve water handling behaviors	– Design related to the culture of the community (materials–labor–techniques) – Respect local and cultural character	– The need for popular participation in the design process – Achieve needs and functions at a lower cost in buildings for the population

The project is divided into (Asmarat 1, 2 and 3). The units of this project are designed in a distinctive architectural style suitable for the Egyptian citizen and provide all services, facilities, recreational units and sales outlets, as well as job opportunities for its inhabitants as shown Fig. 7.

6.3 Third Project: Grape Gheit Project (Bashair Al-Khair) in Alexandria City

It is located in the area of Karmouz—Alexandria City, one of the oldest popular neighborhoods. Recently, it was a slum area and it is now an example of the ability of Egyptians to turn the dream of a decent human life into a reality for the people of this region today. President Abdel Fattah El Sisi signaled the start of the project to transform the slums of grape ghetto into a sophisticated residential and service community. (Ministry of Housing Utilities and Urban Communities, 2020a, b, c).

Bashair Al-Khair consists of three phases completed by the Northern Military Area in cooperation with several civil society and business community the first phase: Bashair Al-Khair 1, which was inaugurated by President El Sisi in 2016; the second phase: Bashair Al-Khair 2, included 37 buildings including 1870 residential units at a cost of LE 400 million, and the third phase: Bashair Al-Khair 3 in 110 buildings including 5840 housing units, a mosque and the development of Qabbari Youth Center at a cost of 2 billion and 325 million pounds as shown Fig. 8 (Simms and Kamal, 2008).

6.4 Fourth Project: "El Mahrousa 1 & 2" Project, Al-Nahda City, Al-Salam Second Neighborhood

The Ministry of Housing, Utilities and Urban Communities is currently completing the work of finishing about 4900 housing units in 2018, shops, offices and administrative units

Fig. 4 Al-Sayeda Zeinab Kindergarten (Tal Al- Aqarib Project)

Fig. 5 Residential blocks are finished with (Hashemite stone and Mashrabiyat)

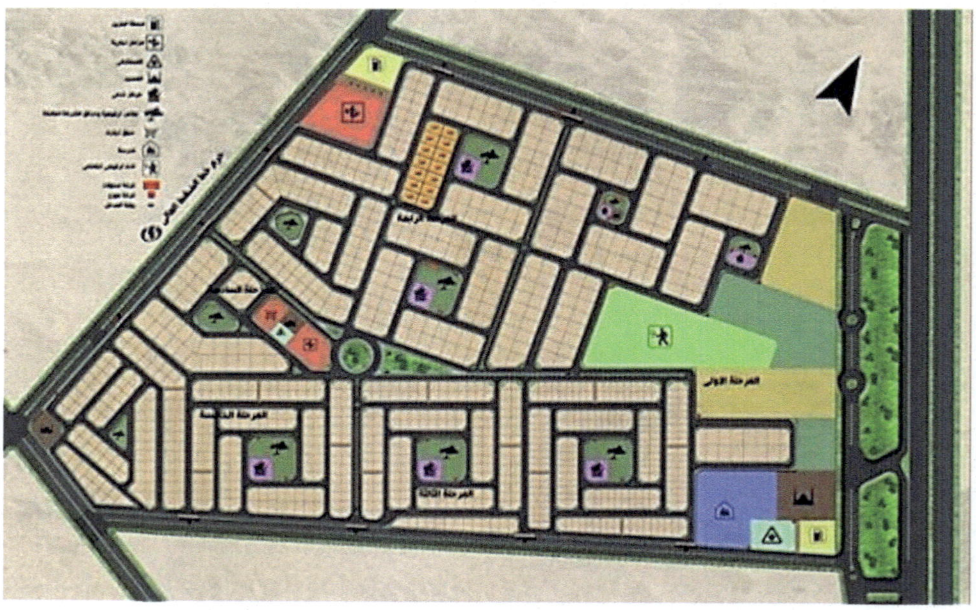

Fig. 6 Layout of Al Asmarat project in Mokattam

Fig. 7 Residential buildings Al Asmarat Project in Mokattam

in the project "Mahrousa 1 and 2" in the city of Al-Nahda, Al-Salam neighborhood in Cairo (Tal Al-Aqrab Project, 2019).

The area of the project is 60 acres including 34.7 acres for buildings and 14.2 acres for services, 8.4 acres for roads and footpaths besides 2.7 acres for green areas and public spaces. Al-Mahrousa 1 project includes the construction of 128 buildings 63 m^2 (114 residential buildings (90 buildings—ground + 5 typical floors), and (24 buildings—basement residential + ground + 5 typical floors) - 14 commercial residential building (basement shops + ground + 5 typical floors) + building women breadwinner. The total number of units is (3157 housing units 63m2 + 72 housing units for breadwinners + 84 stores + 39 administrative offices) as well as the work of connecting the facilities what is currently being completed is the completion of service buildings, paving roads and site finishes as shown Fig. 9 (UN Habitat Housing, 2019).

7 Results and Discussion

There are studies and researches that have analyzed the problems of not applying sustainability standards in architectural housing projects because there are high initial cost of green building projects that use the latest technologies to achieve sustainability.

The classification process is complex, requires time and procedures by using of modern sustainability technologies that are far from the identity and culture of society to

Fig. 8 Residential buildings Grape Gheit project (Bashair Al-Khair) in Alexandria City

Fig. 9 Residential buildings El Mahrousa 1 and 2 project, Al-Nahda City, Al-Salam Second neighborhood

decrease energy saving standard and make more of awareness in prioritizing users, so the constraints are to assessing the sustainability of architectural design in Egypt:

a **Administrative constraints**: The most important administrative obstacles can be insufficient sources of funding needed to achieve sustainable development and lack of support from developed countries by providing support to assist developing countries.
There are non-application and amendment of laws and legislation related to the specifications of green architecture that led to negligence of the state in addressing environmental problems and increasing their accumulation without imposing sanctions on consumers, also the existence of cumulative organizational problems starting with the emergence of the Unified Building Law No. 119 issued in 2008 and did not take any consideration of the concepts of sustainability.

b **Technical constraints**: Many of designers use on two-dimensional schemes such as AutoCAD, which have many design problems and the lack of use of modern technology such as BIM, which reduces errors in designs, which saves cost and time, but did not adopt the state of modern technology and they are high cost of those modern programs and devices
The disproportionate number of special devices and tools required for evaluation with the high building density in Egypt is specialized companies in green construction; also, they have lack of simulation programs to analyze results and develop alternatives and solutions to problems.

c **Artistry constraints**: There are not include the concept of sustainability in the architectural education in the stages of education continuously and focus on environmental design only and lack of attention on the cultural and social dimensions that led to lack of experience sufficient technical expertise did not a day in the process of evaluation of the buildings despite the presence of engineering experts in Egypt for green architecture, but did not benefit them at the design and executive level and lack of availability of architecture skills that are necessary for green architecture techniques.

The following table shows a study of comparison and evaluation of the local examples in Egypt that suffer in achieving the housing projects for sustainable design and the availability of sustainability criteria from its components and services, which are summarized as Table 2 follows.

The analytical study of the four cases and the evaluation of the elements of sustainability show that Al Asmarat and Grape Gheit Projects had poor of public areas like green areas, so green areas are benefit to reduce pollution and high sun rising and provide shading and natural ventilation.

The general form of the housing project in Al Asmarat and Basheer Al-Khair neighborhood is considered of the buildings that were built in the form of blocks without playing in the design of the blocks form that play the important role in providing shade to reduce sunlight and provide nature ventilation, but those buildings were created to sell these residential assets for investment. They were constructed of building materials from concrete and iron, not using recycled, local and renewable materials that caused the high prices of housing units, so in new cities still need more awareness of the principles of sustainability, which consider and develop a vision for sustainability.

Table 2 Comparison and evaluation of the local examples in Egypt that suffer in achieving the housing projects for sustainable design

Project name	Al-Sayeda Zeinab Kindergarten (Tal Al-Aqarib Project)	Al Asmarat project	Grape Gheit project (Bashair Al-Khair)	El Mahrousa 1 and 2 project
Date of establishment	2018	2016	2018	2018
Location	Is located in Al-Sayeda Zeinab Kindergarten, Cairo, Egypt	Is located in Mokattam, Cairo, Egypt	Is located in Karmouz, Alexandria, Egypt	Is located in A1-Nahda City, Al-Salam Second neighborhood. Egypt
Project type	New urban civilization	New urban civilization	New urban civilization	New urban civilization
The project area	The total area of the city is about 7 and a half acres	The total area of the city is about 65 acres	The total area of the city is about 13 acres	The total area of the city is about 60 acres
Project objectives	Provide adequate housing to reduce overcrowding in the capital area. pressure on facilities and improve production	Provide adequate housing to reduce overcrowding in the capital area, pressure on facilities and improve production	Provide adequate housing to reduce overcrowding in the capital area, pressure on facilities and improve production	Provide adequate housing to reduce overcrowding in the capital area, pressure on facilities and improve product»co
Availability of sustainability elements of energy and water recycling	The project use local and renewable materials like Hashemite stone. Mashrabiyat was established in the outlets to take the Islamic form in architecture and not reuse of water	The project did not use recycled, local and renewable materials and not reuse of water	The project did not use recycled, local and renewable materials and not reuse of water	The project did not use recycled, local and renewable materials and not reuse of water
Availability of public services and facilities	They had poor of availability of public services like green areas	They had poor of availability of public services and facilities to the growth of people	They had poor of availability of public services and facilities	They had availability of public services and facilities about 142 acres for services
Use local materials	The project use local and renewable materials like Hashemite stone	The project did not use local and renewable materials	The project did not use local and renewable materials	The project did not use local and renewable materials
Photo of project				

8 Conclusions

The results were obtained that the research concluded studying of local experiences of the problems in housing for the possibility of achieving sustainable construction in housing.

After studying the Egyptian experiences and examples, the research found that the economic housing is present in the existing cities that have been implemented in the projects in the past years; it lacks the principles and criteria of sustainability and the increase of problems in it from the lack of renewable energy sources and increased consumption. However, in the new cities that are still under construction, there are some sustainability criteria, but these projects are aimed at distinguished housing, and there is no economic housing for the purpose of profit and financial gain and investment only.

The selection of materials is carried out in the feasibility stages of the new buildings, especially in the new cities, where the architect must collect information and data for the specifications of the materials and choose them as available in the local market, which suits the cost and environmental sustainability The community did not participate in the planning and design process in the new cities despite the agreement of experts and land owners in the preparation of plans and people did not participate, where the investment and sale of these buildings for profit, where the government and the private sector only provide housing models, but the materials used are high cost and therefore not The ability for families to buy those housing.

9 Recommendations

1. The necessity of the state's interest in the design and implementation of buildings in the consideration of sustainable building materials and taking appropriate decisions after studying them to operate these buildings without a negative impact on users and the environment to achieve high-efficiency buildings.
2. Modify and implement legislation that includes green architecture, where companies working in the field of construction are obliged to apply these concepts with the need to issue codes related to green and local architecture.
3. Provide and develop sustainable building strategies for companies and design offices to implement these standards.
4. Develop a system of sustainable materials for application in existing and new cities in Egypt.
5. Develop an urgent plan to activate renewable energy sources (such as windmills and solar power plants), especially in new cities today.
6. Activate and develop the recycling system for building waste instead of importing materials from abroad, thus reducing costs and maintenance.
7. Respect and take into account the great diversity and variability between the different climatic regions in order to apply the tools of assessment of sustainable buildings and the selection of available local materials.

Acknowledgements First of all, I am thanking *my God* Almighty for accomplishing this modest work, hoping that holds the good for the scientific researchers and students locally and internationally. Deep respect and gratitude to my honorable *Professor Dr.* ***Hossam Eldin Hassan Othman Elborombaly*** for his valuable instructions and for being my ideal ethically and scientifically. I would like to acknowledge to *Assoc. Prof.Dr.* ***Ali Kamal Ali Altwanasy*** under whose supervision this paper was accomplished.

References

Ahmed, K., Samira E., & Shaheen, A. (2012). *Residential satisfaction and its relationship to economic house furnishing for youth at new Damietta city*. In Fourth Annual Scientific Conference, Mansoura University, Folder 3, pp. 1601–1634, Egypt.

Al Asmarat Project. (2019, September 20).Retrieved from http://www.sis.gov.eg/

El Mahrousa Project. (2019, September 24). Retrieved from https://www.mubasher.info/news/

El-Attar, M. E., Mubarak, L. M., et al. (2018). The current methodology for evaluating sustainable buildings in Egypt between possibilities and obstacles. *Journal of Engineering Sciences, 46* (2), 263–281.

Grape Gheit Project. (2019, September 27). Retrieved from http://www.sis.gov.eg/

Hilal, M. M., & Mahdi, K. H. (2014). *Sustainability in architecture: A study on the role of sustainable design strategies in reducing impacts on the Urban environment*. In 13th Al Azhar International Conference, Egypt.

Jomaa, A. A., & Mohammed, S. A. (2017). The mechanism of achieving urban sustainability through the UN treaties applied to some new Egyptian cities. *Journal of Engineering Sciences, 45*(5), 657–687.

Mayson (2014). Mohie Hilal and Hadi Mahdi, *Sustainability in Architecture: A Study on the Role of Sustainable Design Strategies in Reducing Impacts on the Urban Environment*, paper published at the 13th Al-Azhar International Conference, Egypt.

Ministry of Housing Utilities and Urban Communities (New Urban Communities Authority). (2020a, April 3). Retrieved from www.http://www.mhuc.gov.eg/

Ministry of Housing Utilities and Urban Communities (New Urban Communities Authority). (2020b, April 3). Retrieved from www.http://www.mhuc.gov.eg/

Ministry of Housing Utilities and Urban Communities (New Urban Communities Authority). (2020c, April 6). Retrieved from www.http://www.mhuc.gov.eg/

Musa, M. A., Ahmed, M. N., et al. (2014). Study and analysis of the experience of the new city of Assiut in the provision of adequate housing for low-income people (the evaluation of the project of the construction build your home is one of the axes of the National Housing Project). *Journal of Engineering Sciences, 42*(6), 1462–1491.

Qutb, H. H. (2013). Towards new mechanisms to revitalize mortgage finance to resolve the housing problem. *Journal of Engineering Sciences*.

Samir, D. (2018). *Sustainable housing planning*. An-Najah National University, Nablus.

Shawkat, Y. (2014). *A fair policy for housing in Egypt*. Egypt.

Simms, D., & Kamal, H. (2008). *Housing study of Egyptian architecture*. USA.

Tal Al-Aqrab Project. (2019, September 17). Retrieved from http://www.sis.gov.eg/

UN Habitat Housing. (2019, August 19). Retrieved from https://sdg.iisd.org/news/un-habitats-annual-report-spotlights-sdg-11-nua-implementation/

Proposed Systemic Method for Selecting Finishing Materials for Building Flooring Using Building Information Modeling (BIM)

Mohamed A. Alrahhal Alorabi, Khalid S. Al-Gahtani, and Ibrahim A. Alhammad

Abstract

Material selection is a crucial factor in decision-making for any construction project and is viewed as a challenging and complex task given the full range of material selections available. When going through the process of material selection, one must consider the environmental, social, and economic impacts that a specific selection may entail on the sustainability of the project at one hand as well as consider and the value engineering components of the selections in terms of functionality, quality, and cost. Floor finishing materials can have a significant impact on the cost, aesthetics, and quality of any construction project. Therefore, it is necessary to find a systemic approach to handle this process efficiently. One such method can be made using modern and sophisticated engineering software such as building information models (BIM) when considering all the factors listed above. The goal of this paper is to develop a framework for selecting finishing materials for building flooring using building information modeling (BIM) and value engineering concepts with considering a sustainable rating system.

Keywords

Quality • Evaluation • Flooring • Materials • Value • Sustainability

1 Introduction

The construction industry is considered one of the essential sectors in the world. It plays a vital role in achieving economic growth due to its multiple effects on other areas of the economy (Joshi & Murari, 2017). Construction materials are the cornerstone of construction projects, and their consumption value is about 40% of the total cost of any project (Bamigboye, et al., 2019). Floor finishing materials are one of the most critical items in any construction project. They are defined as the material used to cover the floor, as these materials provide the required quality and at a moderate cost (Onochie, Emoh, & Anyanwu, 2017). The material selection process is one of the tasks that engineering design requires (Salit & AL-Oqla, 2017). Due to the wide variety of types of flooring finishing materials, they will be chosen based on the value engineering concept. This selection method will help the decision-makers to select the most valuable materials, which perform the required function with high quality and at the lowest cost, as explained in Eq. (1) (Dell'Isola, 1997).

$$\text{Value} = \frac{\text{Function} + \text{Quality}}{\text{Cost}} \quad (1)$$

While it is easy to quantify the cost factor and compare possible selections based on monetary value, the same cannot be said for quantifying the quality factor. Fifteen years ago, AHP, TOPSIS, and ELECTRE methods were the most used selection methods. In this paper, the AHP method will be used in the materials selection process as it is. The most appropriate method given the limited number of variables, and it provides numerical values for ranking (Salit & AL-Oqla, 2017). Sustainability is the fulfillment of the requirements of people and society by employing technology and practice optimally throughout the building's life cycle (El-Alfy, 2010). Sustainability is evaluated by the rating system through awarding points to meet the parameters of green buildings. Due to the differences in environmental, regional and cultural conditions, these parameters and their weights differ from one system to another (Al-Gahtani et al., 2016). This paper presents an outline model for selecting floor finishing materials using the concept of value engineering and considering the sustainability factor.

M. A. Alrahhal Alorabi (✉) · K. S. Al-Gahtani · I. A. Alhammad
Civil Engineering Department, Collage of Engineering,
King Saud University, Riyadh, Saudi Arabia
e-mail: alhammad@ksu.edu.sa

C. Alalouch et al. (eds.), *Towards Implementation of Sustainability Concepts in Developing Countries*, Advances in Science, Technology & Innovation, https://doi.org/10.1007/978-3-030-74349-9_13

2 The Proposed Method for Selecting Floor Finishing Materials

This paper aims to develop an approach to choose floor finishing materials based on quality and cost considerations. The steps mentioned in Fig. 1 represent the evaluation process that assigns a value for each potential material type. This process will be linked with the BIM model to facilitate data input and automate the output accordingly. Finally, the materials will be selected based on material value factor (*V*) in addition to the sustainability factor (*S*), which will be extracted based on the sustainability rating system. In the following sections, a description of the proposed method steps to select floor finishing materials on any construction project.

Step 1: Choosing the predominant criteria

The purpose of this step is to identify the quality criteria for the floor finishing material. Currently, there are several international quality standards used to determine the required quality of the Material (ISO, SASO, ASTM, and EN). These standards aim to determine the required quality of the materials. Saudi Standards Metrology and Quality Organization (SASO) adapted many of these standards to be used in Saudi Arabia. In this study, the authors studied a long list of criteria used by SASO with different types of floor materials. Examples of these materials are ceramic, marable, and porcelain. The challenge of this study is to identify the most common criteria from this long list and narrow it down to a manageable level. To do this, we consulted with several quality control engineering experts working for several reputable manufacturers. Accordingly, the following sections describe the eight most common criteria for the three types of floors: ceramic, porcelain, and marble:

Criteria 1—Water Absorption: It is one of the fundamental properties of floor finishing materials to determine the effect of water on the flooring surface. The low-water absorption of the floors is an indication of their quality (Gajek et al., 2016).

Criteria 2—Breaking Strength: It is an important criterion that affects the durability of the flooring. The higher the breaking strength, the higher the quality of the tiles (Ferretti, 2013).

Criteria 3—Modulus of Rupture: It is a concept relating to the quantity obtained by dividing the breaking strength by

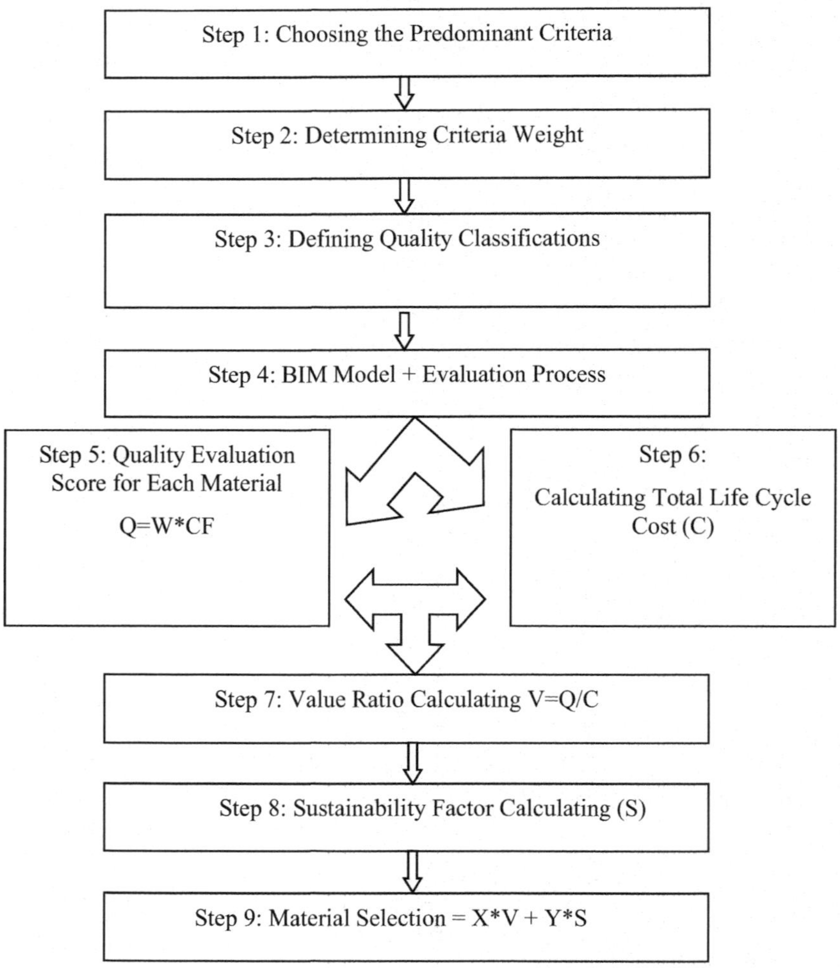

Fig. 1 Steps of the material selection process

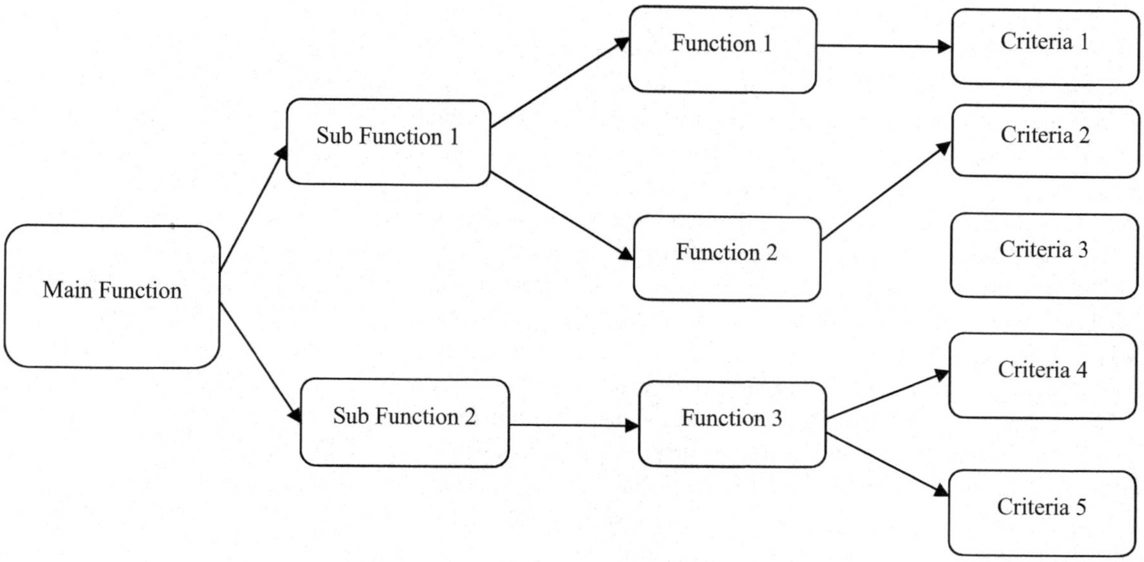

Fig. 2 Main components of FAST diagram for a criteria weight's analysis

the square of the minimum thickness along the broken edge (Fragassa, 2015). The higher the modulus of rupture, the higher the quality of the tiles.

Criteria 4—Abrasion Resistance: Tile surface resistance to abrasion is determined by rotating the abrasive load on the surface and evaluating the abrasion results visually. The abrasion surface can be stained with light oil to ensure that the dirt particles have not covered the place of abrasion (ASTM, 2019).

Criteria 5—Surface Hardness: It indicates the ability of the tile surface to resist abrasion and scratches. It also represents the surface durability and aesthetic properties. It is measured by using a Mohs scale (Bijan et al., 2006).

Criteria 6—Coefficient of Friction: This standard affects building design, as industrial and recreational buildings. It is essential for easy cleaning and draining, and so on (Brough et al., 2001).

Criteria 7—Thermal Conductivity: Low-water absorption floors have high thermal conductivity. Therefore, low thermal conductivity floors (white stoneware) are not suitable for radiant floor heating applications. Al2O3 particles play an important role in increasing the thermal conductivity to 50% when adding 20% to the ceramic paste (Garcia, et al., 2011).

Criteria 8—Stain Resistance: Many materials cause stains on the surface of the tiles. As a result, staining resistance considers one of the most sophisticated quality standards that must be achieved due to the possibility of interactions between these materials and the surface of the tiles (Dondi et al., 2008).

Criteria 9—Aesthetic: Aestheticism is an essential criterion in selecting floor finishing materials. By the visual impression of the types of flooring, their color, and surface finishing quality, aesthetic perception of them will be formed (Sedliačiková & Michalková, 2011). On the other hand, this criterion may significantly increase the cost of these materials.

Criteria 10—Durability: "the ability of a product to perform its required function over a lengthy period under normal conditions of use without excessive expenditure on maintenance or repair" (Cooper, 1994).The concept of durability for flooring can be evaluated through the previous quality criteria 2, 3, 4, 5.

Step 2: Determining the criteria weight

The importance of quality criteria varies depending on the project's function and goal. As we mentioned, the AHP method is a suitable method for this research among MCDM methods, as it depends on the pairwise weighting. For better accuracy, we will establish a correlation between the weights of criteria with the project function and then combine the weight with the function. This process will be done with the function analysis system technique (FAST) diagram (Singh, 2017), as shown in Fig. 2. Each criterion will be attached to its relevant function. Some criteria may be related to more than one function. After criteria integration, all criteria must be evaluated with the assigned weights. Weights will be given at each level of the diagram.

Step 3: Defining Quality Classifications

In this step, the numerical values for each of the eight quality criteria will be determined through specific tests according to international standards, especially Saudi standards (SASO). Accordingly, quality engineers with expertise in this field will be consulted to classify those numerical values into three classifications in terms of quality: high, medium, and low. These classifications will be converted to factors to facilitate the evaluation process.

Step 4: Linking Evaluation Process with the Building Information Modeling (BIM)

BIM is a technology that links the drawings to the project database. Thus, it can be relied upon in project management, which contributes to improving productivity and quality and helps to deliver the project at the lowest cost (Azhar, Nadeem, Mok, & Leung, 2008). The user can, through BIM, get various data related to the project during any stage of the project to improve the design (Eastman et al., 2011). BIM tools can also evaluate building model performance (Nguyen et al., 2016). By using parametric building modeling technology, Autodesk Revit (one of the popular tools of BIM) can represent the building as an integrated database (Dzambazova et al., 2008). Revit ensures the ability to apply external features on the BIM model through what is known as API "Application Programming Interface." From the database, BIM has different dimensions3D, 4D, 5D, … ND and each dimension represent a specific type of data (Cost, Scheduling, Sustainability … etc.) (Czmoch & Pĉkala, 2014). The long future goal of this study is to develop a new dimension of BIM that is related to overall building value engineering. The main objective of the proposed method in this paper is to assist decision-makers in selecting the optimal materials based on function, quality, and cost factors with considering sustainability as an option. For modeling the proposed approach by using BIM, all data relating to the types of materials, their quality criteria, and their values must be included. The above evaluation process will be linked to the BIM model through API.

Step 5: Total Quality Scores Calculation

For each quality criterion, we calculated the function weight and defined the three quality classifications as per the numerical values. Material with high quality will be given a classification factor (CF) of 1. Classification factors (CF) for material with medium and low quality will be calculated by normalizing them with the classification factor of the high quality. Finally, we will get the total quality score by multiplying the function weight for each criterion by classification factor of each material. We will clarify the above calculation steps through the following selected criteria as an example shown in Tables 1 and 2:

In Table 1, the numerical values were obtained from SASO. The materials were classified according to the quality by consulting quality control engineers. A classification factor (CF) has been given for each material based on its quality classification. By multiplying the classification factor by the corresponding function weight of the criterion, we will get the total quality score (Q), as shown in the last row of Table 2.

Table 2 clarifies the calculation: The materials with quality criteria classified as "high" were given a CF of 1, and then the materials with quality criteria classified as "mid" and "low" quality were normalized appropriately with the criteria classified as high.

Step 6: Calculating the Total Life Cycle Cost

This step requires detailed information about materials cost. Materials cost has two parts; the first part is the initial market cost of purchasing the materials. The second part is the cost of operating and maintenance over the life of the building. For illustration purposes, we used an approximate cost factor to reflect the total material cost. The following Eq. (2) can be used to determine the total life cycle of any building materials (Halpin & Senior, 2009):

Table 1 Quality classification as per the numerical values

C	Criteria	Ceramic (BIIA)	Porcelain (BIA)	Marble
C1	Breaking strength	$\geq$ 1000 N	$\geq$ 1300 N	$\geq$ 1850 N
		Low	Mid	High
C2	Modulus of rupture	$\geq$ 22 N/mm2	$\geq$ 35 N/mm2	$\geq$ 6.9 N/mm2
		Mid	High	Low
C3	Abrasion resistance	$\leq$ 345 mm3	$\leq$ 175 mm3	$\leq$ 250 mm3
		Low	High	Mid
C4	Surface hardness	$\geq$ 5 Mohs	$\geq$ 5 Mohs	$\geq$ 4 Mohs
		High	High	Low

Table 2 Total quality scores numerical example

	Criteria	Function weight W	Quality classification factors for floor finishing materials (Normalized) CF		
			Ceramic (BIIA)	Porcelain (BIA)	Marble
C1	Breaking strength	30%	0.54	0.7	1
C2	Modulus of rupture	15%	0.63	1	0.2
C3	Abrasion resistance	35%	0.51	1	0.7
C4	Surface hardness	20%	1	1	0.8
	Q score = W*CF		0.64	0.91	0.74

Table 3 Material values, a numerical example

	Ceramic (BIIA)	Porcelain (BIA)	Marble
Q = material quality score	0.64	0.91	0.74
C (FLCC) = Material normalized cost	0.13	0.25	1
V Material Value	4.92	3.64	0.74

$$\mathrm{FLCC} = \mathrm{IC} * F/P + M\&R * F/A \tag{2}$$

whereas:

FLCC: Future compound amount for life cycle cost.

IC: Initial cost.

M&R: Maintenance and repair cost.

F/P: Future compound amount factor for single payment.

F/A: Future compound amount factor for uniform payment series

$$F/P = (1+i)^n \tag{3}$$

$$F/A = \frac{(1+i)^n - 1}{\mathrm{i}} \tag{4}$$

whereas:

i: Interest rate.

n: Project life span.

Step 7: Value Scores Calculation

This is the final step to get the proposed method result. Table 3 provides a numerical example of the selected three materials. It shows that although ceramic gets the lowest quality score, it gets the highest value, and it should be chosen. The material value will be calculated as the following formula (5):

$$V = Q/C \tag{5}$$

V = Material value.

Q = Material quality score.

C = Material normalized cost = FLCC.

Step 8: Sustainability Factor

Sustainability is an essential factor to consider when selecting materials. However, in this study, this factor remains undetermined because sustainability rating systems are country-specific and vary depending on the standards adopted by each country. Any future studies need to develop a sustainable factor (S) that reflects the sustainability rating system that is the country takes.

Step 9: Final Materials Selection.

As final steps, we need to calculate the final material selection based on both value engineering factor (V) and sustainability factor (S) as shown in Eq. (6):

$$\text{Material Selection } = X * V + Y * S \tag{6}$$

where the summation of the coefficient *X* and *Y* should be 100%, the sustainability factor coefficient (*Y*) need to be identified by local experts to reflect the importance of sustainability in selecting building material.

3 Validation

The study results are validated by submitting a questionnaire to professional architects and design engineers and asking for their opinions on the selected material. Another validation method can be used in future research, namely a survey of building users investigating their satisfaction with the selected material. This can undoubtedly help to validate the results of this study. This can undoubtedly help to validate the results of this study (Chew et al., 2004; Oyedele et al., 2012).

4 Conclusion

The material selection process is a difficult choice for decision-makers due to its complexity. In this paper, we developed an outline approach based on the concepts of value engineering and sustainability. We found the common influencing factors and generated a model by using BIM, expedites the process for decision-making based on those factors and criteria. All the inferred criteria were verified, whether from previous researches or through experienced engineers, suppliers, and others. The result of this research is a great help in improving the current market practice of value engineering. Value engineering is a time consuming and challenging task. This is because the method is not following a systematic approach, and there is no common agreement of material function and quality criteria for building materials. Thus, this study will help the practitioners in this field to determine the value of each flooring finishing material during the design stage by considering the sustainability impact.

Acknowledgements The authors would like to thank Deanship of Scientific Research (DSR), King Saud University (KSU) for funding and supporting this research through the initiative of DSR Graduate Students Research Support (GSR). Also, we would like to thank the Saudi Standards Metrology and Quality Organization (SASO) for their support.

References

Al-Gahtani, K. S., Alsulaihi, I., El-Hawary, M., & Marzouk, M. M. (2016). Investigating sustainability parameters of administrative buildings in Saudi Arabia. *Technological Forecasting and Social Change*.

ASTM. (2019). Standard test method for determining visible abrasion Resistance of glazed ceramic tile. In *ASTM C1027–19.*

Azhar, S., Nadeem, A., Mok, J. Y., & Leung, B. H. (2008). *Building Information Modeling (BIM): A new paradigm for visual interactive modeling and simulation for construction projects.* First International Conference on Construction in Developing Countries (ICCIDC–I)"Advancing and Integrating Construction Education, Research & Practice".

Bamigboye, G., Davies, I., Nwankwo, C., Michaels, T., Adeyemi, G., & Ozuor, O. (2019). *Innovation in construction materials—A review*. IOP Conference Series Materials Science and Engineering.

Bijan, E., Yekta, B. E., Alizadeh, P., & Rezazadeh, L. (2006). Floor tile glass-ceramic glaze for improvement of glaze surface properties. *Journal of the European Ceramic Society*.

Brough, R., Malkin, F., & Harrison, R. (2001). Measurement of the coefficient of friction of floors. *Journal of Physics D Applied Physics*.

Chew, M. Y., Tan, S. S., & Kang, K. H. (2004). A technical evaluation index for curtain wall and cladding facades. *International Journal of Building Pathology and Adaptation (Emerald Group Publishing Limited)*.

Cooper, T. (1994). Beyond recycling: the longer life option. *The New Economics Foundation*.

Czmoch, I., & Pėkala, A. (2014). Traditional Design versus BIM based design. *Procedia Engineering*.

Dell'Isola, A. J. (1997). *Value engineering practical applications for design, construction, maintenance and operations.*

Dondi, M., Raimondo, M., & Chiara, Z. (2008). Resistance to stains of ceramic tiles. *Ceramic World Review*.

Dzambazova, T., Demchak, G., & Krygiel, E. (2008). *Mastering revit architecture*.

Eastman, C., Teicholz, P., Sacks, R., & Liston, K. (2011). *A guide to building information modeling for owners, managers, designers, engineers, and contractors.*

El-Alfy, A. (2010). Design of sustainable buildings through Value Engineering. *Journal of Building Appraisal*.

Ferretti, E. (2013). A cell method stress analysis in thin floor tiles subjected to temperature variation. *Cmc -Tech Science Press*.

Fragassa, C. (2015). Limits in application of international standards to innovative ceramic solutions. *International Journal for Quality Research*.

Gajek, M., Kmita, A. R., Dudek, M., & Partyka, J. (2016). Mechanical properties and microstructure of fast—Fired clinker tiles based on Wierzbka I raw material. *Advances in Materials Science*.

Garcia, E., Pablos, A., Bengoechea, M., Guaita, L., Osendi, M., & Miranzo, P. (2011). Thermal conductivity studies on ceramic floor tiles. *Ceramics International*.

Halpin, D. W., & Senior, B. A. (2009). *Financial management and accounting fundamentals for construction.*

Joshi, A. M., & Murari, S. (2017). Precast construction methodology in construction industry. *International Conference on Emerging Research Trends in Mechanical & Civil Engineering*.

Lee, D. E., & Arditi, D. (2006). Total quality performance of design/build firms using quality function deployment. *Journal of Construction Engineering and Management*.

Nguyen, T. H., Toroghi, S. H., & Jacobs, F. (2016). Automated green building rating system for building designs. *Journal of Architectural Engineering*.

Onochie, I. J., Emoh, F. I., & Anyanwu, C. I. (2017). An evaluation of factors affecting the choice of the different floor finishes in the building industry within Abia State of Nigeria. *International Journal of Latest Engineering and Management Research (IJLEMR)*.

Oyedele, L. O., Tham, K. W., Fadeyi, M. O., & Jaiyeoba, B. E. (2012). Total building performance approach in building evaluation: Case study of an office building in Singapore. *Journal of Energy Engineering*.

Salit, M. S., & AL-Oqla, F. M. (2017). *Materials selection for Natural fiber composites.*

Sedliačiková, M., & Michalková, M. (2011). *Analysis of competitiveness of the chosen company*. Annals of Warsaw University of Life Sciences - SGGW.

Singh, A. (2017). Function analysis system technique—Structuring uncertainty. In *Quantitative risk management and decision making in construction.*

Utilizing Wind-Driven Ventilation Force as a Technique in Adaptive Passive Interior Design

Eman Ahmed Elsayed Mahmoud AlAkaby

Abstract

Sustainability has always been known as the quality of being able to continue over a long time considering environmental, social and economic aspects and its approaches are meant to find solutions not only for the carbon dioxide emissions reduction or facing climate changes that affect the surrounding environment but also to guarantee the interior environment air quality and space thermal comfort. So, wind-driven ventilation considered a natural force that can be used as a technique to respond to climate change in passive spaces for the sake of man health and the society correlation with the surrounding environment. The purpose of this research is to provide empirical study shows how natural ventilation can be used as an interior design solution to save energy and turn the space into a self-sufficient. That can be achieved by following the old examples and discussing how architecture and interior design tried to create the most suitable and useful solutions for users' needs. It will present an analytical study on some of the closest examples of climate change and natural ventilation challenges to our society in Egypt. Those examples focused on environmental forces and predicted the natural environmental effect. Otherwise, it is important to analyze how they can affect the interior design and architectural form. Finally, the research will try to summarize the main interior architectural implications and find some design solutions that can turn the interior space into a passive one to adapt to climatic changes utilizing wind-driven ventilation force as an "adaptive passive space design" technique.

E. A. E. M. AlAkaby (✉)
Faculty of Décor Departement "Interior Architecture", Fine Arts, Alexandria University, Alexandria, Egypt
e-mail: eman.a.akaby@alexu.edu.eg

Keywords

Passive interior space • Natural ventilation • Ventilation strategies • Thermal comfort • Wind-driven airflow

1 Introduction

The natural ventilation in the architectural design depends on the difference of air pressure resistance between the interior and exterior through openings. Likewise, the airflow inside the space is based on the status of the surrounding environment. The influence of the winds depends on the shape of the exterior building, how close other architectural blocks exist with its surroundings, location, the density of the plant elements and their coordination, etc. Here, the question was why the same strategies are not exploited on a smaller scale to operate entirely within the interior space as design solutions to natural ventilation problems. Thus, wind forces are utilized through openings with the plants' interior landscape, or ways of distributing furniture and partitions to control natural ventilation solutions inside instead of excessive consumption of electrical energy and economic waste.

The research deals with the relation between natural ventilation, airflow, and the design of passive interior architecture, that the relationship between natural ventilation and thermal comfort of the interior space is determined. Then, the state of explaining the strategic and technical levels of designing natural ventilation and its impact on interior architecture design philosophically.

1.1 Research Problem

Egypt has always been known for its hot, dry climate in summer and warming rainy one in the winter. Now, it is clear that the climate changes widely that the intensity of heat in summer has been increased, as its duration has too.

C. Alalouch et al. (eds.), *Towards Implementation of Sustainability Concepts in Developing Countries*, Advances in Science, Technology & Innovation, https://doi.org/10.1007/978-3-030-74349-9_14

That necessitated a better use of cooling methods and the search for healthy ventilation methods that would improve the efficiency of the interior space in light of the increase and convergence of the heat-absorbing concrete blocks and the lack of attention to the location of the openings and shading methods.

At present, the idea of ventilation is often dealt with as a design tradition that is limited to ancient architecture, such as Pharaonic or Islamic architecture, and concerning interior architecture about guidance for characterization of marketing only. The handling of ventilation has been absent as a fundamental solution to the good of the internal environment of interior architecture, users' health considerations, and their economic income. As well as, an ideal way to create a sustainable interior space that can adapt to climate changes, and even reduce energy consumed to a minimum to transform into a passive space. That can be achieved by utilizing natural forces such as wind and solar rays, as both affect the thermal state of the interior space.

1.2 Methodology

The research tries to find the main methods in architecture and interior design to create the most suitable and useful solutions for users' needs and to create an adaptive sustainable interior space. So, it can be achieved by using an analytical study for some examples of climate change interior design solutions and natural ventilation challenges using the natural environmental forces like wind and solar rays. Otherwise, it is important to classify how they can affect the interior design and its form.

2 Sustainability for Resistance Ability

The author "Edward Oelson" in his book "The Future of Life" (Wilson, 2002) describes the state of our planet and the pressures that human activity imposes on the environment, such as global warming caused by man, pollution, deforestation, and resource depletion, which led to an environmental crisis threatening the survival of life and the human race. Oelson has cautioned against the attitude of a man who considers himself separate from his environment (Sassi, 2006, p. 3). More importantly, "The natural environment that is dealt with such ignorance and randomness represents the highest levels of development, adaptation, and self-reform, and the design of interior architecture remains influenced by its forces in one way or another, either negatively or positively. While, The principles of sustainability aim to address the problems of environmental degradation and quality of life, by supporting sustainable development and the ability to preserve the benefits of the surrounding natural environment and affecting the interior environment to be in the interest of long-term health stability" (Fig. 1).

3 Natural Ventilation in Interior Space

The origin term of "ventilation" is the Latin word "Ventus," which means air movement. It defines ventilation in interior architecture as "the process of supplying or removing air by natural or mechanical means to and from the interior thermal comfort". Natural ventilation in most cases meets the requirements of interior space users because it provides a healthy environment and energy-efficient.

The term "natural ventilation in interior architecture" (Fig. 2) also refers to systems designed to control the quality and quantity of air movement, and use it to achieve sustainability through achieving thermal comfort or internal air quality, as natural ventilation can be very effective in maintaining interior thermal comfort in an acceptable functional manner, where It is considered a technique of passive cooling, which is suitable in all climatic regions, and an ever-changing adaptive solution and responding to climate changes while emphasizing that well-designed natural ventilation systems can significantly reduce energy consumption (Natural Ventilation, n.d.).

While mechanical ventilation can help overcome the shortfalls that occur from natural ventilation as it can prevent difficulty in Controlling Airflow, It can consume more of electric energy, which would need more economic considerations. It is one of the main driving forces of natural ventilation. Here came the question of how to use the most

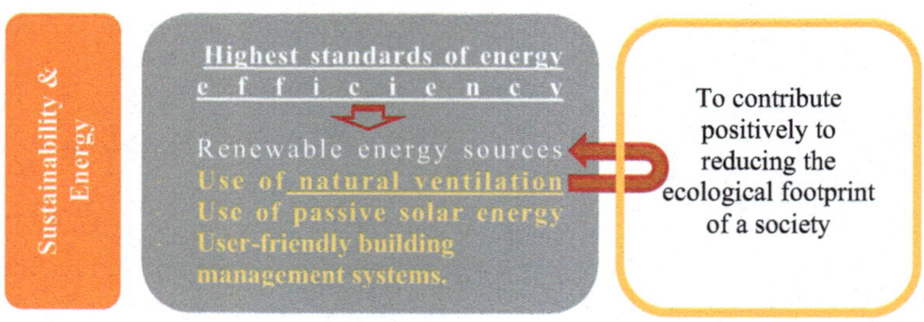

Fig. 1 Diagram shows how the sustainable design for resistance can contribute reducing the ecological footprint of a society (Author)

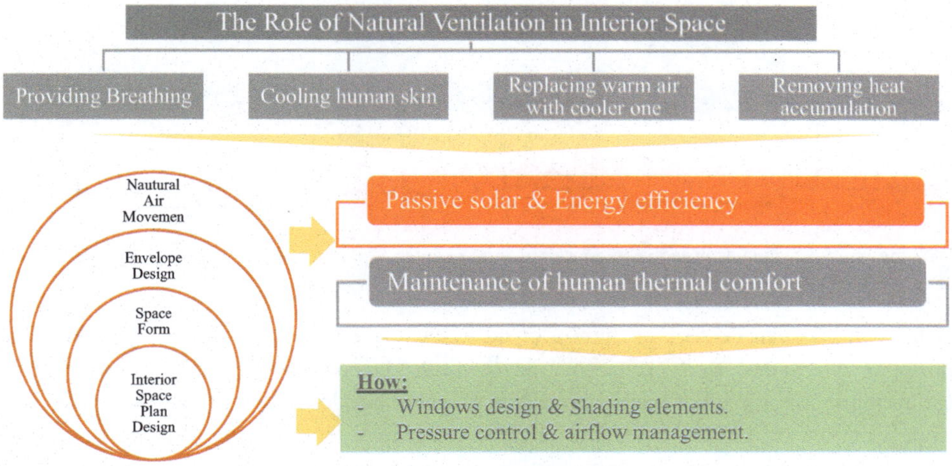

Fig. 2 Diagram shows the role of natural ventilation in interior space (Author)

amount of natural wind force to give the benefits of healthy airflow with a minimal amount of mechanical ways like using small sensing devices instead of using air-conditions or central cooling systems. Depending on wind-driven flow can introduce a balanced amount of fresh air into interior spaces to find users' needs and to create a natural type of adaptive passive interior space (Natural Ventilation vs Mechanical Ventilation, 2014; Energy.gove offices, n.d).

3.1 The Ventilation of Interior Design Over Decades

The wind environment used in regional residences is not limited to interfacial openings such as doors and windows, but rather the overall use of interior and exterior conditions, organizing of ventilation approaches and setting according to wind direction, speed, solar radiation, building design, and space envelope shape. However, in construction means, by designing the open-element response to achieve a wonderful performance ventilation effect. That is why traditional architectural and interior architecture has been succeeded to achieve sustainable function over the decades (Mina & Zhang, 2016).

Traditional architecture has found that its elements and its cultural unity achieve that what humanity needs psychologically and physiologically as comfort throughout the decades is a "natural selection" that we can develop using modern technologies to keep pace with the times. For example, the courtyard as was known in Islamic architecture planning design is the exposed central space that rounded with the other interior spaces of the building from all sides. It is considered a thermostat and moderating agent for the temperature inside the space. It was the ideal solution to allow the airflow and the wind movement through interior spaces and cause its cooling.

Also, airflow paths in interior space create an effective cooling strategy using wind force, which decreases the interior temperature and achieves thermal comfort. Both the courtyard and the airflow paths were found in other designs in earlier decades, like in Pharaonic architecture or later decades (Ancient Egyptian Houses: Domiciles From Pharaohs to Farmers, History on the Net, n.d.), like Hassan Fathy's projects and Renzo Piano's architectural projects (Figs. 3 and 4).

In ancient Egyptian houses and buildings, while the clay and rocks were used, it was not easy to open wide windows. So, there were nets opened at the end of the wall near the

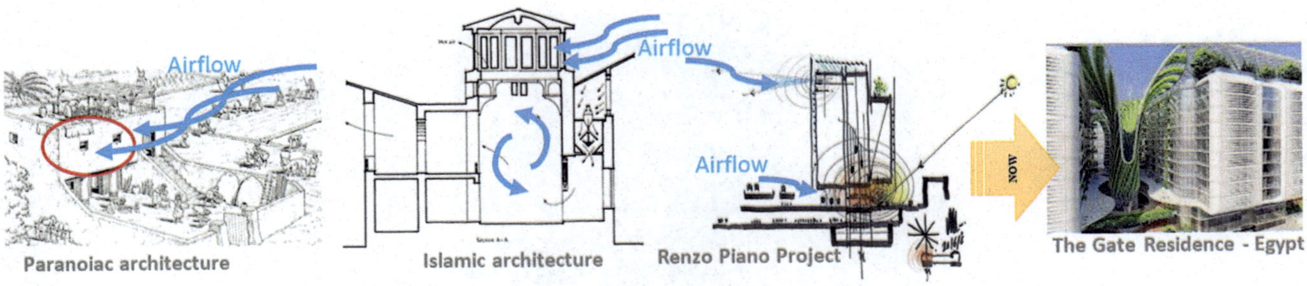

Fig. 3 Sketches show natural ventilation techniques in interior design over decades

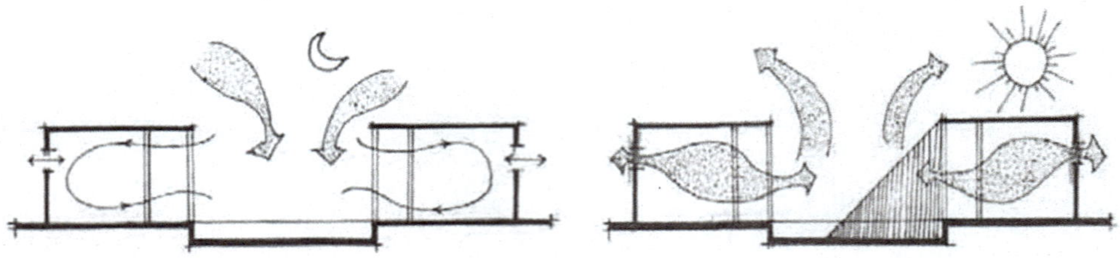

Fig. 4 Sketches show the wind movement through the courtyard in the center of interior space with creating interior shading (Samy, 2013)

ceiling. Those small windows can be seen at the top of the upper rooms on upper floors and they were covered to prevent dust, flies, and heat. On the other hand, a courtyard was a basic design solution at the center of the ancient Egyptian houses of the rich. Also, there is important historical evidence of using wind-catchers in ancient Egypt, dating back to 1300 BCE. It was depicted in a wall painting of Neb Amoun's house, at Tel El-Amarna tombs in Luxor—now in the British Museum. Wind-catchers were an important architectural element to face the current airflow or wind and trap it to get down into the interior space of the buildings to create natural ventilation. (Ramadan, 2016).

Another example is the solar breakers are an ancient technique and method for creating spaces for shading, as in Islamic Mashrabiya, which developed using sun shading systems and solar screening systems, which have become one of the most prominent modern technology in the twenty-first century for the design of an environmentally friendly spaces, where solar shading systems are controlled intelligently and with high accuracy characterized by the presence of sensors that work when the sun's rays fall. These techniques, in response to closing the openings and providing the appropriate shading on the space facade and then on the interior space. It achieves the appropriate savings in the energy needed for heating and cooling and the use of lighting as it works as a borer passes the air nicely into the interior space (Figs. 5 and 6).

After many decades, When the industrial revolution swept through and after that technological progress, the pioneers of the architectural design movements raced intending to break the link with the past and prepare to enter the era of modern technology. Also, new materials were introduced and entered the architecture, so the walls became clear of decorations and the glass was used a lot. For example, Frank Loyd Wright was looking at the light as accompanied by heat, so most of his building openings were shaded with protrusions and balconies to break that heat or allow only the pleasant air to pass to inside the interior spaces.

Another example that comes out for the near future is «The Gate Residence» as an innovative design idea of a green building located in the heart of this New Cairo, Egypt. Its roof provides shadow where necessary like in old traditional architecture where created shadow walkways to reduce the amount of solar heat coming to interior space (Fig. 7). On the other hand, there are several green spaces located in the heart of the building creating void spaces that take the same function of the old courtyard but in a multi-story building (Fig. 8)—to turn the hot airflow to a cool one. Then, It will help to reduce the energy used for cooling interior spaces surrounding the courtyard. So, It can be passive interior spaces. (The Gate Heliopolis | Vincent Callebaut Architectures, 2019; Vincent Callebaut Architectures, n.d).

3.2 The Role of Natural Ventilation in Interior Space

The nature of interior spaces and their envelopes in the past did not use air-conditioners, but rather were designed to become self-adaptive to the temperature of the surrounding environment, whether by cooling or heating using natural

Fig. 5 Islamic Mashrabiya and its work in passing air and light (Jamal, 2007, p. 1588)

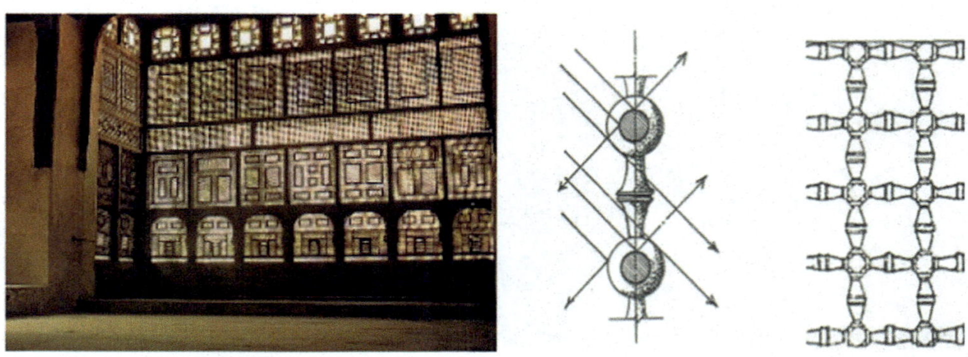

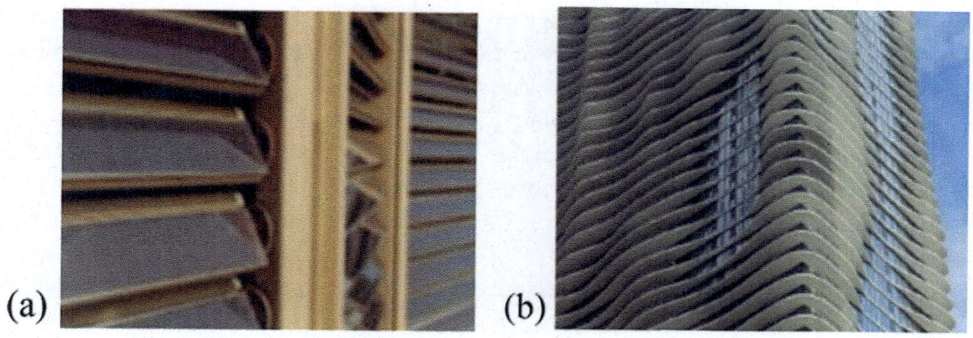

Fig. 6 a House of Di Kathleen who used solar shading from wood frames with photovoltaic cells at the front of the building 2007. b Using sun shading in a residential tower to prevent solar radiation from entering Frank Gehry's project. (Abd ALkader et al., 2017)

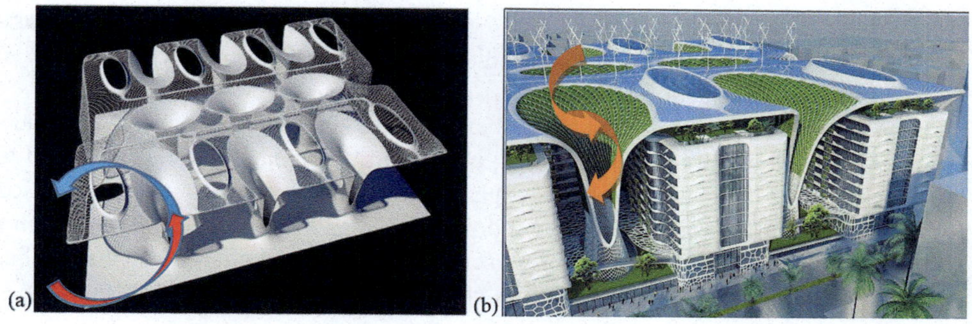

Fig. 7 **a** and **b** Analytical study shows the wind-driven airflow through the space in the center of the building

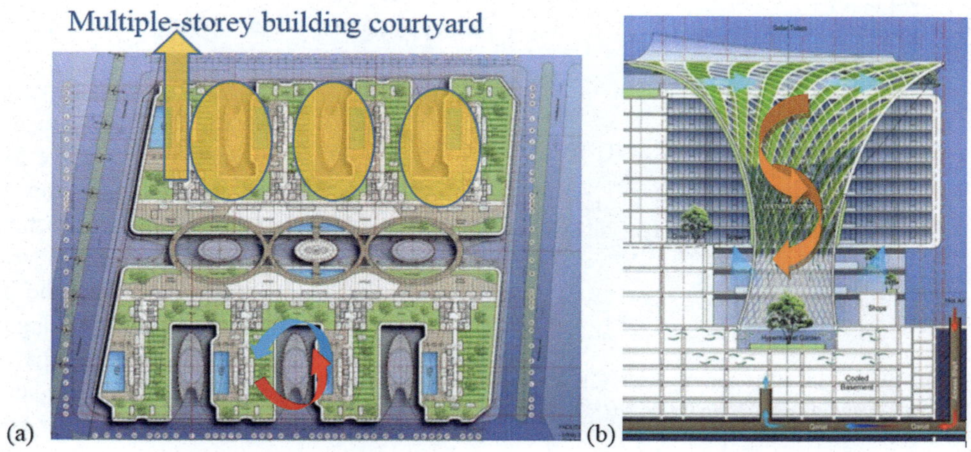

Fig. 8 Analytical study shows the similar function of the courtyard but in a different scale at center of multiple-story building

airflow, as one of the highly influential natural forces, and as a kind of response to that climate and improving thermal comfort—which constitutes Simply the complete physiological and mental sense of comfort—so the promotion of natural ventilation came as a passive cooling strategy in residential blocks, for example, in the climate of Egypt, which is predominantly desert, because the space blocks, and between them and their components are characterized by several methods to achieve this strategy of passive cooling and energy consumption.

This strategy, along with the use of local materials, wall thickness and space plan design, contributes to looking after the rates of air change according to direction and speed, which reduces energy consumption and dynamic ventilation systems, or what is known as active ventilation.

3.3 Climatic Applicability of Natural Ventilation

The applicability of natural ventilation gets the most beneficial function through the interior space form, its envelope, space planning, and even internal furnishing details. The movement of natural air currents is under the influence of direction and speed changes as mentioned before, as well as the change of natural pressure between the interior and exterior due to thermal variation. It may be affected by the design of the architectural cover skin of the interior space, and the location of the openings in it. It may help or prevent climate control, or to address and handle an interior space plan failure (Aynsley, 2014; Sayed et al., 2012). Therefore it was necessary to know some facts about the dynamics of the movement of the airflow, and what might affect the movement of the wind inside the interior architecture:

- **Pressure Forces**: The air moves from the higher pressure area to the lower pressure area where the speed of the airflow in a small space leads to a decrease in air pressure, which allows it to be used in reverse to withdraw the airflow from inside the interior space if it is not desired (Fig. 9).
- **Thermal Forces**: As the airflow of wind-driven through interior spaces transport due to the effect of the temperature difference between the interior and exterior, the hot air rises to its low density according to its physical nature, to be replaced by cold air, and this explains the use of small windows and openings on the top of walls in the old traditional interior architecture as mentioned previously for cooling of spaces in the desert climate, as well as the use of the central courtyard as a moderate, shady and damp area that softens the temperature and distributes it to the surrounding spaces, and maybe further mitigated by the use of the water element during its interior design (About Natural and Mixed Mode Ventilation, n.d.) (Fig. 10).

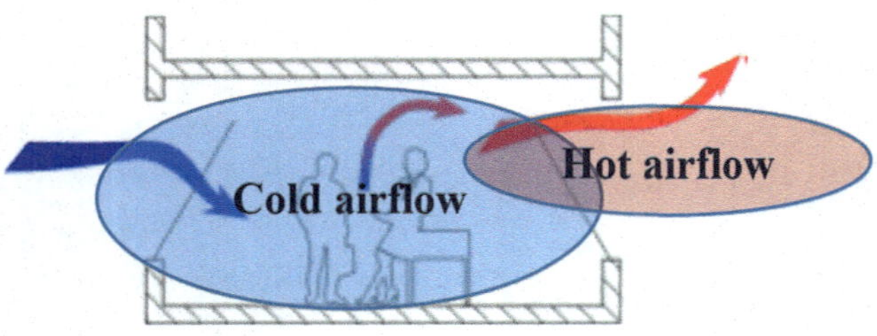

Fig. 9 The sketch shows that Pressure difference can affect Wind-driven airflow through openings to control the passive interior temperature (Osman, 2011, p. 48)

- **Interior Landscape**: The presence of plants or the plant element within the interior space with certain regulations as interior landscape forms enhances the direction of the wind, as it may deflate the airflow and change its speed, or create a pressure zone that operates according to the density, or it may be a breaker of airflow to create positive pressure area (Fig. 11).

3.4 Methods of Natural Ventilation in Interior Space Design

The principle of ventilation refers to how the exterior and interior airflows of wind force are related, and thus how the natural impulses are used to utilize a building ventilate. Moreover, the ventilation principle indicates how airflow is introduced into the building for cooling or heating (Lien and Ahmed, 2011) (Fig. 12).

3.5 Integration of Several Aspects

The utilized wind-driven ventilation force as a technique depends on several aspects. These technique aspects can be either passive or active for providing the energy needed, as natural ventilation can be stimulated by several forms dependent on the direction of the airflow, interior planning and space form to achieve a good ventilation effect (Fig. 13).

When discussing the interior design and cooling factors inside the sunny buildings in Egypt with its dry, hot weather in summer, we find that some thermal blocks absorb heat and trap it inside the space, represented by concrete and bricks. They emit that heat in the winter for heating. But, also release it in the summer, which prevents its cooling

Fig. 10 Sketches show how the window's shape can enhance the thermal forces effect in cooling interior space and turn it to a passive one (Osman, 2011, p. 64)

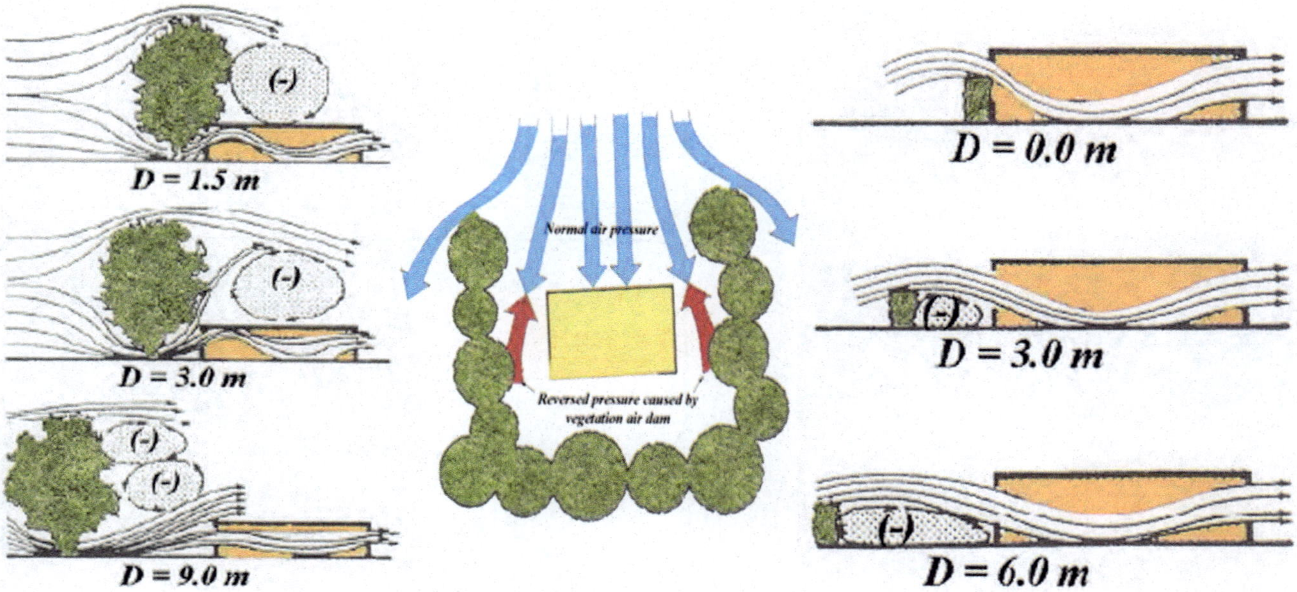

Fig. 11 Sketches show the effect of Plants hedges, Scale, dimensions and location on the airflow direction and deflection at different distances from the building. (Osman, 2011, p. 90–92)

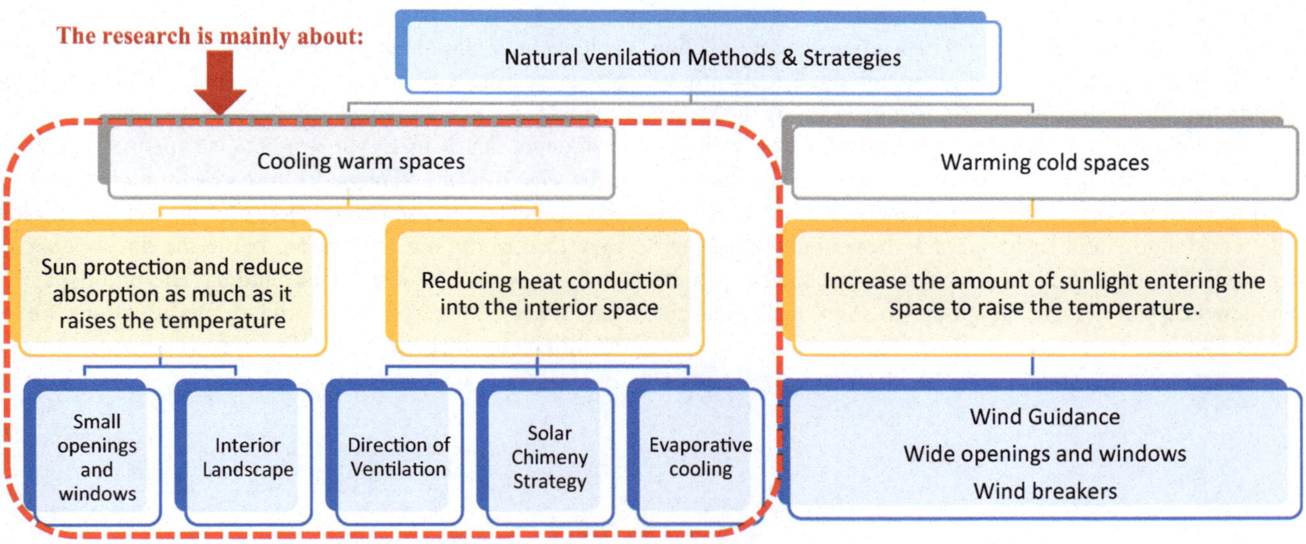

Fig. 12 Diagram shows the natural ventilation methods & strategies (Author)

without consuming a large amount of electrical energy or, more precisely, converting it into a passive cooling space.

In addition to that, there are internal thermal blocks like the walls, interior furniture, and methods of distributing partitions, which we may work to increase the heat and retain it within the close buildings' blocks and not to make optimal use of the windows designed for ventilation mainly. Therefore, it was necessary to follow the strategies and methods used in planning and architectural design to exploit the natural energy of winds-driven to cool interior spaces. That natural force works to reduce the stored heat without relying on energy consumption and find design solutions for interior spaces (Short et al., 2004; Dardir, 2006) (Fig. 14).

4 Case Study

As observed, in traditional techniques of natural ventilation, most of the environmental-vernacular designs depend on cross-ventilation airflow within spaces. As noticed, in

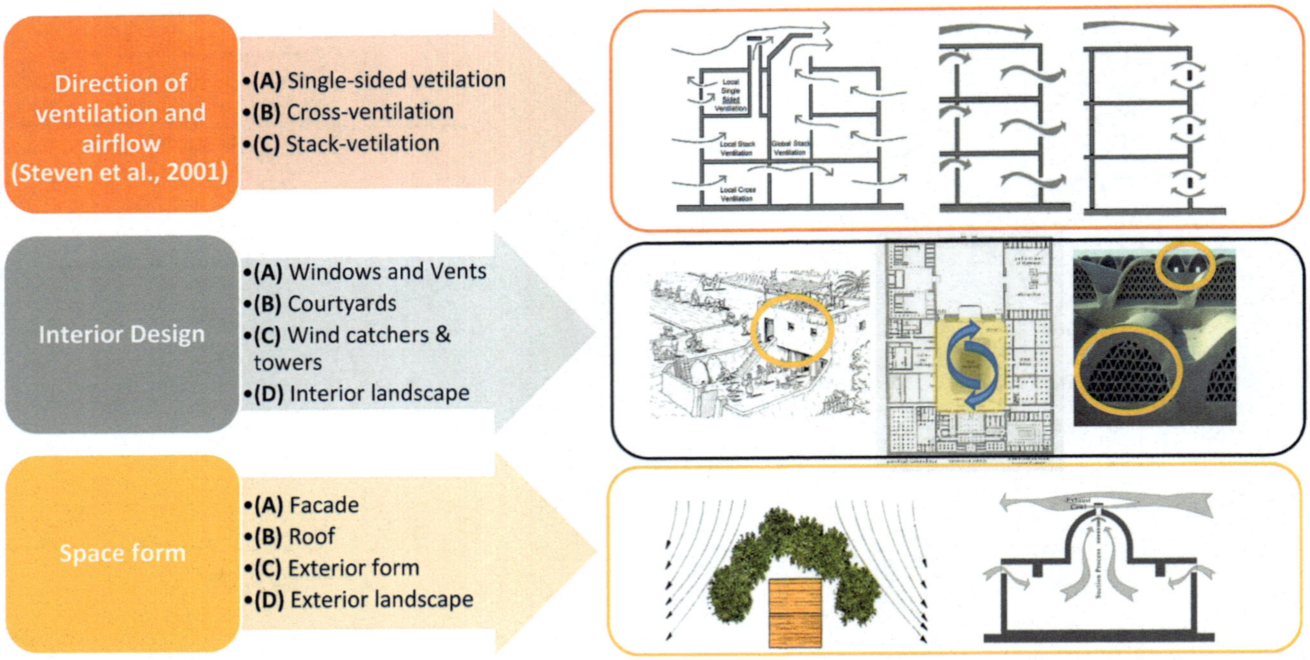

Fig. 13 Diagram shows wind-driven airflow aspects can be either passive or active providing the energy needed in interior spaces (Author)

traditional methods of natural ventilation, most of the environmental-vernacular designs depend on cross-ventilation airflow within the spaces. That is through the actual entry of airflow into the interior space that is influenced by the shape, size, type, location, and material. Then, It becomes to find an interior design solution utilizing wind-driven ventilation force as a technique in an adaptive passive way by simulating the site constraints through monitoring topography and surrounding buildings, building orientation due to airflow path and Landscaping elements design but in minor scale suites the interior spaces.

After a brief study of general principles and forms of architectural and interior design, which essentially cared about the quality of ventilation in proportion to the Egyptian atmosphere, that dominated by high temperatures, a case of residential space in Egypt was studied in one of the new cities east of Cairo. This is for research through monitoring the residents' experience of healthy living and the problems they face from poor ventilation due to the hot weather and the convergence of residential buildings, despite the lack of high floors. Also, trying to find design solutions for the interior space theoretically. This is provided that an applied

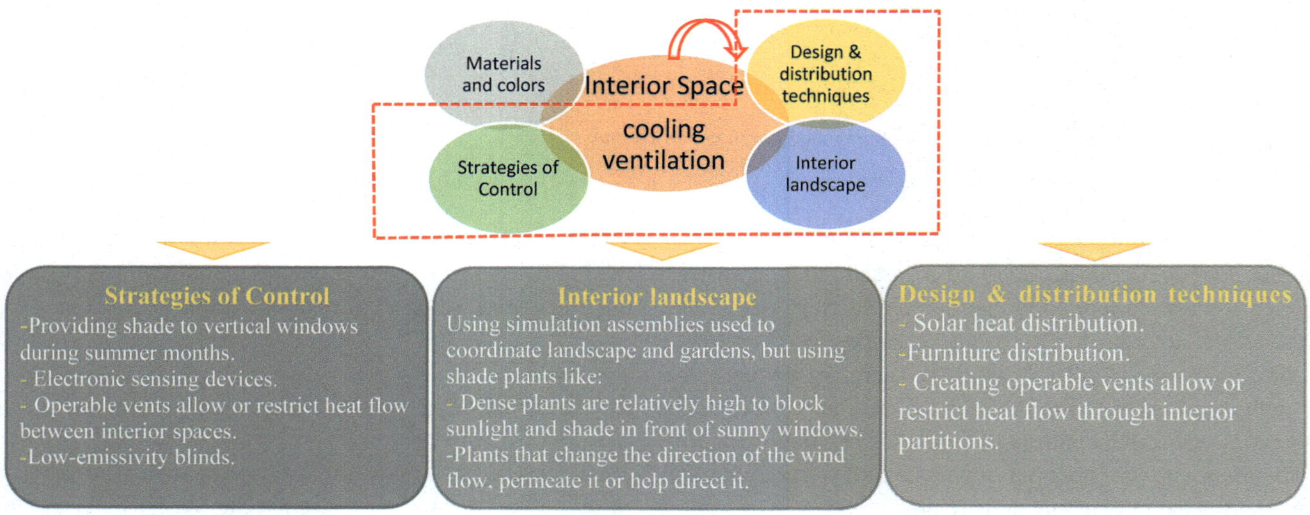

Fig. 14 Diagram shows integration of several aspects to utilize wind-driven force in Interior Design to turn it to be passive (Author)

practical study of these theoretical ideas and solutions will be studied in a more extensive way in independent research.

The residential unit is located in El Obour City, one of the new cities east of Cairo. It is on the first floor of a four-story building. It represents most of the residential units in Egypt surrounded by many concrete buildings comparable to or less than the height. This prevents proper ventilation most of the time as it also impedes the entry of sunlight, especially on the lower floors, despite the building maintaining the high temperature and locking it inside. However, it faces heat retention in it for most of the year, especially in the summer, especially the temperature became higher and its duration is longer. This requires the use of several air-conditioners in all rooms and requires their work most of the time, which leads to the consumption of large electrical energy and affects the family income of the space users (Fig. 15).

4.1 The Current Status of the Housing Unit

It is a middle-income housing unit. It overlooks two interfaces, one north-west, and one south-west.

- The openings on the south-western side, which contain more air currents, have relatively small and openings that do not allow the airflow currents that permeate and cool the interior space, which represented in the reception space.
- The north-eastern side contains the interior rooms of the apartment, which makes it hotter and less ventilated most of the time because no air currents pass through it most of the time.
- In general, the unit contains limited air currents, mostly hot, which leads to a rise in temperature in most of its spaces, and they are trapped most of the time, especially in the summer.

4.2 Suggested Design Solution:

The design solution is based on utilizing wind-driven ventilation force as a technique to turn a traditional space into an adaptive passive one with adding more aesthetic and functional form (Figs. 16 and 17).

- The expansion of some of the openings on the south-western side (the south-western façade). It is legal since there is nothing to prevent changing and modifying the facades' design in that area.

Fig. 15 Two shots show the location of the building that contains the studied residential space and the distance between buildings affects the natural ventilation to work correctly

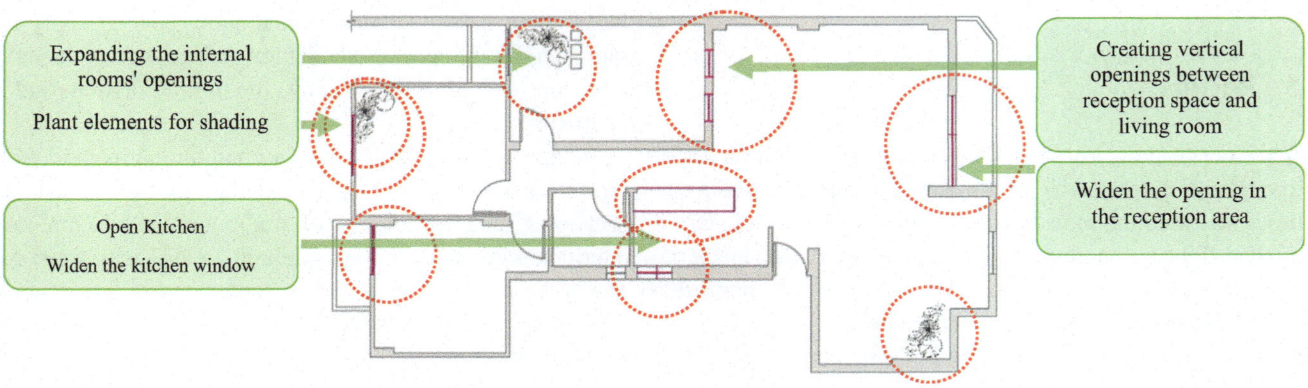

Fig. 16 Plan shows the design solution utilizing wind-driven ventilation force in the residential interior space to Increase the cold airflows to reduce the temperature and allow them to pass through spaces to become renewable and less energy-consuming

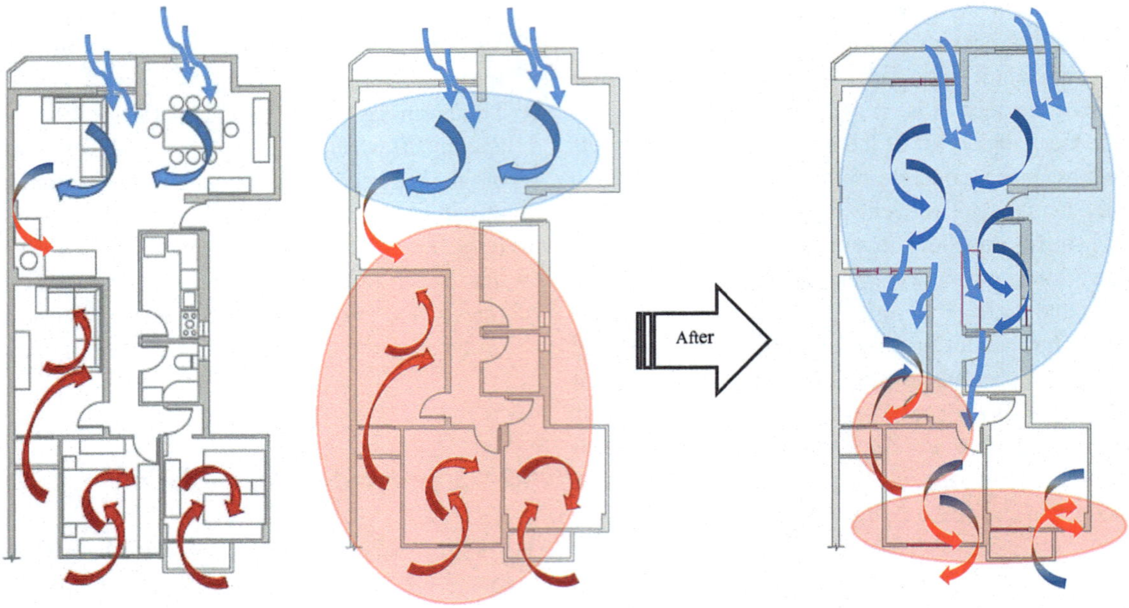

Fig. 17 Apartment plans show the increasing of the cooling airflows through its spaces by using wind-driven ventilation force in order to consume the energy used for cooling spaces

- Make a set of vertical openings between the reception space and one of the rooms that represent the living area to transfer the cold airflow to it, taking advantage of the aesthetic shape of the design and the visual extension between the two spaces.
- Turn the closed kitchen to opened one and removing the wall separating it from the reception space that allows the passage of a cold air stream that reduces the temperature imposed by its functional nature.
- Widen the kitchen window to create airflow with the reception vents and allow hot air to escape.
- Expanding the internal rooms' openings with the addition of some plant elements in a way that creates a shaded entity that reduces the heat of sunlight and does not prevent its penetration, as well as to operate on cold currents within the same space.

5 Results

A review of previous studies of natural ventilation methods throughout the research of different examples at different times reviewed the most important points or ideas of its aims to shed light on passive cooling techniques for architectural envelopes and interior spaces. That should start since the design begins so that the research reaches how to solve problems related to natural ventilation in the present. That helps to turn the interior space into a passive space if it already exists, and its sustainability has not been fully taken care of from the beginning. Mostly, the reliance on natural ventilation is one of the means to reduce energy consumption when the design depends on a force of nature like the wind.

This research didn't mean to go through many details of natural ventilation principles or dissect its techniques as all have been listed and classified in many other researchers. But this paper tended to learn from old traditional eco-design over decades that created passive interior spaces using utilizing wind-driven ventilation force as a technique of adaptability by chance. Also, It suggests some ideas can be reflected as a solution to turn the interior space into an adaptive passive one.

After trying to suggest an experimental solution of utilizing wind-driven ventilation force in interior space, It turns into a passive one taking cooling space as an example. These Ideas come into theoretical points show how to apply the same guides of natural ventilation design but on a more miniature scale. It helps to adjust the utilized driven-wind in interior space. (Table 1).

Finally, This research comes to monitor the main ideas to create a passive interior space simulate the same methods, principles, and techniques of traditional architecture and interior design as they were one unit to deal with from day one of design. So, This research recommends a further study of applying those ideas in the actual life with actual modifications in a separate paper.

Table 1 Author

<table>
<tr><th>Small interior space</th><th>Open/wide interior space</th></tr>
<tr><td>• Using the same opening shapes' types and their location (position) in the details of window's designs (Fig. 18)</td><td>Back to old traditional designs' solutions and principles in interior planning especially using courtyard and interior-landscape treatments (Fig. 21)</td></tr>
<tr><td>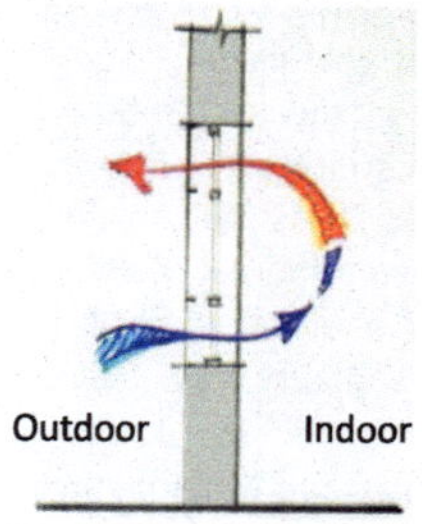

Fig. 18 Wall section shows the idea of how to use the upper opening technique in the window design to cool the interior space</td><td rowspan="5">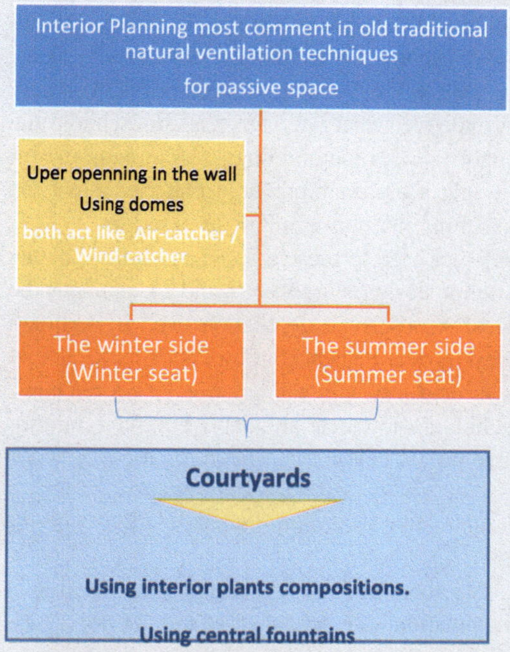

Fig. 21. Old traditional designs' solutions and principles</td></tr>
<tr><td>• Using partitions in some areas to allow the wind-driven airflow to pass between spaces smoothly or to change its direction (Figure 16.19)</td></tr>
<tr><td>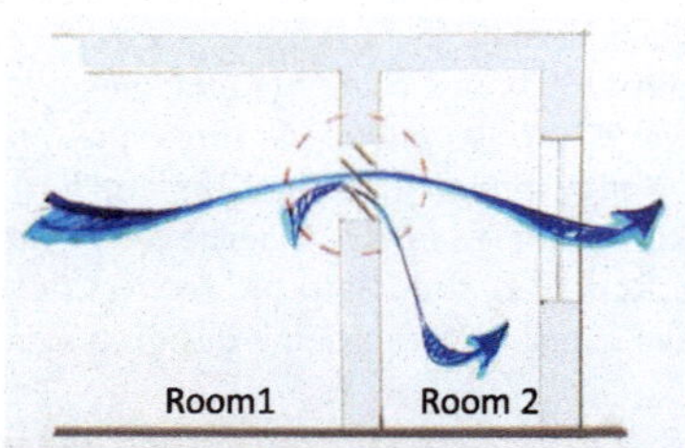

Fig. 19 Interior section sketch shows the idea of using partition design or interior upper wall opening to allow the current airflow to move smoothly between spaces</td></tr>
<tr><td>• Using plants in different compositions and scales to help also in changing the direction of the wind-driven airflow in the interior space (Fig. 20)</td></tr>
<tr><td>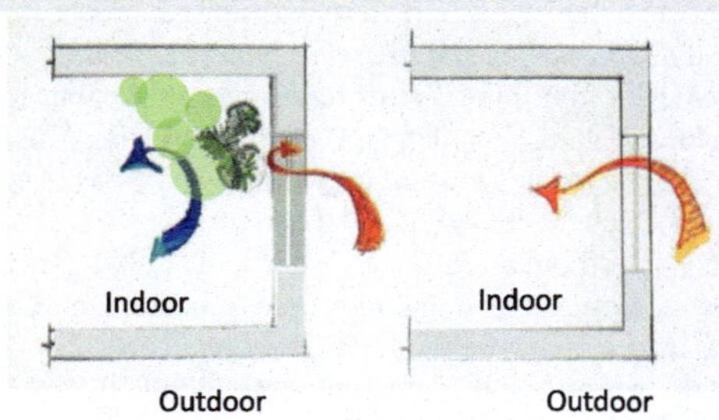

Fig. 20 Plan sketches show the idea of using interior plants to change the direction of wind airflow to maintain the interior cooling.</td></tr>
</table>

6 Conclusions

By the brief analytical study of some old traditional ventilation techniques over different periods, They were all agreed on creating shaded walkways and open small windows facing away from the sun. Also, they used several types of wind-catchers—sometimes in the same building depending on size, wind direction, and dust concerns.

It was clarified that natural ventilation principles were all the same but in different forms, designs and technologies. And, they were able to resist ecological foot-prints of society and to reduce expended energy. Above all, to achieve the goal of sustainable spaces methodology, while natural ventilation took its place as a natural force that can affect the space shape and function either indoor or exterior.

Finally, the paper tried to present some solutions that can be taken as interior design suggestions and ideas using the same methods of wind-driven ventilation used in architectural interior design over decades but on a smaller scale and more details as ideas can suit the interior space contains. They act as techniques to create an adaptive passive interior space to achieve points below:

- Find a fundamental solution to serve the interior environment.
- Fix and handle some defects of interior space form as a result of architectural defects that affect interior space quality and thermal comfort.
- Find solutions can make every interior space as an adaptive passive space, that can adapt to climatic changes and reduce energy consumed to the minimum.
- Utilize the old traditional knowledge of using wind as a natural force and clean source.
- Separate study research is needed to complete what this paper tried to achieve as new fundamental solutions for interior spaces that had problems in ventilation and energy consumption.

References

Abd ALkader, M., Kamal, A., & Magdy, B. (2017, July). Improving thermal performance in the internal environment of residential buildings in Egypt by using smart interfaces. *Journal of Al Azhar University Engineering Sector, 12*(44), 1225–1241, Retrieved from https://jaes.journals.ekb.eg/article_19185_72adf33a8d4466b629460790811b033a.pdf

About Natural and Mixed Mode Ventilation. (n.d.). *Window master*. Retrieved from https://www.windowmaster.com/solutions/natural-ventilation/strategies

Ancient Egyptian Houses: Domiciles From Pharaohs to Farmers, History on the Net. (n.d.). Retrieved from https://www.historyonthenet.com/ancient-egyptian-houses?fbclid=IwAR13w74U228bHkVRbJfgNj4YWSO0MFFsfpEq3CUWMqwvLoIT0fLNH1sWl-8

Aynsley, R. (2014, August). Natural vintelation in passive design. *Environment Design Guide*, Royal Australian Institute of Architects,1–16. Retrieved from https://www.jstor.org/stable/26151921?fbclid=IwAR0kqIV52Kqu3hqhcxVRwYgPEu6m5ojLBjYJ0tzfeVoPhzEQxQ2iICTAppw&seq=1

Dardir. M. (2006), *Natural Ventilation Techniques as a Base for Environmental Passive Architecture* [Master's thesis in Architecture], Faculty of Engineering Ain Shams University.

Energy.gove Offices. (n.d.). Retrieved from https://www.energy.gov/energysaver/natural-ventilation

Jamal. S. (2007), The architectural and constructing treatments for the climatic and environmental factors. *Journal of Engineering, Sciences, 35*(6), 1575–1596, Assiut University. Retrieved from http://www.aun.edu.eg/journal_files/183_J_456.pdf

Lien. J., & Ahmed N. A. (2011, July 27). Wind driven ventilation for enhanced indoor air quality. In N. Mazzo (Ed.), *Chemistry, emission control, radioactive pollution and indoor air quality* (pp. 539–562). School of Mechanical and Manufacturing Engineering University of New South Wales. Retrieved from https://cdn.intechopen.com/pdfs/16338.pdf.

Mina, T., & Zhang, T. (2016), Study on the Climate adaptability of architectural interface opening in Suzhou's regional residences. In N. H. Wong & S. K. Jusuf (Eds.), *Procedia Engineering: Curriculum transformation* (Vol. 169, pp. 108–116), Fourth International

Natural ventilation vs mechanical ventilation. (2014, September 17). Retrieved from https://zehnderamerica.com/natural-ventilation-vs-mechanical-ventilation/

Natural Ventilation. (n.d.). *Green Home Technology Center*, Ohio State University Extension. Retrieved from https://greenhome.osu.edu/natural-ventilation

Osman, M. (2011, May). *Evaluating and enhancing design for natural ventilation in walk-up public housing blocks in the Egyptian desert climatic design region* [Doctoral dissertation], Dundee School of Architecture, College of Art, Science & Engineering, University of Dundee. Retrieved from https://discovery.dundee.ac.uk/en/studentTheses/evaluating-and-enhancing-design-for-natural-ventilation-in-walk-u

Ramadan, S. (2016, July 31). *Air conditioners of ancient times.* Retrieved from https://www.bibalex.org/SCIplanet/en/Article/Details?id=5213&fbclid=IwAR2njWxd_-A3r61DP0FWz0WIt63E_tmhGYRx_r_gVYhWYhGuiDomEJCVjec

Samy, R. (2013). *Assessment of architectural heritage elements using computational fluid dynamic [Master's thesis in Architecture].* Cairo University.

Sassi, P. (2006). *Strategies for sustainable architecture, British library cataloguing in publication data.* Taylor & Francis.

Sayed, A., Hiroshi, Y., Eid, M., & Radwan, M. (2012, June), Indoor natural ventilation using evaporating cooling strategies in the Egyptian housing: A review and new approach. *IACSIT International Journal of Engineering and Technology, 4*(3), 229–233. Retrieved from https://www.researchgate.net/publication/284849581_Indoor_Natural_Ventilation_Using_Evaporating_Cooling_Strategies_in_the_Egyptian_Housing_A_Review_and_New_Approach__, http://www.ijetch.org/papers/355-B1004.pdf

Short, C. A., Lomas, K. J., & Woods, A. (2004). Design strategy for low-energy ventilation and cooling within an urban heat island. *Building Research & Information, 32*(3), 187–206. Retrieved from https://www.tandfonline.com/doi/full/10.1080/09613210410001679875?scroll=top&needAccess=true

Steven, J. E., Dols, W. S., & Axley, J. W. (2001, August), *Natural ventilation review and plan for design and analysis tools.* Architectural Energy Corporation Boulder, CO, NIST Interagency/Internal

Report (NISTIR)—6781, p. 6.Retrieved from https://www.nist.gov/publications/natural-ventilation-review-and-plan-design-and-analysis-tools

The Gate Heliopolis | Vincent Callebaut Architectures. (2019). Retrieved from https://www.arch2o.com/gate-heliopolis-vincent-callebaut-architectures/

Vincent Callebaut Architectures. (n.d.). Retrieved 6 July 2020. Retrieved from http://vincent.callebaut.org/object/141029_thegate/thegate/projects

Wilson, E. O. (2002). *The future of life*. Random House. Retrieved from https://epdf.pub/the-future-of-life.html

Reflections on the Development of Prefabricated Buildings in China from a Historical and Global Perspective

Zhang Junjun, Wang Haining, and Zhang Hong

Abstract

China is experiencing rapid urbanization, which has, and will continue, to yield strong demand for buildings, exerting substantial resource and environmental pressures on the country. Prefabricated overall correlation, parallel design, modular combination, and generalization is the best way to transform the construction industry from a traditional sector to an industrial sector which is of great importance to the sustainability of the whole society. This article firstly analyzes the history of the development of prefabricated buildings in China and then focuses on the different stages of the development of prefabricated buildings in developed countries. With the historical and global perspective, some problems and suggestions for prefabricated buildings in China to achieve sustainability are discussed in the end.

Keywords

Prefabricated buildings • Developing countries • Developed countries • China

1 Introduction

Construction is considered as one of the main factors having an impact on the environment. According to the United Nations Environment Programme, the construction industry consumes 40% of the energy, 30% of the raw materials, and produces 36% of the total greenhouse gas emissions. Carbon emissions generated by building operation and construction account for 50%, 33%, and over 40% of the total social output in the UK, Japan, and the US, respectively (Ikaga et al., 2002). Today, a significant amount of construction is taking place in developing countries, and the right way of construction can reduce carbon emissions. Prefabricated building construction is thought to be more efficient in energy use compared to traditional building construction, which makes it an important way to improve sustainability (Cao et al., 2015).

A prefabricated building, informally a prefab, is a building that is manufactured and constructed using prefabrication. It consists of factory-made components or units that are transported and assembled on-site to form the complete building. Prefabricated overall correlation, parallel design, modular combination, and generalization have transformed the construction industry from a traditional sector to an industrial sector which is of great importance to the sustainability of the whole society (Smith & Timberlake, 2010).

In developed countries, prefabricated buildings are widely used. The prefabrication industry in Japan was technologically advanced in a very large domestic market (Goulding & Rahimian, 2019). In Singapore, the prefabrication technology appeared to be the ideal way to improve the poor image of the construction industry and to resolve the problem of lack of human resources (Ting & Jin, 2000). In Hong Kong of China, prefabrication was considered a good solution for problems such as high waste generation, high risk, and labor-intensive construction (Zhang et al., 2018). Swedish prefabricated housing had been consistently highly represented, particularly in the single-dwelling timber detached housing market where it claimed an 88% share. In Germany, 9% of new residential building permits were for prefabricated buildings, while in the Netherlands 20% of all new housing used wood or concrete prefabrication (Noguchi, 2016). The degree of standardization, serialization, specialization, commercialization, and socialization of components and parts for residential use in the USA was high, reaching almost 100% (Gu, 2017). Although prefabricated buildings are facing a decline in the twenty-first

Z. Junjun (✉) · W. Haining · Z. Hong
Department of Architecture, Southeast University, Nanjing, 210096, China
e-mail: 230179003@seu.edu.cn; ; zjjarch@163.com

C. Alalouch et al. (eds.), *Towards Implementation of Sustainability Concepts in Developing Countries*,
Advances in Science, Technology & Innovation, https://doi.org/10.1007/978-3-030-74349-9_15

century in developed countries, they are still an important way to achieve sustainability (Lukyanov, 2020; Ross, 2006). However, prefabricated buildings are still in the technological development stage in developing countries, such as India (Smith & Narayanamurthy 2009). China has achieved substantial development success since its reform and opening up in 1978 and is the biggest developing country in the world. Today, China is an upper-middle-income country but has remaining challenges, of which the construction industry needs great improvements to achieve sustainability.

2 Methodology

China is experiencing rapid urbanization, which has, and will continue, to yield strong demand for buildings, exerting substantial resource and environmental pressures on the country. The Chinese government is strongly promoting building prefabrication and numerous difficulties have emerged during the process (Chang et al., 2018). It is necessary to analyze the history of the development of prefabricated buildings in China to make clear the gap with developed countries. Researches on the different stages of the development of prefabricated buildings in developed countries are also essential work to do because many experiences could be obtained from them (Kieran & Timberlake, 2004).

The first part of the research will be a study of the industrialization of construction in China. Unlike the Western world, China's scientific and technological progress has made jumps since its founding. As a result, the development of prefabricated buildings in China could be a special case. The second part will be a global study of the industrialization of construction which is accompanied by the gradual escalation of several industrial revolutions. The third part will be the characteristics of prefab technology in various developed countries. Finally, China's current situation will be compared to other countries, problems will be identified and recommendations will be given.

3 Overview of the Development of Prefabricated Buildings in China

As early as the beginning of founding, China had proposed a great target of building industrialization. However due to technical limits and many other reasons, the development of prefabricated buildings twists and turns. An analysis of the historical process of China's prefabricated buildings is conducive to understanding the current status of China's construction industry, which could help to analyze specific issues and work out the best way of construction industrialization in line with China's national conditions.

The overall development of China's prefabricated buildings can be roughly divided into the following four stages (Table 1):

3.1 The Early Stage of the Development of Prefabricated Buildings

From 1950 to 1976 is the early stage of the development of prefabricated buildings in China when they learned advanced technology from the Soviet Union and other countries (Jiang, 2010). The China State Council issued the "Decision on Strengthening and Developing the Construction Industry", which explicitly proposed the goal of "three modernizations", namely design standardization, component production factorization, and construction mechanization (Yi et al., 2017). Thanks to this, the country's concrete prefabrication technology had developed rapidly, and both academic research and architectural practice had flourished unprecedentedly. A group of concrete technical experts emerged in China, and prefabricated component factories were also established. Among them, hollow floor slabs are the main prefabricated products and are widely used in engineering. Unfortunately, in the later period, affected by some special factors, the construction industrialization process developed slowly and basically stagnated. Fortunately, this process had accumulated the material basis of factorization and mechanization for the future development of prefabricated buildings.

3.2 The Ups and Downs of the Development of Prefabricated Buildings

From 1977 to 1995, after the Chinese economic reform, the building industrialization is activated again. During this period, the industry-standard "Design and Construction Rules for Prefabricated Slabs" promulgated by the Ministry of Construction in 1991 was of iconic significance. However, due to technical and construction reasons, a large number of problems such as waterproofing, sound insulation, cold bridges, cracking, and earthquake resistance of prefabricated buildings were exposed. At the same time, China's cast-in-place concrete technology had gradually developed and organically combined with a large amount of cheap labor, the cast-in-place technology quickly occupied the construction market. Both internal and external problems coexisted in the prefabricated buildings, which once again stagnated.

Table 1 Chinese four stages of prefabricated buildings

Time	Stage
1950–1976	The early stage of the development of prefabricated buildings
1977–1995	The ups and downs of the development of prefabricated buildings
1996–2015	The promotion period of the development of prefabricated buildings
2016-	The great development period of prefabricated buildings

3.3 The Promotion Period of the Development of Prefabricated Buildings

From 1996 to 2015, with the commercialization of housing, the demand for housing became higher and higher, and the requirements for quality and diversification became a non-negligible factor. The industrialization of construction entered a period of slow development during which the main construction method is still in-situ technology-based (Yang, 2009). Ready-mixed concrete system developed with the emergence of ready-mixed concrete technology supported by the mixing station equipment, management, and transportation technology. Although the cast had been very mature, its disadvantage was also raised significant out. It took too much time for the manual work of reinforcing steel and the effort to remove the supporting molds. Besides, the pouring process generated serious pollution. As the requirements for a green environment and low carbon are getting higher and higher and the number of people engaged in manual labor is decreasing sharply, the government and enterprises gradually realized the importance of prefabricated buildings, as a result, the construction industrialization was again put on the agenda. The state issued a series of policies to study, support and promote the industrialization of construction, and a number of pilot projects for prefabricated buildings emerged. Enterprises led by Vanke and Yuanda also started the practice of prefabricated buildings. Despite the slow development, China gradually explored building industrialization technology centered on prefabricated structures through the study and reference of advanced foreign technologies. Its main feature is to connect precast concrete components with cast-in-situ concrete or infused mortar. It should be pointed out that due to the high cost of this type of technology and the immature technology, the development of prefabricated buildings is still slow, but it had attracted attention from all sides (Ye, 2016).

3.4 The Great Development Period of Prefabricated Buildings

From 2016 to present, with the active policy guidance, prefabricated buildings show a booming trend (Ye & Li, 2016). In February 2016 the State Council issued "Opinions on Further Strengthening the Urban Planning and Construction Industrialization Work", which clearly stated to promote the prefabricated buildings. China will increase policy support and strive to take about 10 years to make prefabricated buildings account for 30% of newly built buildings. With the unprecedented large support from the government, 2017 witnessed great development of prefabricated buildings. However, the development of prefabricated buildings in China is still in its early stages where are challenges and opportunities.

3.5 Summary

From the above four processes, we can see that due to historical conditions, the earliest development period of prefabricated buildings in China failed to advance effectively. In the second stage, due to the development of cast-in-place concrete technology and cheap labor, the demand for fabricated buildings was not high. Until recent years, with the reduction of labor resource and the demands for protection of the environment, the Chinese government began to vigorously promote prefabricated buildings.

4 History of the Development of Prefabricated Buildings

China's construction industrialization process cannot be separated from the global process, and is deeply affected by the global industrialization process. The global construction industrialization process is closely related to the industrial revolution process. The overall development trend is getting closer to the development process of human civilization and will lead to the development and change of the global construction industry. According to the internal causes and performance of the global construction industrialization development, it can be roughly divided into the following four development stages (Shan, 2018), as illustrated in Table 2.

4.1 Building Industrialized 1.0 Era: Mechanization (1900–1950)

With the rise of the Second Industrial Revolution and the end of the First World War, prefabricated buildings played a

Table 2 Four stages of the development of prefabrication

Time	Stage	Main technical features
1900–1950	Building industrialized 1.0 era	Mechanization
1950–1980	Building industrialized 2.0 era	Standardization
1980–2010	Building industrialized 3.0 era	Informatization
After 2010	Building industrialized 4.0 era	Intelligence

very important role in Western post-war reconstruction and economic recovery. The concept of industrialized construction based on factory production and mechanized assembly began to take shape. However, the technology was immature, the management was rough, the construction cost was relatively high, and it did not have market-oriented conditions. It was basically in a government-led and business-involved mode.

4.2 Building Industrialized 2.0 Era: Standardization (1950–1980)

After the 1950s, with the post-war economic rapid-rate rise, the third industrial revolution (scientific and technological revolution) began to rise, which provided good economic conditions and technological development for prefabricated buildings. The concept of standardization and modularization in prefabricated buildings began to form. The technology had a perfect market foundation so the technical system was gradually improved. As a result, the construction methods were continuously innovated, and prefabricated buildings ushered in a period of rapid development.

4.3 Building Industrialized 3.0 Era: Informatization (1980–2010)

After the 1980s, with the advent of the information age, Auto-CAD software, BIM technology, network technology, and communication technology have been widely used in the field of prefabricated buildings, effectively promoting the improvement and management level of prefabricated building technology systems. The concept of "general system", "open architecture" and "hundred-year house" began to take shape. The prefabricated building industry chains are gradually established and improved in developed countries.

4.4 Building Industrialized 4.0 Era: Intelligence (After 2010)

With Germany leading industrial 4.0 Generation—the arrival of the fourth industrial revolution, people in developed countries put forward higher requirements for the quality of life and the environment. The connotation of prefabricated buildings began to sublimate and develop in the direction of human-centered design, environmentally friendly construction, and intelligent living. The technologies and cultural connotations of prefabricated buildings have been continuously enhanced, and the industrialization process of construction has started simultaneously with the industrial revolution.

4.5 Summary

We can draw a conclusion that the development of construction technology is closely related to the process of the industrial revolution. Technological changes brought by the industrial revolution have given rise to advances in building materials, construction machinery, and other related fields. On the whole, however, the progress of architectural technology has always been slower than the industrial revolution, partly because of the time required to validate the application of the new technologies (Zhang 2017).

5 Overview of the Levels of Prefabricated Buildings in Typically Developed Countries

Learning advanced technology from developed countries and drawing lessons from their experiences can further accelerate the development process of China's construction industrialization.

The following Table 3 shows the status of prefabrications in typically developed countries and China.

Japan's strong role in the factory-built housing industry emerged as a response to World War II. The ministry of international trade and industry established the Japan Prefabricated Construction Suppliers and Manufacturers Association (JPCSMA) in 1964 to promote best practices, develop and disseminate research, and form industry standards (Smith and Quale 2017). Japan frequently experiences large earthquakes so its earthquake-resistant technologies are leading in the world. Most of the houses adopt the framework system and KSI system. Besides, the mature PCa method system has a high level of industrialization.

Table 3 Development status of prefabricated buildings in different countries and areas

Country	Main technical features	Industrialization level
Japan	Main structure: PC structure Good at: earthquake-resistant technology	4.0 era
Singapore	Main type: high-rise residential buildings Good at: standardization	4.0 era
Sweden	Main structure: wooden structure Main type: low-rise residential buildings Good at: dry system	4.0 era
Germany	Main structure: PC structure Good at: energy-saving technologies The earliest initiator of steel structure building	4.0 era
France	Main structure: frame structure Good at: prestressed concrete	4.0 era
USA	Main structure in large cities: PC structure Main structure in small towns: light steel structure and wooden structure Good at: modular technology	4.0 era
China	Main structure: PC structure Good at: construction technology	3.5 era

Singapore has developed up to 30 stories high-rise buildings with prefabrication technology which accounts for more than 80% of the total number of dwellings in the country. The optimization of the layout of the plane, the repeatability of the size of the components, and the standardization of connecting nodes are the keys to achieve a high level of prefabrication (Ting & Jin, 2000).

Sweden's main structure type of prefabrication is the wooden structure, supplemented by the large slab structure system. It is the most developed country for residential prefabrication with the representative technology of dry system and modular coordination integrated into the SIS system (Steinhardt et al., 2020).

Germany is widely recognized as the earliest initiator of steel structure building, the inventor of prestressed concrete and reinforced truss. The emergence and the massive application of early precast concrete construction techniques in Germany were primarily intended to address the social contradictions of the vast lack of housing in the post-war period. At present, Germany is focusing on green and sustainable development and on the use of environmentally friendly building materials and construction systems. Through the detailed optimization process of planning, design, and construction, the projects seek a comprehensive balance of individuality, economy, functionality and ecological and environmental performance. A large number of building components are standardized and modularized for use in buildings, emphasizing the durability of the building, but not pursuing large-scale factory prefabrication rates.

France was one of the first countries in the world to promote prefabricated buildings. The frame structure or slab-column structure is the mainstream, with the steel and wood structure supplemented. The prestressed concrete frame construction system is widely used, with an assembly rate of up to 80%.

USA has the best modular technology with BL quality certification system, making the level of industrialization high. In small towns, it is dominated by light steel structure and wooden structure systems while in large cities it is dominated by PC structure and steel structure systems. In addition to focusing on quality, prefabricated buildings in the USA are now more focused on beauty, comfort, and individuality.

According to Chap. 3, the Chinese government began to strongly promote prefabricated buildings in 2016. In China, the vast majority of buildings are of concrete construction. So the main structure of prefabricated buildings in China is also concrete structure. It is well known that China has the largest population in the world, which has necessitated the construction of a large number of high-rise buildings. Therefore, China is now mainly learning from the experience of Japan and Singapore. China's traditional construction technology is very strong, fully realizing mechanization and standardization, reaching a high degree of information, but almost with no application of intelligence. As a result, China's status of prefabricated buildings is defined in the 3.5 era.

6 Discussion

Through the overview of the development of prefabricated buildings in China, the analysis of the history of the development of prefabricated buildings, and the overview of the levels of prefabricated buildings in typically developed Countries, we can make a discussion in terms of standards, management, connecting nodes, prefabrication type, and structure system respectively.

Building standards and codes are the foundation for the healthy development of the construction industry. The building codes of Western countries have been developed and repaired over nearly a century and are in line with the basic national conditions of each country. Many of the Chinese building codes are borrowed from the Soviet Union, and in the process of use they gradually show various problems. In particular, the Chinese government does not have an effective building code for prefabricated buildings. This has limited the sustainable development of prefabricated buildings.

Building a house is a complex project. Prefabricated construction is more demanding for construction management. Construction companies in Japan usually have a complete approach to project management. Construction management in Singapore is done jointly by the government and the firm. Construction management in European countries relies more on information and intelligent collaboration. China's current construction industry is still dominated by manpower management and is less efficient.

Prefabrication is a construction method where parts of a building are manufactured in a factory and transported to site for assembly which means that the connecting nodes of the prefab components are significant issues. Most countries use dry connection technology, such as Japan, Sweden and France. However, currently in China dry joint technology is not recognized, mainly because it does not meet the Chinese building codes. In China, wet connection technology is still used more often because it is less technically demanding. It is slower to build with the method and is still a traditional construction method.

The type of prefabricated buildings applied varies from country to country. Prefabrication in Singapore is mainly used for high-rise residential construction, while prefabrication in Sweden is mainly used for low-rise buildings. The type of prefab technology in the USA applies differently in large and small cities. Currently, similar to Singapore, prefabrication in China is mainly for high-rise housing. The Chinese government is also seeking a broad application of prefabrication technology in high-rise public buildings. In terms of building type, China should learn standardized prefabricated building design techniques and prefabricated component design techniques from Singapore.

Many highly industrialized developed countries have developed special systems for prefabricated buildings. The main structural system in Japan is PC with earthquake-resistant technology, while the main structural system in Sweden is wood. Germany specializes in PC structure integrated with energy-saving technologies, and France has more research and applications for prestressed concrete. The special system for prefabricated buildings has been proven to be very suitable and technically economical. China's prefabricated concrete single-story industrial plants and large slab buildings for residential use also fall under the category of specialized structural systems. But there is still a lack of a clear choice of structural types for high-rise residential buildings in China.

7 Problems and Suggestions for Prefabricated Buildings in China

In summary, the analysis of the development and significance of prefabricated buildings is carried out above. For now, China still has many problems in prefabrication, mainly in: the lack of unified standards, the low construction management level, the difficulties in node ductility and waterproof issues, the low level of product diversification and standardization, the uncertain main structure system, etc. Details are as follows.

7.1 Unified Standards

Although China has formulated and implemented prefabricated construction standards and codes, from a practical point of view, there is still no unified standard, which results in differences in the standards for the fabrication of prefabricated components, on-site measurement, and connection of nodes. This status results in the lack of scientific standards in engineering construction, which is very unfavorable to the development of prefabricated buildings.

7.2 Construction Management Level

In order to ensure the effective development of prefabricated building construction, a professional prefabricated factory is required to provide technical support in every detail, especially the management. The construction plan design, the component manufacture and transportation, the connecting nodes, and other processes have high requirements. However, in reality, the on-site construction method is still used in the construction of prefab projects. It is urgent to improve the level of management for the prefabricated building firms.

7.3 Connecting Nodes Issues

Connecting nodes are important parts in prefabricated buildings. In order to improve the overall quality of the building, the connecting nodes should be reasonably designed to ensure that the node stiffness meets the relevant requirements. Effective measures should be taken in the specific implementation, otherwise, the project quality will be affected. In the construction of prefabricated buildings, constructors should also pay attention to the problem of joint waterproofing, avoid joint leakage, and ensure the improvement of project waterproof performance.

7.4 Product Diversification and Standardization

It is undeniable that prefabricated buildings are facing a decline in Western countries. A very important reason for this is the lack of diversity in prefabricated buildings, which leads to the perception that prefabricated buildings are boring and unattractive. In fact, on the basis of standardization, it is possible to create a beautiful architectural form through the combination of components. Architects should spend more time on standardization and diversity, and the government should encourage architects and developers to work hard on this.

7.5 Special Structure System

China's construction market is the largest in the world. Mass and marketability are important factors that prefabricated buildings can be implemented. With such an advantage, China should consider inventing its own special structure system and promoting it. Just as Japan focuses on PC structures and Sweden on wood structures, China should focus on a suitable structural form and continue to research and deepen it. On the basis of this special structure, modalized building components and high-performance connection nodes will be developed.

8 Conclusion

In many developed countries, prefabricated buildings are the most important way of construction industrialization. With the advantages to improve quality, shorten the construction period, save energy, reduce consumption, and achieve cleaner production, the vigorously developing prefabricated buildings is an inevitable choice for China. It is a socio-economic method to get a certain level in terms of reducing potential damage to the environment, shortening construction cycles, improving building quality, and reducing labor. As a new building construction method, prefabrication has many advantages that are incomparable with traditional buildings, despite some defects and deficiencies. Therefore, in the era of rapid development of modern construction industry, China should change its concepts and thoughts, invest more manpower and material resources to research and develop prefabricated buildings, make it the mainstream direction of modern construction industry development, and provide the world with safe, efficient, green and sustainable living space.

References

Cao, X., Li, X., Zhu, Y., & Zhang, Z. (2015). A comparative study of environmental performance between prefabricated and traditional residential buildings in China. *Journal of Cleaner Production, 109*, 131–143.

Chang, Y., Li, X., Masanet, E., Zhang, L., Huang, Z., & Ries, R. (2018). Unlocking the green opportunity for prefabricated buildings and construction in China. *Resources, Conservation and Recycling, 139*, 259–261.

Goulding, J. S., & Rahimian, F. P. (Eds.). (2019). *Offsite Production and Manufacturing for Innovative Construction: People*. Routledge.

Gu, T. C. (2017). Current status of the development of assembled buildings in ten countries, including China, Britain and the United States. *Journal of Architectural Design Management, 34*(08), 39–40.

Ikaga, T., Murakami, S., Kato, S., & Shiraishi, Y. (2002). Forecast of CO2 emissions from construction and operation of buildings in Japan up to 2050 *Journal of Asian Architecture and Building Engineering, 1*(2), 149–156.

Jiang, Q. J. (2010). A review of the development of domestic and international assembled concrete buildings. *Journal of Construction Technology., 41*(12), 1074–1077.

Kieran, S., & Timberlake, J. (2004). *Refabricating architecture: How manufacturing methodologies are poised to transform building construction*. McGraw-Hill Companies.

Lukyanov, F. (2020). Prefabricated world order and its decline in twenty-first century. In *Russia in a changing world* (pp. 23–38). Palgrave Macmillan.

Noguchi, M. (Ed.). (2016). *ZEMCH: Toward the delivery of zero energy mass custom homes*. Springer.

Ross, R. S. (2006). Nothing new under the sun: prefab's slow rise from affordable housing to sustainable solution. LA architect 50–56.

Shan, L. L. (2018). Thinking about the development of assembled buildings in China from a historical and global perspective. *Construction Technology, 47*(S4), 595–598.

Smith, R. E., & Quale, J. D. (Eds.). (2017). *Offsite architecture: Constructing the future*. Taylor & Francis.

Smith, R., & Narayanamurthy, S. (2009, July). *Prefabrication in developing countries: a case study of India*. Wood Structures Symposium, p. 5.

Smith, R. E., & Timberlake, J. (2010). *Prefab Architecture: A Guide to Modular Design and Construction* (1st ed.). Wiley.

Steinhardt, D., Manley, K., Bildsten, L., & Widen, K. (2020). The structure of emergent prefabricated housing industries: A comparative case study of Australia and Sweden. *Construction Management and Economics, 38*(6), 483–501.

Ting, S. K., & Jin, H. F. (2000). *Prefabrication in the Singapore Construction Industry*. In 17 th International Symposium on Automation and Robotics in Construction.

Yang, S. (2009). The major development of construction technology in China during the 60 years of the founding of the country. *Journal of Construction Technology, 40*(9), 774–778.

Ye, H. W. (2016). Reflections and countermeasures for the industrialization of new construction. *Journal of Engineering Management, 30*(2), 1–6.

Ye, H. W., & Li, C. S. (2016). New construction industrialization and future construction development. *Journal of Construction Technology, 19*, 45–49.

Yi, Q., Li, C., & Jiang, Q. (2017). A study on the current status of assembled building development in China based on policy analysis. *Journal of Construction Technology, 22*, 50–52.

Zhang, J. J. (2017). *A preliminary exploration of industrialized architectural design method based on assembly-type rigid steel cage technology* [Master Thesis, Southeast University].

Zhang, W., Lee, M. W., Jaillon, L., & Poon, C. S. (2018). The hindrance to using prefabrication in Hong Kong's building industry. *Journal of Cleaner Production, 204*, 70–81.

Toward a Sustainable Design of the Rehabilitation Centers for Addiction

Basma Ibrahem, Doaa Abouelmagd, and Alia Amer

Abstract

The process of applying sustainability in the design of addiction rehabilitation centers is essential. Previous studies can be defined sustainability from two aspects, the first is technical and the second is psychological. Environmental sustainability involves interaction between patients and the surrounding spaces. The literature review reveals that the design of external or internal spaces surrounding the patient has significant influences on the patient's state. There is a lack of researches that informs architects of the design elements and requirements that significantly influence patients' state and show that designing spaces from a healing perspective have a positive effect on patients' state. Those requirements enable the patients to interact with the surrounding environment and do several activities, which improve the healing process and care services for them. This study aims to identify the design requirements from international, national guidelines, and analytical studies of the patient's needs that make them interact with the surrounding environment. It too shows how to apply those requirements in the design process.

Keywords

Rehabilitation centers • Sustainability • Surrounding environment • Patient needs

B. Ibrahem (✉)
Misr University for Science and Technology, 6th of October City, Egypt

D. Abouelmagd · A. Amer
Architecture Department, Faculty of Fine Arts, Helwan University, Helwan, Egypt

1 Introduction

Nowadays, Rehabilitation centers for addiction are essential facilities as almost 9.8% of Egypt's population are addicted to drugs, therefore, it is necessary to establish centers of addiction suitable for the needs of addicted patients as Egypt suffers from a lack of centers of addiction (Hamid, 2014). According to previous studies rehabilitation centers for addiction are considered a category of mental health care (Abdelhay, 2016). The sustainability of health care buildings can be defined from psychological and physical perspectives. A physical perspective cares about energy resources as an ecosystem. On the other hand, a psychological perspective depends on the interaction between the human needs of patients and the surrounding environment (Altman, 1975). The design of rehabilitation centers should include the human, the social, and the psychological needs of the patient. Evidence and researches present that the addicted patient needs special care during the treatment process (Chrysikou, 2014b).

The design process of these centers is one of the most complex design processes. The design represents an efficient role in the external and internal atmosphere of the building. It has positive or negative influences on the state of patients, medical teams, and other users.

Composition of functional programs, proper designing, and social programs are required for the healing environment of health care buildings (Xenopoulous, 2011). Functional programs can assure the sustainable design of the building. The social program is a critical factor that has a positive influence on the state of addiction patients. Supporting social programs can increase the interaction between the patient and the surrounding environment (Lawrence, 2005).

To address the psychological needs of addiction patients and staff in rehabilitation centers, sustainable designers should look to nature and addicted patients' responses to the surrounding environment (Ulrich 1992). The architectural design considers a part of the healing process through the

C. Alalouch et al. (eds.), *Towards Implementation of Sustainability Concepts in Developing Countries*, Advances in Science, Technology & Innovation, https://doi.org/10.1007/978-3-030-74349-9_16

creation of spaces to fulfill the process of rehabilitation (Lundin, 2015). In this paper, we address the effect of sustainable design attributes on the patients of addiction.

2 Literature Review

The literature review shows the definition of addiction, its causes, how to treat it, and emphasized the importance of social connection for preventing addiction. It also considers a rehabilitation center for addiction as a type of building, that have many features and elements which reduce the stress of the patients (Ulrich, 1991), such as home-like characteristics, patient privacy, sports, and recreational facilities, natural daylight, natural views, noise decrease, art, color, dayrooms, accessibility, security, and safety (Dhingra, 2017).

As has been previously reported in the literature the surrounding environment and design of spaces have an effect on the psychology and the state of the patient, psychiatric clinics and rehabilitation centers (Laudet, 2011); therefore, studies were dealing with the surrounding environment and design of spaces have an effect on the psychology and the patient state, psychiatric centers, and rehabilitation centers therefore, studies were dealing with the surrounding environment of patients that have an effect on the healing process and the needs of the patient. It is also well acknowledged that providing the services that help him to return to the community has a positive effect on the patient state (Abdelhay, 2016).

It was reported in the literature that has recognized that most of the rehabilitation centers were focusing on the detention approach of treatment instead of medication, neglecting the patients' psychological, social, and spiritual needs, which are the important elements that achieving effective treatment (Unodc-Who, 2017) Therefore, it is also well acknowledged that the combination between therapeutic techniques in drug addicts center and design process through functions program, connection to nature, adopting the healing gardens, using color therapy and artworks on walls have an effect on the patient state (Habibu, 2015).

Most early studies, as well as current work, focus on health care building sustainability. Environmental sustainability is improved by the interaction between patients and surrounding spaces (Seyun, 2010). Therefore, there is a study to examine how space design interacts with patients. Along with how hospital designs reflect the influence on the patient state (Easy, 2015).

The previous studies reveal that Maslow hierarchy for human needs is a successful theory (Maslow, 1987), it used in many fields such as business according to Idemobi (2011), education according to Adiele and Abraham (2013), Heritage according to McLeod (2007) and medicine according to Zalenski (2006). Also, it is applied in urban design according to Asma Mehan (2016) and architecture design of healthcare buildings according to Abdo (2006) in cancer hospitals and psychiatric hospitals according to Hammad (2014). According to the previous study, that theory of Maslow's hierarchy for human needs didn't apply in rehabilitation centers for addiction so we can use it as a method to conclude addicted patient's needs according to design requirements.

The main problem is that there is a lack of researches in Egypt, addressing the relationship between sustainable design, the surrounding environment, and the satisfaction of addict patients (Hamid, 2014). Also, the relationship between patients' needs in rehabilitation centers for addiction and environmental elements has not been addressed (Abdelhay, 2016). Therefore this paper identifies the most important design requirements that have an impact on the satisfaction of addicted patient needs.

2.1 Definition of addiction

According to the American psychiatric association (APA): "*Addiction is a complex condition, a brain disease that is manifested by compulsive substance use despite harmful consequence*" (Ranna Parekh, 2017). There are two types of addiction (Shaffer, 2018), the first is the behavioral addiction, which is defined as a passionate growth, and realization is making some individuals more liable to "out of control behaviors (Shaffer, 2018). This type is distinguished with the use of repetitive actions, started by a motive that cannot be stopped, causing an individual to escape, drug, soothe, release tension, lessen anxiety or feel euphoric. (i.e., cell phone, social media, exercise, food, eating, gambling, Internet, love and relationships, masturbation, sex, shopping, and video games) (Karim, 2012).

The second type is substance addiction, and it can be defined as "*A complex condition, a brain disease that is manifested by compulsive substance use despite harmful consequence*" (Medicine, 1980). People of addiction focus on using a particular substance, such as alcohol or drugs. We can define drug addiction as a brain disease that leads to changes in the structure and function of the brain (Habibu, 2015).

2.2 Addiction Treatment Stages

Addiction treatment includes medication for depression and other disorders, at first, the caring is intensive. After completing the intensive stage, patients shift to outpatient treatment stage, which for fewer hours per week to help them to enhance their healing process (Nida, 2018a). According to the National Institute on Drug abuse treatment process starts with Detoxification it which is a process, to clear the human body from a toxic substance. The patient needs to isolate in special space with good visibility from the staff to prevent them from self-harm followed by medication, the patient needs private spaces to receive medication with the opportunity to ask and receive answers to questions regarding medication. Then behavioral therapy through group therapy treatment and some activities, the patient needs special spaces for those with comfort colors to reduce the stress during it.

2.3 Rehabilitation Center for Addiction

It is a secure and private institution that renders the treatment of drug addiction. it is considered a category of mental health facilities to treat substance dependency, and it offers 24-h intensive care, including safe housing and medical care. it works to provide medical support for drug detoxification and prevention of retrogression (Nida, 2018a).

Rehabilitation centers for addiction can be worked on an outpatient and Inpatient. Some centers support patients to work on arts and crafts, while others support other methods such as treatment by using music, dance, or drama therapies. Many addiction patients have physical-emotional issues; they are supported to play sports, look after gardens, or take walks around the facility's grounds. The rehabilitation centers for addiction are ready to handle any medical emergencies which may engender during the treatment process. (Nida, 2018b).

3 Research Methodology

According to the study, there are no clear design requirements for addiction centers in national and international guidelines. Therefore this study uses Maslow's Hierarchy of Needs to analyze the needs of addiction patients in rehabilitation centers. The study introduced the effect of a sustainable environment on the state of addiction patients by carrying out analytical studies of national and international rehabilitation centers for addiction. Based on this analysis, the study shows that the application of Maslow's Hierarchy of Needs and elements of healing environment have a positive effect on the state of the addiction patients. The study used El Matar Psychiatric hospital as a national example along with Groot Klimmendaal as an international example.

4 International (Australian & American Guidelines) and National(Egyptian Guidelines)

According to international guidelines, which are the American and Australian, there are no clear design criteria for designing rehabilitation centers for addiction, but there are guidelines for designing psychiatric hospitals. Added to that there are no design standards for rehabilitation centers for addiction in Egypt, but there are general standards for psychiatric hospitals according to the Egyptian guidelines, which are currently used.

All of the above shows there aren't clear standards to design rehabilitation centers for addiction on the local and international level; according to guidelines, there are common elements that must be taken into consideration during design process such as providing privacy to the patient, providing a safe environment that helps the patient to carry out the daily task successfully, flexibility in design, accessibility, using natural light, choosing a calm environment, the connection between the surrounding environment and nature and achieving control of climatic conditions (Ahia, 2010; Institute, 2010).

5 Sustainable Design of the Rehabilitation Centers for Addiction

Sustainability is the main principle of environmental design. It includes preserving natural sources by not exhausting the environment with materials that cannot easily be replaced through the design and construction processes. Sustainability shall ensure the continuity of productivity for many years under optimal conditions (Duzenli, 2017). The concept of sustainability is defined as "*the principle of meeting today's needs without sacrificing the right and the ability of future generations to meet their own needs* Development, 1987)," furthermore, sustainability assures the quality of life by considering biospheric transport capacity, ecosystem, and resources (Smith, 2000). According to the different definitions, sustainability provides the quality of life and interact with natural resources permanent (Ruano, 1998).

The sustainability of healthcare buildings has an augment and can be defined from both physical and psychosocial aspects (Seyun, 2010). we can define the physical perspective, as-built, and natural environment and the psychosocial perspective as the interaction between the patient and the surrounding environment (Easy, 2015). Understanding of rehabilitation centers' design is related to understand patient needs therefore the designer can create a supportive physical environment to provide patients needs according to Maslow's Hierarchy of Needs (Huitt, 2007).

According to the analysis of Maslow's Hierarchy of Needs and attributes of a healing environment, the designer has to take into consideration patients' psychological, social, and spiritual needs. The interaction between the patient and space is important for sustainable design, therefor surrounding spaces in rehabilitation centers have psychosocial effects on the state of patients (Özcan, 2006). Building a psychosocial identity in space can preserve sustainability environment qualities. The designing according to patient needs, social programs, and suitable functional programs and supplying the social programs are prerequisites for the healing environment for the addict patients. Functional programs can assure the sustainability of healthcare buildings. Using design elements suitable for the patient's needs and supporting social programs can increase the interaction between the patient and the surrounding environment. Therefore the designer is focusing on psychosocially supportive design for a sustainable environment (Nguitragool, 2019).

6 Features of Social Sustainable Design for Rehabilitation Centers for Addiction

1. **Meet the basic human needs**: The design should allow the fulfillment of the primary life needs of patients; those needs include food, clean water, and appropriate temperature, in addition to bathing, and sleeping (Zahra Kefayati, 2015).
2. **Respecting and caring for patients:** One of the main features of social sustainability is to protect the patients by surrounding spaces in both interior and exterior areas. It will create a sense of ownership and enforce the feeling of respect (Mohamed Abdel-Raheem, 2016).
3. **Social interaction:** The human need for social communication at different levels is not hidden from anyone, therefor architectural designers take into consideration the design of surrounding space suitable for this interaction (Davoodi, 2013).
4. **Social security:** Creating a sense of security in space is an important indicator of developing social sustainability. Some points should be taken into consideration during the design process are the design of an architectural space that can increase the sense of controllability in users (Jabaree, 2017).
5. **Social participation:** Supporting activities can enhance a user's connection to society (Woodcraft, 2015).
6. **Flexibility:** the building must be in harmony with the natural and surrounding environment. Flexibility can define diversity, adaptability, and variability. It means having a multi-functional space where it is suitable to change the function of the space proportional to requirements (Moberg, 2016).

7 Addiction patient's Needs

The needs of inpatient patients change according to patients' cases, but there are general needs of patients as shown below.

7.1 Medical Needs

The main aim for patients when entering the rehabilitation centers is to receive the necessary treatment according to the patient's state. Therefore the centers provide the necessary treatment for these patients using various treatment methods and rehabilitating them to return them to their normal life again (Asadi-Lari., 2004).

7.2 Human Needs

Many studies dealt with the relationship between human behavior and the surrounding environment, and many researchers divided the basic human needs. The most important of those divisions was what Abraham Maslow (1954) did as a hierarchy in which those needs arose from physiological needs to Self-actualization at the top of the pyramid as shown in Fig. 1 (Uysal, 2017).

7.2.1 Physiological Needs

The design of rehabilitation centers should allow the fulfillment of the basic needs of patients; those needs include Daylight exposure, Fresh air, Natural view, Noise reduction

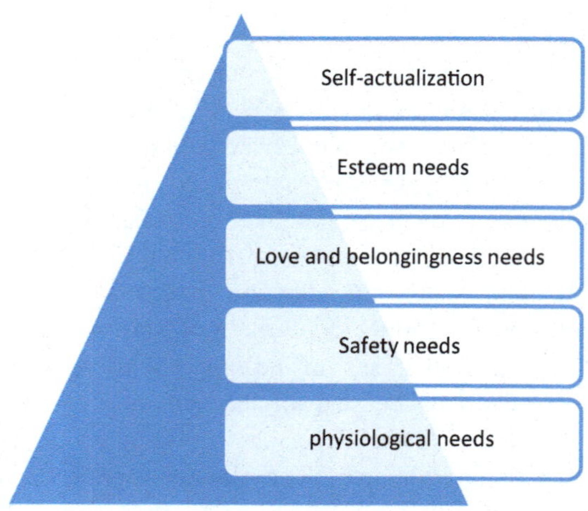

Fig. 1 Maslow's Hierarchy of needs (Huitt, 2007)

in addition to suitable colors to reduce the stress (Polikandrioti, 2011). Those needs could be applied through:

- Using natural light in public zones, group therapy spaces, and inpatient rooms. It has a positive impact on the patient psyche; it decreases depression and promotes shorter inpatient stays as shown in Figs. 3 and 4 (Joseph, 2006). Also Using warm lighting rather than cool in all areas such as corridors, patient rooms, and treatment rooms. It creates a more home-like atmosphere and feeling of harmony (Admin-Hc, 2018).
- Accessibility for patients and visitors as difficult wayfinding increases stress so the design has to support easy wayfinding in addiction centers which may reduce the stress (Alert, 2016).
- The corridors should be designed with appropriate widths that not less than 2.4 m to allow two people from each side to assist the patient as shown in Fig. 2 (Institute, 2010).

7.2.2 Safety Needs

It can be defined as protecting the patient from the risk of suicide, self-harm, and aggressive behaviors (Ranit Mishori, 2016) according to guidelines for design hospitals and healthcare facilities the design achieves this through:

- Use the fences in the exterior spaces which do not allow you to climb up or pass through them to secure the patient protection and preventing their escape. The suitable height for fences ranging from 2.7 to 4.0 m. Also, the entrances and exits must be specific and have control

Fig. 2 Corridors with a suitable width and using soft colors to reduce stress, king country hospital (Afify, 2019)

Fig. 3 Using atrium to provide natural light, warm light and soft colors in patients bedrooms to reduce stress, Östra Psychiatric hospital, Gothenburg (Ulricha, 2018)

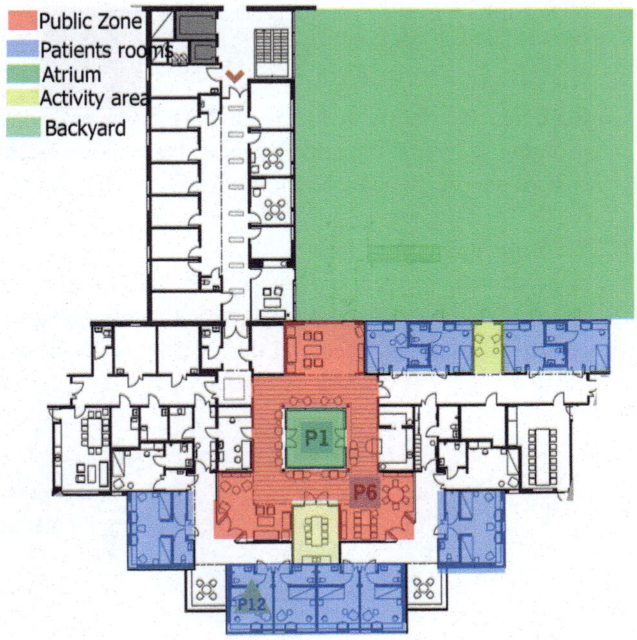

Fig. 4 Providing natural lights in public zones, patient's zones and activities zones by using atriums, Östra Psychiatric hospital, Gothenburg (Ulricha, 2018) edited by author

points that allow clear visibility of anyone entering or exiting.
- Using outdoor fixed furniture to prevent patients from harming themselves and no sharp corners to prevent the patient from harming themselves as shown in Fig. 5.
- Avoid long corridors when designing movement paths because they increase the sense of nervousness.
- Visual access to patients through providing stations in from of dayrooms with large observation windows and good visibility of wards, corridors, and dining rooms and separation between public circulation and patient's

Fig. 5 No sharp corners to prevent patients from harm themselves, king country hospital (Afify, 2019)

circulation by doors. Also, avoid long corridors when designing movement paths because they increase the sense of nervousness as shown in Fig. 6 (Institute, 2010).

7.2.3 Belongingness Need

It can be defined as the need for support and social interaction, it comes important in the behavioral therapy phase to reduce stress. So the architectural designer must consider it when designing rehabilitation centers spaces according to Saul McLeod, it can be achieved through:

- Designing activity spaces that allow a patient to communicate with others to give him a sense of belongingness and make patients don't feel alone and, designing rehabilitation outdoor and indoor spaces for group therapy treatment.
- Allowing patients to regulate their personal spaces and relationships with others in shared spaces such as dayrooms, activities areas, group therapy areas, and eating areas as shown in Fig. 8.
- Providing movable seats in shared spaces promotes the patient‘s ability to organize personal space and interact with others to decrease stress. Also, designing chairs around small tables in the middle of the room to increase interaction and, providing green areas surrounding the patient as shown in Fig. 7 (McLeod, 2018).

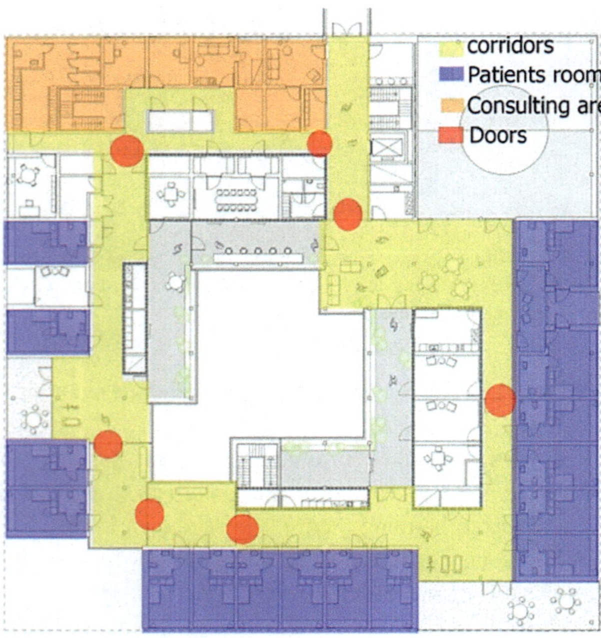

Fig. 6 Using separation doors between departments and avoiding long corridors, Nuuk Psychiatric Clinic, Greenland, (Duddy, 2018) edited by author

7.2.4 Esteem Needs

It can be defined as it is related to patient satisfaction, such as independence, appreciation, dignity, and respect that leads to feelings of self-confidence, value, strength, ability, and sufficiency. According to (Gough Len Doyal, 1984) it can be achieved through:

- Providing patients with single bedrooms to decrease stress. Also, providing patients with sports facilities such as gym hall, swimming pools, living spaces, green

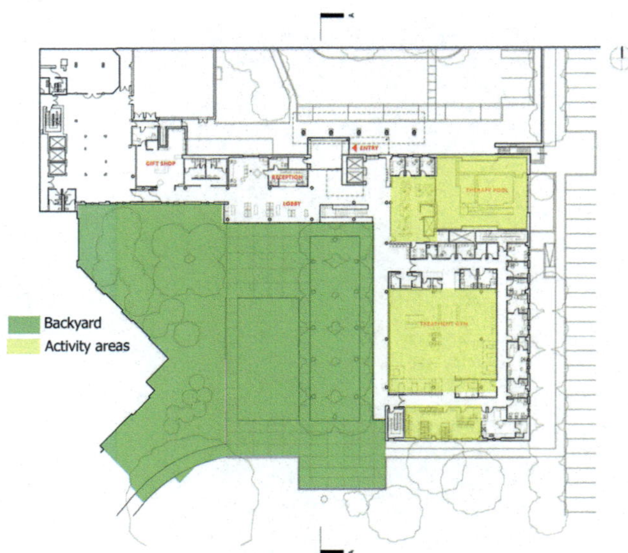

Fig. 7 Green area surrounding the building, activities area (Gym, pool and workshops) to increase interaction between patients, rehabilitation center by St. Johns Rehab / Montgomery Sisam Architects and Farrow Partnership Architects, Toronto, Canada, (Architects, 2012) Edited by author

Fig. 8 Activity spaces with fixed furniture and no sharp corners to protect patients from harm themselves, natural light and soft colors to reduce stress, king country hospital (Afify, 2019)

spaces, they have a positive impact such as achieving satisfaction, happiness, and self-esteem.

- Suitable environment to allow patients to interact with surrounding because it is a sign of respect to the addicts (Len Doyal, 1984).

7.2.5 Self-Actualization

It can be defined as it is realizing personal potential, self-fulfillment, seeking personal growth, and peak experiences. Desire to become the most that one can be. It can be achieved by providing patients with new crafts and activities, which gives them a sense of responsibility and designing workshops to help them try new things and accept themselves (Hoffman, 1999).

8 Healing Environment

It can be defined as creating an environment that makes patients feel good. we generalize by saying that if you feel good, you will "heal" faster, which means as an environment that supports the healing process. Healthy rehabilitation centers design is a compatible mix of nature and architecture that supports health and aids early recovery for those who are sick. Therefore the surrounding environment can play a factor in their sense of well-being and actual recovery. The quality of space in rehabilitation centers buildings affects the outcome of the healing process, and architectural design, so it plays an essential role in the healing process (Chrysikou, 2014a). Rehabilitation centers should provide a cheerful, inviting ambiance, and a caring and healing environment. To achieve a good design, it needs a multi-disciplinary approach involving architects, healthcare professionals, interior designers, landscape architects, environmental scientists, clinical psychologists to form a perfect healing environment (Biley, 1996).

8.1 Benefits of Healing Design (Habibu, 2015)

> Help in the treatment of the physical body, achieve the Physical, Psychological, human, and spiritual needs of the patients, and have positive effects on the patient.

8.2 Elements of a Healing Environment

8.2.1 Privacy, Company, and Dignity

It is essential to give privacy to patients in rehabilitation centers of addiction to improve patient satisfaction levels. This can be achieved by enabling them to be alone and to be with others when they desire. Also, it can be applied to acute bed space and a waiting space (Yen-Ko Lin, 2013).

8.2.2 Views

It is important to provide patients and staff rooms with a window that allowed them to view the scenes outside. The level of anxiety and stress can be reduced among patients and staff by having a window to view the scenes outside as shown in Figs. 9 and 10. Previous research showed that patients who have views outside recover more quickly (Ulrich, 2001).

8.2.3 Nature and Outdoors

The design has to provide patients and staff connection with nature because it is known that connecting to nature improves the patient state (Soh Oi-Zhen, 2015). (Figs. 11 and 12).

8.2.4 Comfort and Control

Design to provide all patient environmental comfort and control. it related to temperature and light. added to that, it also involves acoustics. the designer should organize space to reduce noise transmission. Providing patients with

Fig. 9 Outside scene by using large window to enable them to see the view, Kronstad Psychiatric Hospital (Arkitektgruppe, 2013)

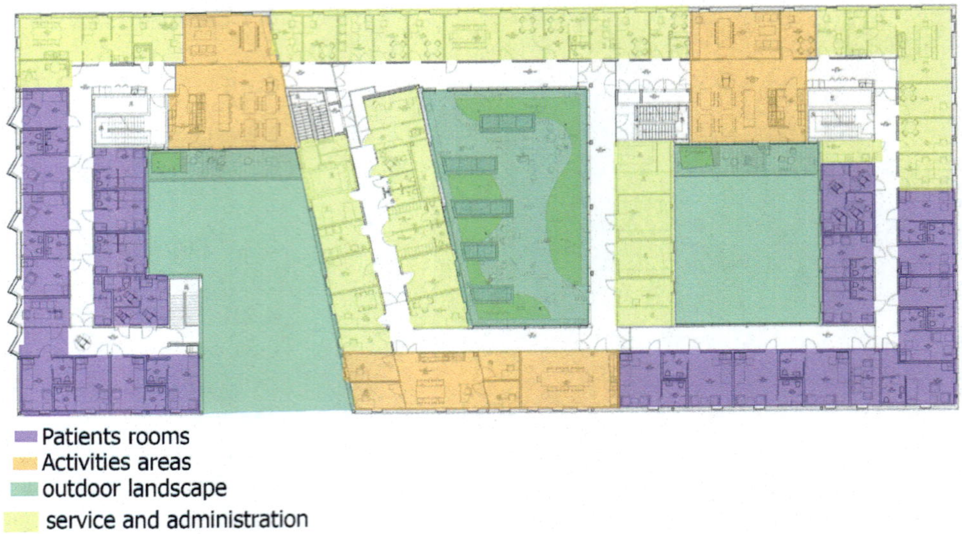

Fig. 10 Show outdoor view from patients zones that be used by patients and visitors, Kronstad Psychiatric Hospital (Arkitektgruppe, 2013) edited by author

Fig. 11 Connect to nature through large windows (Architekten, 2014)

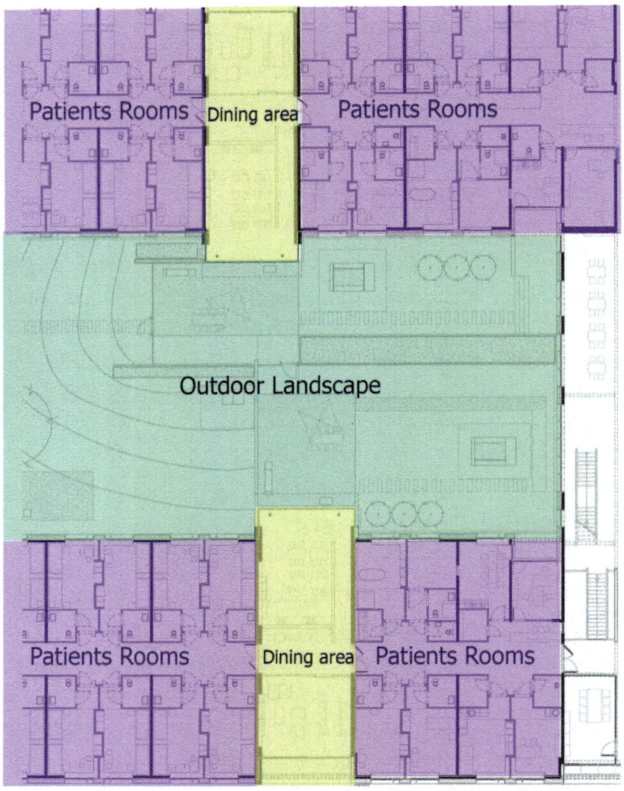

Fig. 12 Outdoor landscape surrounding the building to reduce stress and give patients sense of freedom (Architekten, 2014) edited by author

bedhead controls of lights, blinds, and curtains and doors effective in decreasing stress levels with low-cost impact (Eriksen, 2005).

8.2.5 Legibility of Place

The designer should take into consideration the concept of accessibility to make patients and visitors understand and can find their way inside the center. Places that can't be accessed increase patient's stress levels. Accordingly, public and private places must be clear along with entrances and exits shall be visible (Verderber & Todd, 2012).

8.2.6 Interior Appearance

Addict patients will spend a long time in rehabilitation centers so the design should give them feel homely with comfort colors. Also Using art paintings and sculptures improve the state of the patient (Dissanayake, 2011).

8.2.7 Facilities and Staff

The design should provide addict patients with basic needs in all aspects such as dining area, playing area, group therapy area, and toilets. Finally, the design should provide the staff with a suitable environment as job satisfaction linked with the working environment. Therefore increase staff motivation to work and provide a better quality of the working environment has a positive effect on the patients (Nijhuis, 2017).

9 Design Requirements for Achieving Patients Needs and Healing Environment Elements

We showed that from the analysis of human needs and study healing environment elements, addicted patients need special design in the surrounding environment that covers all their needs. The surrounding environment for addicted patients can play a significant role in the rehabilitation and healing of the patient. Therefore, the architectural design should help to reduce the anxiety and stress by designing spaces similar to home feeling where the extended stay of patients. It also must take into account the internal and external spaces that give a sense of harmony with the surrounding.

Another promising finding shows design requirements for rehabilitation centers for addiction linked with human needs according to Maslow's hierarchy and elements of healing environment as shown in Table 1.

10 Case Study Analysis

we choose these examples to make the analysis of patients' needs and how to achieve these needs through the design. (Table 2).

10.1 Rehabilitation Centre Groot Klimmendaal

Groot Klimmendaal center was established in 2010 by KoenvanVelsen, and it is an international center. It has an

Table 1 Design Requirements to achieve patient needs and elements of healing environment

Patient needs	Elements of a healing environment controlling levels of noise to meet patients need sent	Design Requirements
physiological needs	Nature and outdoor, view, comfort and control, Legibility of place	– Natural daylight is maximized – Buildings are surrounded by landscape and opened to the sky and view – Social and public spaces are located to provide views into external areas – Controlling temperature and ventilation – Controlling levels of noise to meet patients needs – Spaces should locate far away from any sources of noise – Areas are accessible for someone using walking aids/wheelchair – Reception areas are well planned – The signage is clear and visible
Safety needs	Interior appearance	– The personal effects area is safe and accessible – Providing the building with existing alarm
Belongingness need	Company	– Activity areas inside and outside – The building has direct access to exterior space for exercise and access to fresh air – Programmed activities were available for patients during the day and at weekends
Esteem needs	Privacy	– Designing areas and bedrooms according to the gender – Privacy in toilets and bathrooms is secured – Spaces for service users to receive visits – Rooms for interviewing and meeting with individual service users – Privacy in receiving medication with the opportunity to ask and receive answers to questions regarding medication
Self-actualization	Facilities and staff	– Spaces are provided for interactions

Table 2 Comparative table of the cases studied

Patient needs	Elements of a healing environment	How to achiev design requirments in Groot Klimmendaal centre	How to achiev design requirments in El Matar Psychiatric hospital
Physiological needs	Nature and out door, comfort and control, Legibility of place	– The building façade is covered with brown anodized aluminum panels which, gives the building harmony with its surroundings and uses natural light – It is not at a great distance from the city of Arnhem	– Natural lighting is provided in inpatient rooms and public spaces – Using soft colors such as white and begie to reduce patients' stress
Safety needs	Interior appearance	– Providing good visibility of, corridors, dining rooms, and shared areas. – Don't use sharp corners inside the center. – The separation between zones by using doors as shown in Fig. 13 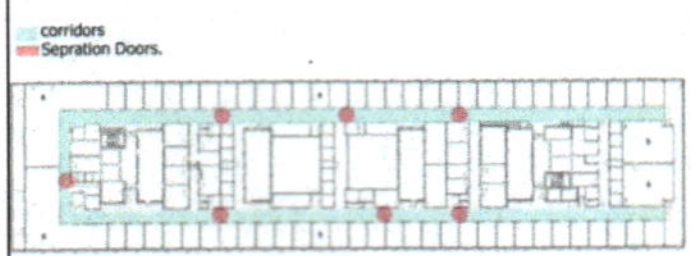 **Fig. 13.** The separation between zones	– The nurse stations cannot observe wards, corridors, and living rooms – Using metal furniture with perpendicular angles
Belongingness need	Company, view	– The building provides living rooms, group therapy spaces, and activity areas to encourage patients to interact with others. – The curtain walls and therapeutic glass use in its façade invite the forest inside the building, giving patients view of nature as shown in Fig. 8.14  **Fig. 14.** The curtain walls and therapeutic glass	– The center consists of activity areas and group therapy spaces – Patient rooms are connected with a single load corridor. It is designed to overlook the outside space as shown in Fig. 8.15  **Fig. 15.** Patient rooms are connected with a single load corridor
Self-actualization	Facilities and staff	– The center doesn't have new crafts and workshops spaces – patients, staff and the local community building to help the patients with their healing process	– The center has workshops to help patients learn new crafts and spend their time on useful things

area of about 14,000.0sqm and is composed of one building with three floors. The center is providing a drug addiction and drug treatment program. The project was awarded Building of the Year 2010 by the Dutch Association of Architects, winner of the first Hedy d'Ancona Award 2010 for excellent healthcare architecture (AlAmeri, 2018). The project presents an unconventional method to design healthcare centers. The building offers a pleasant and comforting experience to its users. It was built on the idea that a positive and stimulating environment increases the well-being of patients and can improve the rate of recovery (Boex, 2016).

10.2 El Matar Psychiatric Hospital

Heliopolis Hospital for Psychiatric was established in 1990, and it is a government and national hospital, located in an area of Heliopolis, next to Cairo International Airport. It has an area of about 5 Feddan, and its buildings are all composed of one floor. The hospital is providing a comprehensive drug addiction and drug treatment program. It considers the first of its kind in Egypt that adopts the policy of "outpatient" in a manner that preserves patient freedom. It considers the Regional Center in Egypt, the Middle East, and North Africa for the International Network for Addiction Treatment. (Hammad, 2014).

11 Discussion

According to the results of the previous table, it is clear that there is a relationship between patient needs and the surrounding environment and patient satisfaction. As it can achieve sustainable design through the interaction between patients surrounding environment, which improves the state of them (Seyun, 2010).

There are some implementations from the previous case studies on the design of the rehabilitation center for addiction:

- Recreational facilities are accessible to the community to support patients'.
- Workshops and classroom to help patients to learn new crafts
- Using natural daylight.
- Using soft colors and a natural view to reduce patients' stress.
- Providing patients with private rooms to reduce their stress and respect their privacy.
- Minimizing sharp corners to protect patients from harming themselves.

Our data suggest that we still have a long way to go to achieve addicted patient needs through the design. From this study, there are some recommendations to achieve patient needs:

Physiological needs by designing glazed corridors, opened to the space to allow the natural light to enter the building, bedroom windows and those in public areas should look out to outdoor space. Also, designing entrances with transitional space to avoid the sudden change in temperature between outdoor and indoor environments, rooms should be provided in areas away from public areas, and there should be a welcoming reception area, with clear wayfinding that accesses the building (health, 1996).

Safety needs by designing bedroom doors to be locked by patients when inside the room and when leaving, appropriate personal attack alarm should be in place and patient-to-staff call systems should be in place in assisted sanitary facilities (Health, Safe management of healthcare waste, 2012).

Belongingness need by designing group therapy spaces areas, areas for activities, with outside space as required. Also, there may be a physical activity area including an internal gym or sports hall (McLeod, 2018).

Esteem needs by designing separate areas according to gender, providing each room with a private toilet, visiting rooms be provided at the entrance to the unit or ward to give privacy to the patients, designing rooms with comfortable seating for interviews and meetings. Also, designing self-medication lockers, a private room located adjacent to either the treatment room or the drug storage room, or medication should be taken to the bedroom (Health, Health Building Note 03–01 Adult acute mental health units, 2013).

Self-actualization by designing spaces for crafts and new activities such as workshops (Hoffman, 1999).

12 Conclusions

This article aimed to identify the design requirements to achieve addicted patient needs according to Maslow's hierarchy. As shown in case studies the surrounding environment plays an important role in enhancing the level of patients satisfaction through enabling them to interact with the surrounding environment. From the previous study, we found that the architecture engineer should understand addicted patient needs to design similar to home feelings to help the patient to interact with the surrounding environment. Thus, this article recommends applying design requirements for achieving addicted patient needs and elements of a healing environment.

In summary, this paper argues that there are design requirements for designing rehabilitation centers to achieve the interaction between patients and the surrounding environment. Architects can design the rehabilitation centers, regarding the psychosocial impacts of design requirements. The design has to combine space and humans need to create a sustainable environment. Enhancing the physical environment is essential for function and Convenience of space, improving the psychological environment is also essential for achieving patient satisfaction. Psychosocially supportive design of a hospital can support the healing environment eventually. On this basis, we conclude that the designer has to consider during the designing process of rehabilitation centers patient basic needs and design requirements to help in the healing process and build a sustainable environment.

It is important that future researches investigate special guidelines for designing rehabilitation centers for addiction especially in Egypt although there are guidelines for designing psychiatric hospitals, but not for addiction centers. Also, we recommend making plans to provide separate centers for addicted treatment because the majority of centers are for mental disease according to guidelines.

References

Abdelhay, M. (2016). Effect of applying therapeutic architecture on the healing of drug addicts. *Second International conference on Sustainability and the future, At the British University*. Egypt.

Abdo, H. M. (2006). *Effect of human needs on the interior design criteria of oncology medical centers.*. Cairo University.

Admin-Hc. (2018, January 5). *Health lighting: how to design Hospital Lighting*. Retrieved from Haichang optotech: https://www.haichanglight.com/health-lighting-how-to-design-hospital-lighting/
Afify, E. S. (2019, November 20). *Hospitals design* (B. Ibrahim, Interviewer)
Ahia. (2010). Environmental considerations. In *Australasian health facility guidelines* (pp. 23–26). Australia.
AlAmeri, S. (2018). *Architecture of drug addiction rehabilitation*. Abu Dhabi University.
Alert, A. P. (2016). *Family Presence during Resuscitation and Invasive Procedures*. Retrieved from American Association of Critical-Care Nurses Practice Alert: www.aacn.org/wd/practice/docs/practicealerts/fam-pres-resusc-pa-feb2016ccn-pages.pdf.
Altman, I. (1975). *The environment and social behavior: Privacy, personal space, territory, crowding*. Brooks/Cole Pub. Co.
Archdaily. (2011, April 8). *Rehabilitation Centre Groot Klimmendaal / Koen van Velsen*. Retrieved from archdaily: https://www.archdaily.com/126290/rehabilitation-centre-groot-klimmendaal-koen-van-velsen
Archdaily. (2013, November 25). Retrieved from Archdaily: https://www.archdaily.com/451158/kronstad-origo-arkitektgruppe
Archdaily. (2013, November 25). *Kronstad Psychiatric Hospital / Origo Arkitektgruppe*. Retrieved from archdaily: https://www.archdaily.com/451158/kronstad-origo-arkitektgruppe
Architects, M. S. (2012, February 24). *St. Johns Rehab / Montgomery Sisam Architects + Farrow Partnership Architects*. Retrieved from ArchDaily: https://www.archdaily.com/211220/st-johns-rehab-montgomery-sisam-architects-farro%25e2%2580%258bw-partnership-architects?ad_medium=gallery
Architekten, H. S. (2014, March 18). *Psychiatric Centre Friedrichshafen*. Retrieved from ArchDaily: https://www.archdaily.com/486389/psychiatric-centre-friedrichshafen-huber-staudt-architekten?ad_medium=gallery
Arkitektgruppe, O. (2013, November 25). *Kronstad Psychiatric Hospital*. Retrieved from ArchDaily: https://www.archdaily.com/451158/kronstad-origo-arkitektgruppe?ad_medium=gallery
Asadi-Lari., M. (2004, June 29). *Patients' needs, satisfaction, and health related quality of life: Towards a comprehensive model*. Retrieved from Health and Quality of Life Outcomes: https://hqlo.biomedcentral.com/articles/10.1186/1477-7525-2-32#citeas
Azmi, H. (1989). *Inpatient human needs in the Egyptian context.*. Cairo.
Azzazi, J. N. (2016). *Psychosocially supportive design within workplace.*. The British University in Dubai.
Biley, F. C. (1996). Hospitals: healing environments?. *Complementary ttherapies in nursing & midwifery*, pp. 111–115.
Boex. (2016, October 4). *An iInclusive rehabilitation center*. Retrieved from Boex: https://www.boex.co.uk/inclusive-rehabilitation-center/
Chrysikou, E. (2014a). *Architecture for psychiatric environments*. IOS Press.
Chrysikou, E. (2014b). *Architecture for psychiatric environments and therapeutic spaces*. IOS Press.
Davoodi, S. F. (2013). *Social sustainability and srchitecture, determining the indicators affecting the development of social sustainability in architecture*. Conference on Architecture, Urban Planning and Sustainable Development with a Focus on Vernacular Architecture to Sustainable City. Mashhad, Iran.
Development, W. C. (1987). *Our common future*. University Press.
Dhingra, S. (2017). Psychology of architecture for the mentally Ill . *International Journal on Emerging Technologies*, 689–692.
Dilani, A. (2007). Psychosocially supportive design as a theory and model to promote health. *International Academy for Design and Health*.
Dissanayake, T. (2011). *Optimal healing Environments characterizing the role of architecture for the recovery of special patent populations*. University of Moratuwa.
Duddy, L. (2018, November 23). *White Arkitekter's design for Nuuk's Psychiatric Clinic emphasizes nature in mental health design*. Retrieved from ArchDaily: https://www.archdaily.com/906440/white-arkitekters-design-for-nuuks-psychiatric-clinic-emphasizes-the-relationship-between-architecture-nature-and-mental-health?ad_medium=gallery
Duzenli, T. (2017). Concept of sustainability and biophilic design in landscape architecture. *The Journal of Academic Social Science*, 43–49.
Easy, F. (2015). A Study on the Effect of the Components of Physical Environment on Patient Satisfaction in Drug Rehabilitation Centers. *Indian Journal of Science and Technology, 8*, 28.
Eriksen CA. (2005). Acoustics and psychosocial environment in intensive coronary care. *Occupational and Environmental Medicine*.
FISCHL, G. (2006). *Psychosocially supportive design in the indoor environment*. Luleå University of Technology.
Frederick, C. (2012, May 18). *The effects of material selections on noise levels in two patient care units*. Retrieved from Healthcare Design: https://www.healthcaredesignmagazine.com/trends/architecture/effects-material-selections-noise-levels-two-patient-care-units/
Gissen, D. (2003). *Big and green: Toward sustainable architecture in the 21st century* . Princeton Architectural Press.
GREGORY, R. (2011, May 27). *Rehabilitation Centre Groot by Architectenbureau Koen Van Velsen, Klimmendaal, Arnhem, The Netherlands*. Retrieved from Architectural Review: https://www.architectural-review.com/today/rehabilitation-centre-groot-by-architectenbureau-koen-van-velsen-klimmendaal-arnhem-the-netherlands/8615224.article
Habibu, S. A. (2015). *Integrating therapeutic techniques in drug addict centre design*. Ahmadu Bello University.
Hamed, R. (2011). *Design Principles of Addiction Treatment Center*. Cairo University.
Hamid, R. S. (2011). *Design principles of addiction treatment centers.*. Cairo University.
Hamid, R. S. (2014). *An analytical study of designing addiction treatment centers in Egypt.*. Cairo University .
Hammad, S. M. (2014). *Design Principles Of Mental Health Care Facilities - A Case Study Of General Secretariat Of Mental Health Hospitals In Greater Cairo.*. Cairo university.
Health, D. O. (1996). *Health Building Note 35 Accommodation for people with mental illness. Part 1: the acute unit.*. The Stationery Office Ltd.
Health, D. O. (2012). *Safe management of healthcare waste*. DH.
Health, D. O. (2013). *Health Building Note 03–01 Adult acute mental health units*.
Hoffman, E. (1999). *The right to be human: A biography of Abraham Maslow*. McGraw-Hill.
Huitt, W. (2007). *Maslow's hierarchy of needs*. Retrieved from . Educational Psychology Interactive.: http://www.edpsycinteractive.org/topics/conation/maslow.html
Institute, T. F. (2010). *guid lines for design hospitals and healthcare facilities.*.: The facility guidlines institute.
Jabaree, E. E. (2017). Social sustainability: A new conceptual framework. *MDPI*.
Joseph, A. (2006). *Impact of Light on Outcomes in Healthcare Settings*. The Center for Health Design.
Karim, R. (2012). Behavioral Addictions. *Journal of Psychoactive Drugs*.

Kholy, M. M. (2011). *the main needs of the elderly during their stay at hospitals and health care facilities.*. Cairo University.

Laudet, A. (2011). The case for considering quality of life in addiction research and clinical practice. *Addiction Science & Clinical Practice.*

Lawrence, D. P. (2005). *Environmental impact assessment: Practical solutions to recurrent problems.*. Wiley.

Len Doyal, I. G. (1984). A theory of human needs. *SAGE Journals.*

Liyin Shen, M. A. (2011). Key Assessment Indicators for the Sustainability of Infrastructure Projects. *Journal of Construction Engineering and Management .*

Lundin, S. (2015). *Healing Architecture: Evidence, Intuition, Dialogue.* Chalmers University of Technology.

Lynn, E. (1995). *Healthcare Architecture Designs for the Future.*

Mansor, D. (2019, November). 10.2. El Matar Psychiatric hospital. (B. ibrahem, Interviewer)

Maslow, A. H. (1987). *A Theory of Human Motivation .*

McLeod, S. (2018, May 21). *Maslow's hierarchy of needs.* Retrieved from Simply psychology:: https://www.simplypsychology.org/maslow.html

Medicine, A. S. (1980, May 1). *A description of addiction.* Retrieved from American Society of Addiction Medicine: https://www.asam.org/resources/resource-links/a-description-of-addiction

Moberg, M. (2016). *Integrating social sustainability within the design of a building.* Chalmers University of Technology.

Mohamed Abdel-Raheem, C. R. (2016). Factors affecting social sustainability in highway projects in Missouri. *International Conference on Sustainable Design, Engineering and Construction.* ScienceDirect.

Nguitragool, M. (2019). *Healing spaces: elements of environmental design that make an impact on health.* University of Strathclyde.

Nida. (2018, January 17). *Principles of Drug Addiction Treatment.* Retrieved from National Institute on Drug Abuse: https://www.drugabuse.gov/publications/principles-drug-addiction-treatment-research-based-guide-third-edition.

Nida. (2018, July 20). *The science of addiction.* Retrieved from National Institute on Drug Abuse: https://www.drugabuse.gov/publications/drugs-brains-behavior-science-addiction.

Nijhuis, J. v. (2017). *Healing environment and patients' well-being.* Wageningen University.

Onaran, B. S. (2009). *Sustainable therapy room surfaces in acute mental health hospitals.* Hacettepe University.

Özcan, H. (2006). Healing design: Aholistic approach to social interaction in pediatric intensive care units in the United States and Turke. *1st International CIB Endorsed METU Postgraduate Conference Built Environment & Information Technologies,* (pp. 244–245). Ankara.

Peters, T. (2016). Architectural interventions for social sustainability: The renovation of modern housing. *Sustainable Built Environment.* Canada.

Polikandrioti, M. (2011). Needs of hospitalized patients. *Health Science Journal.*

Ranit Mishori, A. E. (2016). Viewpoint improving the safety and security of those engaged in global health traveling abroad. *Global Health: Science and Practice,* pp. 522–528.

Ranna Parekh, M. M. (2017, January). *What Is addiction?* Retrieved from The American Psychiatric Association: https://www.psychiatry.org/patients-families/addiction/what-is-addiction

Roger, S., & Ulrich, R. F. (1991). Stress recovery during exposure to natural and urban environments. *Journal of Environmental Psychology.*

Ruano, M. (1998). *Eco-Urbanism: Sustainable human settlements, 60 case studies.* Watson-Guptill.

Seyun, A. (2010). The sustainable space planning of hospital design towards the. *International Conference on Sustainable Buildings, Asia.*

Shaffer, H. (2018). Using the syndrome model of addiction: A preliminary consideration of psychological states and traits. *International Journal of Mental Health and Addiction,* 1373–1393.

Smith. (2000). Sustainbility and Urban Design. In E. H.-W. Chan (Ed.), *Building Hong Kong: Environmental consideration* (pp. 17-42). University Press.

Smith, M. D. (2012). *vista rehabilition.*

Soh Oi-Zhen, C. W.-W.-T. (2015). Quality of Healing Environment in Healthcare Facilities. *Teknologi,* 102–108.

Streifel, A. (1999). *Design and maintenance of hospital ventilation systems and prevention of airborne nosocomial infections.*. Infection Control and Hospital Epidemiology.

Treatment, G. S. (2015). *El Matar Psychiatric hospital.* Retrieved from General Secretariat Of Mental Health And Addiction Treatment: http://www.mentalhealthegypt.com/index.php/ar/2015-10-11-01-10-32/matar/heliopolis

Ulrich, R. S. (1992). How design impacts wellness. *Health care Forum Journal.*

Ulrich, R. S. (2001). Effects of healthcare environmental design on medical outcomes. In Design and Health. *Proceedings of the Second International Conference on Health and Design.* (pp. 49–59). Svensk Byggtjanst.

Ulricha, R. S. (2018). Psychiatric ward design can reduce aggressive behavior. *Journal of Environmental Psychology,* 53–66.

Unodc-Who. (2017). *International Standards for the Treatment of Drug Use Disorders.*. United Nations Office on Drugs and Crime.

Uysal, H. T. (2017). Maslow's hierarchy of needs in 21st century: The examination of vocational differences. *Researches on science and art in 21st century Turke,* pp. 211–227.

Verderber, S., & Todd, L. G. (2012). Reconsidering the semiprivate inpatient room in U.S. Hospitals. *Health Environments Research & Design Journal.*

Vipin Kaushal, P. S. (2004). Environmental control including. *Postgraduate Medical Institute of Education & Research,* pp. 229–232. Chandigarh.

Woodcraft, S. (2015). Understanding and measuring social sustainability. *Journal of Urban Regeneration and Renewal.*

Xenopoulous, S. (2011). *Architectural Spaces and the reformation of drug addicts.* University of Nicosia.

Yen-Ko Lin. (2013, February). Building an ethical environment improves patient privacy and satisfaction in the crowded emergency department: A quasi-experimental study. *BMC Medical Ethics.*

Zahra Kefayati, H. M. (2015). Developing effective social sustainability Indicators In architecture . *Bulletin of Environment, Pharmacology and Life Sciences.*

Zalenski, R. J. (2006). Maslow's hierarchy of needs: A framework for achieving hhuman potential in. *Journal of Palliative Medicine.*

The Architectural Design of Outdoor Spaces in Oncology Hospitals: Toward Achieving Social Sustainability for Oncology Patients

Menna Allah Omar Ramadan, Doaa Abouelmagd, and Alia Amer

Abstract

Cancer is one of the most life-threatening diseases that cause significant psychological consequences among the patients that receive treatment. The oncology patients need to do some activities outdoors with their families that improve their psychological state. A sustainable hospital is a corporate, socially responsible organization that strives to provide a healing environment. Literature review reveals that achieving good "health" contributes to the added values increasing patient satisfaction, healing environment, and supporting their activities. The design of the outdoor spaces plays an essential role in providing a healing environment for those patients. The aim of this research paper is double fold. First, it identifies the most critical design variables in outdoor spaces in oncology hospital and attributes, which link to the aspects and characteristics of social sustainability. Second, the research compares the identified variables and aspects of the social sustainability design of outdoor spaces in oncology hospitals and the real oncology patient's needs. The paper aims at increasing patient satisfaction, efficiency of healing environment, and supporting patient activities.

Keywords

Oncology hospitals • Psychological needs • Social sustainability • Outdoor spaces • Rates of healing

1 Introduction

Cancer is considered an enormous challenge in the medical field that faces the researchers, doctors, and the patients (Bodai & Facs, 2015). It is evident that following an interdisciplinary approach in combining with the research, treatment, and care will help in achieving high healing rates and finding more ways for preventing and treatment methods in the initial stages of this disease (Cameron, et al., 2010).

Although Egypt has not an accurate national cancer registrar yet. Age-standardized incidence rates per 100,000 were 166.6 (both sexes), 175.9 (males), and 157.0 (females) (Zeeneldin et al., 2012). By 2050, a threefold increase in incident cancer relative to 2013 was estimated. The previous data are the only available cancer rates at the national and regional levels of Egypt (Ibrahim et al., 2014). The 57,357 hospital is one of the major oncology hospitals in Egypt as shown in Fig. 1 The reception lobby in 57,357 hospital, which permits the daylight on the space by using the design of the facade with fixed windows. Another hospital in Egypt is The National Cancer Institute in Cairo as shown in Fig. 2. This is a waiting area of the outpatient ward that has a lack of natural ventilation, the agglomeration of the patients in this space influence in the psychological state of the patients, and also, a lack of segregation of the states of patients (Zeeneldin, 2012).

Regarding the medical complications and the long stay of patients in these hospitals, Cancer treatment is a procession involving both the patients and clinicians over a long time in cancer hospitals. This long period harms the psychological state of the oncology patient due to the affection of harshness of the treatment methods as chemotherapy, radiotherapy, and others (Weiderpassa, 2018). Most of rating systems for the evaluation of the sustainability of buildings focus on the environmental and efficiency aspects; however, it neglects the social aspects for the users (Ctca, 2018).

Recently, the researchers focus on the patient's psychological state to achieve high rates of recovery and to raise the patient's psychological state, especially cancer patients

M. A. O. Ramadan (✉) · D. Abouelmagd · A. Amer
Architecture Department, Faculty of Fine Arts, Helwan University, Cairo, Egypt

C. Alalouch et al. (eds.), *Towards Implementation of Sustainability Concepts in Developing Countries*,
Advances in Science, Technology & Innovation, https://doi.org/10.1007/978-3-030-74349-9_17

Fig. 1 Reception space that has a daylight on the space, 57,357 hospital

Fig. 2 Waiting area of the outpatient ward, The National Cancer Institute in Cairo

(Blaschke, O'callaghan, & Schofield, Nature-based care opportunities and barriers in oncology contexts: a modified international e-Delphi survey, 2017). Thus, the architecture design for cancer hospitals must consider the psychological needs of the oncology patients, which affect their healing rates, besides achieving their clinical requirements of diagnosis, treatment, and care at first. The cancer hospital has many important departments for the oncology treatment, such as the emergency ward, chemotherapy units, nursing ward, diagnosis unit, the clinical unit, radiotherapy units, and the outdoor and indoor spaces (Barton, 1995).

This paper will study the outdoor space as one of the crucial components of the oncology hospital that affect the patient's psychological state. The paper will identify the most critical design variables and attributes in outdoor spaces in the oncology hospital, in relation to the characteristics of social sustainability, and comparing these variables with the oncology patient's needs to optimize physical, mental, emotional, and spiritual healing for the individual, and collective needs of patients. Also, this paper is based on the premise that indoor spaces in oncology hospitals can achieve the same positive effect in the psychological state of the oncology patients.

2 Methodology of the Research

The research methodology of this paper is a conceptual study. It provides an overview of the design criteria of the outdoor spaces and social sustainability aspects. Then, make a combination of them to provide design variables or attributes for the outdoor spaces that consider the psychological needs of oncology patients. The research process is to determine the psychological needs of the oncology patients, the criteria of the outdoor space, and the social sustainability aspects. Afterward, make a comparison between social sustainability aspects and patients' needs to achieve the design variables for the outdoor spaces. Finally, trying to implement these variables in the indoor spaces due to the medical state of some patients that require isolation.

3 The Background and Limitations of Previous Researches

Several theories have proved that the outdoor spaces in hospitals contribute to the revival of the body, mind, and spirit. Also, it can function as an expansion of the architectural surroundings of these buildings (Verderber & Refuerzo, 2006). A clear-cut finding in the literature is that the design can improve the environmental quality and social sustainability of healthcare facilities. The social aspect of sustainable design is preferred at the design phase of the hospital to achieve the psychological needs of patients especially oncology patients (Russ, 2018).

There is a lack of research in this field in Egypt. One of these researches aims to prove the relation between the design of the healing garden and hospitalized children's behavior to explore the elements of the design principles which impact the recovery of children (Momtaz & Shaban, 2018). The other aims to revive social contact and interaction with nature by regenerating biophilic design (Abdel-Meguid, 2014). In Egypt, there is a lack of guidelines for the design of the outdoor spaces in the oncology hospitals in the Egyptian code for health care (General Authority for Healthcare Accreditation, 2019).

Scientists mentioned that cancer patients are immune-compromised, due to the illness, the treatments, and thus, they need to be isolated (Torronteguy, Bonorino, & Souza, 2012). Some reveal that achieving social needs for cancer patients in the isolation stage of his treatment is the most problematic, due to the impossibility of dealing with the external environment (Lupinacci, 2012). It collects the social activities required for raising the patient's psychological state as a result of his lack of immunity. Although previous studies have almost exclusively focused on the negative effect from staying in the isolation room, which results in a wide range of emotions and psychological burdens, no systematic review was done to provide design decision to support the psychological needs of these patients (Lee, 2011).

There are some potentially open questions about the validity of the positivity of the outdoor spaces on cancer patients, especially needing to medical isolation, the characteristics of the design of outdoor spaces indoor to be suitable with the isolation stage for cancer patients, and the psychological needs of the cancer patients in the outdoor spaces to achieve in indoor areas (Muzzin et al., 1994). A more systematic and theoretical analysis is required for addressing the characteristics of the design of indoor spaces to supply the same positive influence to the cancer patient in the outdoor spaces, to achieve the psychological needs of cancer patients and high rates of healing.

4 Psychological Needs of Oncology Patients

Medical practice in many oncology hospitals concentrates fundamentally on treating disease but ignores a patient's psychological, social, and spiritual needs. The architecture design must spur and engage people both mentally and socially in integration with the building and try to raise the individual's psychological state (Dilani, 2009). Health processes could be promoted by implementing a Salutogenic design, which focuses on the factors that keep us well, instead of making us unwell. Salutogenic design aims to create an environment that stimulates the mind to create pleasure, creativity, satisfaction, and enjoyment. Also, it strengthens the relationship between the user's health and the characteristics of the surrounded environment (Heerwagen & Associates, 1995). Salutogenic design consists of some factors, which are:

- Connection to nature: Getting in touch with nature generates social and physiological benefits. Therefore, the architect can achieve these benefits through some ways such as direct link (outdoor healing gardens) and indirect link (window view on healing garden) (Croome, 2006).
- Sense of community and belonging: It is required to study in depth the activities that provide a feeling of social connections and healing rates. Specifically, oncology patients belonging to support groups of patients promote the feeling of cheerfulness (Erfani et al., 1 May 2016).
- Behavioral choice and control: Trying to make a strong connection between patients by green space and trees. Social benefits of the healing gardens are to recover from stress and raise moods. (Stacey et al., 2010).
- Opportunity for regular exercise: It is about how the environment can improve the physiological function of the person through relative scaling of sedentary, light, and vigorous activities according to health status for patients. This affords a wide opportunity to assess physical activities for cancer patients and survivors (Schrack et al., 2017).
- Meaningful change and sensory variability: The researchers proved that humans admire levels of sensory variability in the environment, including variation in light, sound, and temperature. The environment freed of sensory stimulus and variability could lead to tedium and negativism (Missel et al., 2018).
- Privacy: The privacy of the patients includes their clinical information, treatment journey, and feeling of secluded, maintaining distances between patients. (Grande, et al., August 11, 2015).

The needs mentioned above could help the architects in identifying the design variables of the outdoor spaces of the oncology hospitals. Also, it could easily be incorporated into sustainable building design to improve the healing rates for patients (Bertucci, et al., February 2019).

5 The Architectural Design of Outdoor Spaces in Oncology Hospitals

The outdoor spaces are essential to the hospital building in general. However, it is necessary to be in the oncology hospital, due to the sensitivity of the patient's state, and his special needs due to the possibility of staying in the hospital for an extended time (Blaschke, The role of nature in cancer patients' lives: a systematic review and qualitative meta-synthesis, 2017). Open space planning and design can play an essential role in improving health and social issues. It helps in promoting healing for oncology patients (Zeliotis, 2017).

In oncology hospitals, the outdoor spaces and indoor spaces are valuable, such as intensive care units, chemotherapy units, and surgery units. Due to their effects on the psychological state for oncology patients, some of the cancer patients in a stage of his treatment must be in a completely isolated section that no direct air exchange or direct contact with the outside world (Block et al., 2002). Thus, this paper aims to address the characteristics of designing a fully isolated space that achieves the practical sense of outdoor space, inside. Moreover, it achieves the psychological needs of cancer patients.

Nowadays, some hospitals try to redesign their buildings by taking the patient's psychological needs into accounts, such as providing fresh air, better views, and building orientation (Aldous, 1992). Evidence-based design often abbreviated to EBD is a new field of study which asserts the prominence of using dependable datum to affect the design process. Nowadays, this approach has become popular in designing healthcare buildings due to improving patient and staff well-being, the patient healing process, stress reduction, and safety (Samimi, 2012).

5.1 The Benefits of Outdoor Spaces

The benefits of outdoor spaces design can be divided into three groups are as following:

Physical benefits: Interaction with a natural ambiance has a positive impact on a patient's feelings, which in turn has a beneficial effect on their physical health, by doing many physical activities between the patients and their families in convenient areas that contributes to improving the psychological state of the patient.

Psychological benefits: When the oncology patient can choose between having privacy and reactive with other outdoor, the patient will be engaged, drawn his concentricity away from disturbing thoughts resulting from his harsh disease, and so participate in his recovery.

Social benefits: Every cancer patient needs recreation and social support. Also, participation in social activities can consider the only means of family, community interaction, and sharing similar experiences. Social support improves immune functions and moods and results in better treatment (Kaplan, july,2001).

5.2 The Design Considerations of Outdoor Spaces

In the design phase for the outdoor spaces' components, several aspects need to be taken into accounts, such as:

Environmental factors (i.e., orientation, insolation, wind exposure, views).

Distance from distractions (i.e., urban noise, parking lots, mechanical systems).

The physical configuration of the complex.

Internal traffic system and its access points.

Specific needs of different user groups (patients, staff, visitors) (Yücel, 2012).

5.3 User Groups and Their Needs

In the design process, the most important group of users is the patients' group, which determines the choices made in the design of the outdoor spaces. The design depends on some elements: the types of cancer patients' cases, their length of stay in the hospital, and their needs for medical isolation or not, besides the physical and psychological needs have cancer patients (Obročníková & Majerníková, June 2017).

On the other hand, the staff is a critical group of hospital users due to spending their working day within the hospital environment. They need outdoor spaces to achieve a restorative escape from the stress in the hospital building (Kabukye et al., April 16, 2018). The visitors are the third user group; they affect the psychological state positively for cancer patients. These visits require a supportive environment to do their social activities and maintain the privacy of patients and their visitors (Sloan Fa, 2007).

5.4 Outdoor Design Criteria

When designing outdoor spaces in the oncology hospital, the design should concentrate on some design variables and aspects, such as:

Accessibility: This is considered an essential demand in the design of hospital outdoor spaces and indoor spaces. Access to these spaces must be easy through entrances and paths (Kent, 2015) As shown in Figs. 3, 4, and 5, the main entrance at University of Arizona Cancer Center is on the west side of the site. It is located on the secondary street to permit the patient to drop off quietly without any congestion. Moreover, there is a good relationship between the welcome desk, the café, the healing garden, and the services core.

Visibility: The outdoor spaces should be visible for all hospital users. Patients' rooms should have views of the garden to enjoy it even if they are unable to visit (Knight et al., 2018).

Fig. 3 Entrance permit the patient to drop off quietly without any congestion, University of Arizona Cancer Center

Feeling of control: Feeling of control leads to improving the psychological state of users, by involving users in the design of the space, for example, a variety of social spots according to users' needs, paths, water views, and quiet nooks (Gascon et al., 2017).

Feeling of security: Due to the oncology patients' physical and psychological state, they need to feel secure. The outdoor areas should have handrails, multiple seating areas, and design elements that make them feeling of the enclosure but without the feeling of monitoring (Burton, 2014).

Physiological comfort: As cancer patients are susceptible due to their weak immune, thus the architecture design of the outdoor spaces should be flexible according to the psychological needs of them. Isolation patients, for example, generally have to keep out of the open spaces. Some solutions, such as locate the isolation rooms on the views of the outdoor space to improve their psyche state, will be required (Consortium, 2019). As shown in Fig. 6, the Cardinals Kids Cancer Center at Mercy Children's Hospital locates the isolation rooms with window's view on the outdoor spaces to improve the patient's psychological state.

Quietness: The official mission of the outdoor spaces is to reduce the stress and the harmful impact on the psyche

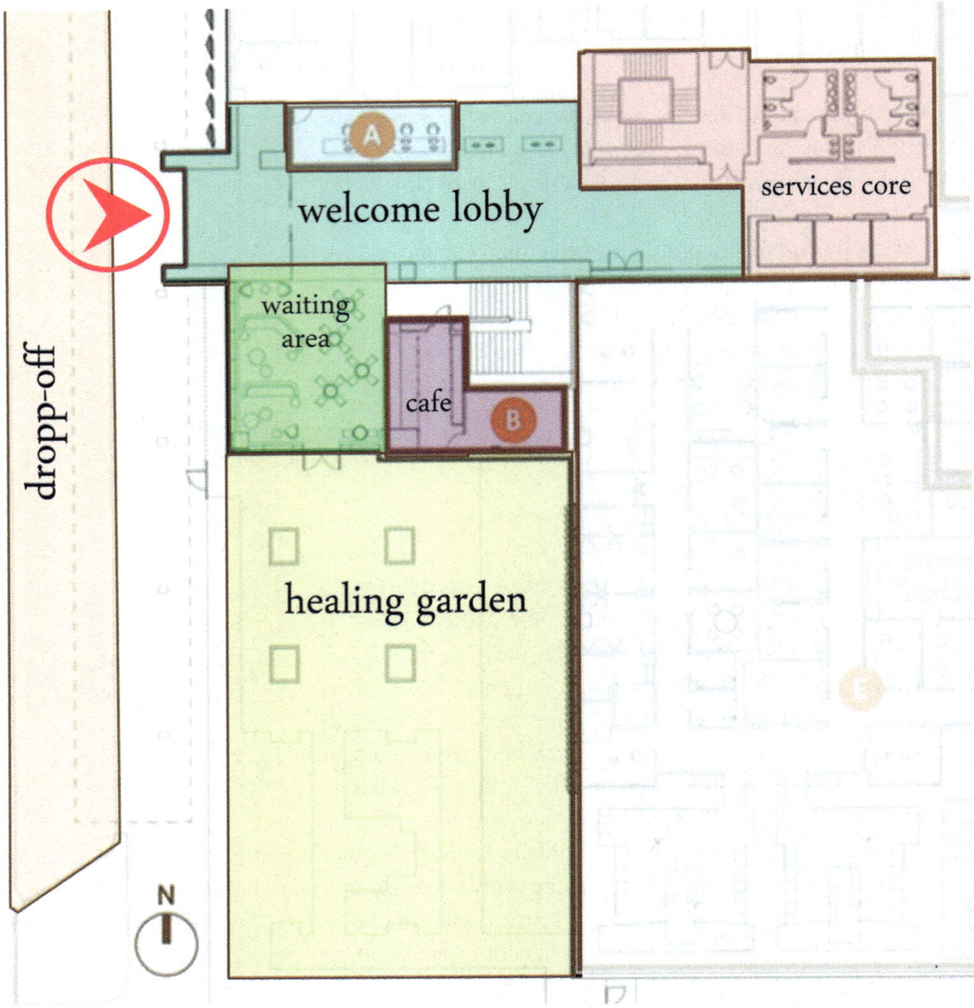

Fig. 4 Relationship between the welcome desk, the café, the healing garden, and the services core, University of Arizona Cancer Center

state of cancer patients, furthermore, to make them feel calm away from the traffic, parking areas, delivery driveways, and helicopter landing pads (Garner & Machin, 2008).

Familiarity: The seriousness of cancer illness causes stress for patients and their families. Thus, they need a familiar environment to be better. Therefore, the design should provide outdoor areas on the human scale, familiar plants, and furniture (Bayır & Fitöz, 2017; Kent, 2015). As shown in Figs. 7 and 8, the infusion area at University of Arizona Cancer Center is located at the north end of the second floor to allow the natural light through floor-to-ceiling windows along the entire north façade.

Sustainability: The usage of sustainability has concentrated on the energy and environment aspects with ignoring the social one. In the design of outdoor spaces in the cancer hospital, the designers should achieve this aspect due to the importance of the patients' psychological state in their treatment (Al-Homoud & Tassinary, 2004).

6 Sustainable Development in Oncology Hospitals

Nowadays, many research papers define the term of sustainability in many forms. However, the term can be summarized as a development that meets the needs of the present without compromising the ability of future generations to meet their own needs (Brundtland, 1987). Sustainable design is also called environmental design, which embraces the social, economic, and environmental facets of any architectural project. It also attempts to design responsible spaces by looking at three facets and determining the best practices for each project (Kras, October 2011).

Further, the sustainable hospital is a corporate socially responsible organization that strives to provide a healing environment that is accommodated by the building (Boström, 2012). Thus, this paper focuses on the social aspect of

Fig. 5 Main entrance is on the west side of the site, University of Arizona Cancer Center

Fig. 6 Cardinals Kids Cancer Center locates the isolation rooms on the views of the outdoor space, Mercy Children's Hospital

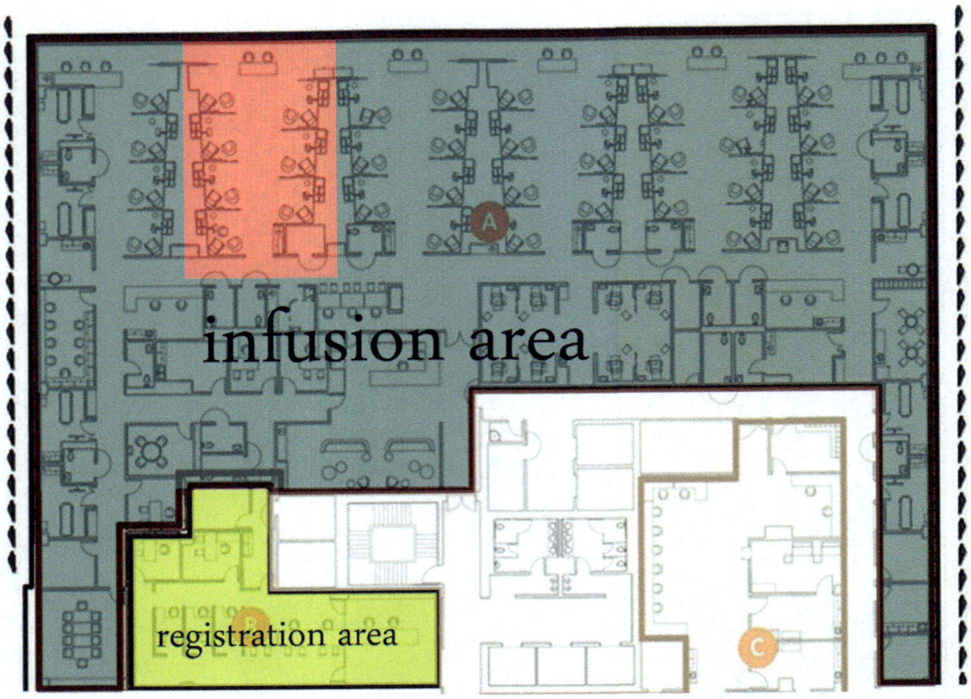

Fig. 7 In the infusion area, it opens up views toward the camelback mountains, University of Arizona Cancer Center

Fig. 8 In the infusion area, there are satellite nurse stations to ensure that nurses all have a direct sightline to each patient, University of Arizona Cancer Center

sustainability in cancer hospitals due to a lack of guidance for this aspect, although it is vital for human life (Kandachar, 2014).

6.1 Social Sustainability Definition

UNESCO project on the "social sustainability of cities," identify social sustainability as: "policies and institutions that have the overall effect of integrating diverse groups and cultural practices in a just and equitable fashion" (Polese & Stren, 2009). Moreover, according to (Dempsey et al., 2011), "social sustainability is neither absolute nor a constant... [It] has to be considered as a dynamic concept, which will change over time (from year to year/decade to decade) in a place." However, it can be defined as specifying and managing both positive and negative impacts of systems, processes, organizations, and activities on people and social life (Balama, 2018). According to (Brain, 2014), the concept of social sustainability is simple; it is about achieving people's needs and desires, considering their actions on the community, and monitoring social change for them to provide their social well-being in the future. Consequently, social sustainability definition could be concluded as the processes that achieve social health and well-being nowadays and futurity.

6.2 Social Sustainability Aspects

Social well-being is one of the essential aspects of social sustainability, especially in oncology hospitals. There are numerous ideas to achieve social well-being; however, no exhaustive list has been listed to cover the factors, which influence social well-being. Nevertheless, it can be summarized in health, safety, equality, community connectedness, personal relationships, environment and services, achievements, and freedom (Rogers, et al., 2012). Nevertheless, the concept of social sustainability is more than just providing our social well-being. Thus, some social factors help in achieving social sustainability (Luoma, 2014). In addition, these factors can be summarized in five dimensions relate to social aspects of sustainability and interrelated measurable aspects of community sustainability, which are:

Social interaction/Social networks in the community: These indicate that there are some people we know and depend on them can influence some aspects of life such as feelings of safety and sense of well-being (Fischer, 1984).

Participation in collective groups and networks in the community: It is evident that participation in organized activities is considered to contribute positively to community sustainability (Skidmore et al., 2006). As shown in Fig. 9, the design of family lounge at Walker Center for Cancer Care provides the comfort and well-being of family and visitors of the patients by adding; armchairs, end tables, and folding chairs for larger groups.

Community stability: This represents the user's connection to their stay place, and interaction with the nature of it and its surroundings (Taub & Wilson, 2007). As shown in Fig. 10, the wellness room at Seattle Children's hospital provides the quietness and comfort to the patients.

Pride/sense of place: A definite sense of correlation to the place is assumed as an aspect of social sustainability. Also, it can be defined as a mix of emotional contact through friction with others and place attachment (Fukuyama, 2000). As shown in Fig. 11, the facility entrance flows a generous atrium that surrounded the reception area and waiting room. Also, the wall of glass that leads the eye into a four-season healing garden. These features create a calming environment for patients in the Roswell Park Cancer Institute.

Safety and security: Providing security and safety is related to community sustainability (Barton, 1999).

These aspects can affect the psychological state of oncology patients positively if the architecture design achieves it correctly as shown in Table 1.

7 Discussions and Results

From the review above, key findings emerge that the psychological needs of the patients affect clearly on the rates of healing. Also, the design of outdoor spaces in the hospital buildings helps in improving the patient's psychological state. On the other hand, social sustainability achieves the user's needs and desires, considering the user's actions and monitoring the social change of him to provide his social well-being in the future.

From the above, it is found that oncology patients need to improve their psychological state, which affects their illness state positively. They are the most in need of developing their hospitals' buildings by implementing social sustainability concepts to design the healing spaces, especially outdoor spaces. The outdoor spaces at Pamela Buffett Cancer Center are designed with smooth surfaces and variable green areas to raise the healing rate for the patients as shown in Fig. 12, the outdoor spaces at Princess Máxima Centre for Child Oncology have variable furnishing such as tables, umbrellas, and guard supports as shown in the outdoor spaces at Princess Máxima Centre for Child Oncology have variable furnishing such as tables, umbrellas, and guard supports as shown in Fig. 13, and the outdoor spaces provide multiple floors types such as rock, grass, and brick wood carpet as shown in Fig. 14.

Fig. 9 Design of the family lounge puts into consideration the well-being of family and visitors of the patients, Walker Center for Cancer Care

Fig. 10 Wellness room that provides the quietness to the patients, Seattle Children's hospital

Fig. 11 Generous atrium surrounded the reception area and waiting room that create a calming environment for patients, Roswell Park Cancer Institute

Table 1 identifies social sustainability factors, features, outdoor spaces characteristics, and design variables that support the oncology patients' needs and achieve these needs.

Regarding the limitations that may force the medical isolation of some oncology patients due to the weakness of their immune system, it could be argued that the friction with the outdoor spaces in hospital building is limited to the stage of treatment. Thus, the designers should try to apply some of the characteristics of the outdoor spaces in indoor social spaces as well (CBE & Dangoor, 2017). Therefore, there is a need to study the psychological needs of patients and social sustainability aspects in indoor spaces.

The results of Table 1 conclude that the application of social sustainability in the design of the outdoor spaces in oncology hospitals to achieve oncology patients' is essential. As it can reduce stress, increased social support, and raise the ability to focus attention on social activities between oncology patients, which raise the psyche state of them (ElMashharawi, 2019).

The components of indoor spaces can contribute to achieving the psychological needs of isolated oncology patients who are forbidden from the friction outdoor spaces by putting into consideration social sustainability features and factors. The indoor spaces consist of some components such as main lobbies, elaborate artwork, corridors, waiting rooms, infusion bays, the bays, resource libraries, appearance boutiques, and wellness rooms. In this paper, I will try to apply the characteristics of outdoor spaces in the components of the indoor space to achieve oncology patients' needs as we can.

Following are the paper recommendations to apply social sustainability features to the components of indoor spaces to achieve the oncology patients' needs:

Connection to nature: By designing the atriums with daylight with non-open fixed windows and orient the patients' rooms to the view of the healing gardens (Andritsch, et al., 2013). As shown in Fig. 15, the patient's room is oriented to the healing garden to permit the daylight to pass through it.

Table 1 Features of social sustainability and outdoor spaces characteristics linked to oncology patient's needs

Oncology patients' needs	Social sustainability factors	Social sustainable features	Outdoor spaces characteristics	Design variables and / or attribute
1. Connection to nature	Community stability	Daylight; views of outdoor natural spaces; views of the sky and weather; interior plantings; interior atria with daylight and vegetation; water features	Visibility, familiarity	Design variable outdoor furnishing such as; barbecues, tables, umbrellas, supports and guards, and plant containers Provide a variety of planting such as; Cochlospermum tinctorium A. Rich, Cissus ibuensis Hook, Desmodium velutinum Willd, etc. (Abubakar, Musa, Abubakar, & Hussaini, 2007) Design inpatient rooms with windows accessing to natural daylight and controlling glare. It is recommended that window area should occupy 20–30% of the window wall (Arneill & Devlin, 2003) Design patient's room orientation on garden view to the early morning sun and light exposure. It needs to be visible from the view of a patient in bed
2. Sense of community and belonging	Social interaction/ social networks in the community	Multiplicity of meeting spaces and centrally located meeting and greeting spaces	Physiological comfort, Quiet	Design a community space, family lounge, and quit room for relax Provide family areas including sleeping facilities Design a space that has floors that include earth, rock, grass, low planting, concrete, asphalt, stone, brick wood carpet, tiles, and so on
3. Behavioral choice and control	Pride/sense of place	Ability to adapt the environment to suit the user needs; ability to move easily between the isolated spaces and social spaces	Quiet, Sustainability, Accessibility	Enable individual control such as; volume controls, light switches, etc., consider providing single rooms as possible due to it helps patients avoid noise and feel privacy Design a line of the landscape that may be as sharp paving slabs, rock, shrubs, hedges, shadows, scent, grass, and ivy provide prayer/spiritual areas
4. Opportunity for regular exercise	Participation in collective groups and networks in the community	Attractive outdoor walking paths; indoor exercise facilities; open interior stairways	Accessibility	Design a therapy gym include basketball, track, etc. Design open activity areas that can reduce disorientation for the patients Design garden rooms that contain the various activities for the isolated patients Construct paths with smooth surfaces
5. Meaningful change and sensory variability	Community stability	Daylight; window views on the outdoor spaces; variation in lighting levels	Visibility, physiological comfort	Design shelter structures to protect from sun, rain, or wind. Such as; pergolas and solid opaque or translucent roofs Design the seasonal plants
6. Privacy	Safety and security	Enclosure; Maintenance of the desired distances between the users and the strangers around them; save the privacy of the public spaces for anonymity	Feeling of control, feeling of security	Design enclosed structures that control vision, movements or both with various heights Incorporate private, and semi-private spaces Place comfortable chairs with a home-like feel

Source Authors' work

Fig. 12 Passes with smooth surfaces and green areas in the outdoor spaces, Pamela Buffett Cancer Center

Fig. 13 Outdoor spaces furnishing such as tables and umbrellas, Princess Máxima Centre for Child Oncology

Fig. 14 Space that has floors that include rock, grass, and brick wood carpet, Smilow cancer hospital

Sense of community and belonging: By designing the wellness rooms, which held the support groups, reading corners, yoga corners, and other passive social activities (Chid et al., 2015). These rooms should overlook the healing gardens. Also, design communal space that like a social hub, it supports many activities; going online, playing games, watching TV, reading, etc. As shown in Fig. 16, the community space of Seattle Children's hospital achieve what the oncology patients' need to feel belonging at the indoor spaces.

Opportunity for regular exercise: Design an ample indoor space to help them work out some non-harsh activities such as relaxation exercises, in-house exercise facilities, and so forth (Paraskevopoulou & Kamperi, 2018). As shown in Fig. 17, basketball court at Seattle Children's hospital help them to workout some activities.

Privacy: By designing private family treatment rooms that offer plenty of space for interaction, lounging, playing games, or sharing a snack. Also, it allows patients to feel a sense of privacy (Dubose et al., 2018).

Behavioral choice and control: The architectural design must control ventilation, reduce infection among patients by attempting an automatic purification of internal air as the weakened immune systems of patients. The architectural design must control ventilation, reduce infection among patients by attempting an automatic purification of internal air as the weakened immune systems of patients. (Bosia et al., 2016). Also, it must reduce the noise by isolating walls to keep calm in the buildings and make a design space for users to move quickly between solitude and social engagement spaces.Also, it must reduce the noise by isolating walls to keep calm in the buildings and make a design space for users to move quickly between solitude and social engagement spaces..

Meaningful change and sensory variability: By maintaining the arrival of the daylight to the indoor spaces with keeping in mind the visual change, pleasant sounds, and lighting levels (Ulrich et al., 2004). As shown in Fig. 18, patients and visitors can relax in The Park at MD Anderson hospital for cancer treatment, a comfortable indoor setting with tropical palms and vibrant plantings, and in Fig. 19, the building meets the public with a transparent facade, welcoming space, views, and thermal comfort in Cafeteria dinning at Winship at Midtown hospital.

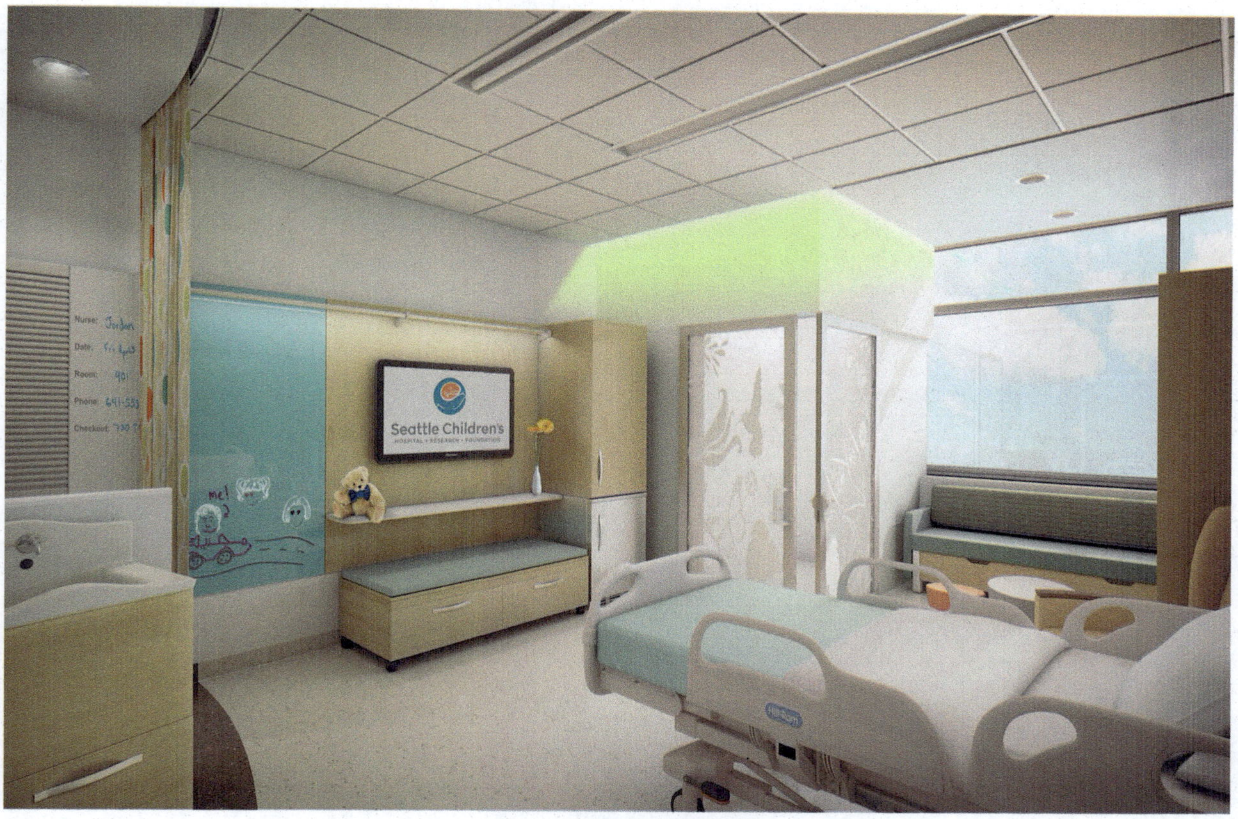

Fig. 15 Patient's room is oriented to permit daylight to pass through the room, Seattle Children's hospital

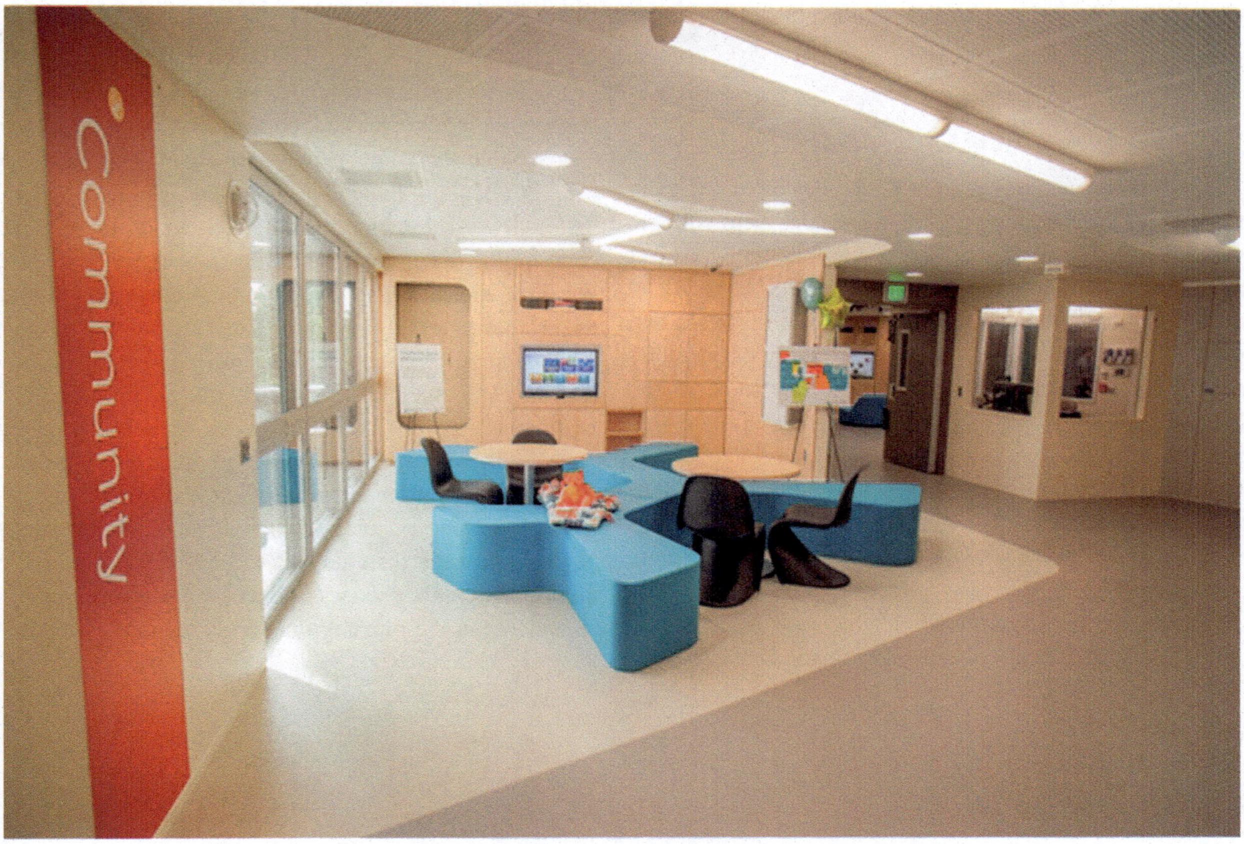

Fig. 16 Community space helps the patients feel belonging at the indoor spaces, Seattle Children's hospital

Fig. 17 Basketball court as a sample to help them to work out some relaxation activities, Seattle Children's hospital

8 Conclusion

The aim of this paper was to identify the design variables in outdoor spaces in oncology hospitals and link them with the aspects and characteristics of social sustainability. Also, it compares the identified variables and aspects of the social sustainability of outdoor spaces in oncology hospitals and the oncology patient's needs. Through the study of oncology patient's psychological needs, we found that some of the oncology patients' state requires the medical isolate that forbids any friction with the outdoor spaces, due to the oncology patients are immune-compromised. Thus, we suggest applying the outdoor characteristics and variables on the indoor spaces to achieve the oncology patient's needs in the oncology hospitals.

In summary, this paper concludes that applying the factors and features of social sustainability on the characteristics of the outdoor spaces can achieve oncology patients' needs. Also, the premise that indoor spaces in oncology hospitals can achieve the same positive effect in the psychological state of oncology patients is accepted.

Future research is needed to delimitate an architectural code for the oncology hospitals to achieve the psychological needs of the oncology patients as references to the architecture designers. It is also required to apply these codes in practice in the design of a hospital to be an operational reference for hospitals. It is recommended to apply these findings on the design of the oncology hospitals in Egypt. Some of these hospitals are under construction now such as; The New Egyptian National Cancer Institute and others already existing oncology hospitals such as; The Egyptian National Cancer Institute. As a lack of guidelines for the outdoor and indoor spaces of oncology hospitals putting into consideration the oncology patients' needs.

Fig. 18 Park can make the patients and visitors relax, MD Anderson hospital for cancer treatment

Fig. 19 Cafeteria dinning with the shaded roof to apply the thermal comfort for the patients, Winship at Midtown hospital

References

AbdelMeguid, S. M. (2014). Biophilic design (Strategies for hospitals retrofit).

Abubakar, M. S., Musa, A. M., Abubakar, A., & Hussaini, I. M. (2007). The perception and Practice of Traditional medicine in the treatment of cancer and inflammations by the Hausa and Fulani tribes of Northern Nigeria. *Journal of Ethno pharmacology*, 625–629.

Aldous, T. (1992). *Urban villages : A concept for creating mixed-use urban developments on a sustainable scale*. Urban Villages Group.

Al-Homoud, M., & Tassinary, L. G. (2004). Social interactions at the neighborhood-level as a function of external space enclosure. *Journal of Architectural and Planning Research*, 10–23.

Andritsch, E., Stoger, H., Bauernhofer, T., Andritsch, H., Moser, R. S., Kasparek, A. K, Samonigg, H. (2013). *The ethics of space, design and color in an oncology ward*. Cambridge University Press (pp. 215–221).

Architects, K. (2004). *The park of MD Anderson for cancer treatment [Photograph found in MD Anderson for cancer treatment, Houston, Texas]*. Retrieved February 12, 2020, from https://www.kmdarchitects.com/md-anderson (Originally photographed 1939).

Arneill, A. B., & Devlin, A. S. (2003). Health care environments and patient outcomes: A Review of the Literature. *American Psychological Association Journal*, 665–694.

Balama, Ş. Y. (2018). *Decision-making for biomass-based production chains*. Acadmic Press.

Barton, H., Davis, G., & Guise, R. (1995). *Sustainable settlements : a guide for planners, designers and developers*. Bristol. Severnside Research and Consultancy Unit.

Barton, H. (1999). *Sustainable communities: The potential for eco-neighbourhoods*. Routledge.

Bayır, E., & Fitöz, İ. (2017). Investigating the effects of cancer disease specific data on the interior satisfaction in oncology services. *Tasarım Kuram*, 114–137.

Bertucci, F., Soriano, A. G., Miramon, A. M., Moulin, J. F., Fluzin, S., Maraninchi, D., & Gonçalves, A. (2019). Outpatient cancer care delivery in the context of e-oncology: A French perspective on "cancer outside the hospital walls". *Cancers*, 1–19.

Blaschke, S., O'callaghan, C. C., & Schofield, P. (2017). Nature-based care opportunities and barriers in oncology contexts: a modified international e-Delphi survey. *BMJ Open*, 1–11.

Blaschke, S. (2017). The role of nature in cancer patients' lives: a systematic review and qualitative meta-synthesis. *BMC Cancer*, 1–13.

Block, K. I., Boyd, D. B., Gonzalez, N., & Vojdani, A. (2002). The immune system in cancer. *Integrative Cancer Therapies*, 294–316.

Bodai, B. I., Facs, M. D. (2015). Breast cancer survivorship: A comprehensive review of long-term medical issues and lifestyle recommendations. *The Permanente Journal*, 48–79.

Bosia, D., Marino, D., & Peretti, G. (2016). Health facilities humanisation: design guidelines supported by statistical evidence. *Ann Ist Super Sanità*, 33–39.

Boström, M. (2012). A missing pillar? Challenges in theorizing and practicing social sustainability: introduction to the special issue. *Sustainability: Science, Practice, & Policy*, 3–14.

Brain, J. (2014). The social side of sustainability. *2018*. planning.org.nz.

Brundtland, G. H. (1987). *Report of the world commission on environment and development: Our common future*. The World Commission on Environment and.

Burton, A. (2014). Gardens that take care of us. *The Lacent Neurology*, 447–448.

Cafeteria Dinning of Winship at Midtown Hospital [Photograph found in Winship at Midtown hospital, Atlanta, Georgia]. (2015). In 1006325230 775386347 S. Architects (Author). Retrieved February 20, 2020, from https://winshipcancer.emory.edu/images/location-imgs/midtown-renderings/cafeteria-dining.jpg (Originally photographed 1937).

Cameron, C. G., DeMille, D., Lynch, M. P., Huntzinger, C., Alcorn, T., Levicoff, J., Mintzer, D. (2010). An interdisciplinary approach to manage Cancer Cachexia. *Clinical Journal of Oncology Nursing*, 72–80.

CBE, N. D., & Dangoor, R. (2017, December 15). *The immune system and cancer*. Retrieved from cancerresearchuk : https://www.cancerresearchuk.org/about-cancer/what-is-cancer/body-systems-and-cancer/the-immune-system-and-cancer#what

Chid, G. I., Edac, & Asic. (2015). Interior designs for cancer care; A patient-focused approach for selecting materials and finishes. *Health Facilities Management*, 1–15. Retrieved from American Academy of Healthcare Interior Designers.

Croome, D. C. (2006). *Creating the productive workplace*. Taylor & Francis.

Ctca. (2018, 10 24). *What's the difference: Oncology specialties*. Retrieved from cancer treatment centers of America: https://www.cancercenter.com/community/blog/2018/10/whats-the-difference-oncology-specialties

Dempsey, N., Bramley, G., Power, S., & Brown, C. (2011, December 19). *The social dimension of sustainable development: Defining urban social sustainability*. Wiley Online Library.

Dilani, A. (2009). Psychosocially supportive design: A salutogenic approach to the design of the physical environment. *SHB2009—1st International Conference on Sustainable Healthy Buildings* (pp. 55–65). Seoul, www.researchgate.net.

Dubose, J., Macallister, L., Hadi, K., & Sakallaris, B. (2018). Exploring the concept of healing spaces. *Health Environments Research & Design Journal*, 33–56.

Dudley, P. (2010, June 15). Basketball court of Seattle Children's hospital [Photograph found in Seattle Children's hospital, Laurelhurst]. Retrieved February 13, 2020, from https://www.flickr.com/photos/seattlechildrens/4705707974/in/album-72157624286247720/ (Originally photographed 1980).

ElMashharawi, A. (2019). Biophilic design for bringing educational spaces to life. *Journal of Design Studio*, 16–21.

Erfani, S. S., Blount, Y., & Abedin, B. (2016, May 1). The influence of health-specific social network site use on the psychological well-being of cancer-affected people. *American Medical Informatics Association* 23(3), 467–476.

Family lounge [Photograph found in Walker Center for Cancer Care, Watertown, New York.]. (2019, October 17). In 1005816176 775082643 R. Studio (Author). Retrieved February 10, 2020, from https://www.healthcaredesignmagazine.com/projects/cancer-care/photo-tour-walker-center-for-cancer-care-at-samaritan-health/#slide-8 (Originally photographed 2019, October 17).

Fischer, C. S. (1984). *To dwell among friends: Personal networks in town and city*. University of Chicago Press.

Fukuyama, F. (2000). *The great disruption: Human nature and the reconstitution of social order*. Simon & Schuster New York.

Garner, D., & Machin, K. (2008). Investigation and management of an outbreak of mucormycosis in a paediatric oncology unit. *Journal of Hospital Infection*, 53–59.

Gascon, M., Zijlema, W., Vert, C., White, M. P., & Nieuwenhuijsen, M. J. (2017). Outdoor blue spaces, human health and well-being: A systematic review of quantitative studies. *International Journal of Hygiene and Environmental Health*, 1207–1221.

General Authority for Healthcare Accreditation and. (2019). GAHAR Handbook for Hospital Standards. Egypt.

Grande, D., Asch, D. A., Wan, F., Bradbury, A. R., Jagsi, R., & Mitra, N. (11/August/2015). Are Patients With Cancer Less Willing to Share Their Health Information? Privacy, Sensitivity, and Social Purpose. *Journal of Oncology Practice*, 378–383.

Grassl, A. (2013). *The setting areas of Smilow cancer hospital* [Photograph found in Smilow cancer hospital, New Haven]. Retrieved February 17, 2020, from https://www.towersgolde.com/smilow (Originally photographed 1974).

Hamed, S. (2019, August 9). *The waiting area of the outpatient ward* [Photograph found in The National Cancer Institute in Cairo, Cairo]. Retrieved December 31, 2019, from https://www.elwatannews.com/news/details/4292478 (Originally photographed 2019, August 9)

Heerwagen, J. H., & Associates. (1995). Investing in people: The social benefits of sustainable design. *Investing In People: The Social Benefits of Sustainable Design*. Seattle, USA.

Hoffman, S. (2017). *The passes and green areas of Fred & Pamela Buffett Cancer Center* [Photograph found in Fred & Pamela Buffett Cancer Center, Omaha, Nebraska]. Retrieved January 28, 2020, from https://www.chihuly.com/exhibitions/fred-pamela-buffett-cancer-center/chihuly-sanctuary (Originally photographed 2017).

Ibrahim, A. S., Khaled, H. M., Mikhail, N. N., Baraka, H., & Kamel, H. (2014). Cancer incidence in Egypt: Results of the national population-based cancer registry program. *Journal of Cancer Epidemiology*, 1–18.

Institute, T. N. (n.d.). *NCI dictionary of cancer terms.* Retrieved from cancer: https://www.cancer.gov/publications/dictionaries/cancer-terms/def/comprehensive-cancer-center

Kabukye, J. K., Koch, S., Cornet, R., Orem, J., & Hagglund, M. (2018, April 16). User requirements for an electronic medical records system for oncology in developing countries: A case study of Uganda. *AMIA Annual Symposium Proceeding Archive*, 1004–1013.

Kandachar, P. (2014). *Materials and social sustainability.* Butterworth-Heinemann.

Kaplan, R. (2001, July). The nature of the view from home psychological benefits. *Environment and Behavior*, 507–542.

Kent, J. L. (2015). *Therapeutic landscapes: an evidence-based approach to designing healing gardens and restorative outdoor spaces.* Routledge.

Knight, A., Black, R., Whitsed, R., & Harvey, R. (2018). Enhancing the usability and benefits of open space for older people in regional Australia. *Australian Planner*, 1–12.

Kras, I. (October 2011, October 1). *Sustainable hospital buildings.* Twynstra Gudde.

Lee, Y. M. (2011). The experience of being a neutropenic cancer patient in an acute care isolation room—A systematic review of qualitative evidence. *Joanna Briggs Institute International Convention*, 309.

Luoma, H. (2014, January 13). *Fundamentals of cancer center design: The patient*. Retrieved from Healthcare Design Magazine: https://www.healthcaredesignmagazine.com/architecture/fundamentals-cancer-center-design-patient/

Lupinacci, M. (2012). *On the importance of social sustainability.* billpeduto.com.

Merrick, N., & Blessing, H. (2015a). *The layout of University of Arizona Cancer Center* [Photograph found in University of Arizona Cancer Center, Phoenix]. Retrieved May 2, 2020, from https://www.archdaily.com/797911/university-of-arizona-cancer-center-zgf-architects?ad_medium=office_landing&ad_name=article (Originally photographed 2015).

Merrick, N., & Blessing, H. (2015b). *The ground floor of University of Arizona Cancer Center* [Photograph found in University of Arizona Cancer Center, Phoenix]. Retrieved May 2, 2020, from https://www.archdaily.com/797911/university-of-arizona-cancer-center-zgf-architects/580aef52e58ece84440000df-university-of-arizona-cancer-center-zgf-architects-plan-1?next_project=no (Originally photographed 2015).

Merrick, N., & Blessing, H. (2015c). *The first floor of University of Arizona Cancer Center* [Photograph found in University of Arizona Cancer Center, Phoenix]. Retrieved May 2, 2020, from https://www.archdaily.com/797911/university-of-arizona-cancer-center-zgf-architects/580aef40e58ece84440000de-university-of-arizona-cancer-center-zgf-architects-plan-2?next_project=no (Originally photographed 2015).

Merrick, N., & Blessing, H. (2015d). *The main entrance of University of Arizona Cancer Center* [Photograph found in University of Arizona Cancer Center, Phoenix]. Retrieved May 2, 2020, from https://www.archdaily.com/797911/university-of-arizona-cancer-center-zgf-architects/580aee20e58ece84440000dc-university-of-arizona-cancer-center-zgf-architects-photo?next_project=no (Originally photographed 2011).

Merrick, N., & Blessing, H. (2015e). *The infusion area of University of Arizona Cancer Center* [Photograph found in University of Arizona Cancer Center, Phoenix]. Retrieved May 2, 2020, from https://www.zgf.com/project/ua-cancer-center/ (Originally photographed 2015).

Missel, M., Borregaard, B., Schoenau, M. N., & Sommer, M. S. (2018). A sense of understanding and belonging when life is at stake—Operable lung cancer patients' lived experiences of participation in exercise. *Europeon Journal of Cancer Care*, 1–8.

Momtaz, R. I., & Shaban, R. (2018). The impact of healing gardens on improving psychological recovery of children—Application on pediatric cancer hospital in Egypt. *Journal of Engineering Sciences*, 333–345.

Muzzin, L. J., Anderso, N. J., Figuered, A. T., & Gudelis, S. 0. (1994). The experience of cancer. *Social Science & Medicine*, 1201–1208.

Consortium, N. C. (2019, March 25). *cancer*. Retrieved from NCI (National cancer institute): https://www.cancer.gov/research/areas/causes

Obročníková, A., & Majerníková, Ľ. (2017, June). Patient satisfaction with health care in an oncology setting. *Nursing in the 21 century*, 20–24.

Paraskevopoulou, A. T., & Kamperi, E. (2018). Design of hospital healing gardens linked to pre-orpost-occupancy research findings. *Frontiers of Architectural Research*, 395–414.

Polese, M., & Stren, R. (2009). *The social sustainability of cities: Diversity and the management of change*. Routledge.

Reception lobby in 57357 hospital [Photograph found in 57357 hospital, Cairo]. (2019, December). Retrieved February 3, 2020, from https://www.57357.org/en/category/gallery/the-present/#pid=22 (Originally photographed 2019, December)

Rogers, D. S., Duraiappah, A. K., Antons, D. C., Munoz, P., Bai, X., Fragkias, M., & Gutscher, H. (2012). A vision for human well-being: transition to social sustainability. *Current Opinion in Environmental Sustainability*, 61–73.

Russ, J. (2018). *Sustainability and design ethics.* CRC Press Taylor & Francis Group.

Samimi, K. (2012, May 1). Children's cancer and transplant Hospital: A micro town within a bubble. scholarworks.umass.edu. Retrieved from https://scholarworks.umass.edu/theses/869

Schrack, J. A., Gresham, G., & Wanigatunga, A. A. (2017). Understanding physical activity in cancer patients and survivors: new methodology, new challenges, and new opportunities. *Molecular Case Studies*, 1–7.

Skidmore, P., Bound, K., & Lownsbrough, H. (2006). *Community Participation: Who benefits.* Joseph Rowntree Foundation.

Sloan FA, G. H. (2007). *Cancer control opportunities in low- and middle-income countries.* National Academies Press. Retrieved from https://www.ncbi.nlm.nih.gov/books/NBK54037/#ch8.s1

Stacey, D., Paquet, L., & Samant, R. (2010). Exploring cancer treatment decision-making by patients: A descriptive study. *Current Oncology*, 85–93.

Staff, M. (2012, November 29). *Interior images of the Cardinals Kids Cancer Center at Mercy Children's Hospital* [Photograph found in The Cardinals Kids Cancer Center at Mercy Children's Hospital, St. Louis]. Retrieved October 14, 2019, from https://www.flickr.com/photos/mercyhealth/8536629617/in/album-72157632943218488/ (Originally photographed 1897).

Stuhaug, E. (2011, March 15). *Patient's room view 1 of Seattle Children's hospital [Photograph found in Seattle Children's hospital* Laurelhurst]. Retrieved February 13, 2020, from https://www.flickr.com/photos/seattlechildrens/5590477162/in/photostream (Originally photographed 1980).

Stuhaug, E. (2013, March 16). *Wellness room view of Seattle Children's hospital* [Photograph found in Seattle Children's hospital, Laurelhurst]. Retrieved January 31, 2020, from https://www.flickr.com/photos/seattlechildrens/8660197723/in/photostream (Originally photographed 1980).

Stuhaug, E. (2014, October 7). *Community space of Seattle Children's hospital [Photograph found in Seattle Children's hospital*, Laurelhurst]. Retrieved February 13, 2020, from https://www.flickr.com/photos/seattlechildrens/15294480479/in/album-72157648131522880/ (Originally photographed 1980).

Taub, R. P., & Wilson, W. J. (2007). *There goes the neighborhood: Racial, ethnic, and class tensions in four Chicago neighborhoods and their meaning for America.* Vintage books.

Tilleman, R. (2018). *The outdoor spaces of Princess Máxima Centre for Child Oncology* [Photograph found in Princess Máxima Centre for Child Oncology, Utrecht]. Retrieved February 12, 2020, from https://www.archdaily.com/899421/princess-maxima-centre-for-child-oncology-liag-architects/5b61df58f197cc027e00006d-princess-maxima-centre-for-child-oncology-liag-architectsphoto?next_project=no (Originally photographed 2018).

Torronteguy, C., Bonorino, C., & Souza, A. P. (2012). The immune system of cancer patients. *Anti-Inflammatory & Anti-Allergy Agents in Medicinal Chemistry*, 1–32.

Ulrich, R., Quan, X., Zimring, C., Joseph, A., & Choudhary, R. (2004). *The role of the physical environment in the hospital of the 21st century: a once-in-a-lifetime opportunity.* Texas A&M University and Georgia Tech: The Robert Wood Johnson Foundation.

Verderber, S., & Refuerzo, B. J. (2006). *Innovations in hospice architecture.* Taylor & Francis e-Library.

Watson, T. (2005, August 31). *The generous atrium of Roswell Park Cancer Institute* [Photograph found in Roswell Park Cancer Institute, Buffalo, New York.]. Retrieved January 27, 2020, from https://www.healthcaredesignmagazine.com/architecture/cayuga-medical-center-ithaca-radiation-medicine-center-ithaca-ny/?hilite=%27Roswell%27%2C%27Park%27%2C%27Cancer%27%2C%27Institute%27 (Originally photographed 1898).

Weiderpassa, E. (2018). Cancer research for better public health policy—A personal view on the role of the International Agency for Research on Cancer in cancer control. *Journal of Cancer Policy*, 70–72.

Yücel, G. F. (2012). *Hospital outdoor landscape design*. IntechOpen.

Zeeneldin, A. A. (2012). *Research profile of The Egyption National Cancer Institute on the PubMed.* PubMed.

Zeeneldin, A. A., Taha , F. M., & Moneer , M. (2012). Past and future trends in cancer and biomedical research: a comparison between Egypt and the World using PubMed-indexed publications. *Springer Nature*, 1–10.

Zeliotis, C. (2017). Where to next for cancer centre design? *Future Healthcare Journal*, 142–145.

Visual Comfort in Elder Care Facilities: Promoting Environmental Gerontology Theory

Alaa M. Edrees, Shaimaa Kamel, Hanan Sabry, and Ashraf Nessim

Abstract

Population aging is an inevitable universal demographic process. Aging exposes people to physical, sensory, and psychological changes that often lead to health declination and cause dependency. In developed countries, elder care facilities emerged widely to ensure the aging perception of home as a safe and comfortable environment when family members are not around to assist. The elderly spend most of their time indoors; thus, indoor lighting design and visual comfort are vital for their safety. Comfort contributes to the well-being of the elderly as a part of the social sustainability context. The aim of this research is to set visual comfort design guidelines for elder care facilities in developing countries. To achieve this, a methodical review was carried out of published literature on lighting, environmental gerontology, and enhancement of visual comfort for the elderly. Then, an analysis of the environmental gerontology approach application in elderly built environments was carried out. Finally, a summarization of preliminary guidelines for visual comfort that can be applied in developing countries to form aging-friendly space was made. The findings establish recommended guidelines and proposed solutions that would improve the daylighting environment in elder care facilities from the gerontology perspective. This research highlights the important role of environmental gerontology in creating an age-friendly environment that fulfills the physical and psychological needs of the elderly, especially in daylighting aspects, thus promoting social sustainability.

Keywords

Elderly • Daylighting • Visual comfort • Elder care facilities • Environmental gerontology

A. M. Edrees (✉) · S. Kamel · H. Sabry · A. Nessim
Faculty of Engineering, Ain Shams University, Cairo, Egypt

1 Introduction

Recently, the United Nations (UN) claimed that population is aging due to the decrease in fertility and the increase in life expectancy (UN, Department of Economic and Social Affairs, Population Division 2017). In Africa, the population of people who are over 60 years old is expected to increase from 5% in 2017 to 11% 2050 (UN, Department of Economic and Social Affairs, Population Division 2017). This demographic change is occurring worldwide, as shown in Fig. 1, due to the accomplishment of socioeconomic development in the past 50 years (WHO, 2015). This demographic change plays an important role in the sustainable development 2030 agenda (UN, Department of Economic and Social Affairs, Population Division 2017).

In most developed countries, the eligible elderly services start at the age of 65 (UN, 2019). In Africa, elderly age is connected to retirement age criteria in each country as the case of most developing countries (WHO, 2002a, b). In Egypt, the retirement age is 60 years for men and 55 years for women (Ministry of Social Solidarity [MOSS] 2019; Essam-Eldin, 2018). This elderly population is rated as 6.7% in 2017 but expected to increase in 2050 to 11.5% of the total Egyptian population (Central Agency for Public Mobilization and Statistics [CAPMAS] 2019).

Arab families' tradition assumes older people are taken care of by their own family members. However, due to changes in lifestyles, social behavior, and demography, it is becoming hard for the family alone to provide full care (WHO, 2019). Elderly care is needed, and it should be a mutual responsibility of not only families but also governments and other sectors to preserve and ensure the rights of elder people in health quality and human dignity (WHO, 2019). Consequently, developing age-friendly cities and communities is one of the 10 priorities to promote a decade of healthy aging and one of the Sustainable Development Goals' agenda in 2030 by the United Nations (WHO, 2017; UN, Department of Economic and Social Affairs, Population

C. Alalouch et al. (eds.), *Towards Implementation of Sustainability Concepts in Developing Countries*,
Advances in Science, Technology & Innovation, https://doi.org/10.1007/978-3-030-74349-9_18

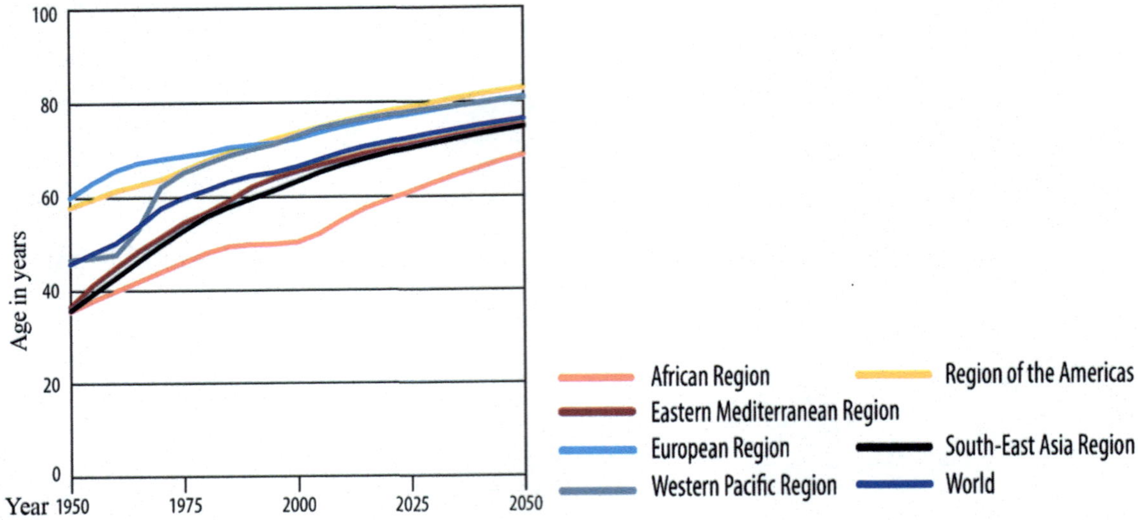

Fig. 1 Life expectancy in world regions through 100 years. *Source* WHO (2015)

Division, 2017). This will be achieved by inspiring change, connecting cities and communities to allow the exchange of information, and supporting finding solutions through innovative and evidence-based technical guidance (WHO, 2015, 2017). Therefore, special care is needed to be applied to elderly care built environments and their design to ensure elderly well-being.

Aging subject human to physical, sensory including hearing and vision, and psychological changes (Grayston, 2018; WHO, 2015, 2018). The decline in physical conditions and loss of abilities may hinder the participation of the elderly in their daily physical and social activities in interior spaces and built environments (Clarke & Nieuwenhuijsen, 2009; George, 2017). These activities are also affected by the barriers in the built environment such as functional vision barrier, which can be described as the inadequacy of lighting and visual aspects (Clarke & Nieuwenhuijsen, 2009; Colombo et al., 2013). This can be a major barrier as the elderly spend the majority of their time indoors, especially residential spaces (Institute of Medicine of the National Academies, 2011).

Accordingly, lighting is considered an important element in the physical environment of the elderly. Appropriate lighting can contribute to the enhancement of visual environment by preventing fall risks and sleeping disorders that highlight its significant role in the health and safety of occupants (Lu et al., 2019; Illuminating Engineering Society (IES) 2017). Well-being is the satisfaction of human needs that represent one of the key themes of social sustainability, as shown in Fig. 2 (Liu et al., 2017). The important relationship between the elderly and the built environment is highlighted to promote their well-being in Fig. 3.

As a result, environmental gerontology theory emerged to develop the connection between elderly persons and their

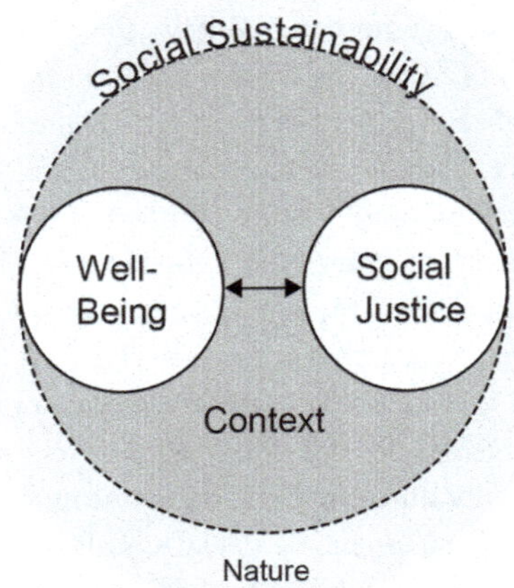

Fig. 2 Social sustainability framework diagram. *Source* Liu et al. (2017)

environment to achieve the well-being and quality of life through the means of intervention (Scheidt et al., 2006). Therefore, setting guidelines to achieve elderly visual comfort through environmental gerontology is needed in elderly care facilities to promote well-being in age-friendly space.

2 Research Objective and Methodology

This research aims to establish visual comfort design guidelines in elder care facilities through the role of environmental gerontology to be applied in developing countries' regulations as a part of social sustainability keys. To achieve this aim, a literature review on elderly visual functions and comfort in elder care facilities will be followed. The systematic literature review will focus on addressing the

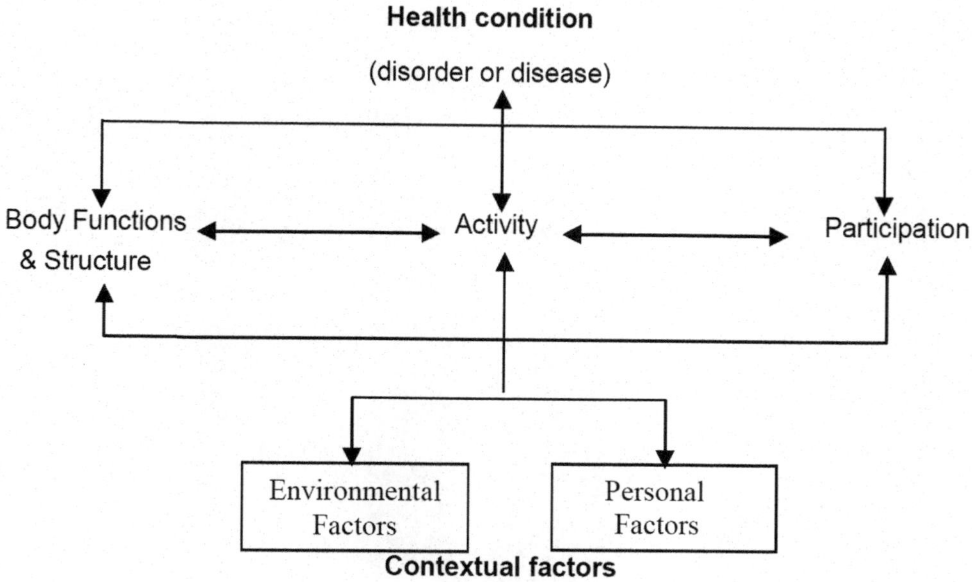

Fig. 3 Diagram of International Classification of Functioning, Disability, and Health (ICF) describing the interaction between disabilities, and social and environmental factors. *Source* World Health Organization [WHO] (2002)

social sustainability aspect of well-being in elder care facilities through improving the visual comfort guidelines for the elderly.

The searching process was based on peer-reviewed journals, books, articles, governmental guidelines, and reports. The databases searched were PubMed, Web of Science, Google Scholar, Scopus, ScienceDirect, and Governmental Websites. They were selected as they provide the different aspects of the topic including keyword set lighting conditions, population, and the built environment. The keywords in lighting conditions were daylighting, illumination, lighting color, light distribution, and glare. The keywords included under the population were elderly, senior, aging, and older adults, while the keywords included for the built environment were assisted living, house, senior living, nursing home, and retirement community. In addition, the environmental gerontology part keyword with visual comfort was searched. The search resulted in 160 searches related to visual comfort in the elderly built environment and 50 searches for environmental gerontology.

These criteria were used for screening the results to select the relevant topics:

- The publishing period was after 2009.
- Published articles should be in English.
- Specific diseases related to elderly studies were excluded such as dementia and Alzheimer.
- Studies should include lighting quality aspects in elderly care facilities.

The eligible studies reviewed for visual comfort in elder care facilities were 30, while the eligible articles for environmental gerontology were 24 studies after adding 4 articles to describe the history of the approach and the origins.

3 Literature Review Results

The literature review is divided into four parts. The research begins with a brief description of the eyes' composition and processing of light. It focuses on eye changes that occur with aging to review the elderly perception of light. Then, the research elaborates on international efforts for creating codes and standards about elderly visual comfort. Moreover, the research investigates the environmental gerontology theory as an age-friendly approach in designing elder care facilities and eliminating visual barriers. It also focuses on the achievement of visual comfort from the point of view of the environmental gerontology theory.

3.1 Aging Eye Changes and Perception of Light

The eye is described as a complex body organ that humans use for vision to gain information of surrounding environments, as shown in Fig. 4 (Dockter & Freed, 2019). Sensory cells and photoreceptors that are divided into rods and cones are formed in the retina (Dockter & Freed, 2019). The rods are concerned with the scotopic vision, which is the vision in low light levels and sensitive to shapes and movements (Dockter & Freed, 2019). The cone function is to process colors (Dockter & Freed, 2019). There is a third type of photoreceptors—photosensitive retinal cells—which are concerned with signal transmission to the brain. These cells help in regulating circadian rhythm, pupil constriction, mood, alertness, and more body functions. Color sensing and visual perception start after light enters eyes and make the lens change its shape to focus light on the retina (Dockter & Freed, 2019).

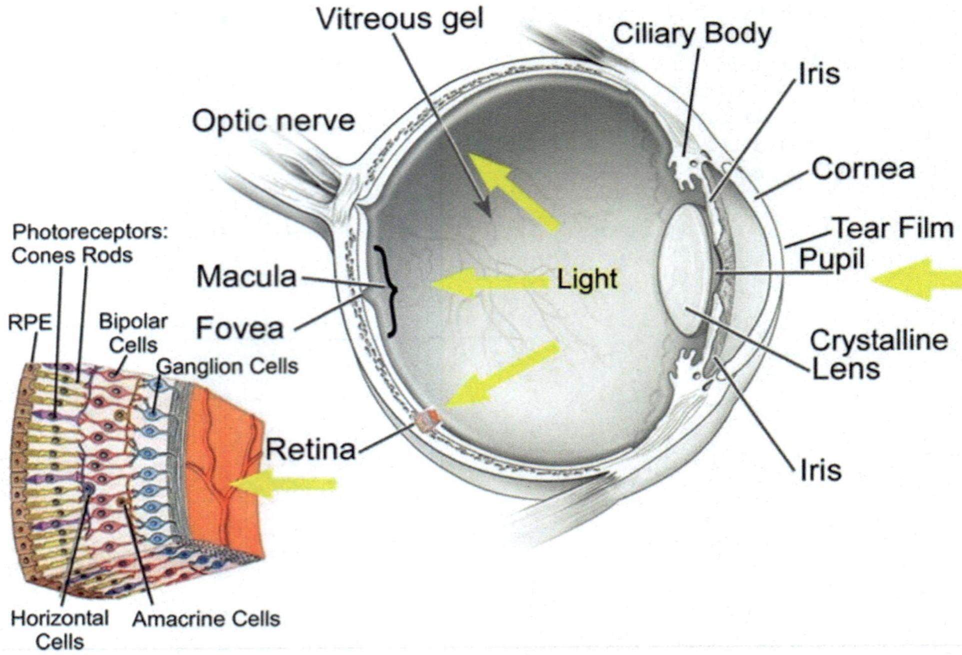

Fig. 4 Description of light filtration through lens and focusing on retina photoreceptors. *Source* Haney (2015)

So when eye ages, it is subjected to various changes including (Besdine, 2017):

- Eyes are subjected to dryness due to eye cells' decreased production of fluid.
- The focus on near objects becomes harder due to lens stiffness which is a condition called presbyopia.
- Seeing in dim light becomes harder due to lens thickening, so retina sensors become less sensitive to light.
- Sudden dark or light transition becomes a problem for the elderly due to the slower reaction of the pupil.
- Color perception changes due to lens yellowing.
- Recognizing fine details and judging distances become difficult due to the decrease in nerve cells count.

In addition, visual impairments due to illness such as macular degeneration, glaucoma, cataracts, retinitis pigmentosa, and diabetic retinopathy are described in Fig. 5 (Derungs Licht, 2011; The German Federation of the Blind and Partially Sighted [DBSV], 2018).

A research by Owsley (2011) discussed scotopic contrast sensitivity, which describes older adults' difficulty to see in low illumination levels. In addition, the research analyzed visual function where visual acuity and contrast sensitivity decrease that makes it harder to distinguish details also the adaptability to light time decrease. Then, it elaborated on the slowed-down visual cycle that is the slowing down of visual speed processing. This causes an increase in dark adaptation time and reduces sensitivity to motion. From the examples derived is that elders cannot distinguish between two moving objects simultaneously. The research conclusion summarized the visual perception process and recommended further research on the vision function process that can help the elderly performing everyday tasks.

Another research by Dupuy et al., (2013) analyzed visual functions that occur in the aging eye that are considered as keys in lighting system design understanding. These visual functions were explained as light amount perceived by eyes is reduced, light scattering produces higher sensitivity to glare, and blue wavelengths reduced perception due to eye lens yellowing.

A research by Wijk et al., (2002) focused on how elders perceive colors. This included color discrimination, color recognition, naming, and preferences related to visual and cognitive functions in old age. The researchers conducted a comparison between older adults' 3 groups 80 years old, 90 years old, and Alzheimer's disease (AD). The results concluded that red and yellow were more discriminated than blue and green, as shown in Fig. 6 (Wijk et al., 2002). As color perception changes with aging due to the yellowing of lenses, the filtration of blue light becomes more difficult, as shown in Fig. 7 (Dockter & Freed, 2019).

Therefore, eye changes that occur when people age affect visual functions and perception of light that can be summarized as shown in Fig. 8. These changes can affect design decisions in visual comfort.

3.2 Elderly Visual Comfort

Researches from various disciplines studied the built environment and visual comfort needed for elderly daily

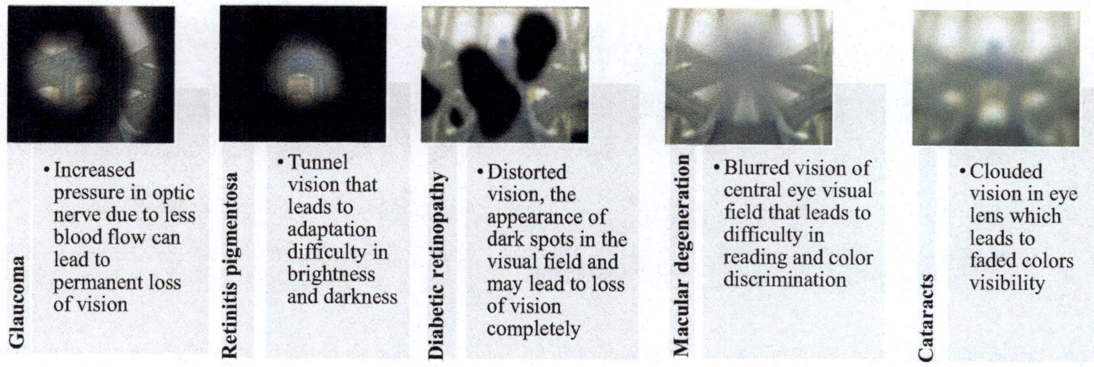

Fig. 5 Visual impairments due to aging. *Source* By authors based on (Derungs Licht, 2011; The German Federation of the Blind and Partially Sighted [DBSV], 2018)

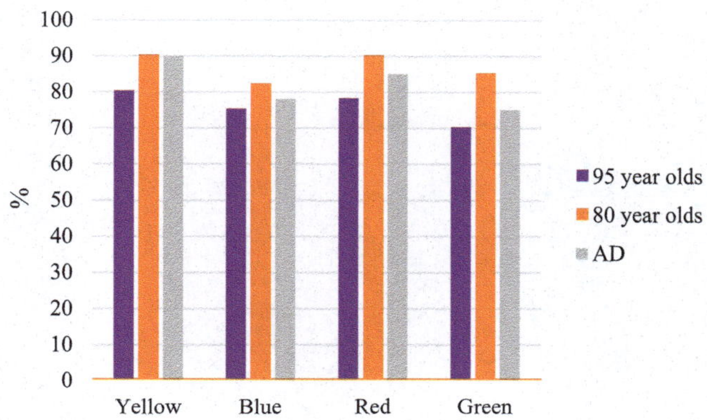

Fig. 6 Color discrimination by elderly groups. *Source* By authors based on Wijk et al. (2002)

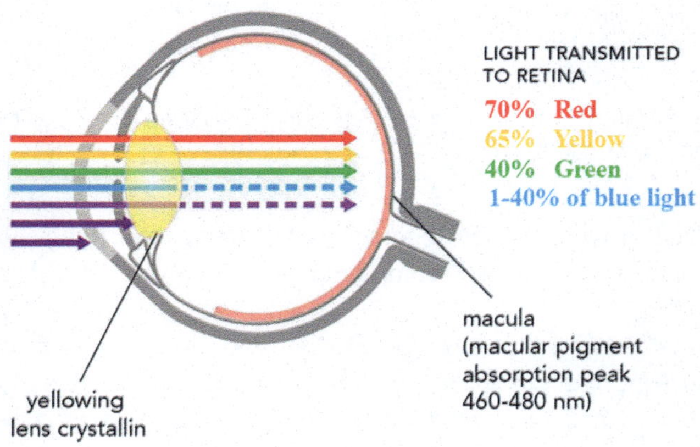

Fig. 7 Light color wavelength filtration through eyes. *Source* Dockter and Freed (2019)

activities and tasks (Lu et al., 2019). The built environment for the elderly they spend most of their time is residential such as houses, assisted living, independent living, nursing homes, and retirement communities; all can be included under the elder care facilities (Lu et al., 2019). The Egyptian General Organization for Physical Planning (2015) created a guide for social facility standards including elder care facilities. Elder care facility should be on the ground floor with accessible ramps, include outdoor space with area minimum 20 m^2/person, and be close to services such as banks, supermarkets, and hospitals. There is a lack of data of spaces included inside the building, and there is no code for visual comfort aspects for the elderly. In developing countries, the sustainable development of elderly facility codes and standards is not common and needs further institutional support (Hannula, 2012).

However, in developed countries as stated by the United States Department of Veterans Affairs, Office of Construction and Facilities Management (2019), the design of elderly care facilities was inspired by hospital care with minimum consideration to person life involvement. Then, they were upgraded to a residential community living center to allow participation of residents in a healing environment. The community living center was evolved into a resident care center involving residents with their families to a new design called small house design model.

According to the United States Department of Veterans Affairs, Office of Construction and Facilities Management (2019), the small house model consists of three levels to convey residence, neighborhood, and community center group. The elder care facility is the home unit that should present familiar surroundings. The housing unit consists of residential rooms as private quarters and then alcove settings as semiprivate areas then moving to public spaces, i.e., living room, kitchen and dining room, utilities, lounges, and open spaces as shown in Fig. 9. This activity room has direct access to external court or garden.

A research by Schneck et al., (2014) conducted an optometry testing for elderly groups of 865 individuals aged 58 to 102 years old to differentiate hues and colors acceptable for this age group. The results of the research stressed on that older people could not distinguish pastel hues of colors, and this causes confusion between different colors. Then according to an interview with Marylin E. Schnek one of the authors and elder vision expert—Ph.D. and a scientist

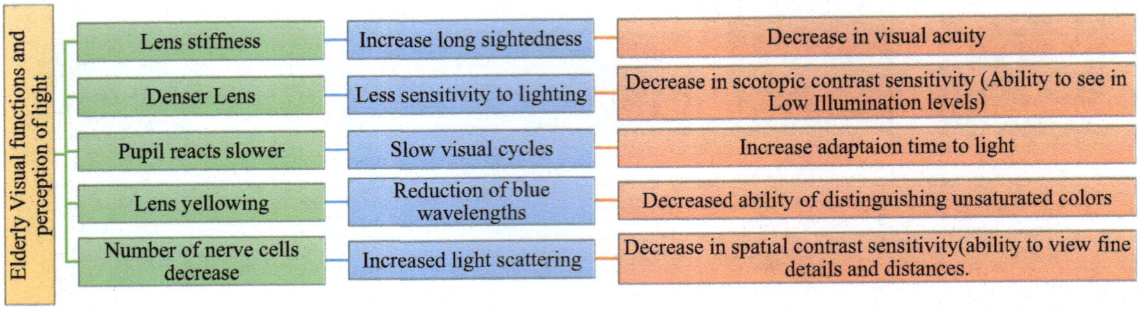

Fig. 8 Summarization of elderly visual functions and perception of light. *Source* By authors based on previous researches mentioned above in Sect. 1

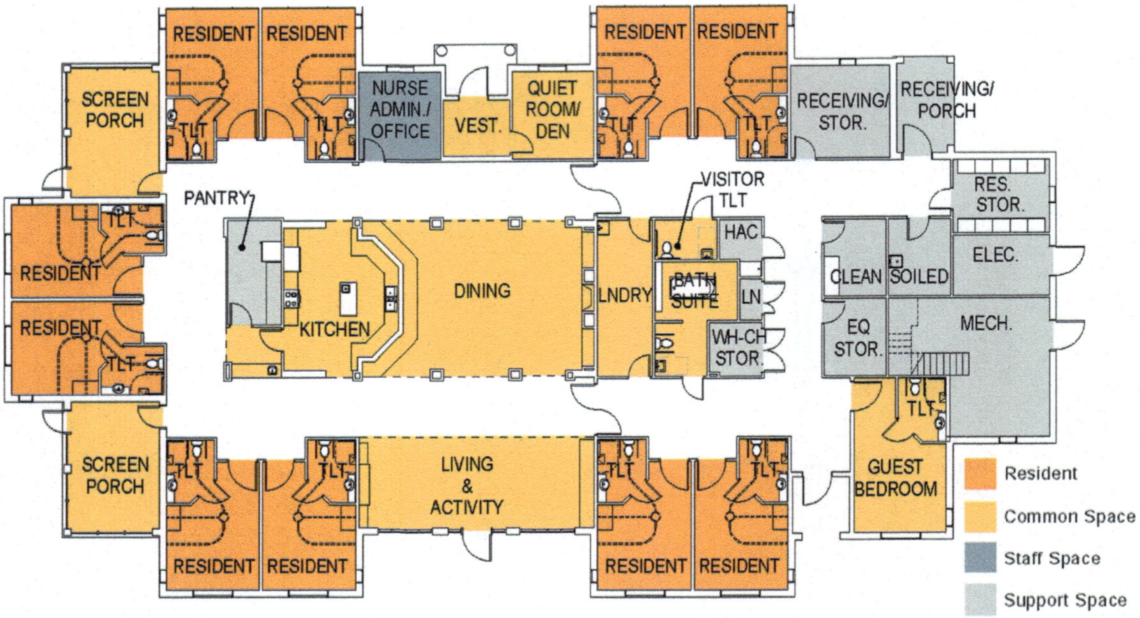

Fig. 9 Floor plan of VA Tuscaloosa Community Living Center Cottage (Building 146) showing the space design and hierarchy. *Source* United States Department of Veterans Affairs, Office of Construction and Facilities Management 2019

at the Smith-Kettlewell Eye Research Institute in San Francisco—to explain these results as applications and findings, the research emphasized on the yellowing of lens that causes hardship in distinguishing shades of blues and greens (Shea, 2019). Nonetheless, the elderly still can perceive blue color but darker than usually seen (Shea, 2019). The results should be useful in color designing for elderly homes and facilities (Shea, 2019).

The conclusion of the interview stated that designer should consider the following points when applying colors to achieve visual comfort such as using contrast between dark and light hues and when using pastel colors—not preferable—consider not shifting from one color to another especially in transitional spaces (Schneck, et al., 2014). In addition, the designer should use saturated and rich colors and provide ambient lighting, task lighting, and dimmable lighting to enhance color renders and perception (Shea, 2019).

Moreover, George (2017) discussed the factors of the design interior spaces in living facilities that affect the elderly and enhance quality of life including physical, psychological, and performance aspects. The research elaborated further on accessibility factors which include orientation of space and hierarchy of spacing providing secured access through solving the relationship between inner and outer spaces, etc. Improving these factors enhances social interaction, activities, and independence which creates a barrier-free design leading to safety and quality of life. The research also recommended the following for enhancing mobility and visual comfort for the elderly in interior spaces:

- Adequacy of lighting levels: Illumination that can help elders perceive visual details.
- Eliminating glare: Direct glare from light sources and daylight rays entering rooms and reflected glare from the light on reflective surfaces.
- Consistent contrast levels: Gradual change in lighting levels especially in transition. Spaces between outdoor daylit areas and indoors must be considered because the uneven changes in light levels cause disorientation and confusion (George, 2017).

The research concluded that solving design factors leads to the enhancement of physical factors and psychological factors to create age-friendly facilities as shown in Fig. 10. In addition, the author recommended studying the problem facing the elderly to develop a guideline for age-friendly facility design and be a chance for future development in rural and urban areas (George, 2017).

However, a research by Kunduraci (2017) emphasized on the role of lighting in residential houses for elderly as poor lighting design in residential settings may lead to fall and make elderly feel unsafe and less socially active. The aim of the research was literature review on applied visual comfort residential spaces occupied by elderly. The methodology was to analyze visual comfort in these spaces considering the visual task use of space and visual requirements without considering architectural features such as windows and shading techniques. The research recommended illumination in entrance and circulation between 200 and 300 lux. The recommended lighting for task surface lighting in bedrooms and living rooms is 500 lux while 100–300 near floors. The results concluded that there was a gap in the mentioned literature studies to provide lighting design considering the relation between elderly satisfaction and using architectural features. The results were qualitative and simple quantitative. He recommended further research to include perception, privacy, the satisfaction of elderly, and architecture design of features within lighting and daylighting design.

Another research by Lu et al. (2019) created literature on lighting effects on older adults' visual and non-visual performance. The literature included evidence-based studies and showed that inadequate lighting conditions contribute to the elderly risk of fall but studies did not mention detailed lighting parameters such as illuminance and light color or strategies. The research recommended:

- Further studies in design practices and changes in residential facilities for older adults regarding the consistency in actual illumination levels or contrast levels.
- Collaborative research among multi-discipline team including professional designers, medical, health researchers, and senior housing organizations to create a theoretical model more advanced with evidence-based lighting standards in the field of lighting and aging eyes.

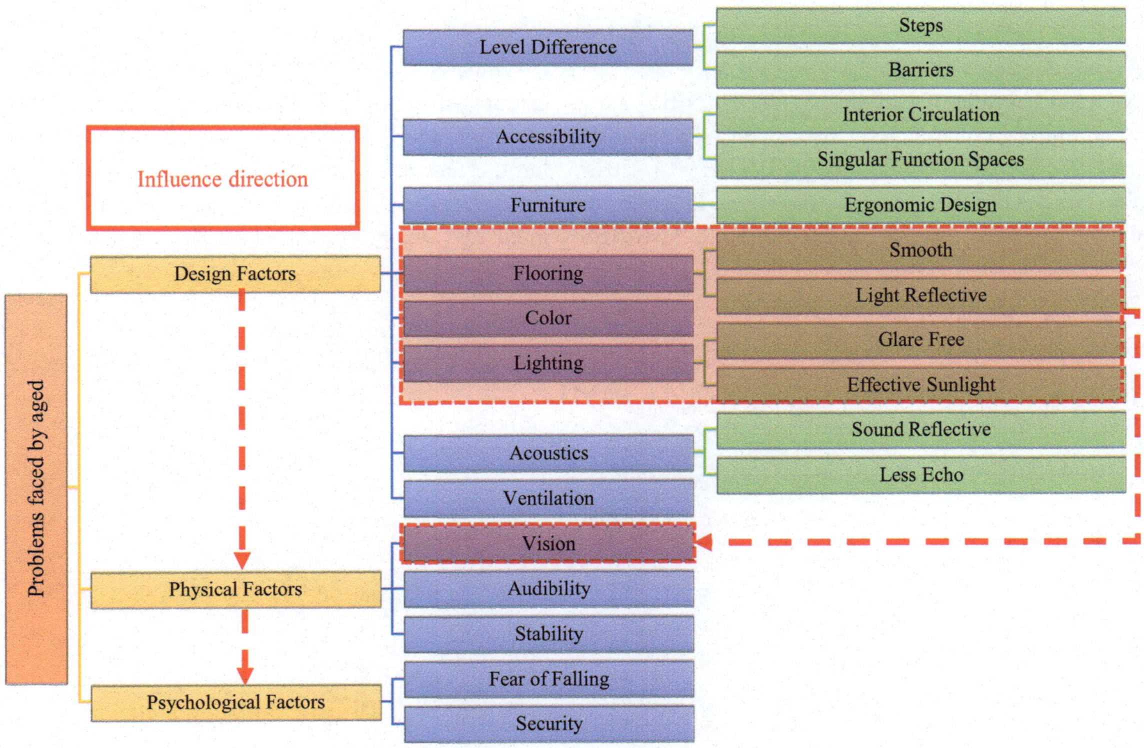

Fig. 10 Design factor influence on physical and psychological factors for the elderly highlighting vision design aspects. *Source* By authors based on George (2017)

As age-friendly environments and elderly care are not only a family responsibility but also governmental, as mentioned by world health organization WHO, creating age-friendly cities and setting aging accessibility guidelines become a world initiative purpose. In the MENA region, the resources provided by governmental guidelines and health-care policies for elder care are scarce and negligible to the emotional and physical problems and leisure criteria for dependent elderly (Sibai & Yamout, 2012). Contrarily, the developed countries such as the USA, Finland, Great Britain, and the Netherlands are considered leading countries in design guidelines for accessibility and elderly care institutional codes under the supervision of governments (United States Department of Veterans Affairs facilities, Office of Construction and Facilities Management, 2011).

However, the majority of codes resulted from governmental parties focused on quantitative energy efficiency rules for typical design excluding low vision people and elderly (which also considered as low vision) (Dupuy et al., 2013). In Great Britain, a handbook was developed in 1995 by the Royal National Institute for the Blind (RNIB) called "Building Sight: Handbook of Building and Interior Design Solutions to Include the Needs of Visually Impaired People" which discussed the modifications and design recommendations in interior spaces including lighting (Dupuy et al., 2013). Then, the British Standards Institute (BSI) developed code "BS 8300: Design of Buildings and their approaches to meet the needs of disabled people. Code of practice" in 2001 and updated it in 2009 (Dupuy et al., 2013). It discussed the partially sighted built environment design to include them in the code (Dupuy et al., 2013).

The National Institute of Building Sciences (NIBS) Low Vision Committee collaborated also with Illuminating Engineering Society (IES) Committee/ASHRAE 90.1 -2014 to develop the ANSI/RP 28–07 (National Institute of Building Sciences, 2014). It was updated 2016 to ANSI/IES RP-28–16 "Lighting and Visual Environment for Seniors and the Low Vision Population" with updated recommendations for low vision, traumatic brain injury, Alzheimer's disease, and lighting required for commercial and public areas such as offices, museums, and libraries (Sanford & Yoshida, 2017). The Illuminating Engineering Society (IES) (2017) set the values of task and accent lighting for elders in public facilities and residential facilities including elder care homes that highlight the quantity of lighting needed as shown in Table 1. It is the most adequate code to address elderly lighting requirements, especially in living communities (Dupuy et al., 2013).

The daylighting quality depends on illuminance relationship between openings, and adjacent and surrounding tasks' surfaces through different times of day and seasons (Low Vision Design Committee, 2011). For low vision and elderly, daylighting improvement depends on three lighting concepts:

- Task Lighting: Illuminating specific surfaces that require high levels of lighting to perform tasks such as reading and writing.
- Accent Lighting: Controlled ratios of color, illuminance, luminance, and contrast.
- Exterior Lighting: Contrast values—affects contrast sensitivity—between elements and adequate illumination levels in transition areas to achieve security and ease in mobility (Low Vision Design Committee, 2011; Illuminating Engineering Society (IES) 2017).

Thus, accent lighting depends on surfaces and materials characteristic for compensating the loss of contrast sensitivity for elderly and low vision. The factors affecting accent lighting are described in Fig. 11 (Dupuy et al., 2013). The characteristics depend mainly on the color's hue, value, and intensity (Dupuy et al., 2013). The color values of surfaces

Table 1 Minimum required illuminance for elderly residential spaces

Space	Ambient light (lux)	Task light (lux)
Interior entry	1000	
Exit stairways and landings	100	
Visitor waiting	200	500
Indoor circulation/lobby/lounge areas	200	500
Activity/meeting/common rooms	300	500
Entrance	100	
Living room	200	750
Bedroom	200	750
Bathroom	200	
Countertop	300	
Sink	500	
Dining	200	500

Source Illuminating Engineering Society (IES) (2017)

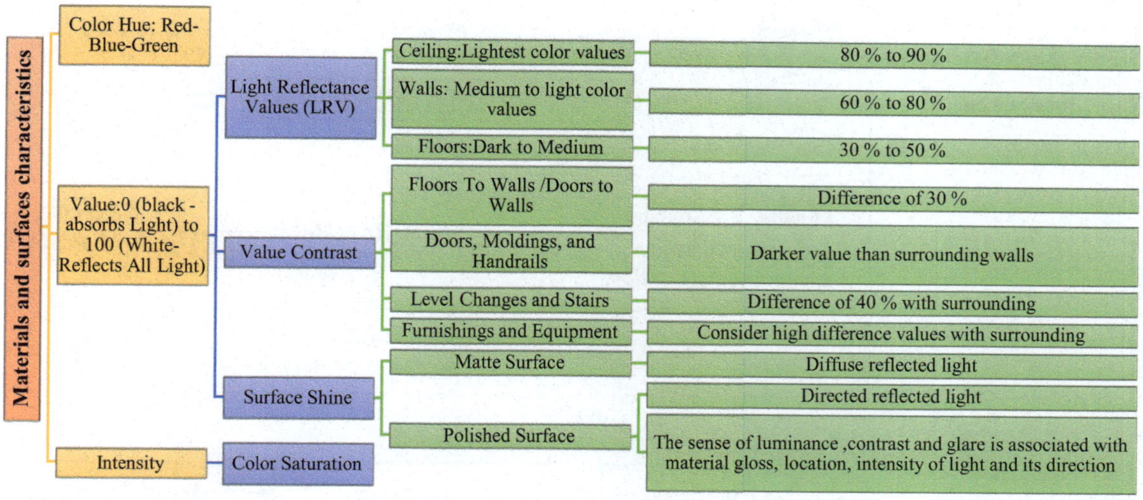

Fig. 11 Materials and surfaces' characteristics including aspects and values that affect the visual environment. *Source* Based on Dupuy et al. (2013) by authors

are represented numerically on a scale from zero to 100 where zero is black color and 100 is white to present lightness or darkness of material (Dupuy et al., 2013). Light reflectance values (LRVs) represented the percentage of color lightness or darkness, while value contrast explained the difference between materials colors to enhance architecture surfaces and details (Dupuy et al., 2013).

Moreover, the National Institute of Building Sciences (NIBS) Low Vision Committee collaborated with FGI—Former AIA Guidelines—to produce coordinated Guidelines Handbook that determines the visual environment recommendations in the facilities for aging and low vision (National Institute of Building Sciences, 2014). The Handbook is divided into two volumes:

- Guidelines for Health Care Facilities: Design and Construction of Hospitals and Outpatient Facilities.
- Guidelines for Residential Care Facilities: Design and Construction of Residential Health, Care, and Support Facilities (National Institute of Building Sciences, 2014).

4 Environmental Gerontology Approach

According to the World Health Organization (WHO) framework of age-friendly cities and communities, countries including Ireland, Canada, Australia, and the USA adopted age-friendly concepts to promote elderly participation, health, security, and independence (The Queensland Government, Department of Communities, Disability and Seniors 2017; WHO, 2010). These can be achieved through accessible and barrier-free built environments, and good environmental quality in indoors and outdoors.

According to the social production theory, subjective well-being in social sustainability is divided into physical well-being and social well-being (Lindenberg, 2001). Providing comfort for human is considered one of the physical well-being aspects. It is reflected in satisfying physiological needs and creating a pleasant and safe environment (Lindenberg, 2001). Sustainable aging is connected to elderly institutions and physical living environment that is adaptable and supportive of their needs (With Seniors in Mind, 2011; Wu & Kaushik, 2015). So, creating a sustainable environment for elderly can contribute to their comfort, thus creating a healthy aging environment contributing to their social sustainability, as shown in Fig. 12 (Parra-Ullauri & Rodrigues, 2017).

Environmental gerontology theory is one of the age-friendly approaches that emerged between the 1960s and the 1980s as a subfield from gerontology (Wahl & Weisman, 2003). This science discussed the relation between the elderly and built environment (Lawton & Nahemow, 1973). It focused on developing person–environment (p–e) relationship that contributes to the well-being and quality of life through the means of intervention in elderly institutions and housing, as shown in Fig. 13 (Lawton, 1977, 1999; Wahl, 2001). The process of p–e relationship is divided into belonging, which is related to elderly experience in built environment while agency related to goal-directed p–e cognitions, behaviors, and social practices, as described in Fig. 13 (Wahl & Oswald, 2010).

4.1 Parameters of Environmental Gerontology

Environmental gerontology is a considered field with input from different disciplines including occupational and physical therapists, human service professionals, nursing or other

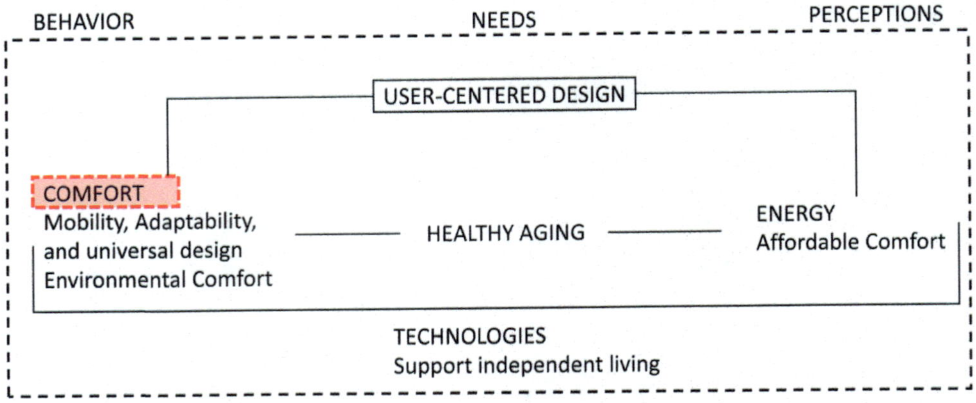

Fig. 12 Sustainable elderly housing framework. *Source* Based on Parra-Ullauri and Rodrigues (2017) by authors

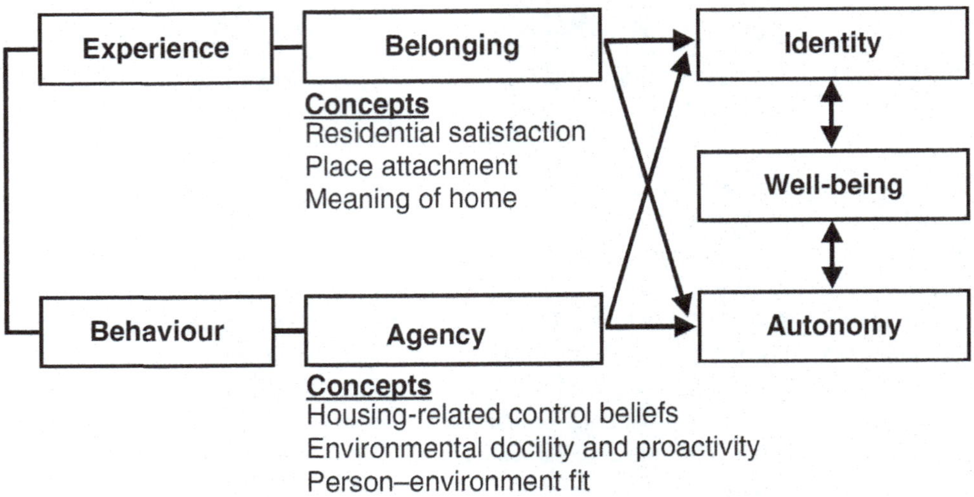

Fig. 13 Conceptual framework on p–e relationship for elderly. *Source* Wahl and Oswald (2010)

medical professionals, and architects and other environmental designers (Stichler, 2013). The environmental gerontology contributions appeared through theoretical background and research findings in social and behavioral gerontology and fields like geriatrics, geropsychiatry, and nursing sciences (Wahl & Weisman, 2003). It developed more to include environmental proactivity that assumes the elderly can change and adapt to any environment other than housing to their abilities and needs (Lawton, 1989). The application of environmental gerontology field extended from housing design at the micro-level to communities and cities at the macro-level (Wahl & Weisman, 2003). Bronfenbrenner (1999), and González and Rodríguez (2016) explained application levels of environmental gerontology as follows:

- The macro-level for building age-friendly communities, urban and rural environments, regions, and landscapes including evidence-based housing and elderly institutions (regulations and values of society).
- The exosystem (linkages between subsystems that indirectly influence the individual).
- The mesosystem as in the neighborhood and public spaces (two or more microsystems combined that affect the developing individual directly).
- The micro-level as personal housing and institution modifications (the interaction of persons directly with the immediate environment).

Qualitative research on environmental gerontology discussed experiences of space for the elderly from a physical and social context, the identity, and attachment to space points of view (Sánchez-González, 2014). Then, studies evolved to widely analyzing environmental dimensions that affect the elderly including environmental attributes and environmental functions. Environmental attributes that affect the elderly included comfort, privacy, accessibility, guidance, control, security, and dignity, while environmental functions for the interaction with elderly were divided into:

- Maintenance: cognitive–affective feelings in later life affecting the continuity and predictability of built environments as in private home environments and planned environments.

- Stimulation: concerned with the departure of the usual environment and the appearance of new social or leisure areas that affect the behavior.
- Support: concerned with the environment (home or planned) role in balancing a reduction or loss in abilities, which reflected in creating accessible and barrier-free environments (Kahana et al., 2003; Kurniawati, 2012; Lawton, 1989).

In addition, Taneli and Kovach (2015) analyzed quantitative methods of space including environmental variables, such as crowding and its psychological effect, illumination importance to circadian rhythm, treatment of agitation, lack of daytime alertness, and sleep problems of individuals with dementia to enhance safety, mobility, visual comfort, and acoustic comfort. The research concluded that these measures could provide a holistic assessment that can affect the influence of decision-making.

4.2 Visual Comfort Through Environmental Gerontology

A research by Pinto et al., (1997) discussed the application of researches involving gerontology in creating living environments and care settings for the elderly through providing techniques and products to compensate sensory and perceptual losses of the elderly. These promoted the improvement in mobility and strength of the elderly. The research analyzed visual acuity that causes environmental risks for the elderly due to inadequate lighting, high illumination-level brightness, different lighting levels between indoor and outdoor, lighting transition between different spaces, and insufficient lighting to recognize objects' details. The research recommended solutions to reduce the risks involved:

- Illumination of floor $\geq$ 300 lx; illumination of the work surface of 500–800 lux.
- Reducing the effect of glare and shadows through locating windows on the sides or at 1.70 cm in height or more and using sun shading systems on east and west windows.
- Using uniform illumination in the transition between indoors and outdoors, and rooms and corridors.

Another research by Lasagno et al., (2014) studied transitional spaces from the exterior to the interior as it is considered a functional vision barrier for three age groups older adults—above 60 years old, middle group—from 33 to 52 years old, and younger group—25 up to 30 years old-. The study examined two factors visual performance and visual discomfort due to rapid change in illumination in the city of Mendoza. The results confirmed that transitional spaces should be considered in the design of the built environment as functional vision barriers considering aspects of age, climate, spatial complexity, and building orientation. The research also suggested the design of appropriate shades and eaves to control daylight and improving visual comfort for users.

Moreover, Engineer et al. (2018) discussed environmental interventions in interior spaces for the elderly including findings of design specifically for elderly users that can promote environmental gerontology. The research discussed daylighting intervention strategies to produce visual comfort by analyzing architect Judith Chafee's house for her elderly mother. She used daylighting strategies including overhangs and clerestory windows with light shelves to reduce the effect of glare and increase visual acuity by distributing daylight. The reduced visual acuity recommended the reduction of using pastel cool colors such as blues, greens, and purples that appear gray for the elderly. In addition, the research recommended minimizing the contrast on floors to prevent falls but it is acceptable in differentiating steps. Increasing contrast on walls is required to illuminate certain features in design as the location of switches and others.

Thus, the elderly encounter visual problems due to the sudden change in illumination levels from inside to outside. According to the environmental gerontology approach, space should be free of barriers and support loss in functions. Focusing this research on visual comfort aspects, the Walumba elders' center contains a roof like a great shade over the building to reduce direct daylight on common areas and public spaces (Fig. 14a) (ArchDaily, 2015). Another daylighting strategy is a semitransparent shading device extending across the roof eaves to allow natural light and reduce glare from direct illumination (Fig. 14b) (ArchDaily, 2015). Another example is Peter Rosegger's nursing home where every transition area is designed with solid overhead on balconies and entrances, as shown in Fig. 15a (ArchDaily, 2014). In addition, some galleries and stairs have horizontal and vertical wooden beams as shading techniques in transition spaces from inside to outside, as shown in Fig. 15b (ArchDaily, 2014).

5 Findings

Achieving visual comfort is an important aspect for the elderly in the design of housing and elderly care facilities. By analyzing the environmental gerontology approach, it emphasized on the role of the physical environment in supporting loss in functions to achieve elderly well-being. The research focused on visual comfort for the elderly in elder care facilities. The environmental gerontology is divided into dimensions, which is highlighted in green, and

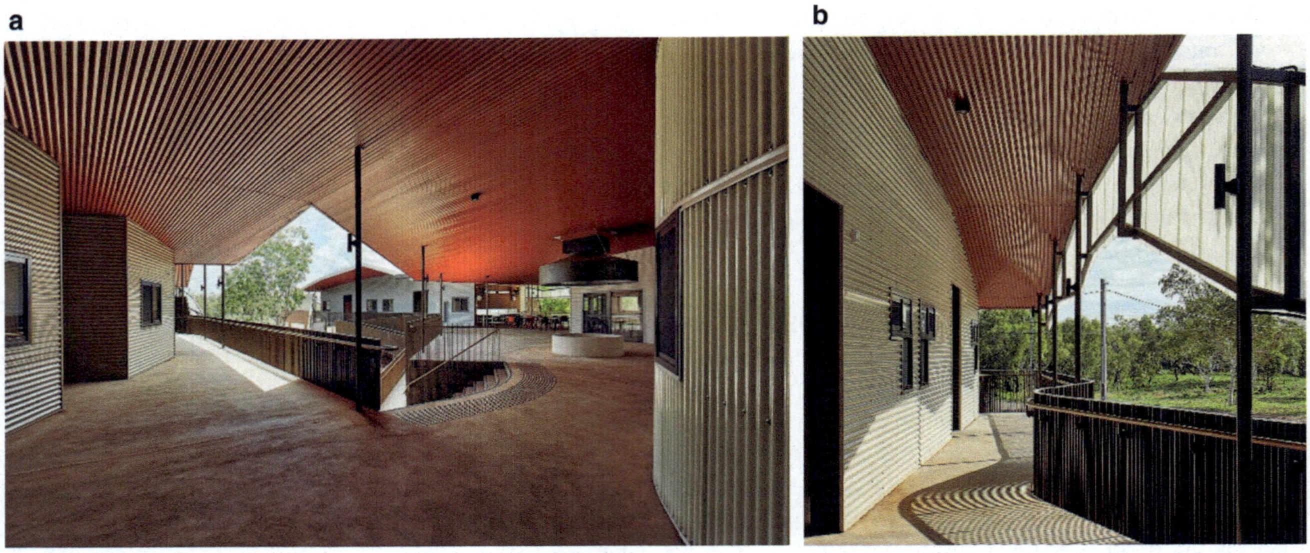

Fig. 14 **a** Roof shading and **b** semitransparent roof eaves. *Source* ArchDaily (2015)

Fig. 15 **a** Solid overhang on balconies and corridors. **b** Horizontal and vertical wooden beams used as shading techniques over transition at the entrance. *Source* ArchDaily (2014)

Table 2 Environmental gerontology matrix conclusion to promote visual comfort

Environmental gerontology matrix		
Environmental dimensions	Environmental functions	Supporting Visual Function
Level of application	Environmental attributes	comfort, accessibility, control and security
Micro Level	Elder care facilities modifications	

Source Based on previous by authors

the level of application is highlighted in blue in Table 2. Through the environmental gerontology approach, the environmental dimension used is supporting visual function to create a comfortable, accessible, controlled, and secured built environment for the elderly. These can be applied through modifications in elder care facilities, which is an intervention in the micro-level of the environment.

Moreover, the international codes and standards for lighting represented the quantity of light required for elderly to perform certain tasks as minimum illumination required. These levels were considered insufficient from the point of view of environmental gerontology researchers, as discussed by Pinto et al., (1997). Moreover, IES 2016 code provided nearly equal illumination levels as suggested by environmental gerontology researches. In addition, after analyzing case studies, transitional spaces became important spaces for the elderly as they encourage mobility and accessibility for elderly from interiors to exteriors and connecting them with daylighting and environment. The deduced daylighting strategies are mainly overhang shading devices, solid or perforated. Therefore, Fig. 16 summarizes the recommended visual comfort guidelines for the elderly in elder care facilities.

The proposed guidelines for elderly visual comfort discussed the quantity of lighting mainly in active areas such as living and dining room, activity room, the transition from interior spaces to exterior ones, terraces, and open corridors. In addition, the proposed quality of daylighting proposed architectural features and daylighting strategies to solve glare and direct sunlight. It is divided according to the three daylighting aspects: task lighting, accent lighting, and exterior strategies. These guidelines can be used as regulations for elder care facilities in developing countries.

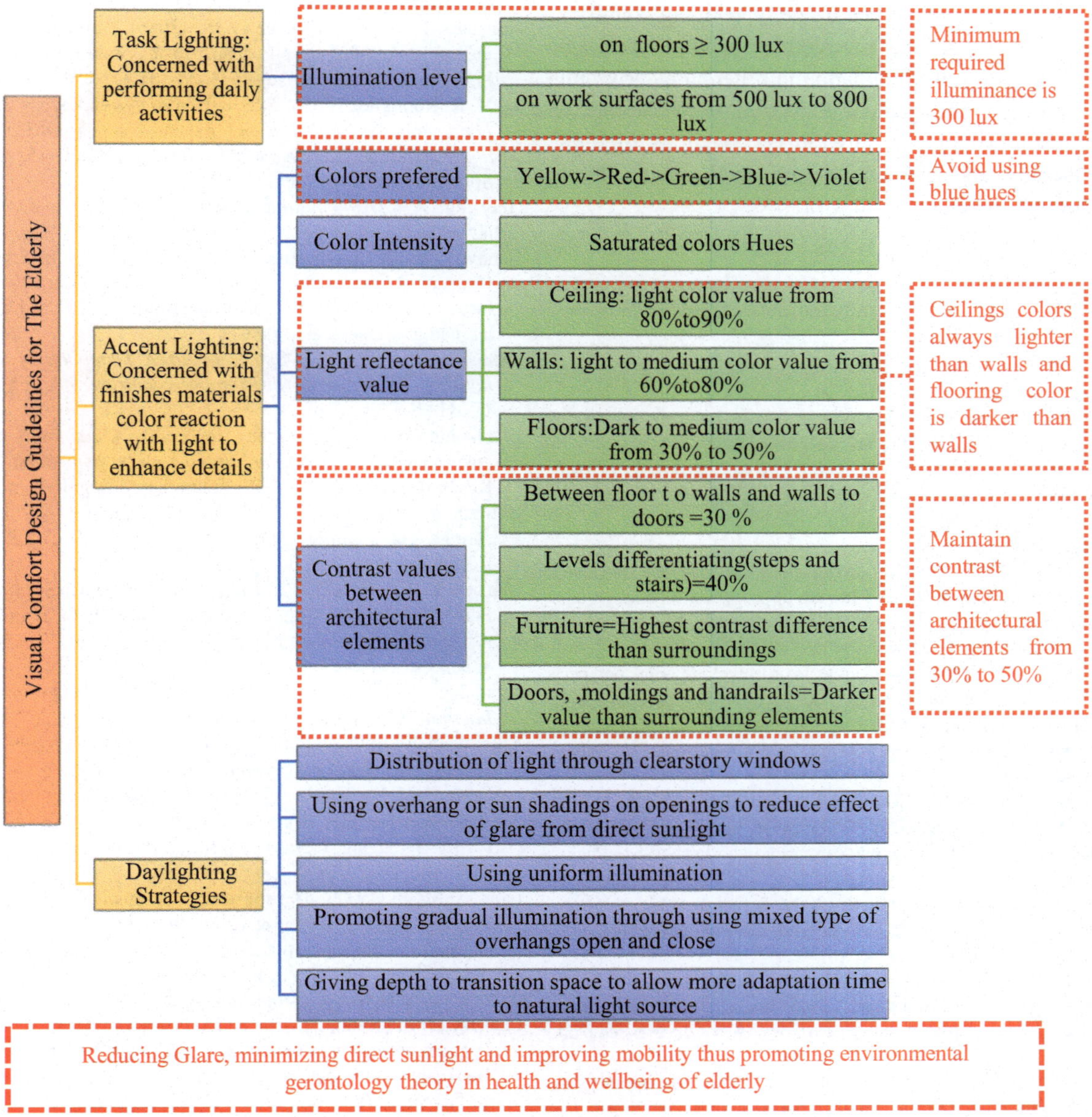

Fig. 16 Deduced visual comfort design guidelines for elderly in elder care facilities. *Source* By authors

6 Conclusion

The built environment for the elderly is considered a crucial aspect of their physical and psychological well-being. Moreover, the need to upgrade elderly care facilities emerged due to the demographic and social changes where elderly spend most of their time. The research focused on the visual comfort aspects of space to reduce environmental risks. To overcome visual acuity and comfort problems, the research elaborated on the multidisciplinary field of environmental gerontology. The research emphasized the role of the built environment in the design approach for elderly physical, psychological, and social health especially housing and care facilities, thus contributing to the social sustainability aspect.

The resulted guidelines are focused mainly on the quality of lighting that can be used as governmental codes and policies for elder care facilities in the developing countries. Therefore, further research is needed on the application and measurable quantity of illumination in different elderly built spaces, especially the gradual transition from indoors to outdoors using simulation programs as well as different shading designs and adjustments to evaluate glare, illumination through the elderly behavior.

References

ArchDaily. (2014). *ArchDaily: Peter Rosegger Nursing Home/Dietger Wissounig Architekten.* Retrieved December 27, 2019, from https://www.archdaily.com.br/br/760936/lar-de-idosos-peter-rosegger-dietger-wissounig-architekten?ad_medium=gallery

ArchDaily. (2015). *ArchDaily: Walumba Elders Centre/Iredale Pedersen Hook Architects.* Retrieved October 27, 2018, from https://www.archdaily.com/625274/walumba-elders-centre-iredale-pedersen-hook-architects

Besdine, R. W. (2017). *Changes in the body With aging.* Retrieved February 2, 2019, from Merck Manual: https://www.merckmanuals.com/home/older-people%E2%80%99s-health-issues/the-aging-body/changes-in-the-body-with-aging#v8967858

Bronfenbrenner, U. (1999). Environments in developmental perspective: Theoretical and operational models. In S. L. Friedman & T. D. Wachs (Eds.), *Measuring Environment across the life span: Emerging methods and concepts* (pp. 3–28). American Psychological Association.

Central Agency for Public Mobilization and Statistics[CAPMAS]. (2019). *Statistical yearbook 2019.* Retrieved January 4, 2020, from Egypt Statistics: https://www.capmas.gov.eg/Pages/StaticPages.aspx?page_id=5034

Clarke, P., & Nieuwenhuijsen, E. R. (2009, September 20). Environments for healthy ageing: A critical review (Elsevier, Ed.) *Maturitas, 64*(1), 14–19.

Colombo, E. M., O'Donell, B. M., Santillán, J. E., & Issolio, L. A. (2013). Functional vision barriers: A new concept analyzed in terms of human visual performance. *Psychology & Neuroscience, 6*(2), 219–226.

Dockter, C., & Freed, C. (2019). Chapter 1: A close look at our eyes. In J. Houston & J. Barrett (Eds.), *How to save Your eyes in the digital age: The handbook for eye care and electronics.* California: eyesafe. Retrieved December 2, 2019, from https://eyesafe.com/handbook/chapter-1/

Dupuy, R., Guarnaccia, G., Noell-Waggoner, E., & Center of Design for an Aging Society. (2013). *Lighting, low vision & building codes.* The National Institute of Building Sciences Low Vision Design Committee Symposium.

Egyptian General Organization for Physical Planning. (2015). دليل المعدلات والمعايير التخطيطية للخدمات معورية مصر العربية *[Guide to rates and planning standards for services in the Arab Republic of Egypt].* Cairo,Egypt: Egyptian General Organization for Physical Planning.

Engineer, A., Sternberg, E. M., & Najafi, B. (2018). Designing interiors to mitigate physical and cognitive deficits related to aging and to promote longevity in older adults: A review. *Gerontology, 64*, 612–622.

Essam-Eldin, H. (2018). سن المعاش في مصر مقارنة بالدول الأخرى *(Reports and Investigations, Pension age in Egypt compared to other countries).* Retrieved May 31, 2019, from Albawabh News: https://www.albawabhnews.com/3028437

George, A. (2017, 12). Influencing factors on age-friendly interiors. *Forensic Science and Addiction Research, 1*(4), 39–45. https://doi.org/10.31031/FSAR.2017.01.000517

González, D. S., & Rodríguez, V. R. (2016). Approaches to environmental gerontology approaches to environmental gerontology: Policy and practice on ageing and place. In D. S. González & V. R. Rodríguez (Eds.), *Environmental gerontology in Europe and Latin America: Policies and perspectives on environment and aging, international perspectives on aging* (Vol. 13, pp. 11–44). Springer International Publishing.

Grayston, F. (2018). Normal changes of ageing. *InnovAiT, 11*(11), 627–633.

Haney, L. V. (2015, November 20). *Things I learned at Science Camp (Part 2).* Retrieved December 28, 2019, from Laura Vican Haney: http://laura-vican-haney.squarespace.com/blog/2015/11/20/things-i-learned-at-science-camp-part-2

Hannula, E. -L. (2012). *Going green: A handbook of sustainable housing practices in developing countries* (M. French, Ed.). UN-HABITAT.

Illuminating Engineering Society (IES). (2017). *Lighting for the visual environment for seniors and the low vision population* (ANSI/IES RP-28–16, ERRATA #1 ed.). Illuminating Engineering society.

Institute of Medicine of the National Academies. (2011). *Climate change, the indoor environment and health.* Tha National Academies Press.

Kahana, E., Lovegreen, L., Kahana, B., & Kahana, M. J. (2003). Person, environment, and person-environment fit as influences on residential satisfaction of elders. *Environment and Behavior, 35*(3), 434–453.

Kunduraci, A. C. (2017). Lighting design for the aging eyes. *MATTER: International Journal of Science and Technology, 3*(3), 185–194.

Kurniawati, W. (2012). Public Space for Marginal People. *Procedia - Social and Behavioral Sciences, 36*, 476–484.

Lasagno, C. M., Issolio, L. A., Pattini, A. E., & Colombo, E. M. (2014). Transitional spaces from exterior to interior as functional vision barriers in ageing. *Lighting Research and Technology, 46*(6), 706–715.

Lawton, M. P. (1977). The impact of the environment on aging and behavior. In J. E. Birren & K. W. Schaie (Eds.), *Handbook of the psychology of ageing* (pp. 276–301). Van Nostrand Reinhold.

Lawton, M. P. (1989). Three functions of the residential environment. *Journal of Housing for the Elderly, 5*(1), 35–50.

Lawton, M. P. (1999). Environmental taxonomy: Generalizations from research with older adults. In S. L. Friedman & T. D. Wachs (Eds.), *Measuring environment across the life span: Emerging methods and concepts* (pp. 91–124). American Psychological Association.

Lawton, M. P., & Nahemow, L. (1973). Ecology and the aging process. In M. P. Lawton & C. Eisdorfer (Eds.), *The psychology of adult* (pp. 619–674). American Psychological Association.

Licht, D. (2011). *Visual timing light: Biodynamic lightin elder care*. Retrieved January 7, 2020, from Derungs Licht,Catalogues: https://goo.gl/wCbCA5

Lindenberg, S. (2001). Social rationality versus rational egoism. In T. J. H. (Ed.), *Handbook of sociological theory* (pp. 635–668). Springer.

Liu, Y., Dijst, M., Geertman, S., & Cui, C. (2017). Social sustainability in an ageing chinese society: Towards an integrative conceptual framework. *Sustainability, 9*(4), 658. https://doi.org/10.3390/su9040658

Low Vision Design Committee. (2011). *Environmental design and operations for people with low vision* (pp. 26–28). NIBS.

Lu, X., Park, N.-K., & Ahrentzen, S. (2019). Lighting effects on older adults' visual and nonvisual performance: A systematic review. *Journal of Housing for the Elderly, 33*(3), 298–324.

Ministry of Social Solidarity [MOSS]. (2019). دور المسنين *(Elderly facilities)* . Retrieved May 30, 2019, from Ministry of Social Solidarity [MOSS]: http://www.moss.gov.eg/sites/mosa/ar-eg/Pages/sector-service-detail.aspx?sid=34

National Institute of Building Sciences. (2014). *2014 Annual report for the president of the United States*. National Institute of Building Sciences.

Parra-Ullauri, A., & Rodrigues, L. (2017). Sustainable homes for the elderly. *The International Journal of Aging and Society, 7*(4), 55–71. https://doi.org/10.18848/2160-1909/CGP/v07i04/55-71

Pinto, M. R., De Medici, S., Zlotnicki, A., Bianchi, A., Van Sant, C., & Napou, C. (1997). Reduced visual acuity in elderly people: The role of ergonomics and gerontechnology. *Age and Ageing, 26*(5), 339–344.

Sánchez-González, D. (2014). Place identity, aging and environmental pressures of the city.Reflections from environmental gerontology. In D. Sánchez-González, & L. Á. Domínguez, *Identidad y espacio público: Ampliando ámbitos y prácticas [Identity and public space: Expanding areas and practices]* (pp. 141–168). Editorial Gedisa.

Sanford, L., & Yoshida, Y. (2017, April 29). *RP-28 Update:Lighting and the visual environment for seniors and the low vision population*. Retrieved March 22, 2019, from http://iessf.org/archives

Scheidt, R. J., Windley, P. G., Salthouse, T. A., Schaie, K. W., & Gatz, M. (2006). Environmental gerontology: Progress in the Post-Lawton era. In J. E. Birren, R. P. Abeles, T. A. Salthouse, K. W. Schaie, & M. Gatz (Eds.), *Handbook of the Psychology of Aging* (6th ed., pp. 105–125). Elsevier Inc.

Schneck, M. E., Haegerstrom-Portnoy, G., Lott, L., & Brabyn, J. (2014). Comparison of panel D-15 tests in a large older population. *Optometry and Vision Science, 91*(3), 284–290.

Shea, K. (2019). *Sherwin-Williams: Color Perception and the aging eye*. Retrieved July 15, 2019, from https://www.sherwin-williams.com/architects-specifiers-designers/inspiration/styles-and-techniques/sw-art-stir-color-aging-eye

Sibai, A. M., & Yamout, R. (2012). Family-based old-age care in arab countries: Between tradition and modernity. In H. Groth & A. Sousa-Poza (Eds.), *Population Dynamics in Muslim Countries* (pp. 63–76). Springer.

Stichler, J. F. (2013). Design considerations for aging populations. *Health Environments Research & Design, 6*(2), 7–11.

Taneli, Y., & Kovach, C. R. (2015). Advancing theory and practice through collaborative research in environmental gerontology. *Research in Gerontological Nursing, 8*(2), 58–60.

The German Federation of the Blind and Partially Sighted[DBSV]. (2018). *Augenkrankheiten (Eye diseases)*. Retrieved February 19, 2019, from The German Federation of the Blind and Partially Sighted[DBSV]. https://www.dbsv.org/augenkrankheiten.html

The Queensland Government, Department of Communities, Disability and Seniors. (2017, March). *Age-friendly communities: good practice review*. Retrieved December 5, 2019, from Queensland Government, Department of Communities, Disability and Seniors: https://www.communities.qld.gov.au/resources/dcdss/seniors/age-friendly-community/qafc-good-practice-review.pdf

UN. (2019). *Global issues: Ageing*. Retrieved May 30, 2019, from United Nations: https://www.un.org/en/sections/issues-depth/ageing/

United Nations (UN), Department of Economic and Social Affairs, Population Division. (2017). *World Population Prospects: The 2017 revision, key findings and advance* (Working Paper No. ESA/P/WP/248 ed.).

United States Department of Veterans Affairs facilities, Office of Construction & Facilities Management. (2011, June). *United States Department of Veterans Affairs Facilities: The Community Living Centers Design Guide*. Retrieved December 24, 2019, from http://www.bidnet.com/bneattachments?/365422700.pdf

United States Department of Veterans Affairs, Office of Construction & Facilities Management. (2019, March). *United States Department of Veterans Affairs facilities: Small House (SH) Model Design Guide*. Retrieved December 25, 2019, from https://www.cfm.va.gov/til/dGuide/dgSHModel.pdf

Wahl, H. W. (2001). Ecology of aging. In N. J. Smelser & P. B. Baltes (Eds.), *International encyclopedia of the social and behavioral sciences* (Vol. 6, pp. 4045–4048). Elsevier.

Wahl, H. W., & Oswald, F. (2010). Environmental perspectives on ageing. In D. Dannefer & C. Phillipson (Eds.), *The SAGE Handbook of Social Gerontology* (pp. 111–124). Sage.

Wahl, H. W., & Weisman, G. D. (2003). Environmental gerontology at the beginning of the new millennium: Reflections on its historical, empirical, and theoretical Development. *The Gerontologist, 43*(5), 616–627.

World Health Organization [WHO]. (2002a). *Towards a common language for functioning, disability and health: international classification of functioning, disability and health*. World Health Organization.

WHO. (2002b). *Health statistics and information systems: Proposed working definition of an older person in Africa for the MDS project*. Retrieved May 30, 2019, from https://www.who.int/healthinfo/survey/ageingdefnolder/en/

WHO. (2010). *World Health Organisation, ageing and life-course: Global Age-Friendly Cities Project*. Retrieved May 15, 2019, from https://www.who.int/ageing/projects/age_friendly_cities/en/

WHO. (2015). *World report on ageing & health*. WHO.

WHO. (2017). *10 Priorities towards a decade of healthy ageing*. WHO.

World Health Organization[WHO]. (2018). *Ageing and health*. Retrieved January 28, 2019, from https://www.who.int/en/news-room/fact-sheets/detail/ageing-and-health

WHO. (2019). *Ageing and life course*. Retrieved January 12, 2019, from https://www.who.int/ageing/features/misconceptions/en/

Wijk, H., Berg, S., Bergman, B., Hanson, A. B., Sivik, L., & Steen, B. (2002). Colour perception among the very elderly related to visual. *Scandinavian Journal of Caring Sciences, 16*, 91–102.

With Seniors in Mind. (2011). *The senior living sustainability guide*. Retrieved May 5, 2020, from With Seniors in Mind: https://www.withseniorsinmind.org/slsg-download

Wu, W., & Kaushik, I. (2015). Design for sustainable aging: improving design communication through building information modeling and game engine integration. *Procedia Engineering, 118*, 926–933.

Sustainable Living Fences (SLF): To Develop the Function and Form of Universities' Fences. (Based on Value Architecture)

Basma S. Kassem, Essam Eldin Badran, Ibrahim Abdel Rashid, and Faysal Abu ElAzm

Abstract

Sustainable facilities have a considerable positive impact on the sustainability of universities. So Egyptian universities should strive to become more sustainable by providing sustainable facilities along with a suitable learning environment for students. There are shortages in financial resources for governmental universities; thus, the Egyptian government calls every university for achieving its additional financial resources to provide full support for scientific research. The traditional universities' fences achieve main physical values as securing the university, ignoring some social and environmental values; therefore, the research presents a solution "added Values" to save universities' operating expenses due to the efficient use of materials, providing more funding for research, and providing a better environment for students, staff, and other academic endeavors. The research proposes a model for sustainable practices that introduces designing "sustainable living fences (SLF)" instead of the traditional universities' fences to develop its function and form. This new term seeks to avoid its traditional problems, adding new visual value, promoting its functions, presenting services, and creating more enjoyable spaces for students, staff, and residents of the surrounding areas.

Keywords

Living fences • Universities' fences • Sustainable values • Function • Form • Portable unites • Value architecture

1 Introduction

The impact on the environment resulting from today's activities inside the universities and the availability of resources for future generations has become an important concern for many researchers, managers, and policymakers toward sustainability (Hodges, 2005; Kirk, 2003; Zhao, 2014). We cannot create sustainability in urban areas alone. A net productive landscape complementing the city, as its border can offer a strategy for deep sustainability. So, the number of sustainable facilities continues to grow (Logan, 2019).

1.1 Research Problems

- The main research problems are as the following:
- The traditional governmental universities' fences are placed in a monumental context, and it is surrounded by special outdoor urban spaces, ignoring some social and environmental values.
- The shortcomings of financial resources for governmental universities lead the Egyptian government to call every university for achieving its additional financial resources to provide full support for its scientific research. This will create a challenging atmosphere to build and remodel facilities that meet the demanding needs of the universities (Elmualim et al., 2010; Sinclair, 2009).
- The shortcomings in marketing the Egyptian governmental universities' special units, as research centers as well as a productive and service sector.

B. S. Kassem (✉) · E. E. Badran · F. Abu ElAzm
Architecture and Urban Planning Department, Faculty of Engineering, Suez Canal University, Ismailia, Egypt
e-mail: Basma.kassem@eng.suez.edu.eg

F. Abu ElAzm
e-mail: Faysal.arch@icloud.com

I. Abdel Rashid
Faculty of Engineering, Ain Shams University, Cairo, Egypt

C. Alalouch et al. (eds.), *Towards Implementation of Sustainability Concepts in Developing Countries*, Advances in Science, Technology & Innovation, https://doi.org/10.1007/978-3-030-74349-9_19

1.2 Research Objective

The main research objective is the adaptive reusing of the traditional governmental Universities' fences, to develop its function and form by adding social, environmental, and economic values. This will increase the percentage of energy supplement to its proposed designed spaces, and to its surrounding areas, and market the local governmental universities, achieving its additional financial resources to provide full support for scientific research.

1.3 Research Question

How can we develop the function and the form of universities' fences toward sustainable living ones?

1.4 Research Structure.

The research is divided into four main parts, as shown in Fig. 1.

2 Theoretical Approach

Sustainable campuses are becoming a target for public higher education universities. Facilities can play a significant role in the sustainability of the campus by achieving the resources savings in addition to the improved workspaces. The research will focus on the universities' fences as a start toward the sustainability of universities.

2.1 Universities' Fences

The adage says, "*Good fences make good neighbors*" (Schwartz, 2019). Fencing defines the smaller blocks, zones like college buildings, perimeters, pedestrian paths, and direct traffic. Building new universities' fences for higher education campuses require planning and scheduling giving a very tight academic schedule (Jamison, 1996). There are a lot of considerations about fencing and related systems with college planning and management, which are as the following:

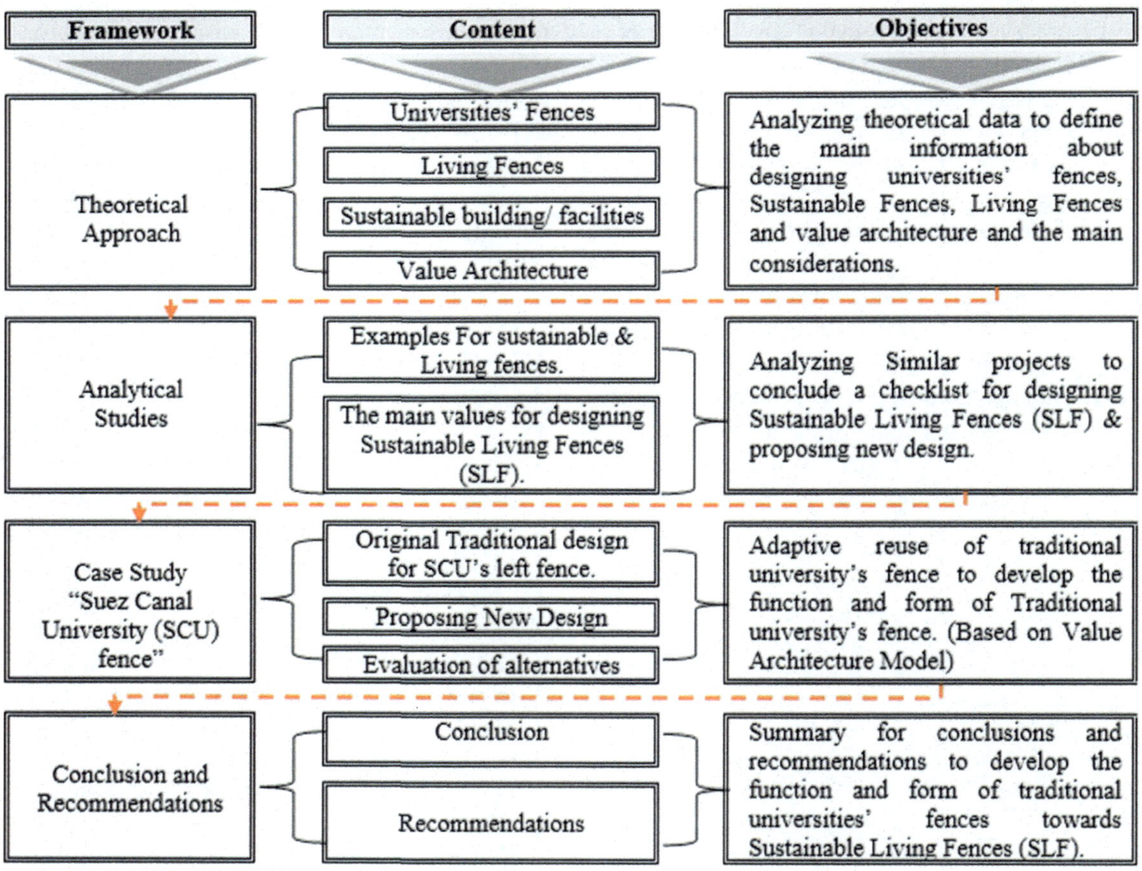

Fig. 1 Research structure. *Source* Researchers

- The costs, materials, environments, and users' needs.
- Defining what kind of fencing system will be used for what areas of campus.
- Defining the fencing system which is informed by the kinds of security and safety scenarios on campus.
- Addressing police and fire department access issues.
- Providing a balance between campus' branding and being a physical component to provide safety for students.

2.2 Living Fences

Living fences are classified into the following:

A **Natural Living fences**

These kinds of fences are constructed of shrubbery, trees, or flowering bushes which are attractive and environmentally friendly (Schwartz, 2019), as shown in Fig. 2. The points needed to be taken into consideration from natural living fences are as the following:

- Defining the purpose of the natural living fence.
- Defining the boundary or location to place the fence.
- Evaluating context/site conditions.
- Determining the desired length, width, and height to achieve the purpose goal.
- Extension of tree and shrub roots, as it is not preferred to plant too close to structures.
- Checking for utility lines as overhead power lines, and underground irrigation heads and lines.
- Installed into the ground or in a raised planter.
- Plant selection regarding the "plant toxicity, soil kind, how much light and moisture it needs to receive."

B **Man-Made Living Fences**

These kinds of fences are constructed of multi-purposes fences. It transforms an ugly and bare chain link fence into marketing, advertising, and promotions opportunity to get target messages and draw the attention of passersby (Wraps, 2019), as shown in Fig. 3. "*The significance of projects isn't only based on their technological innovations or how much water, energy, or other resources they conserve*" says David Goldberg, a partner with the Seattle design firm, San Francisco. The great impact comes from the thousands of students and visitors who take their lessons out to the world (Jamison, 1996). So, the campus should be highly productive including marketing shops, native plantings, a field lab, a café, a dining commons, housing, classrooms, and offices. These multi-purposes designed spaces can be supported by a photovoltaic array catches for rainwater, outdoor furniture, and shelter to students waiting for public transportations. The points needed to be taken into consideration from artificial/man-made living fences are as the following:

- Campus' gates, to market adjacent designed spaces.
- Managing traffic and designing parking lots.
- Kinds of adjacent furniture as just one piece of furniture can define a space and establish a safe and functional perimeter on an educational campus; furniture can be benches, bike racks, trash cans, and table sets.
- Marketing research to gather information about users' needs and understand the cultural, society, socioeconomic, geographic, and personal makeup of them. This

Fig. 2 Hedera Ivy screening "natural living Fence." *Source* https://www.artigreen.co.uk/hedera-ivy-screening?lightbox=dataItem-jl5cnzn2

Fig. 3 Boxpark, London "man-made living fences." *Source* https://beebreeders.com/shipping-containers-re-purposed-but-never-redesigned

will help the designers to create a project that provides solutions, increase their chances for success (Sheehan, 2017).

- An awareness of architectural elements serving multiple purposes.

2.3 Sustainable Building/Facility

Constructing sustainable facilities can reduce the water consumption of a building by 40% and energy use by as much as 50% (Holowka, 2008). Sustainable facilities are marketable to potential students who are choosing which university to attend (Attaran & Celik, 2015). Attaran and Celik determined—in an online survey of 162 students in a university in New England—that students are willing to pay more for sustainable facilities. Constructing sustainable facilities will improve the work environment and reduce campus energy consumption (Attaran & Celik, 2015). The points needed to be taken into consideration from sustainable building/facility are as the following:

- Project location and land utilization "Context analysis."
- Materials and resources selection (water and energy).
- Materials and resources conservation and management.
- Environmental regulations and laws.
- The costs of executing sustainable construction during construction phases, from inception through completion, use, re-use, and demolition.
- Addressing initial costs, running costs, maintenance costs, and future modification costs and ensuring that projects make a profit and return on investment.
- Many sustainable systems and component options for new facilities will cost more initially. However, over the life cycle of the facility, there will be benefits from reduced energy costs and improved work environments for its occupants (Buchanan, S. 2011). If the long-term financial return is more than the initial cost to install, then the more sustainable option should be installed in the project (Abidin & Pasquire, 2007; Nalewaik & Venters, 2009; Wao, 2014).

2.4 Value Architecture

Value architecture (VArch) is finding the right balance between architectural investment resources like materials, structure, energy, time, cost … etc., by appropriate technology to meet today's needs and requirements, unignoring the future needs (Kassem & El-Azm, 2017). It is concerned with various forms of financial implications associated with construction resources based on value methodology. It is also concerned with enhancing the quality of people's life, meeting the expectations of people, and satisfying their needs. Using value architecture methodology correctly will provide the ability to achieve a sustainable building. It can also provide the university with the best possible value for money invested. The following considerations should be considered in the evaluation and selection of alternatives:

- Evaluating alternatives to meet local users' needs.
- Minimizing the idle time by optimum use of resources.
- Optimum use of local and available materials, manpower, machines, and equipment with optimum cost.

- Optimum use of recyclable, renewable, and reusable resources.
- Complying with environmental regulations.
- The right mix of team members who evaluate the alternatives of designs.

3 Analytical Studies

The research will display similar project and conclude the research's checklist which is as the following:

3.1 Similar Project

A. The Cohon University Center's Fences "a Hub of the Carnegie Mellon campus, USA"

The Cohon University Center's Fence is designed to support community engagement, wellness, and health. It offers some special spaces as conference space, meeting rooms, studios, fitness rooms, multiple dining locations, and studying spaces (Jared, 2019), as shown in Fig. 4. It supports opportunities for undergraduate and graduate students in which they can connect, explore, and engage with one another to create a set of experiences (Carnegie Mellon University, 2019a). Therefore, a positive change in the community around will be created.

The rain garden, as shown in Fig. 5, acts as a holding tank separated from the city's water lines. So that the filtrated water and sewer systems do not mix. Indigenous plants and vegetation can grow with fresh water. The crated rock formations offer plenty of seating for eating lunch outdoors (Carnegie Mellon University, 2019b).

B Egyptian Sporting Clubs' Man-Made Living Fences

For sports organizations, coaches say that stadium fences are not only beautifying their outer fences but also help in branding their programs for fans and fulfill their needs (Wraps, 2019).

Some Egyptian stadiums' fences, as shown in Figs. 6 and 7, were converted into commercial stores and the advertising boards with a huge amount of rents to be paid in advance. Its profits are for developing the services inside the club. However, this project faced some obstacles, as organizing traffic and shortage of parking.

3.2 Checklist for Designing Sustainable Living Fences (SLF)

From the theoretical approach and the similar project, the research proposes the following checklist for designing sustainable living fences (SLF), as in Table 1.

Fig. 4 Student leadership, involvement, and civic engagement. *Source*https://www.cmu.edu/student-affairs/slice/#slice, analyzed by researchers

Fig. 5 Cohon University Center's fences. *Source* https://www.cmu.edu/environment/university-commitments/index.html

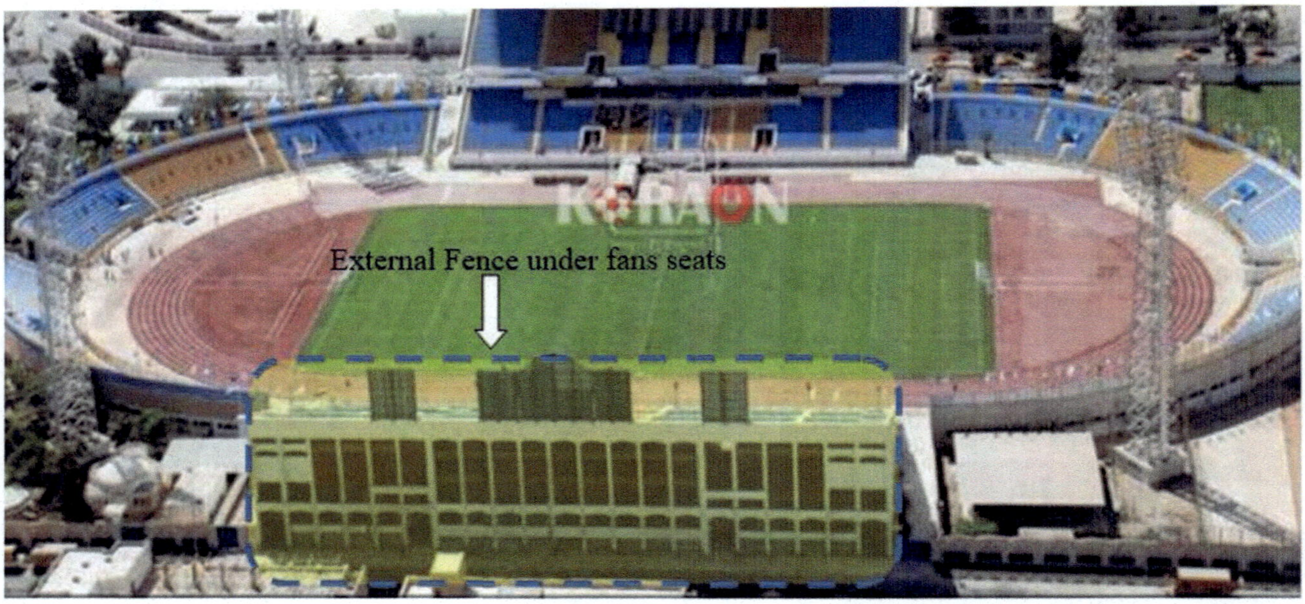

Fig. 6 Ismaily Sports Club's Commercial fence. *Source* https://www.koraon.com/, analyzed by Researchers

Fig. 7 El-Zamalek Sports Club's Commercial fence. *Source* https://4rent-apartmnt.blogspot.com/2018/04/mohandeseen.html

Table 1 Research's checklist for designing sustainable living fences (SLF)

	Target Values		
University Fence	The main values needed to be achieved from Universities' Fences are as the following:		
	• Define space. • Direct traffic.	• Provide security. • Reflect the university image.	• Enhance safety. • Provide aesthetics.
Living fences.	**Natural Living fences' Values:**		
	• Provide marketability. • Respect Environment. • Provide natural privacy. • Make inner boundaries greener.	• Increase productivity. • Prevent pollution. • Improve air quality.	• Provide aesthetics. • Attract passage. • Increase biodiversity.
	Man-made Living fences' Values:		
	• Increase productivity. • Define boundaries. • Provide messages.	• Encourage gathering. • Prevent pollution. • Provide marketability and branding.	• Refine functionality • Attract passage.
Sustainable building/Facility.	**A. Environmental Values:**		
	• Reduce Waste. • Eliminate pollution. • Access to daylight • Control access. • Reduce Water usage • Re-use and recycle wastes and materials.	• Reduce energy consumption. • Conserve water and energy. • Provide natural ventilation. • Use local resources. • Minimize the building's impact.	• Increase productivity • Reduce CO_2 emissions. • Connect with nature. • Eliminate toxic materials. • Use renewable energy. • Promote indoor air quality.
	B. Social Values:		
	• Respect legislation. • Satisfy users' needs and expectations. • Create Jobs for the local community.	• Respect governmental regulation	• Enhance corporate image • Respect organization ethos
	C. Economic Values:		
	• Provide profitability • Drive investment • Support of the local economy	• Minimize Lifecycle cost • Provide financial gain.	• Remain competitive • Market expansion

Source Researchers

4 Case Studies

The research will study the ability to re-design governmental universities' fences and will focus on Suez Canal University's left fence in Ismailia, Egypt, as a case study.

4.1 General Information

Suez Canal University is an Egyptian governmental university, distinguished by its mission directed to all members of society. It was established by the presidential decree no. 93 for 1976 (Suez Canal University, 2019).

4.2 University's Aim "main Targets"

Suez Canal University aims to occupy a prominent place among other Egyptian higher education institutions based on its contribution to develop education and extent its interaction with its surrounding community through pioneering community-related research. It also aims to provide opportunities to its students for learning and working after their graduation, responding to the needs of the Egyptian society and the labor market. Suez Canal University had established special units as shown in Fig. 8, to activate its role in providing distinctive social services to the university and the surrounding community, for social and economic development. Special units depend on two pillars, which are as the following:

- Providing services, consultations, and putting the experiences of the faculty members and their research capabilities to achieve the university's mission in community service.
- Providing financial resources to operate the units and financing some of its projects. The units of a special nature at the University of the Suez Canal provide many services.

Fig. 8 Suez Canal University's special units. *Source* Google Earth & Engineering Administration, Suez Canal University, analyzed by researchers

4.3 Original Design

Suez Canal university's left fence lies at the entrance of Ismailia city, Egypt. It is about 1000 m long and has a buffer zone about 16 m wide as the following Figs. 9, 10, and 11:

4.4 Function Analysis

The research seeks to develop the values within Suez Canal University without compromising to abandon its identity. The research suggests that it is better to make universities' fences efficient with considering renewable energy. Governmental universities can save money by making these fences as efficient resources as possible. The following Fig. 12 shows the (physical and non-physical) function analysis for SCU's left side Fence.

4.5 Re-design

After a market research and function analysis, the proposed design for developing the function and the form of Suez Canal Universities' fences consists of representative units for its special units, offices, advertisements, open cafes, closed restaurants, and an open area for settings which can be used as an observation deck, regarding its identity, as shown in the following Figs. 13, 14, 15, and 16:

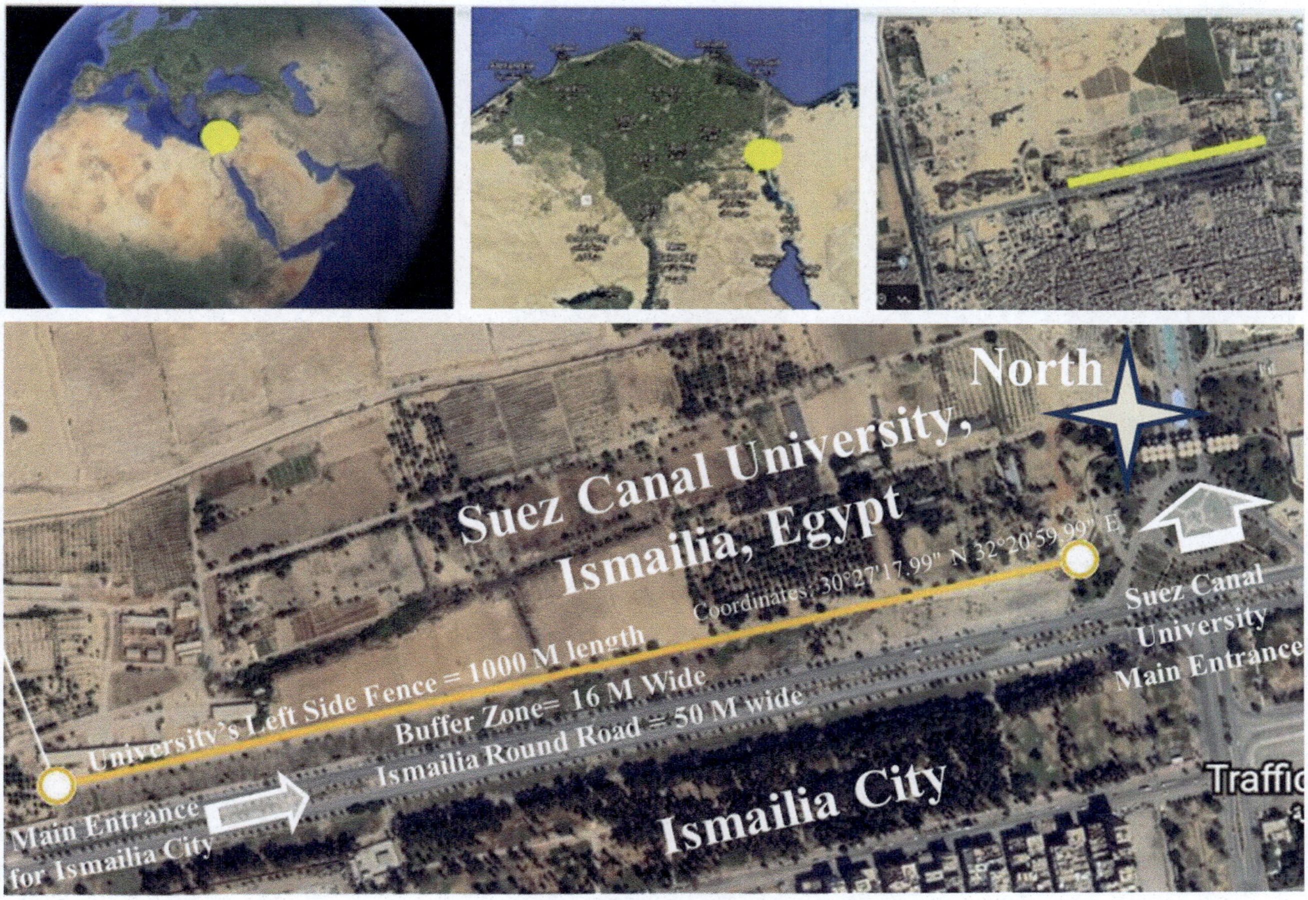

Fig. 9 Layout for Suez Canal University's left side fence, Ismailia, Egypt. *Source* Google Earth, reproduced by researchers

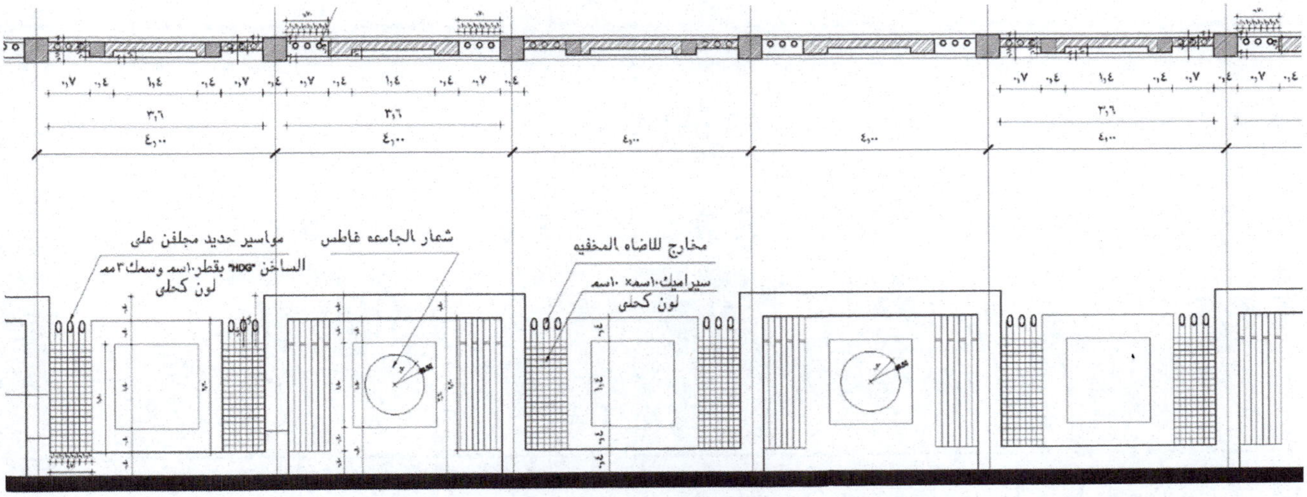

Fig. 10 Working drawings for Suez Canal University's left side fence. *Source* Engineering Administration, Suez Canal University, Ismailia, Egypt

Fig. 11 Real photo for Suez Canal University's left side fence elevation, Ismailia, Egypt. *Source* Captured by researchers, October 2019

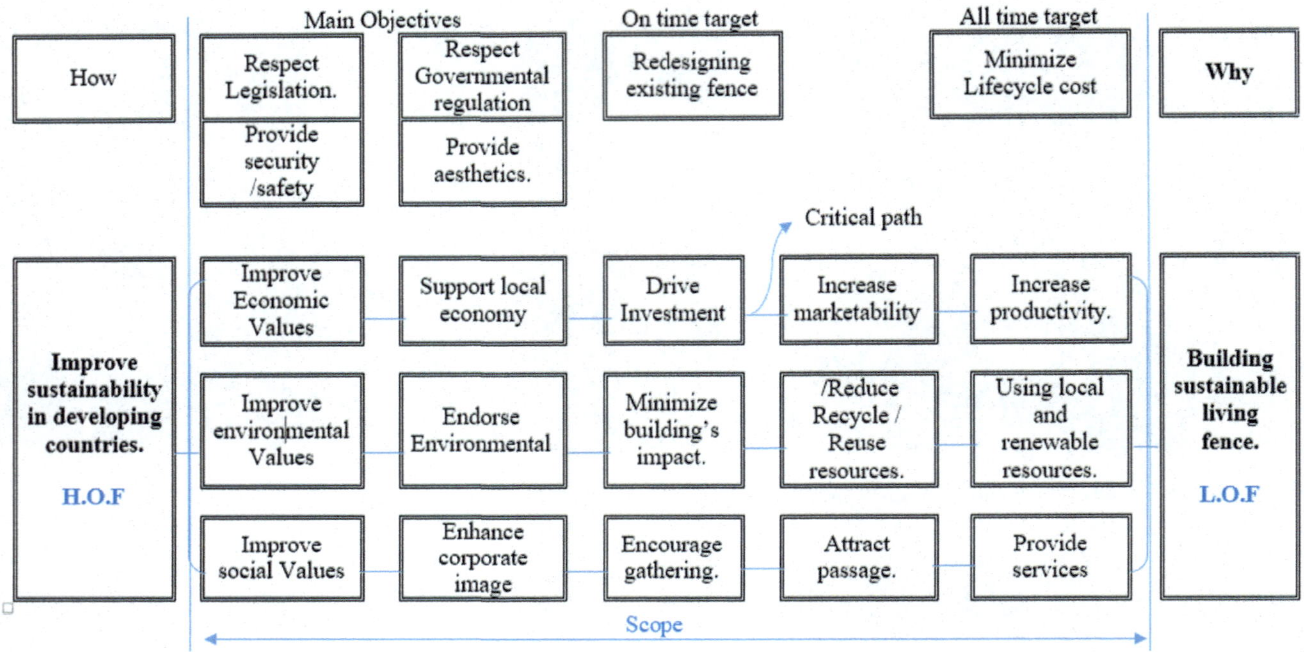

Fig. 12 (Physical and non-physical) function analysis for SCU's left side fence. *Source* Researchers

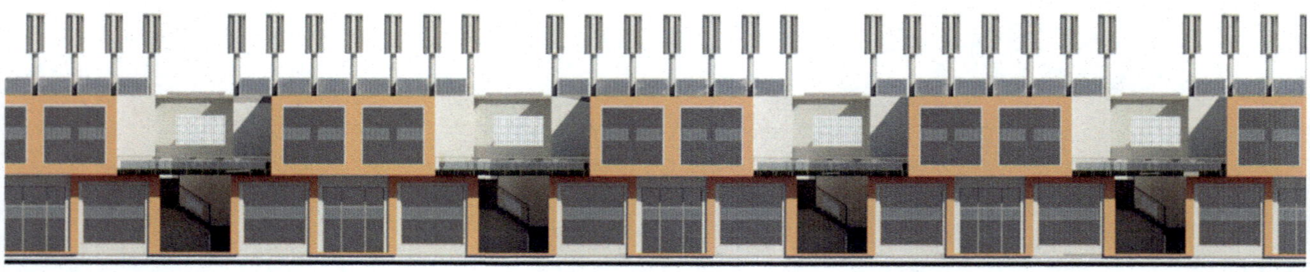

Fig. 13 Elevation for the proposed design. *Source* Researchers

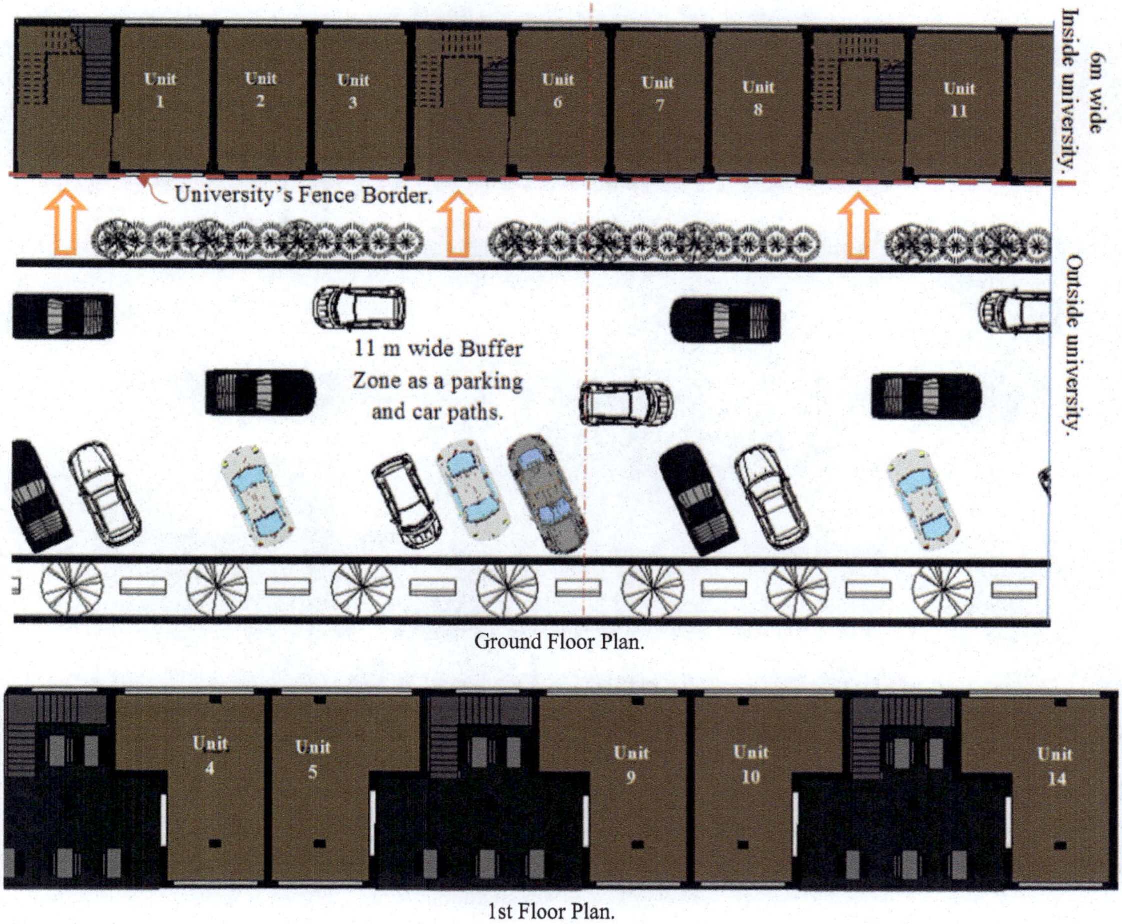

Fig. 14 Proposed plans for Suez Canal University's left side fence. *Source* Researchers

4.6 Creativity for Alternatives

According to previous studies, it is concluded that the most term that affects the costs of the project is the construction materials. Therefore, the research team proposes two alternatives for the traditional construction material "reinforced concrete and bricks," for one unit—a repetitive building unit—with dimensions "3 m width × 6 m length × 4 m hight," as shown in the following Fig. 17.

4.7 Evaluation for Alternatives

The research team with the help of some specialists conducted a meeting to evaluate the traditional construction material and the other two alternatives for the original building material, based on the evaluation matrix as in Table 2, and they deduced the following:

After evaluation, it is concluded that traditional construction materials have the greatest quality points and alternative (1) portable unit "Container" will raise the value of sustainable living fence and decrease the time needed for construction. Hence, in the developing phase, the research recommends to put the two proposals into consideration and chooses the lowest life cycle cost to minimize resource depletion.

5 Conclusions and Recommendations

From all the above, we can deduce the following.

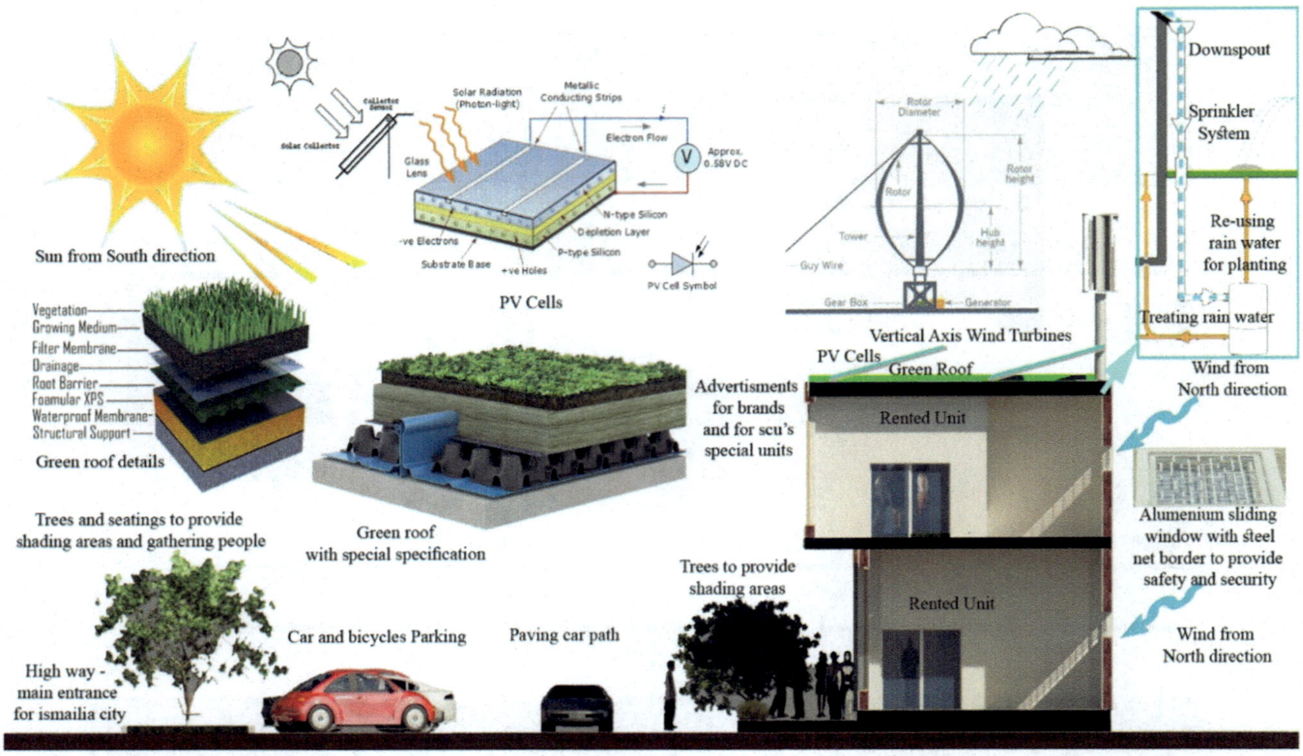

Fig. 15 Section studies for the proposed design. *Source* Researchers

5.1 Conclusions

Value architecture methodology can be applied through all phases of construction projects toward sustainable buildings in Egypt, as in Fig. 18. Sustainable living fence (SLF) can be defined as: "A multi-functional landmark Fence in which people can work and enjoy their entertainments. It conserves the use of resources and produces energy to satisfy the building electricity demands plus the additional electrical energy for the surrounding areas."

Sustainable living fence (SLF) can supply the governmental universities with a lot of money, which will affect the economy of the developing countries. There are millions of University's fences all over the world, so the development of its function and form will change the global urban context to a sustainable one.

5.2 Recommendations

The research chooses Suez Canal University's Fence as a case study for opening the field for other designers to re-design other universities' fences toward sustainable living ones. Regardless of the research's University's Fence design, the research opens the field for other designs that could have added values. Therefore, the research recommends that the proposed "functions analysis" could be followed while designing a new university's fence to achieve

Fig. 16 External shots for the proposed design for Suez Canal University's left fence. *Source* Researchers

(a) "Reinforced concrete and bricks" with finishings. US $5000/ unit | (b) Alternative (1) : Portable unit "Container". US $2000 | (c) Alternative (2) : Precast unit cuboid. US $3000

Fig. 17 Traditional construction material and its alternative ideas. *Source* **b** and **c** https://www.alibaba.com/onlinestore

Table 2 Evaluation matrix for the proposed alternatives

A. Durability	A								
B. Safety	2B	B							
C. Non toxic	A/C	B/C	C						
D. Aesthetics	A/D	2B	2C	D					
E. Maintenance	A/E	2B	C/E	D/E	E				
F. Construction Time	A/F	2B	C/F	D/F	2F	F	Q	C	V= Q/C
Weight	4	9	6	3	3	5	**Quality Points**	**Cost per unit**	**Value**
Weight Percentage	13.5%	30%	20%	10%	10%	16.5%			
Traditional construction materials	5	5	4	5	4	3	460	$5000	0.181
	67.5	150	80	50	40	49.5			
Alternative (1)	3	5	4	5	3	5	415	$2000	0.2075
	40.5	150	80	50	30	82.5			
Alternative (2)	3	5	4	5	5	5	445	$3000	0.148
	45	150	80	50	50	82.5			
Rates --------	**5= Excellent 4= Very good 3= Good 2 = Fair 1 = Poor**								

How Important
2 = points of major preference.
1 = points of minor preference.
1 = points each for same preference.

Source Group of specialists with researchers

Fig. 18 Value architecture framework. *Source* Researchers

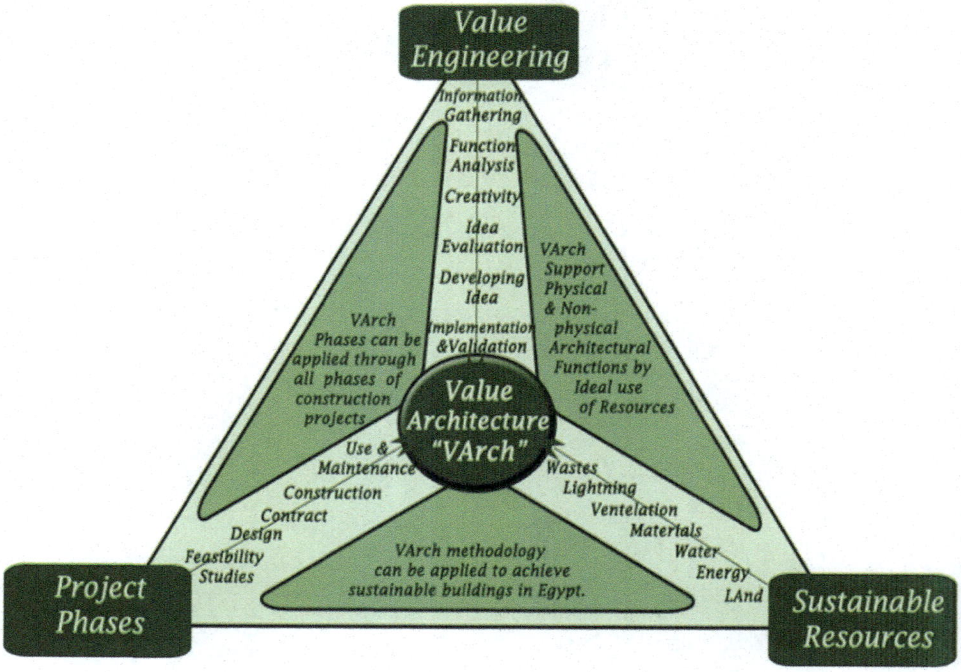

sustainable living fence (SLF), which could be followed on constructing any new building to reach sustainable living buildings (SLB) as well.

References

Abidin, N., & Pasquire, C. (2007). Revolutionize value management: A mode towards sustainability. *International Journal of Project Management*.

Attaran, S., & Celik, B. (2015). Students' environmental responsibility and their willingness to pay for green buildings. *International Journal of Sustainability in Higher Education*.

Buchanan, S. (2011). *Integrating sustainability into the facilities capital planning process*. Facilities Manager. Retrieved December 1, 2019, from the World Wide Web: http://files.eric.ed.gov/fulltext/EJ920383.pdf

Carnegie Mellon University. (2019a). Retrieved December 1, 2019, from the World Wide Web: https://www.cmu.edu/student-affairs/slice/#slice

Carnegie Mellon University. (2019b). Retrieved December 1, 2019, from the World Wide Web: https://www.cmu.edu/environment/university-commitments/index.html

Elmualim, A., Shockley, D., Valle, R., Ludlow, G., & Shah, S. (2010). Barriers and commitment of facilities management profession to the sustainability agenda. *Building and Environment Journal, 45*(1), 58–64. https://doi.org/10.1016/j.buildenv.2009.05.002

Hodges, C. (2005). Facility manager's approach to sustainability. *Journal of Facilities Management*. https://www.emerald.com/insight/content/doi/10.1108/14725960510630498/full/pdf

Holowka, T. (2008). LEED and higher education. *Environmental Design & Construction Journal*.

Jamison, A. (1996).*Guidelines for developing a campus master plan document for small colleges and universities*. Andrews University.

Jared, L. (2019). *Cohon University Center*. Retrieved December 1, 2019, from the World Wide Web: https://www.cmu.edu/cohon-university-center/

Kassem, B., & El-Azm, F. (2017). *Value architecture (VArch): As another approach of Value Engineering (VE)*. London International Conference for Sustainable Design of the Built Environment (SDBE).

Kirk, C. (2003). Sustainability: Taking the long view. *Planning for Higher Education*.

Logan, L. (2019). Sustainable campus development—Campuses go green: Colleges and universities take environmentally responsible design to new levels. *Architectural Record*. Retrieved December 1, 2019, from the World Wide Web: https://continuingeducation.bnpmedia.com/courses/multi-aia/sustainable-campus-development/

Nalewaik, A., & Venters, V. (2009). Cost benefits of building green. *Cost Engineering*.

Schwartz, D.B. (2019). *11 Living fences that look better than chain link*. Retrieved December 1, (2019), from the World Wide Web: https://www.bobvila.com/slideshow/11-living-fences-that-look-better-than-chain-link-47520

Sheehan, A. (2017). *The complete guide to market research for your retail business*. Retrieved December 1, 2019, from the World Wide Web: https://www.shopify.com/retail/how-to-do-market-research-the-retailer-s-guide-to-doing-homework-for-your-niche-business

Sinclair, B. (2009). Culture, context, and the pursuit of sustainability: Contemplating problems, parameters, and possibilities in an increasingly complex world. *Planning for Higher Education*.

Suez Canal University. (2019). Official online site, Retrieved December 1, 2019, from the World Wide Web: http://scuegypt.edu.eg/en/?page_id=13

Wao, J. (2014).*Value engineering methodology to improve building sustainability outcomes* (Ph.D thesis). University of Florida.

Wraps, C. (2019). *Construction and fence mesh graphics*. Retrieved December 1, 2019, from the World Wide Web: https://www.capitalwraps.com/mesh-graphics/

Zhao, X. L., Li, W., & Stanbrook, J. (2014). A framework for the integration of performance based design and life cycle assessment to design sustainable structures. *Advances in Structural Engineering, 17*(4), 461–470.

Sustainable Development Management, Impact, and Education

A Procedural Paradigm for Green Project Management of Sustainable Development

Dina Khater

Abstract

The author discusses that project managers (PMs) are generally liable for the actualization of the construction management objectives, as they have effective roles in developing a sustainable building process. The project management practices and their performance in the currently on-going Egyptian national housing projects from initiation and planning stage through to construction and operation stages will be analyzed and evaluated in an effort to adopt a paradigm shift of communities' project management to a sustainable path. The understanding of green building projects should be based on the management of the project life cycle stages as an integrated process. Project Management Body of Knowledge (PMBOK), on the contrary, introduces the project management execution process as a repeated process in each of the project life cycle stages. Therefore, the consideration of sustainability fundamentals in project management practices requires a parametric shift. This shift is discussed in this paper as the successful integration of project management aspects through project life cycle stages. Hence, the author utilized the existing knowledge areas in PMBOK with further adjustments, added new knowledge areas and modified project management processes flow to go linear with project life cycle stages (PLC). This has resulted in having sustainability indicators that can be monitored and assessed through the project management processes of housing projects. The paper results in the formation of a guide model that is introduced as a procedural paradigm that helps project managers to plan, monitor and assess sustainable-based housing projects from inception phase to operation stage and involve the appropriate corrective action at the proper time.

D. Khater (✉)
Faculty of Engineering, Architectural Engineering Department, Cairo University, Cairo, Egypt

Keywords

Green project management • Integration concept • Guide model • Critical success factors • Performance index formula

1 Introduction

The similarities between the principles of sustainable development and sustainability offers a basis for understanding the interrelationships between them. Both principles can be featured simply as the process and the goal. Sustainable development is a way to attain the goal of sustainability, where sustainability assessment (SA) is the mechanism that ensures the accomplishment of this purpose and provides the basis for evaluation during the implementation of that process. The framework for delivering sustainable construction projects should be fulfilled with a procedural process that is validated by SA process as a comprehensive type of assessment designed to support decision-making and planning in the economic, environmental and social context. This procedural process is literally based upon the three fundamentals of sustainability (economic, environmental and social) dimensions and the achievement of the necessary credit points required by several green buildings rating systems. Nevertheless, this procedural process has not yet formalized the integration with the project management aspects because the project management scope needs to be extended to incorporate PLC stages as an integrated process. The paper presents a procedural paradigm for green project management of housing projects with a series of statements that have a common and well-defined aim for the long-term goal of sustainability.

The project management processes and knowledge areas are the basis of this paradigm which will be applied through PLC stages to accomplish an integrated process for the delivery of sustainable housing relevant goals. The project

C. Alalouch et al. (eds.), *Towards Implementation of Sustainability Concepts in Developing Countries*,
Advances in Science, Technology & Innovation, https://doi.org/10.1007/978-3-030-74349-9_20

management aspects in Egyptian housing sector will be discussed with the assumption about project management to be a tool for an enhanced planning and sustainable performance of housing projects. Where, it is important to recommend a better approach, when considered, critical success factors can bring sustainable housing to an efficient level. The deployment of such key factors in the current practiced project management would help resolving the delay in completion, enhancing the poor quality of finishing, minimization of capital investment and ensuring the sustainability of housing projects within the local economy. The paper introduces a guide model validated as a future paradigm for a sustainable-based project from inception stage to operation stage from project management vision, green project management (GPM) critical success factors and green project management (GPM) performance index formula in housing projects.

2 The Contrast Between Sustainability and Project Management Principles

Increasing awareness of an adaptive management framework for sustainable buildings is pushing project management process and areas of knowledge toward sustainability. The paper discusses that sustainability contributes greatly to the method of project management and the ability to guide the construction sector on a more prosperous path toward sustainable development. In construction sector, its development is largely focused on the current relationship between clients, economic benefit, the built environment and management of these projects. Projects and people, who are managing these projects, operate in a wider sense than the project itself. Project managers are in an exclusive role to improve the construction sector in order to support continuous growth, a long-term vision for the construction industry and to be the agents of transformation beyond their traditional role, i.e., "Integrating sustainability goals can assist to accomplish cost, time and quality goals and help in sustainable delivery of projects" (Hills et al., 2008).

2.1 Project Management Critical Success Factors

Organizations have recently recognized that project management is needed to put corporate strategies into actions; correspondingly, it is necessary that projects would be successful (Baccarini & Collins, 2003). Critical factors that contribute to the success of the project are significant considerations and are therefore a set of circumstances, evidence or causes that relate to the results of the project. Thus, in order to increase the chances of a successful project, according to Mobey and Parker (2002), it is important for the organization to consider the critical success factors. Traditionally, time, cost and quality are the key criteria for assessing the performance of the project and are regarded as the "iron triangle" for the project to be successful. In this sense, the paper questions whether the performance of a sustainable project could be accomplished and the sustainability goals of the project can only be met by project management key criteria or not.

2.2 Key Elements and Principles of Sustainable Housing

The development of the project is sustainable where the full breakdown of the job operations is oriented in such a manner as to maximize the elimination of the environmental and health impacts arising from the construction projects. Therefore, it can be argued that it is necessary to establish a problem-free management for housing projects that will enable housing to become sustainable. The conceptual framework for identifying and assessing key components for achieving sustainable housing will show that this framework is made up of three aspects for a sustainable-based development, including socio-economic and environmentally sustainable programs. The expansion of sustainable strategies does not actually transform into appropriate housing without the engagement of housing developers and project managers to successful management plans (Ibem & Aduwo, 2015). This underlines the vital role of effective management and organizational capability in effective introduction of sustainable housing criteria.

3 Attempts to Integrate Between Project Management and Sustainability

In 2009 book "Sustainable Development & Project Management" (Eid, 2009), Mohamed Eid introduced the incorporation of sustainable development into building project management (Eid, 2002). Here are some of the conclusions from his research:

- Project management is an efficient vehicle for bringing a more profound change not only in the operation of the construction industry, but also, most significantly, in the culture of the company.
- Project management processes and knowledge areas are not devoted to a balanced strategy.
- Mapping sustainable development through project management and field experience identifies possibilities for the implementation of sustainability concepts in all project management processes.

In its efforts to link sustainable development to project management, researchers distinguished content-related concepts of sustainable development from process-related ones, while highlighting challenges and potential for their application (Gareis et al., 2010). Table 1 shows definitions of a sustainable-based management for projects. The PMI research focused on determining how eight environmental concepts can be used in order to increase the standard of project performance and reliability of project management procedures. A simultaneous and interconnected economic, ecological and social orientation, as well as a temporal, spatial and value-based orientation, will create possibilities and limitations for sustainable development that have been referred to by other researchers (Messikomer et al., 2011). Further researchers (Gareis, 2013; Goedknegt & Silvius, 2012; Økland, 2015; Silvius et al, 2012; Turner, 2010) continued upon the principles identified by pioneer researchers (Messikomer et al., 2011) and stressed on the fundamentals that literature finds to be essential for achieving sustainable-based projects and their management processes. The exception to this provision was when balancing and aligning human, ecological and profit factors, as well as legal and ethical considerations as essential to Sustainable Project Management overlooked (Økland, 2015).

The previous illustration explained the researchers' efforts to define the expected enhancement in project content and outcome when project management aspects are practiced with sustainability principles consideration. Nevertheless, this practice requires a guide model with defined inputs and expected outputs to integrate between project management traditional plan and project sustainability management plan. This guide model is designed to shift the traditional scope of project management. Accordingly, project managers will be able to attain the controllability of various activities, management of different stakeholders and predictability for a better opportunity when project planning comprises sustainable objectives through project life cycle stages and is able to put the deviated activities into a corrective action plan.

4 The Design of the Procedural Paradigm

The paper tackles the basic tools of project management which are project management processes and knowledge areas. In the Project Management Body of Knowledge (PMBOK) (Project Management Institute, 2017), the process groups are the chronological phases that the project goes through in each of its stages, and the knowledge areas occur throughout any time during the process groups. These knowledge areas are the core technical subject of the project management profession, and they bring the project to life. The following discussion will provide ground for understanding the integration of sustainability in project management and the places of impact in project management processes and knowledge areas. These implications will first relate to the processes, areas of knowledge and accordingly the methodology for the overall project management plan and its subsidiary plans. Hence, the paper introduces another concept; that is different from PMBOK; as the project management processes go chronologically in linear relationship with PLC stages. A new version for mapping the project management process groups, project management knowledge areas through PLC stages is illustrated from sustainability perspective by the author which contradicts with that one in PMBOK. This contradiction will be validated at the results section of this paper.

Table 1 Definition of sustainability in project management (Gareis et al., 2009)

Sustainability in project management
"Sustainable Project Management is minimizing the resources you and your team use to work a project from project initiation through close"
"Sustainable Project Management aims to apply the principle of meeting the needs of the day without compromising the benefits of future generations, to the construction industry by providing ways of buildings that use less virgin material and less energy, cause less pollution and less waste but still provide the benefits that construction projects have brought us throughout history"
"Sustainable Project Management is the promotion of positive and minimizing of negative sustainability impacts (economic; environmental; and social) within the process by which projects are defined, planned, monitored, controlled and delivered such that the agreed benefits are realized and contributing to a sustainable society."
"Sustainability in projects and project management is the creation, implementation and management of project-organized improvement in strategies, procedures, services, properties or organizations, taking into account the six sustainability values of the project, its outcome and its effects"

4.1 Factors of Sustainability in Project Life Cycle

Wu and Low (2010) stressed on a comprehensive approach for the entire life cycle of the project, from planning till operating for the development of a green building. An integrated solution is also essential for the life cycle of green building due to the insufficiency to create a green building with only modern environmentally sustainable materials and technology. The introduction of an integrated design approach includes the collaboration of architects, engineers, property managers, building owners, operators and construction industry stakeholders in the design of the project (Kubba, 2010). It involves a continuous project management work flow throughout the life path of green building with the view of encouraging the coordination between practitioners, and the most important thing is to orchestrate the flow of information and the scheduling of overlaps between multiple consultants / experts by project managers. Indeed, the entire life cycle of a building should be viewed by sustainable architects and contractors rather than by initial capital investment. The constructed world can, therefore, be viewed as part of a natural environment where resources and energy are shared between them.

4.2 Sustainability in Project Management Knowledge Areas

The author introduces new knowledge areas in the way to formulate the guide model for the procedural paradigm of sustainable developments. These knowledge areas are illustrated to emphasize the integrated vision of project management for the whole project life cycle in order to build a green building. The knowledge areas well-known from PMBOK will remain as they are, however, looked differently toward the incorporation of sustainability fundamentals. Additionally, the author introduces new knowledge areas which are concluded from the literature review and seen as crucial for having project management practices on the sustainable path.

4.3 The Structured Questionnaire for Experiencing Project Management Current Practices

A structured questionnaire was designed to experience the current practices of project management in the framework of sustainability principles, taking the housing projects as case study. It will make use of the basic tools of project management, project management processes and knowledge areas, as well as factors of sustainability in PLC retrieved from literature review as tools for the design of the paradigm. The structured questionnaire was presented in interviews, site visits and correspondences directed to project managers, design and construction engineers in addition and online forms directed to end user. The conclusion of the questionnaire identified the places to intervene in the existing project management aspects for the incorporation of sustainability principles in its processes and knowledge areas through PLC stages.

5 The Development of the Procedural Paradigm (Structured Questionnaire)

Sustainable performance (SP) of a construction project during its life cycle (LC) is a main objective to achieve sustainable development (SD). Factors impacting the SP of a construction project can be explored in three major categories: economic sustainability factors (ESF), social sustainability factors (SSF) and environmental sustainability factors (EnSF). These factors are classified in relation to the PLC stages; inception phase, design phase, construction phase and operation phase (Enshassi et al., 2016). Based on this line of reasoning, studies that promote the integration of the sustainability concept into project management were deployed (Enshassi et al., 2016; Martens & Carvalho, 2017; Sudha & Timea, 2017). The author elaborated these studies to address the sustainability dimensions through PLC stages and relate them with the basic tools of project management which are project management processes and knowledge areas. A paradigm is proposed, which is designed to discuss factors of sustainability in PLC and sustainability in project management knowledge areas. The designed paradigm was developed through a structured questionnaire to survey the current status of project management practices. The questionnaire aimed to accomplish the paradigm development that maintains the linkage between the time-dependent nature of project management and the long term of sustainability through (2) sections, Tables 2, 3, 4 and 5:

- Section (1): Factors affecting SP of construction projects throughout PLC stages from project management perspective have been investigated on Likert scale rating. The key variables were coded and clustered into three sets aligned with the sustainability triple bottom line and the basic project management process aspects (Enshassi et al., 2016; Gareis, 2013; Gareis et al., 2009; Kubba, 2010; Martens & Carvalho, 2017; Økland, 2015; Project Management Institute, 2017; Silvius et al, 2012; Wu & Low, 2010). In the operation stage, the end users gave their input.
- Section (2): Check listing the level of application of project management practices in housing projects on a

Table 2 Questionnaire section (1), sustainability factors (Economic Factor "EF," Social Factor "SF" and Environmental Factor "EnF") in project inception stage [developed by author]

Analysis on likert scale		Strongly disagree (%)	Disagree (%)	Neutral (%)	Agree (%)	Strongly agree (%)
Sustainability factor/project stage (inception stage)						
EF	1. Was the capital budget defined to plan and control project total cost (life cycle cost analysis)? Has it been extended to consider not only elementary cost but total cost for building-up, operating project over its life cycle?	9	55.5		35.5	
	2. Was the planned profit extended beyond focusing on stage or sectional profits and considered total profit from operating a construction project across its life cycle?	9	64.50	12	11.50	3
SF	3. Was land selection for project site based on cropland and natural resources protection?			18.75	81.25	
	4. Were negative impacts avoided from project development on any cultural and natural heritage?			18.75	81.25	
	5. Was the project able to provide local employment?				100	
	6. Has the project improved local infrastructure capacity: drainage, sewage, power, roads … etc.?		18.75	37.50	43.75	
	7. Were end users cultural aspects considered (cultural background, financial category and their identity?	18.75	25	25	31.25	
EnF	8. Were potential air pollution from the proposed project and its impact on local climate examined?				100	
	9. Was waste generation at both project construction and operation stages examined?				100	

Likert scale rating (Enshassi et al., 2016; Sudha & Timea, 2017).

The government recent national housing developments were surveyed in four different cities and for four different levels of income. The questionnaire was discussed in interviews, site visits and correspondences to monitor and report the current status of project management practices from sustainability perspective with a total of 115 engineers working as design project managers, construction managers, construction supervision engineers, architects, owner representatives and PMO managers in addition to contractors in order to monitor and report the current status of project management practices from sustainability perspective. A total of 106 engineers has responded to this questionnaire, Table 6. Additionally, in the operation stage, a total of 168 end users, who have moved to live in their units and those who are waiting to move, gave their responses through a questionnaire that was available on google documents, and its link was announced on social media where end users were informed and easily responded, Table 6. On a Likert scale that varied from 1 = strongly disagree; 2 = disagree; to 3 = neutral; 4 = agree; and 5 = strongly agree, interviewees from design and construction professions as well as project managers and end users rated the questionnaire topics, in addition to open questions that revealed some important details. These details have been standardized and classified under the main questionnaire categories.

The results of the questionnaire revealed the gaps in the existing project management processes to perform sustainably. The author then identified the positions of influence in the present system. Therefore, potentials of project management to ensure green housing projects delivery were concluded as will be illustrated hereafter.

The survey has monitored comprehensively the categories of housing classified according to the level of income as per the (*) code in previous table:

- Low Income: Social Housing (*)
- Average Income: Sakan Misr (**)
- Above Average Income: Dar Misr (***)
- High Income 1: R3 District (****)
- High Income 1: R5 District (****)

5.1 Questionnaire Conclusions (Concluding the Current Gaps)

The analysis of the received responses for the questionnaire resulted in supporting the author's designed paradigm and revealed the gaps in the efficiency of the current projects

Table 3 Questionnaire section (1), sustainability factors (Economic Factor "EF," Social Factor "SF" and Environmental Factor "EnF") in project design and construction stages [developed by author]

Analysis on likert scale		Strongly disagree (%)	Disagree (%)	Neutral (%)	Agree (%)	Strongly agree (%)
Sustainability factor/project stage (design stage)						
EF	1. Was the total cost involved in PLC, i.e., site formation, construction, operation, maintenance cost considered?		46.25	11.25	42.50	
	2. Were economic consideration given for durability and availability for material selection?		16.75	34.75	41	7.50
	3. How far clustering and prototype have been followed?			33.50	32	33
	4. How much was the compliance with the site conditions considered (topography and site survey)?		18.50	33	48.50	
	5. How do you evaluate the mistakes and discrepancies in delivered design documents?		15	15	70	
SF	6. How do you evaluate the considerations in designing process for life safety and emergencies?				100	
EnF	7. Is the designer knowledgeable of energy savings and environmental issues?		25	16.50	58.50	
	8. Was modular and standardized components to enhance build ability and to reduce waste generation utilized?		25	16.50	58.50	
Sustainability factor/project stage (construction stage)						
EF	1. How far do you agree that the following expenses / cost is planned and well managed?					
	1a. Materials cost (concrete, steel, timber, brick … etc.)?		37.5		62.50	
	1b. Energy cost?		37.50		62.50	
	1c. Water resources cost?		37.50		62.50	
	1d. Equipment costs for using various equipment, tools?		37.50		62.50	
	2. How far do you agree that the following expenses / cost is planned and well managed?					
	2a. Labor cost?		32.50		67.50	
	2b. Professional fees paid to various professionals?		31.25	15	53.75	
	3. The afforded site security (various types of measures for protecting the site safety)?		37.50	6.25	56.25	
SF	4. Are there standardized measures during for on-site health, site hygiene, safety measures and insurance?		50	43.75	6.25	
	5. Is there a provision for public safety (warning boards and signal systems, safety measures)?		87.50	6.25	6.25	
EnF	6. Is the application of renewable materials and materials reuse (rubble, earth, concrete, steel and timber) applied?	72.50	24.50		3.75	
	7. Are there a policy conditions / ISO conditions applied to manage the following:					
	1a. Air emission and pollution?				100	
	1b. Waste produced from project operation?		80	20		
	8. Are consideration being given to the reduction of earthwork and excavation, formwork, reinforcement, concreting and waste treatment during structural operation?			20	80	

Table 4 Questionnaire section (1), sustainability factors (Economic Factor "EF," Social Factor "SF" and Environmental Factor "EnF") in project operation stage [developed by author]

Analysis on likert scale					
Sustainability factor/project stage (operation stage)					
Validity evaluation	No (%)	Don't know (%)	Yes (%)		
1. Were you informed about the units handing over date?	61.25	13.75	25		
2. Were you periodically notified about any delay in work progress and modified delivery date?	61.25	13.75	25		
3. Were you able to follow up on the work progress?	41	12.50	46		
4. Did you notice a change in master planning, facades design, residential units plan design, the finishing model you chose?	31.25	12.50	56.25		
Satisfaction evaluation	Very dissatisfied	Dissatisfied	Neutral	Satisfied	Very satisfied
5. How satisfied are you with the unit that was allocated to you by lottery system?	15	13.75	27.50	25	18.75
6. How do rate the following (Price / m2)?	12.50	33.75	5	26.25	22.50
7. How do you rate (Flat Area) and (Flat Interior Design)?	25	25	10	30	10
8. How do you rate advantage of reserving at this sector?	16.25	30	10	30	13.75

management performance to be sustainable due to the hereafter analysis conclusions:

(i) The fragmentation in the management of PLC stages and the diversity of stakeholders causes negative implication in time, cost and risks evaluation as a consequence of conflicts in decision-making.
(ii) Sustainability in project management processes should ensure the comprehensive planning for PLC stages, while monitoring and controlling should overview the design, tender and construction stages. Sustainability performance can be assessed through sustainable project planning (SPP) which a new introduced project management knowledge area by the author.
(iii) The lack of the integration between the management of PLC stages supports the project management capacity to bridge this gap and have PLC stages streamlined with defined sustainable goals. Hence, this stresses the importance of the chronological linear relationship between project management processes and PLC stages.
(iv) Project managers should be responsible to plan for the mechanism for SP achievement, facilitate the collaboration among various professionals through PLC stages and have them consistent and coordinated.

Hereafter, is the author diagraming for (3) important constructs retrieved from above conclusions, Figs. 1, 2 and 3 and are interpreted in knowledge areas as discussed in Sect. 3.2 to complete the development of the designed paradigm.

5.2 Guide Model and Implementation Guidelines

The author at this stage is looking to establish the comprehensive perspective to deliver the proper project management for sustainable performance and proposed a model which was based on mapping the project management process groups, project management knowledge areas through PLC stages, i.e., linear form not fragmented stages with isolated project management processes. The project stages (inception, design, tendering, construction and operation) are considered collectively as one work breakdown structure (WBS) linked linearly on the horizontal line with the project management processes in a chronological sequence, while the project management areas of management on the vertical line have been divided into sub-groups with their activities defined through PLC stages. This perspective is introduced as a procedural paradigm to ensure the delivery of sustainable-based construction projects. This necessitated the inevitability to validate this paradigm which is performed with another structured questionnaire, as discussed in Chap. 5.

The developed paradigm through mapping the project management process groups, project management knowledge areas through PLC stages, illustrated in Table 7. This mapping is traditionally defined in PMBOK by mapping of the project management expertise to the five project management process groups of initiating, planning, executing, controlling and closing and the nine project management knowledge areas for each of the PLC stages. The level of conformity with the paradigm is defined by the author by how well the new defined mapping of new project

Table 5 Questionnaire section (2), project management knowledge areas [developed by author]

Analysis on llikert scale	Strongly disagree (%)	Disagree (%)	Neutral (%)	Agree (%)	Strongly agree (%)
Project budget					
1. Is the project budget according to contract?		63		37	
2. How much is the total actual project budget, the expenses incurred, sum m^2 of construction floor area, the rework costs?		75		25	
3. Do you think green buildings cost more than traditional buildings?		75		25	
4. Do you think these are of financial benefit for green buildings: 4a. Lower energy/water usage, less waste disposal?					
4b. Durability of building materials?				100	
Contract conditions					
5. Are there any "green" requirements in the contract?	100				
6. Are you satisfied with incentive/penalty clauses in contract?			25	75	
Project team characteristics					
7. Do you agree about the following:					
7a. Importance of organizing team relationships in RFP?				100	
7b. The contractual terms for project team members?				100	
7c. The team member's experience in similar facilities, green buildings and delivery systems adequate?		75	25		
7d. The team communication of this project?		25	50	25	
Design and construction process					
8. How is the timing of communication within team members? Is it satisfactory?		50	25	25	
9. Do you agree about presence of a GB consultant? What do you think about his contractual position?			25	75	
10. How is the design charettes and commitment level of team members?		25	50	25	
Analysis on likert scale	Very Unsatisfactory	Unsatisfactory	Neutral	Satisfactory	Very satisfactory
Project schedule					
11. How do you evaluate the variance between the planned project duration and the actual project duration?		50	25	25	
12. How far is the reflection of the speed of the following on project schedule:					
12a. Construction (Actual duration/floor area)?			25	75	
12b. Material availability (time delay because of supplying materials)?		70	15	15	
12c. Equipment availability (time delay because of lack of equipment)?		70	15	15	
12d. Labor availability (time delay because of lack of labor)?		75	25		
Project quality					
13. How do you valuate the difference in level between quality expectation of owner and real project quality?	15	10	15	60	
14. How do you evaluate the budget and time required to rework unsatisfied quality requirement works?	3.75	30	2.5	63.75	
Design and construction integration					
15. How are you satisfied with the consultants' / contractors' work integration?	10	28.75	37.50	23.75	
16. How do you evaluate the involvement of a "green" consultant?		87.50	6.25	6.25	
17. How do you evaluate decision-making process in this project?	10	33.75	17.50	38.75	

Table 6 Number of interviews and questionnaires responses [developed by author]

Number of interviews /questionnaires responses	Sherouk city		New Cairo city		6th October			New administrative capital	
	(***)	(*)	(***)	(**)	(*)	(***)	(**)	(****)	(****)
City authority engineer (owner representative)	1	1	1	1	1	1	1	5	5
Architects	3		3	5		3	5	5	3
Supervision consultant				2			2	3	2
Owner's management consultant									1
City authority engineer (supervision)		1		1	1		1		
Contractor	1	3	1	2	4	1	2	4	3
Technical office	1	1	1	1	1	1	1	1	1
Contractor project management consultant	2		2			2			
Sub-contractor	2		2			2			
City authority engineer (external works supervision)	1		1			1			
Contractor (external works)	1		1			1			
End user	25	20	20	14	25	30	19	10	5
Total (%)	23		33		27			18	

Fig. 1 Long-term orientation

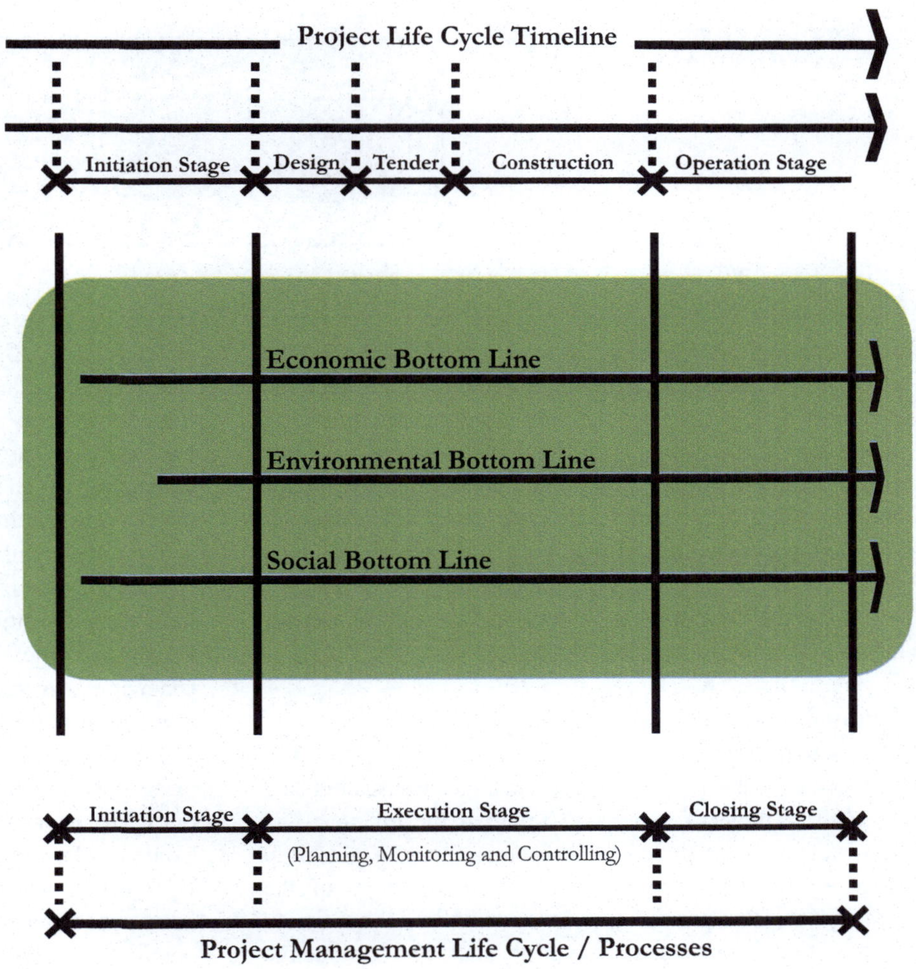

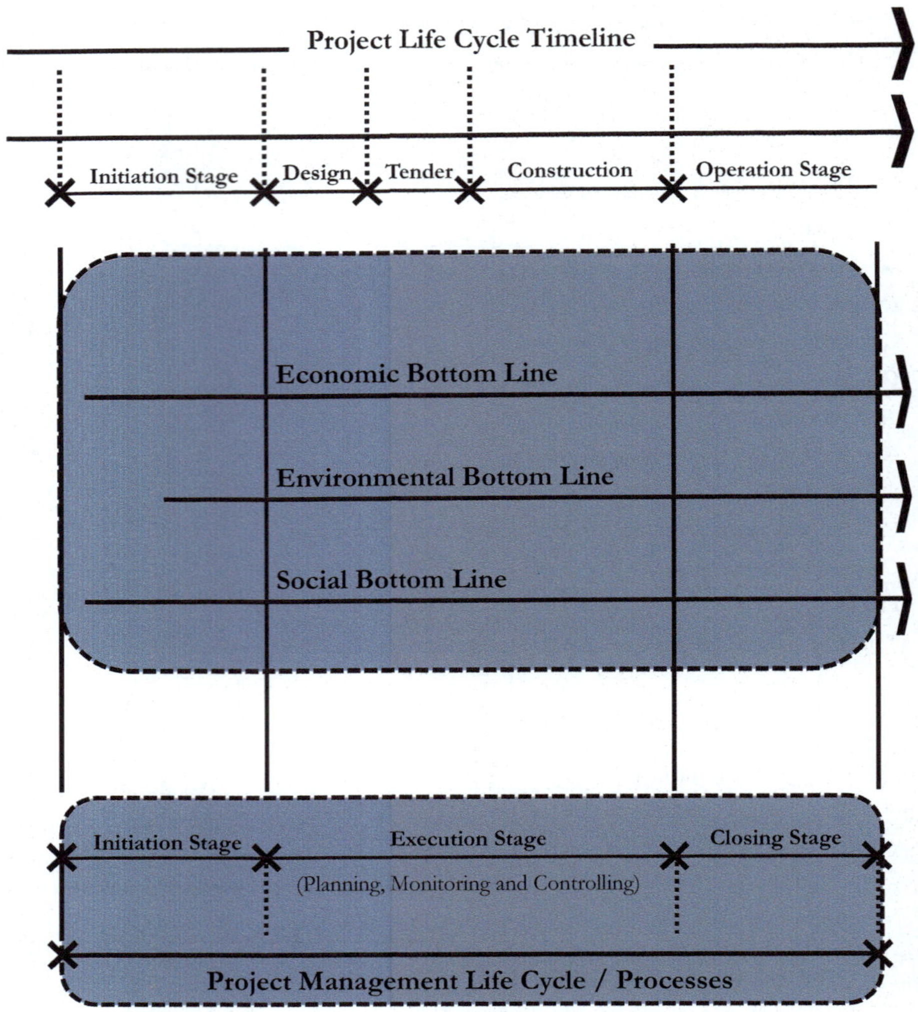

Fig. 2 Integrity concept

management knowledge areas to project management processes through PLC stages is maintained. Therefore, the place of the deficiency can be highlighted. The deficiency in performing any of the mapping components means a deduction of the value of this knowledge area in green project management (GPM) formula coming hereafter. This deduction will be relevant to the range, each of the areas of knowledge has recorded, Table 11.

Here is the illustration for knowledge areas introduced by the author in the procedural paradigm:

1. Sustainable Project Planning
   Sustainable project planning can provide the appropriate tools to make priorities more concrete, transparent and guide a project's actions to improve the efficiency of sustainable execution. By comparing sustainability objectives with actual progress and specifiers of the program, the project can be accomplished by timely monitoring, economic, environmental and productive delivery throughout its life cycle driven to completion.
2. Integration Management
   The value of using evidence and experience to make sustainability-related decisions should be emphasized. Project managers require a combination of soft skills and hard skills to carry out a comprehensive review and to understand the effect of actions in the PLC processes that their ability requires to assess capital, to make trade-offs and to cope with competitive activities. This include (planning, organization, networking, leadership, interaction management, critical thinking capability, data processing, impact evaluation, scheduling, budgeting, change management and risk management). Creating transparent contact networks with the project team and partners is essential to help ensure that knowledge is exchanged and that proper impact analyses are carried out to determine alignment points or dependencies.
3. Contract Management
   Performance-based service contracting depends on the customer company to delineate the result or desired outcome and leaves it to the contractor to determine how the desired outcome is best achieved. Contracting shares

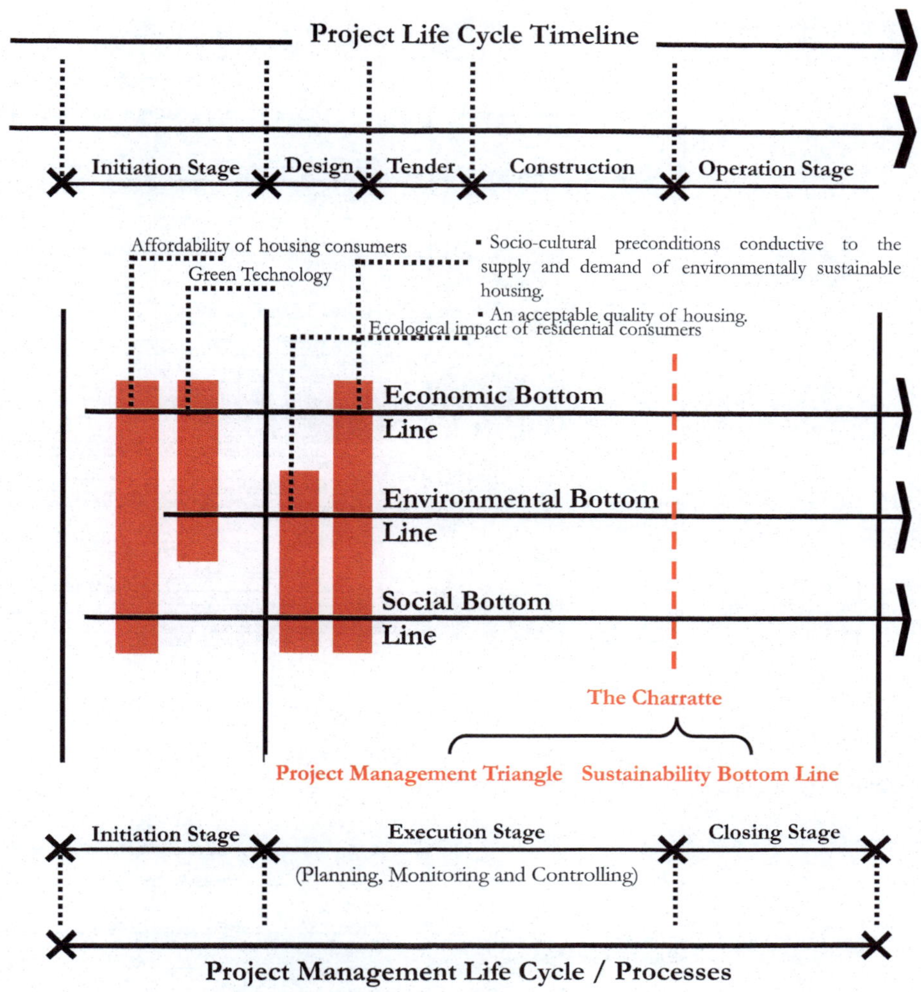

Fig. 3 Sustainability principles activation through project management processes

Table 7 GPM procedural paradigm (the guide model) [developed by author]

Project Management Areas of Knowledge		Project Life Cycle Stages / Project Management Processes				
			Planning, Execution and Monitoring & Controlling			
		Initiation	Design	Tender	Construction	Closing
1. Sustainable Project Planning	1.1 Plan Development	- Develop projects strategic objectives - Develop sustainability parameters				
	1.2 Plan Evolution		- Developing Project Management Plan - Develop the framework for integrating sustainability principles through project plan	- Direct and manage project work - Communication of project management plan and Sustainability Assessment System report		
	1.3 Assess & Control		- Develop sustainability assessment system	- Evaluate and update Project Management Plan - Evaluate the integration of sustainability goals		- Project Learned Lessons
2. Project Integration Management	2.1 Planning	- Develop project charter	- Develop IGBP-PDRI Model - Develop procurement method - Develop quality standard (QMIF) - Identify project constraints			
	2.2 Execution		- Identify stakeholders' interests - Develop stakeholders' requirements			
				- Maintain procurement method		
			- Ensuring quality management integrated framework (QMIF)			
	2.3 Monitoring & Control		- Monitor and control project work - Perform integrated change control			
3. Contract Management		- Development of contractual relationship guidelines (CRG)	- Contract drafting			
			- Issue green requests for proposals - Contract management			- Closing contracts
4. Scope Management	4.1 Planning		- Collect stakeholders' requirements - Define scope - Plan scope management			
	4.2 Execution		- Create overall WBS for project milestones			
			- Create work breakdown for stage milestones			
	4.3 Monitoring & Control		- Scope validation and control - Reporting changes			
5. Communication Management		- Plan communication	- Manage and control communications			

Table 7 (continued)

Project Management Areas of Knowledge			Project Life Cycle Stages / Project Management Processes				
				Planning, Execution and Monitoring & Controlling			
			Initiation	Design	Tender	Construction	Closing
			management				
			– Manage integrated change control				
6.Sustainability Principles Activation (Economic Fundamental)	Projects Constants:	8.1.1 Time	– Initiating baseline time schedule				
				– Planning integrated time schedule			
				– Updating time schedule and include dated corrective actions			
		8.1.2 Cost	– Early stage design decisions – Identify target group capabilities				
			– Develop project budget scale with respect to sustainability				
				– Green building life-cycle analysis			
				– Direct and manage life-cycle analysis recommendations			
	Projects Variables (housing projects): Determining housing affordability		– Identify target group				
			– Performing feasibility study				
				– Direct and manage feasibility study recommendations – Implement sustainability assessment, evaluate and decide corrective actions			
6.Sustainability Principles Activation (Social Fundamental)	Projects Constants:	8.2.1 Stakeholders	– Identify Stakeholders				
			– Plan project team management – Manage stakeholder engagement and develop project team				
				– Control stakeholder's engagement and project team management			
		8.2.2 Culture	– Identify target group requirements				
			– Identify sustainability decisions orientation				
				– Sustain sustainability decisions orientation			
	Projects Variables (housing projects): Implement housing social criterions		– Determining density and urban form – Determining dwelling size				
				– Developing density and urban form – Developing dwelling size			
					– Direct and manage sustainability decisions		
						– Ensuring adaptability and Ensuring social acceptability	
				– Implement sustainability assessment, evaluate and decide corrective actions			
6.Sustainability Principles Activation	Projects Constants:		– Determining water conversation and energy efficiency methods				
				– Determining construction			

Table 7 (continued)

Project Management Areas of Knowledge			Project Life Cycle Stages / Project Management Processes				
				Planning, Execution and Monitoring & Controlling			
			Initiation	Design	Tender	Construction	Closing
(Environmental Fundamental)		8.3.1 External Policy		materials			
					– Determining construction methods		
		8.3.2 Internal Procedures	– Identify sustainability decisions orientation				
				– Develop internal procedures			
				– Implementing process and procedures regulations – Sustain sustainability decisions orientation			
	Projects Variables (housing projects): Maintaining quality of life criterions		– Developing housing indicators reflecting the acceptable quality of life				
				– Maintaining humanization in design development			
					– Direct and manage sustainability decisions		
				– Implement sustainability assessment, evaluate and decide corrective actions			
7. Quality Management				– Plan quality management			
				– Perform Quality Assurance – Control Quality			
8. Procurement Management				– Plan Procurement Management			
				– Conduct Procurements – Control Procurements			
							– Close Procurements
9. Risks Management			– Plan risk management – Plan preventive technique				
				– Identify risks – Perform qualitative & quantitative risk analysis – Plan risk responses	– Identify risks – Perform qualitative & quantitative risk analysis – Plan risk responses	– Identify risks – Perform qualitative & quantitative risk analysis – Plan risk responses	
				– Apply integrated remedial technique			

in savings, a form of performance-based contracting, can be helpful when cash flow is an issue. The contractor finances a plan in exchange for a portion of the savings actually made by the authority (Improving Services Acquisitions, 2001).

4. Scope Management
Egyptian green building rating was the green building evaluation handbook which is relevant to the Egyptian constraints in terms of sustainability. The integration between Egyptian green building rating and Egyptian building law will evolve the parameters of the Egyptian IGBP-PDRI evaluation model. Taking this into implementation will result learned lessons for green building policy which will be subject to assessment and evaluation for further enhancement.

5. Communication Management
Communication is the role that the project manager must apply in the sustainable integration of cost, range and time to produce a quality product and can therefore be considered as a core knowledge field of project

management in the project management sustainable approach. Therefore, successful execution of sustainable projects depends heavily on the construction project manager's ability as communicator to lead the team and manage a construction project successfully.

6. Sustainability Principles Activation
The architecture of housing buildings is a task for architects to be responsive to the needs of consumers that are unique to the context and, at the same time, to provide them with comfortable conditions. In other words, representing the character of the location and many other elements, such as the hierarchy of spaces, the various degrees of transparency and enclosures, the courtyard, the use of soft and hard landscaping, the particular uses of spaces, patterns of activity and the schematic properties of spaces, could allow the architect to identify social space and stress national identity as well as cultural identity. This should be incorporated into the planning framework of sustainable affordable housing and viewed as a conceptual priority at an early stage of design and backed up during the design, tendering and construction stages.
7. Quality Management
The advantages of the integrated design process are necessary for a cost-effective green construction project to be implemented successfully. The successful implementation of the model requires various measures on all specified levels (construction products, processes/project and building). Green procurement today has high priority, and the proposed model can be an important part of green procurement schemes for both public and private customers. National sustainable development policies must allow the integrated quality to be fully implemented in construction.
8. Procurement Management
It is inevitable that the involvement of the decision-making represented in the developer effective role in procurement processes is necessary to avoid any deviation from the sustainability path planned for the project. Nevertheless, the engagement of the stakeholders themselves (basically the developer) toward sustainability is the tool to ensure that. The way to resolve both constraints is the full engagement between stakeholders' management and project integration management. The tool to execute that is the contract, which should be the document reflecting the sustainability strategic goals into detailed standards and procedures to be ensured its execution by all project stakeholders including the developer.
9. Risk Management
The only way to deal with the risk is to stop it. Risk can be eliminated at the planning stage prior to the launch of the project by making the required improvements arising from the evaluation of the necessary risks. Two types of risk control methods should be considered: forecasting approaches that can be used to minimize the dangers predicted to occur during the completion of the project before the launch of the project; and remedial techniques that are used during the development process after a vulnerability has emerged (Iqbal et al., 2015). Preventive risk control methods do not mitigate risks; thus, they will arise from time to time during the execution of the project, and remedial management approaches must be used to assess their effect and, if possible, to minimize them.

6 The Validation of the Procedural Paradigm (Validation Questionnaire)

Ensuring the streamline of sustainability vision in project management processes and practices through PLC stages can be achieved when maintaining the long-term vision which requires the project managers engagement to sustainability. This revert us back to the paper focal point which is to integrate the project management aspects and the sustainability principles in a procedural paradigm. To introduce the new shift needed to apply and integrate sustainability and project management, the correlation between project management processes and project management adjusted knowledge areas through PLC stages was introduced in another questionnaire (validation questionnaire) performed with earlier interviewees. In the validation questionnaire, interviewee will first rate the linear relationship between the stages of PLC and PM processes with the new alignment of PM knowledge areas as illustrated in Sects. 6.1 and 6.2.

6.1 New Alignment of PM Processes and Adjusted PM Knowledge Areas Through PLC Stages (Correlation Coefficient Method)

In the statistics science, correlation coefficients are used to determine how strong a relationship is between two variables, its value varies from + 1 (means strong relationship) to -1 (means negative relationship), with a neutral relation at Zero (means no relation). The correlation coefficient of each knowledge area with PLC stages and project management processes was concluded from the ranges of respondents' responses. The correlation matrix between knowledge areas through PLC stages and project management processes was determined by correlation statistic method performed by Microsoft Excel developed by author. This has refined the author designed paradigm and turned it from a draft version to a developed (final) version that will be considered as the guide model. Conclusions confirmed the interviewee's agreement about the new alignment of PM processes and adjusted PM knowledge areas through PLC stages.

6.2 Critical Success Factors for Green Project Management (Ranking Method)

The need to identify the dominant factors and the less dominant factors in the correlation matrix was also a main pillar for getting the paradigm from draft version into the final version. The conformity of the project performance with the guide model reflects that the project management practices are on the right track toward success. Therefore, guide the way for enhancing the Sustainable Project Management performance. The procedural paradigm was discussed in this part of the validation questionnaire, and (3) questions were asked after the interviewee finalizes his review for the guide model about the following:

1. Defining GPM critical success factors (CSFs), Table 8.
   A crucial step to reach this was a question about green project management (GPM) success criteria, where respondents replied with their level of agreement by ticking on ten-point Likert scale, 0 (not important) to 10 (very important) for the areas of knowledge the respondent thinks they directly relate to the success of sustainable-based project management (CSFs), Table 8. The areas of knowledge which had below 5 grades were then excluded.
2. Identification of CSFs for GPM, Table 9.
   Respondents are then requested to re-assess their replies by identifying the critical factors influencing the success of GPM where respondents replied with their level of agreement by ticking on the ten-point Likert scale, according to the respondent believe about the areas of knowledge contribution toward the success of sustainable-based project management, Table 9. The result of this question was then computed to reflect the scores each area of knowledge has achieved.
3. Ranking the Critical Factors Influencing the Success of GPM, Table 10.
   The third question was about ranking the critical factors influencing the success of GPM in an ascending order starts from (1) to (9), Table 10, where ranks equal to 5 grade and above were considered as one factor, Table 11. The respondents assumed the weight of each factor toward the success of GPM according to its importance based on their experience and their vision for sustainable-based performance of sustainable development projects.

7 Results

The green project management performance success index is a systematic way of prediction developed by the range formula analysis which is based upon a mathematical series performed by Excel. The highest ranked knowledge areas (dominant critical success factors), i.e., contribution toward green project management success are interpreted in range formula analysis which enable to monitor and predict the performance the project management practices on the

Table 8 GPM success criteria [developed by author]

Level of agreement	Sustainable project plan (SPP)	Integration management	Contract management	Scope management	Communication management	Sustainability principles activation	Product management	Risk management
0	0	0	0	0	0	0	0	0
1	0	0	0	0	0	0	0	0
2	0	0	0	0	0	0	0	0
3	0	5	0	0	0	0	0	0
4	5	5	0	7	0	0	5	17
5	5	25	10	25	7	7	25	25
6	15	10	10	5	7	7	17	5
7	30	28	5	35	28	11	35	35
8	15	20	20	20	22	20	10	10
9	15	3	17	5	18	22	5	5
10	12	1	35	0	15	30	0	0

Table 9 Identifying the critical factors influencing the success of GPM [developed by author]

Level of agreement	Sustainable project plan (SPP)	Integration management	Contract management	Scope management	Communication management	Sustainability principles activation	Quality management	Procurement management	Risk management
0	0	0	0	0	0	0	0	0	0
1	0	0	0	0	0	0	0	0	0
2	0	0	0	0	0	0	0	0	0
3	0	0	0	0	0	0	0	0	0
4	0	7	0	7	0	0	5	17	17
5	10	25	10	25	7	7	25	25	25
6	15	5	10	5	7	7	17	5	5
7	25	35	5	35	28	11	35	35	35
8	20	20	20	20	22	20	10	10	10
9	15	5	17	5	18	22	5	5	5
10	12	0	35	0	15	30	0	0	0
Total Scores	97	97	97	97	97	97	97	97	97

Ten-point Likert scale 0 (not important) to 10 (very important)

Table 10 Ranking the critical factors influencing the success of GPM [developed by author]

	Scores	Rank	Average rank	Weight factor
Sustainable project plan (SPP)	60	3	3.5	0.23
Integration management	55	5	6	0.04
Contract management	60	3	3.5	0.20
Scope management	55	5	6	0.04
Communication management	65	1	1.5	0.21
Sustainability principles activation	65	1	1.5	0.18
Quality management	55	5	6	0.04
Procurement management	35	9	9	0.03
Risk management	50	8	8	0.03

Table 11 Range formula output [developed by author]

Sustainable Project Planning (SPP)		Contract Management (CM)		Communication Management (CMM)		Sustainability Principles Activation (SPA)		Other Knowledge Areas	
Mean	18.03	Mean	16.81	Mean	17.53	Mean	13.65	Mean	10.53
Standard Error	0.26	Standard Error	0.29	Standard Error	0.25	Standard Error	0.33	Standard Error	0.32
Median	20.00	Median	15.00	Median	18.00	Median	13.00	Median	10.00
Mode	20.00	Mode	15.00	Mode	15.00	Mode	10.00	Mode	8.00
Standard Deviation	2.59	Standard Deviation	2.83	Standard Deviation	2.45	Standard Deviation	3.25	Standard Deviation	3.12
Sample Variance	6.72	Sample Variance	8.01	Sample Variance	6.00	Sample Variance	10.54	Sample Variance	9.75
Kurtosis	-1.67	Kurtosis	-0.50	Kurtosis	-1.99	Kurtosis	-1.65	Kurtosis	-0.19
Skewness	-0.23	Skewness	-0.33	Skewness	-0.03	Skewness	0.05	Skewness	1.09
Range	9.00	Range	10.00	Range	5.00	Range	8.00	Range	10.00
Minimum	15.00	Minimum	10.00	Minimum	15.00	Minimum	10.00	Minimum	8.00
Maximum	24.00	Maximum	20.00	Maximum	20.00	Maximum	18.00	Maximum	18.00
Sum	1749.00	Sum	1631.00	Sum	1700.00	Sum	1324.00	Sum	1021.00
Count	97.00	Count	97.00	Count	97.00	Count	97.00	Count	97.00
Confidence Level (95.0%)	0.52	Confidence Level (95.0%)	0.57	Confidence Level (95.0%)	0.49	Confidence Level (95.0%)	0.65	Confidence Level (95.0%)	0.63

sustainable path through the performance index equation concluded from this analysis.

The application of the green project management performance index formula is explained herewith:

1. Highest ranked knowledge areas (dominant critical success factors) and other knowledge areas are concluded with minimum and maximum values and range to be as inputs in the formula shown its details in Table 11.
2. The values of the knowledge areas are determined by the project manager according to the concluded minimum and maximum contribution to success as well as the success range for each of them based on the analysis of interviews responses. The responses about the contribution percentage of GPM critical success factors toward success of GPM were grouped into the range formula which provided minimum, maximum and the range of contribution, where it can be decided from project manager monitor process the level of conformity with the procedural paradigm of green project management, Table 11.
3. The success range interprets the level of conformity to the specified activities in the paradigm and the successful conformity of green project management practices.
4. The formula represents the performance index for the proposed paradigm of green project management for sustainable development projects.

8 Conclusion

The vital difference between the guide model is introduced in this paper, and the Project Management Body of Knowledge (PMBOK) were explored. The author utilized the existing knowledge areas in PMBOK with further adjustments, added new knowledge areas with the modification of project management processes flow. This has resulted in having sustainability indicators that can be monitored and assessed through the project management processes of sustainable development projects. The paradigm is expected to be the tool to control and involve the appropriate corrective action at the proper time to guarantee the sustainable delivery of the end product. It will enable the executive management and the project managers to have specified measures of performance in sustainable developments. This guide model was validated and is ready for the future studies for the verification. It is recommended by the researcher to be verify the guide model in order to take the validated theory into a verified application. It recommended as well to study the mechanism to resolve obstacles which is important for the paradigm verification.

Green Project Management Performance Index (GPMPI)

Green project management performance index will be calculated as follows:

GPMPI = 0.23 SPP + 0.20 CM + 0.21 CMM + 0.18 SPA + 0.18 Other

Maximum Value of GPMPI = 0.23 (24) + 0.20 (20) + 0.21 (20) + 0.18 (18) + 0.18 (18) = 5.52 + 4 + 4.2 + 3.24 + 3.24 = 20.20

Minimum Value of GPMPI = 0.23 (15) + 0.20 (10) + 0.21 (15) + 0.18 (10) + 0.18 (8) = 3.45 + 2 + 3.15 + 1.8 + 1.44 = 11.84

Hence, Index Value lies between 11.84 $\leq$ GPMPI $\leq$ 20.20

Where:

- **SPP (sustainable project planning):** its successful performance range is (9) with minimum value (15) and maximum value (24)
- **CM (Contract management):** its successful performance range is (10) with minimum value (10) and maximum value (20)
- **CMM (communication management):** its successful performance range is (5) with minimum value (15) and maximum value (20)
- **SPA (sustainability principles activation):** its successful performance range is (8) with minimum value (10) and maximum value (18)
- **Other (Other knowledge areas management):** its successful performance range is (10) with minimum value (8) and maximum value (18)

References

Baccarini, D., & Collins, A. (2003, December 2–5). Critical success factors for projects. In A. Brown (Ed.), Fremantle, Western Australia. Surfing the waves: Management challenges; Management solutions, Proceedings of the 17th ANZAM Conference. http://hdl.handle.net/20.500.11937/14127.

Eid, M. (2002). *A sustainable approach to the project management Odyssey*. In PMI Research Conference (Frontiers of Project Management Research and Application), Seattle. Philadelphia. PMI Institute, Philadelphia, Pennsylvania, USA. University of Edinburgh, School of Arts, Culture and Environment, Department of Architecture, Scotland, UK.

Eid, M. E. M. (2009). *Sustainable development & project management: rethinking relationships in the construction industry: integrating sustainable development (SD) into project management (PM) processes*. LAP Lambert Academic Pub.

Enshassi, A., Kochendoerfer, B., & Al Ghoul, H. (2016). Factors affecting sustainable performance of construction projects during project life cycle phases. *Factors Affecting Sustainable Performance of Construction Projects during Project Life Cycle Phases, 7* (1), 50–68.

Gareis, R. et al. (2010). *Relating sustainable development and project management: A conceptual model.* In PMI Research Conference (Defining the Future of Project Management), Washington, DC, Newtown Square, PA: Project Management Institute.

Gareis, R. (2013). Re-thinking project initiation and project management by considering principles of sustainable development. In *Sustainability integration for effective project management* (pp. 129–143). IGI Global.

Gareis, R., Heumann, M., & Martinuzzi, A. (2009). *Relating sustainable development and project management* (p. 52). IRNOP IX.

Goedknegt, D., & Silvius, A. J. G. (2012, October). *The implementation of sustainability principles in project management*. Proceedings of the 26th IPMA World Congress, pp. 875–882.

Hills, M., Fox, P., Skitmore, M., Hon, C., & Fong, P. (2008). *The role of project managers in construction industry development*. Proceedings of the AACE International's 52nd Annual Meeting and ICEC's 6th World Congress on Cost Engineering, Project Management and Quantity Surveying 2008, pp. 1–10. AACE.

Ibem, E., & Aduwo, E. (2015). *A framework for understanding sustainable housing for policy development and practical actions*. Architects Colloquium, College of Science and Technology, Covenant University.

Improving Services Acquisitions, GAO-02–179T. (2001). *Seven steps to performance-based services acquisitions*. U.S. General Services Administration, October 1994

Iqbal, S., Choudhry, R. M., Holschemacher, K., Ali, A., & Tamošaitienė, J. (2015). Risk management in construction projects. *Technological and Economic Development of Economy, 21*(1), 65–78. https://doi.org/10.3846/20294913.2014.994582

Kubba, S. (2010). Green construction project management and cost oversight. Architectural press 2010, an imprint of Elsevier, pp. 438–484, ISBN 9781856176767, https://doi.org/10.1016/B978-1-85617-676-7.00012-9

Martens, M. L., & Carvalho, M. M. (2017). Key factors of sustainability in project management context: A survey exploring the project managers' perspective. *International Journal of Project Management, 35*(6), 1084–1102.

Messikomer, C., Huemann, M., Dominguez, O., Gareis, R., Navarro-Flores, O., & Morgese, P. (2011). Sustainability and project management: The Future is Now. *Project Management Institute, 10*, 1–137.

Mobey, A., & Parker, D. (2002). Risk Evaluation and Its Importance to Project Implementation. *Work Study, Modern Economy, 51*(5/4), 202–208. https://doi.org/10.1108/00438020210430760

Økland, A. (2015). Gap analysis for incorporating sustainability in project management. *Procedia Computer Science, 64*, 103–109.

Project Management Institute. (2017). *A guide to the project management body of knowledge* (PMBOK Guide), Vol. 6

Silvius, A. J. G., et al. (2012). *Sustainability in project management*. Gower Publishing.

Sudha, R., & Timea, K. (2017). *Sustainability in Project Management*. UMEA University, School of Business and Economics, Europe.

Turner, R. (2010). Responsibilities for sustainable development in project and program management. In H. Knoepfel (Ed.), *Survival and sustainability as challenges for projects, international project management association, Proceedings of IPMA Expert Seminar.*

Wu, P., & Low, S. P. (2010). Project management and green buildings: Lessons from the rating systems. *Journal of Professional Issues in Engineering Education and Practice, 136*(2), 64–70. https://doi.org/10.1061/(ASCE)EI.1943-5541.0000006

Can Developing Countries Use Global Systems Priorities for Neighborhoods Sustainability Certification? Case Study: Asunción, Paraguay

P. Argüello Meza, J. Fariña Tojo, and E. Román López

Abstract

The BREEAM-CM and LEED-ND Neighborhood Sustainability Global Certification Systems have been developed in first world countries, where sustainable construction has been debated and implemented for years. Currently, they are very present in Latin America, even in countries like Paraguay, where there is still no solid framework on which to anchor the sustainable development of cities. Consequently, this paper is the result of the analysis of these global systems priorities in contrast with the most urgent sustainability needs to attend in Asunción, as part of a doctoral thesis research that the author is developing at the Universidad Politécnica de Madrid. For this purpose, several qualified professionals were consulted in Asunción through interview. As a result, significant discrepancies between these priorities have been identified, as well as some of these global systems weaknesses when contextualized in a developing country. This has led to some recommendations for Paraguayan public institutions in order to develop sustainable urban planning.

Keywords

BREEAM communities • LEED for neighborhood development • Universal methodologies • Ecological • Green building

Nomenclature

GS-CSN	Global Systems for the Certification of the Sustainability of Neighbourhoods
BREEAM-CM	Building Research Establishment Environmental Assessment Methodology Communities
LEED-ND	Leadership in Energy & Environmental Design for Neighborhood Development
AMAS Action Plan	Action Plan for Sustainability to Asunción Metropolitan Area

1 Introduction

According to the United States Environmental Protection Agency—USEPA, the current concern for sustainable construction has its origins in the energy crisis of the 1970s, where the oil price increase stimulated the development of renewable energy sources in order to reduce dependence on fossil fuels, as well as promoted the research on energy efficiency. All together, including the environmental movement, produced the first studies on contemporary sustainable construction (USEPA, 2016). From then on, energy efficiency has always been one of the most important criteria to consider in a building that aspires to be sustainable.

One of the consequences of this phenomenon was the creation of methods to certify the sustainability of homes, offices, hospitals, neighborhoods, etc. The leadership of the United Kingdom and the USA was confirmed by the creation of the systems: **BREEAM** and **LEED**. Both are today the most used systems in the world (Kyrkoua & Karthausa, 2011) and, therefore, in Latin America.

In 1972, the Building Research Establishment—BRE was created: a research laboratory funded by the United Kingdom government whose main objective was to investigate the behavior of materials and construction methods, as well as fire prevention in the country. Although it was privatized in 1997, in order to keep the authority and independence that it had when it was financed with public funds, it created the Foundation for the Built Environment—FBE (later called The BRE Trust), with members from all construction

P. Argüello Meza (✉) · J. Fariña Tojo · E. Román López
Department of Urban Planning and Lan Management, Universidad Politécnica de Madrid, Madrid, Spain

C. Alalouch et al. (eds.), *Towards Implementation of Sustainability Concepts in Developing Countries*, Advances in Science, Technology & Innovation, https://doi.org/10.1007/978-3-030-74349-9_21

industry sectors. Certifications to buildings are initiated in 1999 through the development of the BREEAM certification system, and later, specialized certifications will be issued for residences, shops, schools, and neighborhoods (BREEAM communities or BREEAM-CM focuses on the sustainability of neighborhoods). In addition, they will develop a global application system by exporting certification to other countries. The BRE Global is created in 2006 to manage the certifications issued outside the UK, as well as the BRE Group born to mediate between BRE, BRE Global, and The BRE Trust (BRE, 2015).

In 1993, the Green Building Council—GBC was created in the USA: a non-profit association whose mission was to promote sustainability in construction, embracing a coalition with the entire construction industry and intending to create a system to classify ecological buildings according to the American context. In 1995, the USGBC partnered with the Natural Resources Defense Council—NRDC to create the LEED sustainable building classification system, and in 1999, the first building certifications were issued. Subsequently, partnered with the congress for the New Urbanism and launched LEED for Neighborhood Development or LEED-ND (USGBC, 2009) and, in that same year, the World Green Building Council was founded with the mission of expanding and issuing certifications globally (WGBC, 2018).

Due to the global extension of both systems (BREEAM-CM and LEED-ND) from now on, they will be named in this investigation with the acronym **GS-NSC** which means: *Global Systems for Neighborhood Sustainability Certification.*

Even though both systems have global aspirations, several authors maintain that local characteristics are crucial for neighborhoods certification. Originally, studies were focused on the general framework (Haapio, 2012; Sharifi & Murayama, 2013; Rueda, 2012), and they evolved into feasibility studies for different contexts, offering very interesting conclusions for this study. According to Garde, both strengths and the limitations of the LEED-ND system must be considered in order to only apply what may be useful for each location (Garde, 2009). Edward Ng stated that these systems were formerly designed just for single buildings, and they still embrace an individualistic perspective which is not aligned with neighborhoods sustainability (Ng, 2010). Kyrkoua and Karthausa pointed out that sustainability goes beyond energy efficiency, since it seeks to endure meeting people's needs, and universal methodologies have the great challenge of creating criteria that prioritize the particularities of each site. They claim that no organization can create a monopoly on sustainability and that the market must allow licensing and cross-certification, based on a set of requirements that can evolve and adapt to each city (Kyrkoua & Karthausa, 2011).

Regarding the importance that these universal tools attach to each requirement[1] (whether it is compulsory or voluntary, and if grants 1 or 9 points), several authors reached similar conclusions. Humbert et al. studied the weight or importance that LEED requirements have in office buildings in California, where this system was born, and concluded that a higher rating is not always synonym of low environment impact (Humbert et al., 2007).

Egyptian researcher Wael Aboneama compared the global LEED system with a local tool called the Green Pyramid, where he found that even though his country suffers from water scarcity, in LEED water use efficiency represents only 9% of the global score, while in Green Pyramid this represents up to 27%, three times more. He also criticized LEED's strategy of capturing rainwater for on-site reuse, which is not very efficient in a country with very little rains. Therefore, he considered that the importance given to each requirement should be determined according to the most urgent local needs, since each part of the world suffers from the lack of one or more resources. For instance, we can see the contrast between the Middle East countries and North America, where the former suffer from a lack of water resources and an extremely warm climate that makes them dependent on refrigeration devices, while the latter main concern is the lack of energy and the cold weather, needing heating devices. However, he also emphasized that the whole world suffers from the consequences of global warming as well as from the ozone depletion, and therefore, he considered that every region of the world should also take into account general requirements that address these global challenges (Aboneama, 2018).

In South Africa, Gibberd compared certification tools with local needs, highlighting the significant differences between countries, exposing that excessive consumption in developed countries should be reduced, while basic human needs in developing countries still have to be met. And for this reason, he considered that sustainability promotion will depend on its context. However, it also highlights that environmental problems require an internationally integrated effort (Gibberd, 2002).

Valentin and Spangenberg also promoted the idea that each community develops its own sustainability requirements but within a common or global structure, in their case Agenda 21. In other words: common structure, different requirements, and priorities. This will, thereby, allow comparison between cities without ignoring their specific needs and situations. He argued that local requirements development provides an opportunity to visualize the singularity of each community, which creates local/regional identity. (Valentin & Spangenberg, 2000).

[1]Requirement: It is the title or nomenclature of what is required (e.g., TM 02 - Safe and appealing streets). It can be easily identified because it is followed by a code (TM 02). The sum of all the requirements is the check list.

Cole recognized that these tools can help to define what constitutes a green building. However, regarding its use in developing countries, he maintained that social and economic concerns are qualitatively different from those of developed countries, since all global certification systems contain their authors' values and priorities, as well as their circumstances, recognizing as problematic the development of a standardized and universally applicable method. Furthermore, he considered that the development of a certification tool is only a way to achieve a goal end but not a goal in itself. Following this and to avoid "continuous improvement" by technical characteristics comparison instead of dealing with equally important questions just as the context in which they operate (Cole, 2005), this research focuses on researching the effectiveness of Global Systems priorities BREEAM-CM and LEED-ND in a very different context from what was thought when they were created, taking as a case of study the capital of Paraguay: Asunción.

In Paraguay, in 2013, the Green Building Council Paraguay—GBC-PY was created, and since then, it has promoted the debate on sustainability in construction and has certified five buildings and registered another six that are currently in certification process.[2] Furthermore, it has collaborated with the *National Institute of Technology and Standardization—INTN*[3] in the publication of five national regulations related to sustainable construction, which contributes to the LEED system implementation in the country. Additionally, Paraguay has the largest hydroelectric power plant in the world, and it bases most of its consumption on this clean energy. Despite this, the country still does not have a solid framework on which to anchor the cities sustainable development, and Asunción is the only one with a *Action Plan for Sustainability to Asunción Metropolitan Area,*[4] *to be named in this investigation AMA Action Plan.*

Taking into account all of the above, this research objective is to analyze and confirm whether the main sustainability requirements for neighborhoods treated in the GS-NSC, such as the production of clean energy, matches the most urgent needs of Asunción and, at the same time, verify the viability of their application in the capital.

2 Methodology

To achieve this research objectives, semi-structured *interviews* were conducted with local experts in Asunción, the pre-established questions for the interview were collected in a questionnaire, thus ensuring that all questions were resolved during the interviews and avoiding excessive duration. In other words: the questionnaire is rather a guide that served to set the topic of the interviews, not being a sampling method. It served to obtain deep and substantial information on the priorities that must be assumed in the capital to achieve its sustainability and to know the justifications for the infeasibility of applying certain requirements of BREEAM-CM and LEED-ND in this city. It is a method frequently used in Anthropology, where the researcher is immersed in the context of interaction that he wants to investigate; in this particular case, the author is from Paraguay.

Therefore, interviews were conducted in this way: the researcher read the questions of the questionnaire while recording the participants' reflections when answering them. When a greater knowledge on certain matter was observed, the participants were asked to delve further into their answers by open-ended questions.

For the target group selection, their knowledge of the BREEAM-CM and LEED-ND systems has been taken into account, thus including: members of the GBC-Paraguay (entity that promotes the LEED methodology in the country), professionals who achieve or at least try to obtain certification with these systems, technicians who prepared the five Paraguayan regulations for sustainable construction whose references are LEED and BREEAM and also students who investigated these methodologies in their final degree projects. Finally, their willingness to be interviewed was also essential.

A total of 34 interviews were obtained and 33 questionnaires were considered valid and will be fully published on the website of the Universidad Politécnica de Madrid, together with the researcher's Ph.D. thesis.

Finally, the limitations present in this study are: on the one hand, the different duration of each interview, where the average was 20 min and some participants could only dedicate 10 min, while others extended up to 40 min of interview, which limited or facilitated in each case the development of the arguments with which they supported the answers to the questionnaires. On the other hand, the high material cost and time invested should also be mentioned, since it was necessary to travel to Asunción, make individual appointments, generally in the participants' offices, since the Paraguayan idiosyncrasy responds better to a personal conversation.

2.1 Questions Set up

The three questions prepared for the questionnaire that served as a guide for the interviews are described in Table 1. *Questionnaires set up.*

[2]Source: https://www.usgbc.org/resources/country-market-brief accessed December 11, 2019.
[3]Instituto Nacional de Tecnología y Normalización – INTN.
[4]Plan de Acción Área Metropolitana de Asunción Sostenible.

First question

This question aims to determine which are the urban sustainable requirements that must be a priority in Asunción; since in the research work of the author's doctoral thesis, the following has been detected.

That there are discrepancies on what is considered a priority in BREEAM-CM, where *Energy Strategy* is emphasized, compared to LEED-ND, where the *Smart Location* stands out (location in areas with urban services such as: public transport, potable water, wastewater infrastructure, in addition to locating them in areas with a certain number of crossings roads). Besides this, in the *AMAS Action Plan,* the priorities described are: mobility and transport, land use, as well as sanitation and drainage.

Therefore, respondents were asked to indicate with an X the requirements that they considered a priority for Asunción. Thanks to this it will be possible to verify why and whether the local professionals are in greater or lesser agreement with BREEAM-CM, and/or with LEED-ND, and/or with the AMAS Action Plan.

Second question

In *SLL Prerequisite: Smart Location* of the LEED-ND system, it was detected as a weak point the fact that it is requested to locate every project on a plot with infrastructure, at least planned, for sewage collection and water supply. Therefore, it was decided to inquire about the reliability of the public organizations responsible for these infrastructures.

Third question

The specifications related to *Public Transport* were reviewed in this question; since on several occasions[5], the LEED-ND system proposes as a valuable option to locate the buildings next to a public transport service, if not existing at least *planned*. As in the following requisites:

Obligatory requirements: SLL Prerequisite Smart Location; NPD Prerequisite Walkable Streets; and NPD Prerequisite Compact Development.

Voluntary request: LT Credit Access to Quality Transit; NPD Credit Walkable Streets; NPD Credit Mixed-Use Neighborhoods; NPD Credit Transportation Demand Management; and NPD Credit Transit Facilities.

Given the various occasions where public transport projects were planned, budgeted, issued as public tenders, such as the *Metrobus*[6] and *Renfe Train* cases, to end up without being even built or without meeting the stipulated deadlines, it was decided to inquire with the local professionals on the effectiveness of public transport *planning*.

3 Case of Study: Asunción, Paraguay

Asunción is the capital and the most populous city in Paraguay. In 2018, it has 523,184 inhabitants (DGEEC, 2015), in an area of 117 km^2, of which 10% are wetlands, since it has approximately 20 km of coastline on the Paraguay River (DGEEC, 2002). The urban dynamic links it with the *Gran Asunción* or the Asunción Metropolitan Area—AMA, composed by the capital and 11 municipalities of the Central Department, forming part of a continuous urban sprawl, which represents 52% of the urban population of Paraguay, explaining why the sustainable development of the region represents a significant impact to improve the quality of life of the country's population.

Due to the horizontal population growth model, the urban footprint expands increasingly in the territory leading to a low density, in Asunción it is only 46 inhabitants/ha, and in the *Gran Asunción* 43 inhabitants/ha. This is due to the large part of urban land which has been developed, and no densification policies have been implemented to promote height models and take advantage of idle floors and abandoned buildings (BID, 2014).

All this resulted in a quantity and quality reduction of the city infrastructures, as well as its equipment and urban furniture. At the same time, land use changes not only pushed the colonization of rural areas, but also created settlements in areas of high ecological value, especially: wetlands and flooding areas, thus improving the vulnerability of the metropolitan region to natural disasters (ASU, 2016).

As an aggravating factor in this situation, strong population growth has not been accompanied by an adequate provision of urban infrastructure and services, especially in the provision of housing, potable water, sanitation, storm drainage, urban transport, and urban solid waste management sectors. Highlighting in particular: the low sewerage coverage (21%), the practically non-existent treatment of collected wastewater (4%), the inadequate management of urban solid waste (only 68% of the metropolitan area), and the low coverage of storm drainage systems (19% in Asunción) (BID, 2014).

[5]Obligatory request: SLL Prerequisite: Smart Location, and NPD Prerequisite: Compact Development. Voluntary request: SLL Credit: Preferred Locations, LT Credit: Access to Quality Transit, NPD Credit: Compact Development, NPD Credit: Mixed-Use Neighborhoods, NPD Credit: Transit Facilities, and NPD Credit: Transportation Demand Management.

[6]Urban bus that runs on an exclusive route, or bus-only lane.

Table 1 Questionnaires set up

(1) Which of the main Urban Sustainability problems in Asuncion do you consider the most urgent to address? Mark with an X	
( ) Traffic congestion	( ) Densification of areas near public transport
( ) Public transport	( ) Wastewater Management
( ) Energy strategy	( ) Urban Solid Waste Management
( ) Potable water	( ) Street connectivity (crossings roads, avoid privatization, etc.)
( ) Densification	
(2) Is the planning of the MOPC[a]/ESSAP[b] reliable to know the areas of Asunción where a new wastewater collection[c] or water supply system will be built? ( ) Yes ( ) No	
(3) Do you think it is reliable to use MOPC planning to know the areas of Asunción where the new infrastructure for public transport will be provided? ( ) Yes ( ) No	

[a]MOPC: Ministerio de Obras Públicas y Comunicaciones (Ministry of Public Works and Communications)
[b]ESSAP: Empresa de Servicios Sanitarios del Paraguay (Health Services Company of Paraguay)
[c]All types of water contaminated by its human use

In addition, there is an economic, labor, and productive centralization in Asunción that relegates neighboring cities to the dependent dormitory villages of this, which generates an important daily influx to and from Asunción (ASU, 2016). However, its rapid urban expansion took place without the necessary planning, leaving an inefficient and unsustainable *público transport* network that has contributed to the increase in traffic congestion in the city (BID, 2014).

However, despite the fact that city sustainability does not yet have a solid framework in which to anchor, the country is the only one in the region that has a surplus of clean energy thanks to its 3 hydroelectric plants, one of them ITAIPU is the one with highest energy production in the world (SEAM & ONCC, 2015).

4 Results of the Interviews

The comments obtained during the interviews will be described below, as well as the results of the 33 valid questionnaires. The total number of mentions received will be placed in parentheses of Table 2. *Questionnaires results*.

First question

It should be noted that all respondents answered this question, and in the light of these results, we can say that for professionals in the Asunción area, the main urban sustainability problems are: first (26 consider it) and tied in votes: *Public Transport* and *Wastewater Management*. And in second place (23 mentions) and also tied: *Traffic congestion* and *Urban Solid Waste Management.*

Second question

In the second question, the participants mainly considered that the MOPC/ESSAP planning is not reliable (23 votes) to know the areas of Asunción where a new wastewater collection or water supply system will be built, compared to a minority that considered it trustworthy (8 votes). Only two respondents did not answer the question.

In order to justify this lack of reliability, the engineer María José Palacios revealed that the works that are currently under construction (sewerage and hydraulic works) were planned 10 years ago, and some have been executed in spaced time periods, without ever offering guarantees on its completion. And, although there are many active works, she standed that sewer coverage in the city is still very low, and there is still no wastewater treatment plant. Therefore, she said that private investment will not be sustainable without a public investment in sewage networks and a treatment plant for the city. In addition, she commented that the lack of these services discourages real estate development (M. Palacios, personal communication, April 4, 2019).

On the other hand, for the architect Alejandra Escauriza, the projects discontinuation after each government change is what makes the planning of the MOPC/ESSAP unreliable (A. Escauriza, personal communication, April 1, 2019). The engineer and businessman Juan José Barrail denounced the lack of inter-institutional coordination, since it is very common for a road to be paved or repaired and, shortly afterward, ESSAP or another institution breaks the new asphalt for infrastructure installation, usually for provision of water or pipes repair (J. Barrail, personal communication, April 3, 2019).

However, the engineer Jorge Abbate considered that this problem is even greater since it is conditioned from its ground, arguing that political decisions are usually prioritized over technical criteria, preventing him from trusting this organism projects (J. Abbate, personal communication, April 2, 2019). The renowned architect Solano Benítez

Table 2 Questionnaires results

(1) Which of the main Urban Sustainability problems in Asuncion do you consider the most urgent to address? Mark with an X

(23) Traffic congestion	(12) Densification of areas near public transport
(26) Public transport	(26) Wastewater Management
(8) Energy strategy	(23) Urban Solid Waste Management
(6) Potable water	(6) Street connectivity (crossings roads, avoid privatization, etc.)
(12) Densification	

(2) Is the planning of the MOPC[a]/ESSAP[b] reliable to know the areas of Asunción where a new wastewater collection[c] or water supply system will be built? (8) Yes (23) No

(3) Do you think it is reliable to use MOPC planning to know the areas of Asunción where the new infrastructure for public transport will be provided? (11) Yes (19) No

[a]MOPC: Ministerio de Obras Públicas y Comunicaciones (Ministry of Public Works and Communications)
[b]ESSAP: Empresa de Servicios Sanitarios del Paraguay (Health Services Company of Paraguay)
[c]All types of water contaminated by its human use

recommended to establish general and non-partial strategies (S. Benítez, personal communication, March 26, 2019).

The engineer Carlos Re commented on his experience in the construction of a 30-storey residential building in the new financial area of the city (Santa Teresa Ave.), where the construction company worked in coordination with ESSAP, ANDE, and the municipality of Asunción in the arrangement of the asphalt and the central promenade of the public thoroughfare, including gardens, bike paths, sidewalks, curbs, as well as the construction of sewage and pluvial channels, as well as the underground electrical cables, to subsequently recover the investment with the obtaining credits or tax exemptions (C. Re, personal communication, March 28, 2019).

Third question

Approximately two thirds of the participants (19 votes) did not consider the MOPC planning reliable to know the areas of Asunción where the new infrastructure for public transportation will be built. Only two professionals did not answer this question, and one registered: «I do not know the scope of the MOPC planning».

Most respondents answered *no* (19 votes out of 33), arguing their response in various ways. The architect Mauro Remonato gave as an example the current delay in the works carried out in Eusebio Ayala Ave. to complete bus-only lane. In addition, he mentioned having worked on his planning, and, even though he considers the MOPC project reliable, this does not include its execution, especially with regard to compliance with schedules (M. Remonato, personal communication, April 1, 2019). For the architect Annie Granada, the lack of reliability is due to the different changes of government that does not respect the pre-existing MOPC planning, under continuous modifyication (A. Granada, personal communication, March 26, 2019). According to Barrail, the lack of precision of the tendered projects is what makes it difficult to meet the deadlines, besides facilitating corruption (Barrail, 2019).

The lack of coordination between the territory planning and urban transport is, according to the architect Luis Silvio Ríos, the main obstacle to sustainable development, and he set as an example the location of the new corporate axis of Asunción for which no provision for public transport system improvement has been planned. He also mentioned the commuter train project that would link the city center with the airport, which unfortunately has not been able to get ahead to date (S. Ríos, personal communication, April 2, 2019). At the same time, the Vice Minister of Urban Planning, Housing and Habitat, the architect Jorge Bosch, mentioned that they have started working with the newly created Vice-Ministry of Urban Planning—MUVH on a comprehensive planning that includes the territory and transporation (J. Bọsch, personal communication, April 2, 2019).

5 Discussing the Results

First question

In order to determine which are the requirements of urban sustainability that should be a priority in Asunción, a comparative Table 3 will be made between what is considered relevant by the respondents, Fig. 1, which is essential for BREEAM-CM and LEED-ND, as well as the imperative for the AMAS Action Plan:

According to these results, Table 3, the respondents agreed more with the *AMAS Action Plan* priorities than with the global systems ones: BREEAM-CM and LEED-ND. Both

Table 3 Comparison between the priorities of the respondents, BREEAM-CM, LEED-ND, and AMAS Action Plan

Options offered in the questionnaire	Priorities of			
	Respondents	BREEAM-CM	LEED-ND	AMA Action Plan
Traffic congestion	X			X
Public transport	X		X	X
Energy strategy		X		
Potable water			X	
Densification				X
Densification of areas near public transport				X
Wastewater management	X		X	X
Urban solid waste management	X			X
Street connectivity (crossings roads, avoid privatization, etc.)			X	

Fig. 1 Asunción main urban sustainability problems identified by local professionals

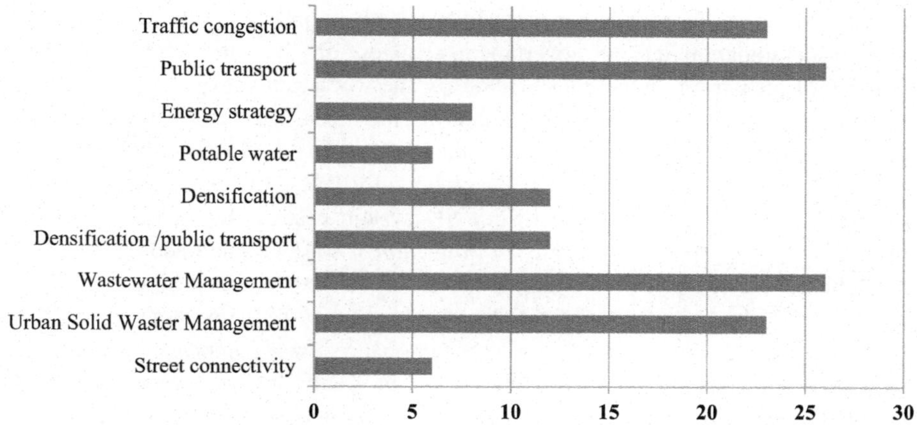

wastewater management and *public transport*, as well as *traffic congestion* and *urban solid waste management* were also proposed by the *AMAS Action Plan* as high priorities. However, the *Energy Strategies* principle of the BREEAM-CM system was not considered as a priority for Asunción according to local professionals. In the LEED-ND system, even though *public transport* and *wastewater management* priorities coincided with interviewees answers, the *potable water* and the *connectivity of the streets* were not in that line.

It is not surprising that the interviewees agree more with the AMAS Action Plan since this is an exhaustive study of the reality of Asunción and its metropolitan region. These results help to demonstrate the initial hypothesis of this research, which considers essential to prioritize local characteristics in the certification of the neighborhoods sustainability. In the same way, these results reinforce Aboneama's claims, who also demonstrated with clear examples that regional challenges are the most important elements to consider in a sustainability certification system (Aboneama, 2018). So it was to be expected that not all of the LEED-ND and/or BREEAM-CM priorities coincided with the Asunción ones. The results also contribute to the conclusions of Cole, who maintains that all global certification systems reflect the values and priorities of their authors, as well as their circumstances (Cole, 2005), making it very difficult to match the main needs of other sites.

However, this does not mean that the *Energy Strategies* or the other requirements should not be taken into account in Asunción, but considering the magnitude of the economic effort required to solve all the urban sustainability problems and the institutional weaknesses of a developing country like Paraguay, and it is necessary to prioritize some requirements over others. The results show that they are not currently a high priority, so it is suggested that BREEAM-CM reduces the scores of this requirement when certifying in Asunción, in order to offer greater relevance to the most urgent needs in this city such as the *management of sewage water*. Having said that, it is necessary to continue working to improve the *Energy Strategies* in Paraguay since, as Aboneama mentioned, everyone suffers from global warming and the depletion of the ozone layer, requiring certain general requirements to meet global challenges (Aboneama, 2018).

In addition, the proposal to reduce the score of the *Energy Strategies* when certifying in Asunción reinforces the conclusions found by Aboneama, who revealed in his research that the importance given to a requirement must be governed according to the most urgent local needs (Aboneama, 2018). It also helps to justify the conclusions of Humbert et al., when analyzing the scores that the LEED requirements offer, who concluded that, even in the USA where this methodology was created, a higher rating is not always synonym of low environment impact (Humbert et al., 2007).

Second question

The problems reported by the participants mainly focused on the wastewater collection system and reliability of the MOPC planning for a new system construction. The professionals unanimously agreed on the impossibility for these public institutions to complete a work in time, on the projects discontinuation due to government changes, on the absence of inter-institutional coordination for the works, on the existing institutional weakness, and on the need to establish general and non-dissociated strategies.

And despite the fact that the LEED-ND system establishes as a priority that the *wastewater collection service* is available (SLL prerequisite: smart location, a obligatory requirement), the fact of offering that these services may be in the stage of *planning* facilitates its non-compliance, since as it was demonstrated by the interviews, in Table 2. *Questionnaires results* that the MOPC planning in Asunción is not reliable. This would lead to the certification of neighborhoods without the minimum sustainable conditions and would also worsen the current situation as the impact of large companies on the current wastewater collection system would be enormous.

However, the situation of the BREEAM-CM system is even worse, since it does not directly address the problem of *wastewater* throughout the manual. Perhaps, it could be included as a community need in the "*SE 02—Priorities and demographic needs*" requirement, but only if is requested by the community.

Therefore, due to the lack of commitment in the construction of new networks, the poor sewerage system coverage (21%), the absence to date of an Effluent Treatment Plant (there is a modern ongoing project, but today the wastewater that is collected by public networks is still throw at the bottom of the Paraguay River), and, mainly, due to the foreign investors demands, large private enterprises have created alternatives for this situation. Specific cases can be mention, such as the buildings of *Paseo de la Galería* and the *Sheraton Hotel*, where they built small plants for the treatment of effluents in situ before returning them into the public network; the UNILEVER Distribution Center went even further and built an effluent treatment plant with sedimentation, flocculation, filtration, etc., where treated water is also used in toilets and gardening.

Third question

Regarding the reliability of the MOPC in *public transport planning*, as initially predicted, professionals offered negative point of views. The arguments held were also very similar to the second question answers in Table 2. *Questionnaires results*: the impossibility of complying with the stipulated deadlines, the lack of projects continuity due to government changes (many quoted the Metrobús example, which works were paralyzed at the time of the interview), and the absence of inter-institutional coordination. Corruption was also mentioned as another factor that hindered the reliability of the MOPC.

Therefore, the fact that LEED-ND offers as an option that the public transport service may be in the *planning* stage would lead to the failure to accomplish a much-needed criterion in Asunción; since as it was shown with the interviews, this *planning* is not reliable in Paraguay. In addition, the LEED-ND system does not consider public transport as a high priority, even though there are various requirements that address this issue; they are voluntary or optional which means that the application of this criterion is left to the investor's discretion. Making them mandatory would be much more effective, remembering that this change would only be for certification in Asunción.

On the other hand, in the BREEAM-CM system, public transport is not treated as a high priority, and the only mandatory requirement that indirectly mentions it is the "*TM 01—Accessibility to transport,*" also offering a very low score within the system (2 points, 11 being the highest). It treats it directly but in two voluntary requirements.[7] However, unlike LEED-ND, it is worthwhile that this system treats transport as a transversal criterion, that is, all the factors that make up urban mobility are treated in a single and mandatory point that serves as a guide and basis for all its components (which will be individually detailed later in voluntary requirements, such as bike lanes or public transport). According to this requirement, an evaluation of the state of transport must be carried out with the local authorities, to elaborate a Travel Plan whose objective is to encourage more sustainable transport alternatives, proposing the reduction of transfers, shortening times and avoiding unnecessary commuting (especially in cars), reducing the distance between buildings and public transport stops,

[7]BREEAM-CM Voluntary requirements: TM 04—Access to public transport, and TM 06—Public transport facilities.

improving walking and cycling trails, as well as public transport facilities, etc. A great achieve for this system, since it is well known that mobility is a broad field and its fragmentation would only give a partial view of the situation. In addition, this global approach is very necessary for Asunción, as traffic congestion was the second most urgent point of attention according to the participants and the AMAS Action Plan.

6 Conclusions and Recommendations

This research main objective was to analyze and confirm whether the main sustainability requirements for neighborhoods treated in the global BREEAM-CM and LEED-ND systems coincided or not with the most urgent needs to attend in Asunción, after consulting it with local professionals through interviews in Table 2. *Questionnaires results*, the following has been concluded:

The results obtained evidenced that, in accordance with Kyrkoua & Karthausa, 2011; Aboneama, 2018; Gibberd, 2002; Valentin & Spangenberg, 2000; Cole, 2005, before importing global systems from outside, its adaptation to the local context is crucial. Because what may be a priority in a first world country, such as the USA or the UK, probably will be faced in a very different way in a developing country such as Paraguay, where sustainability must keep working on the foundations, attending to basic needs such as proper *management of wastewater* or *public transport.*

In addition, these systems could contribute to the city's sustainability objectives, such as the *Action Plan for Sustainability to Asunción Metropolitan Area.* Making individual work contributes to the collective and avoiding the excessive individualist pointed out by Edward Ng. Therefore, it is proposed that private actions from the neighborhood must be coordinated with the public actions taken from the municipality. This is because in developing countries, where public institutions still do not have the necessary strength to promote sustainable development, and the private initiative contribution could serve as a great impulse. And, although public–private alliances are scarce, work must be done to increase this reality, providing information, reducing the process bureaucracy, establishing fiscal insertions, and among other measures.

While it is true that, in accordance with Cole, these universal tools offer clarity and definition of what constitutes a green construction (2005), and it is also important to recognize that environmental problems require an internationally integrated effort, just as Gibberd concluded (2011), and that a common or global structure that allows comparison between cities is needed (Valentin & Spangenberg, 2000). As well as confirming that its insertion in developing countries has positively driven the practice of sustainable measures in construction. However, BREEAM-CM and LEED-ND must work to establish requirements that best contribute to the context in which they are inserted.

In view of the fact that wastewater management is a high priority in Asunción, and after checking the deficiencies found in both international systems on this point, it is recommended that: the LEED-ND system no longer considers *planning stage* as valid for the wastewater management service, and that the BREEAM-CM system treat this problem directly. In addition, both systems must address this issue in a much more precise manner as proposed below:

Urban developments should coordinate the wastewater management policies with the competent authorities, since they will depend on the capacity available in the public collection network, the future city effluent treatment plant and the size of urban development. In addition, it is proposed to offer two alternatives: the connection to the public network (avoiding mere *planning state*) and in situ treatment (since the current coverage is very low and almost no certifications could be given in Asunción otherwise).

On the *connection to the public network*, due to the lack of its update/renovations, as well as due to the impact that large urban developments mean for the cities infrastructures, large enterprises should be required to carry out an impact study, which would generate the development on available public infrastructure for the collection and treatment of wastewater. And, if necessary, on-site treatment before returning them into the public network, or the establishment of public–private partnerships for the renovation/expansion of the wastewater collection and treatment system, so that the new development does not collapse the existing public infrastructure, all this under the guidelines of the public agency in charge, of course.

In those places where there is *no access to the public network*, the construction of an effluent treatment plant must be required if the necessary space is available, or at least the use of small plants, such as the Rotating Biological Contactors—RBC, depending on the size of the development. Small urban projects could establish partnerships with other neighborhoods (new and/or existing) for the construction of these infrastructures. So private enterprises contribute to the urban infrastructure, vital to the city, as some projects have already done. In other words, every great development must be responsible for the impact it generates in the city.

Additionally, in a city like Asunción where the *public transport system* has great deficiencies and is one of the crucial factors to be treated, measures that contribute to increasing the use of public transport must be established as mandatory. This is essential in the BREEAM-CM system, which considers them voluntary. For its part, the LEED-ND system should no longer consider valid that the public transport service is just in a planning state, as well as treat mobility as a whole as the BREEAM-CM system does, since public transport is integral part of the urban mobility system, and for this, it is necessary to work in coordination with the competent authorities, contributing from the neighborhoods (the private) with the sustainable mobility strategies of the city (the public).

Likewise, public entities responsible for wastewater management and public transport must improve the reliability of these services planning. Mainly in the fulfillment of the deadlines of the works, the continuity of the projects despite the government changes (the metrobus works were canceled and dismantled during the writing of this article after the change of government), as well as combating corruption through transparency and technical precision in the projects launched to tender. Inter-institutional coordination and projects development based on integrative strategies must also be improved. In addition, the public transport system needs to be improved, the sewerage coverage in the city must be increased, and is crutial to compleate the construction of the city's effluent treatment plant.

Not forget to mention that improvements must be achieved in the coordination between the territory planning (especially when the densification capacity of a neighborhood is increased) with public services and infrastructure, such as those related to wastewater management and public transport. As the interviewee Palacios mentioned, the lack of these basic services in the city discourages real estate development, as well as hinders to achieve real sustainability in buildings.

Of course, all these measures must change as sustainability evolves in the city of Asunción, fulfilling the aforementioned precept of being adaptable to a context that is always mutable, as Kyrkoua & Karthausa pointed out in their investigation (2011).

Finally, it is necessary to point out that in order for neighborhoods to truly contribute to urban sustainability, the government must first establish its goals, regulations, urban projects, etc., regardless of private entities interests, such as the World Green Building Council or the BRE Group. Avoiding the formulation of national regulations that exclusively seek the implementation of their systems in the country, as happened in Paraguay with the five national regulations on sustainable construction launched by the INTN in collaboration with the Green Building Council Paraguay.

References

Aboneama, W. (2018). Protect the Planet through sustainability rating systems with local environmental criteria—LEED in the Middle East. In *Asia Pacific International Conference on Environment-Behaviour Studies.* 2012 (pp. 752–766). https://doi.org/10.1016/j.sbspro.2012.12.264

ASU. (Ed.) (2016). *Plan de Desarrollo Sustentable Asunción 2030.* Gobierno Municipal de Asunción. Asunción.

BID. (Ed.) (2014). *Plan de Acción, Área Metropolitana de Asunción Sostenible.* Banco Interamericano de Desarrollo. https://webimages.iadb.org/PDF/Plan+de+Acci%C3%B3n+ICES+Asunci%C3%B3n.pdf

BRE GLOBAL. (Ed.) (2012). *BREEAM Communities Technical Manual: SD 202. Version: 2012 Issue: 1.2.* https://www.breeam.com/communitiesmanual/

BRE. (Ed.) (2015). BRE: Our history. https://www.bre.co.uk/history

Cole, R. (2005). Building environmental assessment methods: Redefining intentions. In *The 2005 World Sustainable Building Conference.* Tokyo. https://doi.org/10.1080/09613210500219063

DGEEC. (Ed.) (2015). *Paraguay Proyección de la población por sexo y edad, según Distrito, 2000–2025.* Dirección General de Estadística, Encuesta y Censos.

Garde, A. (2009). Sustainable by design: Insights from U.S. LEED-ND pilot projects. *Journal of the American Planning Association, 75*(4), 424–440. https://doi.org/10.1080/01944360903148174

Gibberd, J. (2002) The sustainable building assessment tool assessing how buildings can support sustainability in developing countries. *Built Environment Professions Convention.* http://citeseerx.ist.psu.edu/viewdoc/download?doi=10.1.1.197.7550&rep=rep1&type=pdf

Haapio, A. (2012). Towards sustainable urban communities. *Environmental Impact Assessment Review, 32,* 165–169. https://doi.org/10.1016/j.eiar.2011.08.002

Humbert, S., Abeck, H, Bali, N., & Horvath, A. (2007). Leadership in energy and environmental design (LEED): A critical evaluation by LCA and recommendations for improvement. *International Journal of Life Cycle Assessment, 12*(1), 46–57. https://escholarship.org/uc/item/01n0q8bx

Kyrkoua, D., & Karthausa, R. (2011). Urban sustainability standards: Predetermined checklists or adaptable frameworks? *Procedia Engineering, 21,* 204–211. https://doi.org/10.1016/j.proeng.2011.11.2005

Ng, E. (2010). *Designing high-density cities: For social and environmental sustainability* (p. 279). Earthscan.

Rueda, S. (2012). *Guía Metodológica para los Sistemas de Auditoría, Certificación o Acreditación de la Calidad y Sostenibilidad en el Medio Urbano.* Ministerio de Fomento.

SEAM & ONCC. (Ed.) (2015). *Primer Informe Bienal de Actualización de la República del Paraguay. Ante la Convención Marco de las Naciones Unidas sobre el Cambio Climático 2015.* Secretaría del Ambiente y Oficina Nacional de Cambio Climático.

Sharifi, A., & Murayama, A. (2013). A critical review of 7 selected neighborhood sustainability assessment tools. *Environmental Impact Assessment Review, 38,* 73–87. https://doi.org/10.1016/j.eiar.2012.06.006

USEPA. (Ed.). (2016). Green building History in the U.S. United States Environmental Protection Agency. https://archive.epa.gov/greenbuilding/web/html/about.html

USGBC. (Ed.) (2009). *LEED 2009 for neighborhood development, technical manual.* US Green Building Council.

USGBC. (Ed.) Country market brief. https://www.usgbc.org/advocacy/country-market-brief

Valentin, A., & Spangenberg, J. (2000). A guide to community sustainability indicators. *Environmental Impact Assessment Review, 20,* 381–392. https://doi.org/10.1016/S0195-9255(00)00049-4

WGBC. (Ed.). About us, our-story. http://www.worldgbc.org/our-story

Impacts of Inaccessibility to Clean Cooking Fuels: Global Versus Regional Perspective

Ifeoluwa Garba

Abstract

Energy accessibility is fundamental in addressing many of the present universal development issues which impact on people's physical, socio-economic well-being. However, especially in developing countries where modern energy system accessibility is mostly embryonic, energy poverty remains a severe development problem – despite the amplified attention from numerous agencies and governments. To date, most studies have concentrated only on accessibility to electricity: failing to tackle the fundamental inaccessibility to clean energy for cooking and/or heating. In this paper, we address this gap by reviewing the issue of inaccessibility to clean cooking fuel. In addition, using the fixed effect and random effect methodologies, we present empirical evidence on the impact of inaccessibility to clean fuels for cooking on social development (using education, life expectancy and labour force participation as indicators). We analyse these impacts from both a global and regional perspective: comparing the magnitudes of the impacts. The results obtained show that the household use of solid fuels has a statistically significant negative effect on primary and secondary education, as well as life expectancy. To the best of our knowledge, the results obtained from these analyses are the first available evidences on the dynamics of solid cooking fuels and social development.

Keywords

Cooking fuels • Developing countries • Social development • Sustainable development

1 Introduction

Modern forms of energy are fundamental for sustainable social development. They are a cornerstone for industrialisation and urbanisation: crucial elements in the development of any economy Lin and Ouyang (2014), Fouquet and Pearson (2012), Rühl et al. (2012). In addition it plays a role in enabling a sustainable environment. However, the ongoing reality is that over a third of the world's population live without access to modern forms of energy for basic day-to-day activities such as cooking, lighting and heating World Bank Group (2016).

The last decade has seen increased movements towards the attainment of universal access to modern energy services. However, much of the efforts have been limited. Up to date, household access to electricity for lighting has been the primary focus of most governments and agencies whilst household access to clean[1] fuels for cooking and heating has been neglected World Bank Group (2014). The neglection of this aspect of energy poverty can be exemplified by governmental goals, targets and documents which tend to incorporate household accessibility to electricity but omit accessibility to clean cooking fuels United Nations (2017). A cause and effect of more people across the globe experiencing inaccessibility to clean cooking fuels than electricity World Bank Group (2014).

Further implications include the substantial gap in literature analysing the effects of household inaccessibility to modern/clean cooking fuels on social development, especially in developing countries. The limited existing studies which have investigated effects of inaccessibility to modern cooking fuels have focussed on the environmental impacts of the phenomenon, such as global and/or local pollution, whilst little-to-no focus have been placed on the socio-economic effects of the phenomenon on people's lives.

I. Garba (✉)
Department of Electrical and Electronic Engineering, Centre for Doctoral Training, University of Strathclyde, Glasgow, Scotland
e-mail: ife.garba@strath.ac.uk

[1]Clean in this paper refers to fuels which produce little-to-no harmful pollutants when combusted.

C. Alalouch et al. (eds.), *Towards Implementation of Sustainability Concepts in Developing Countries*, Advances in Science, Technology & Innovation, https://doi.org/10.1007/978-3-030-74349-9_22

At most, some studies have considered some health effects of heavy reliance on traditional fuels for cooking. Across these studies, there is an assumption that continued lack of access to modern energy is possibly detrimental to socio-economic development. Yet, there is limited literature that methodically analyses or addresses issues relating to cooking energy for developing regions Garba and Bellingham (2018).

Therefore, in this paper, we attempt to address some of these gaps and further existing literature by presenting empirical results on the effects of inaccessibility to modern/clean cooking fuels on some indicators of social development: primary education, secondary education, employment and life expectancy. To the best of our knowledge, this paper is amongst the first studies looking into this topic.

Table 1 Population without electricity and relying on traditional use of solid fuels, 2014 (millions)

Region	Population without electricity	Population relying on traditional use of biomass for cooking
Africa	634	793
Sub-Saharan	632	792
Other	1	1
Developing Asia	512	1875
China	0	453
India	244	819
Europe	0	0
Latin America	22	65
Middle East	1	8
World	1186	2742

2 Wider View of Energy Issue

Looking at the total energy utilised in 2016 across the globe, the largest consumptions were observed in Asia (45%), America (27%) and Europe (15%) whilst the lowest consumptions were seen in Oceania (1%), Middle East (6%) and Africa (5.89%). In the context of the fuel sources constituting the energy production and consumption, contributing about 80%, fossil fuels were the biggest source of energy. Despite the push for a move towards cleaner energy sources, coal, crude and natural gas featured as the dominant sources across all regions (excluding Africa). In addition, although the global focus is being placed on increasing the share of renewable energy in the global energy mix and thus shifting towards cleaner energy sources United Nations (2017), with a few exceptions, the energy mix across regions remains significantly imbalanced.

When considering only biomass consumption, upon further exploration into the energy mix within the regions, it is observed that Asia (39.36%) and Africa (28.65%) account for almost 60% of the global biomass consumption, and when considering sectoral consumption within these regions, a further breakdown of data would show that in Africa and Asia, about 56% and 28%, respectively, of the total biomass are utilised by households. The implications of these can be anticipated to be significant—especially considering that at household levels, the utilised biomass fuels are generally unprocessed and thus, inefficient and harmful Garba and Bellingham (2018).

According to the International Energy Agency, around 70–90% of the total energy used in developing countries are consumed by households. Yet, the appalling reality faced by billions of people is inaccessibility to necessary energy services such as, clean cooking and/or heating provisions, lighting, amongst others. Currently, across the globe, nearly 2.8 billion people have to rely on solid fuels for cooking and/or heating. Truly, more people across the world experience inaccessibility to clean cooking fuels than to electricity. The population affected by this issue are concentrated in two main regions: developing Asia, where more than 1.8 billion people still rely on traditional cooking fuels and sub-Saharan Africa, where nearly 800 million people rely on solid fuels for cooking World Bank Group (2016). What's more, the majority of the affected population reside in rural parts of these developing regions. However, this situation is not exclusive to only developing regions. A few developed or transitioning countries, such as China, are observed to have large proportions of population utilising traditional solid fuels for cooking—see Table 1.

To further illustrate the current distribution of this issue, using data from the World Bank, Figs. 1 and 2 demonstrate the proportions of households with access to clean cooking fuels across the aforementioned regions. Starting with the African region (Fig. 1), only three countries (Algeria, Egypt, Tunisia) which are located in the northern part of the region have achieved universal access. In addition to these countries, all countries but South Africa, Gabon, Botswana and Cape Verde in the African region have less than 50% access levels. In the case of the countries within the sub-Saharan region, less than 10% of the population within the countries are seen to have access to clean cooking fuels. Many countries, such as Liberia, Mali and Uganda, are reported to have less than 2% access levels.

In the case of south Asia (Fig. 2a), the highest level of access to clean cooking fuels across the south Asian region is found in Bhutan (68%). Across the region, even countries such as Sri Lanka which have relatively high electrification

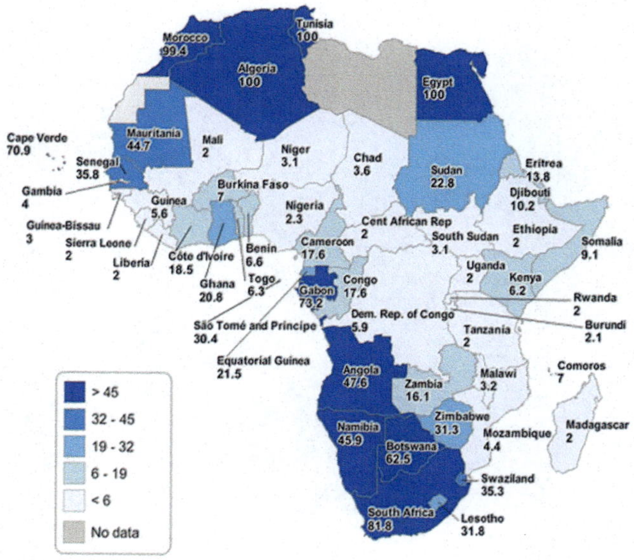

Fig. 1 Distribution of household use of solid fuels across Africa

rates have significantly low portion of their population with access to clean cooking fuels. Lastly, when considering the East Asian region (Fig. 2b), countries such as Brunei Darussalam and Singapore, have attained universal access. Unlike previously discussed regions, only few countries in East Asia have below 50% access levels. The lowest levels of access to clean cooking fuels are Lao People Democratic Republic (4.6%), Myanmar (9.1%) and Cambodia (13.4%).

3 Data and Methodology

With regard to the aspects of social development investigated, we analyse impacts on education, employment and life expectancy. Using the focus areas of the United Nations (UN) Sustainable Development Goals (SDGs) United Nations (2017), these components of social development have been chosen.

Data

The analyses performed in this paper are based on the global aggregate data, as well as, data collated for the forty-six sub-Saharan Africa; seven South Asian as well as twenty-four East Asian and Pacific countries. The variables used in this study were developed based on the combination of the current focus of the UN SDG, as well as, from studies in the literature on energy and social Development. As such, these selected variables are based on the annual time series on solid fuels use; primary education enrolment rate; secondary education enrolment rate; life expectancy and employment rate. For these analyses, based on data from the World Bank (WB); World Health Organisation (WHO); United Nations Educational, Scientific and Cultural Organisation (UNESCO) and the International Labour Organization (ILO), we construct a panel dataset of variables for the global and regional aggregate.

The panel data used is for the period of 15 years, spanning across 2000–2015. In terms of classification of countries for regional aggregates, the World Bank's regional categorisation approach has been used. For all variables, excluding life expectancy, the variables used in the analyses are deprived from the 100% census, kept in their natural log forms to reduce heteroscedasticity and as such, are unweighted.

Table 2 shows the descriptive statistics of the key variables used for both the global and regional panels. As shown in Table 2, a substantial (a little over a third) amount of households across the world rely on solid fuels. Focusing on the three energy most deprived regions, on average, every 7 in 10 households rely on solid fuels. Table 2 also showcases the differences across the other key variables, between the global picture and the three regions, taken together. The average rate of enrolment in primary education across the three regions is 85.43% in comparison to a global average of 90.93%. However, in the secondary enrolment, the average difference becomes more pronounced, with an almost 20% difference on average. For life expectancy, there is an average difference of 5 years. Implying that on average, a person residing on one of the three regions can be expected to live 5 years less than the global average of approximately 70 years.

Finally, considering the literature surrounding employment levels in developing regions, it is striking to see that compared against the global average. There are approximately 5% more people employed in these regions.

Methodology

To investigate the effects of solid fuel use on social development, we utilise two econometric panel models using the data described above. For these analyses, we designed three (3) bivariate estimator models—the least squares, fixed effect and random effect models—to test the effects on each indicator. However, we present the results from only two models.

The fixed effect (FE) model explores the relationship between the independent and dependent variables within the examined entities (countries) using a transformation to eliminate the unobserved effects within the entities (countries), before estimating the relationship between the variables. The most common types of unobserved variables are fixed effect and as such, tend to be time invariant characteristic of the entities (countries), for example, race, culture, gender, etc. Therefore, due to the model controlling for these time-invariant differences between the countries, the obtained estimated coefficients of the models are unbiased by omitted time-invariant characteristics. On the other hand,

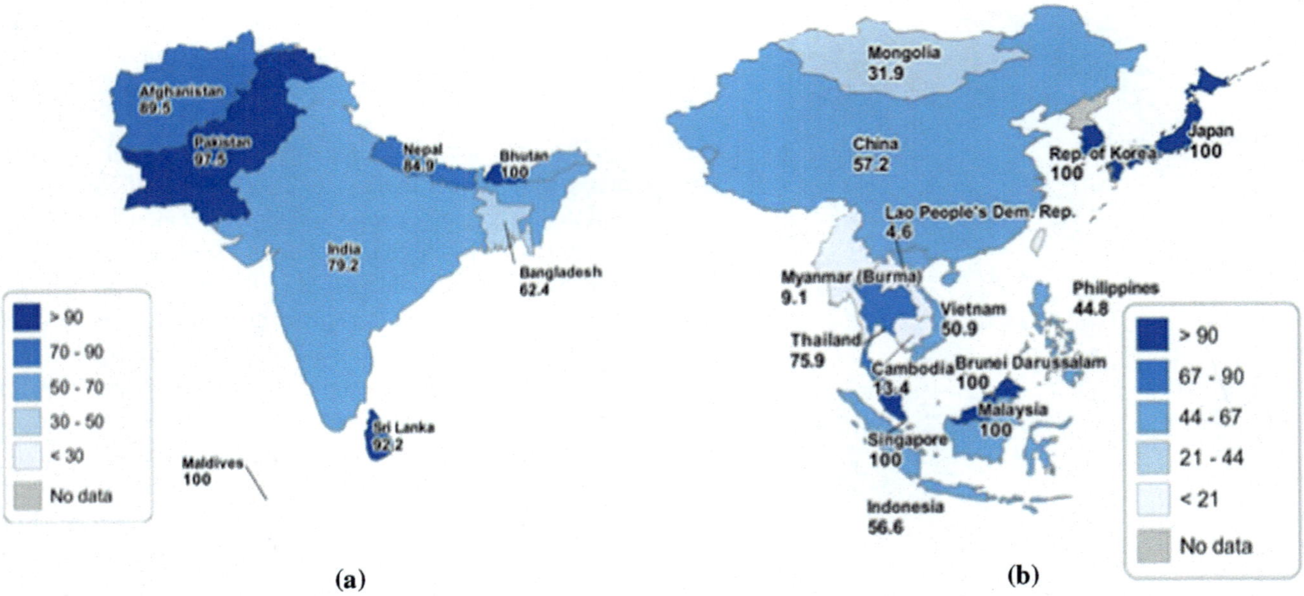

Fig. 2 Distribution of household use of solid fuels across south (**a**) and east (**b**) Asia

Table 2 Descriptive summary of variables

	Summary of the variables							
	Min		Max		Mean		Std. dev	
	Glo	Reg	Glo	Reg	Glo	Reg	Glo	Reg
Solid fuels	0.00	40.78	88.79	88.79	33.82	69.48	33.30	16.85
Primary E	60.23	60.23	97.97	97.49	90.93	85.43	9.04	11.41
Secondary E	20.70	20.70	90.85	77.17	65.50	48.72	20.34	18.27
Employment	40.87	52.76	70.48	70.48	57.47	62.31	7.80	5.29
Life expectancy	50.51	50.51	79.06	75.10	70.29	64.65	7.47	8.09
Labour	47.02	54.88	73.89	73.89	61.97	66.02	7.48	5.96

E—education; Glo—global; Reg—regional

the random effect (RE) model explores the relationship between the independent and dependent variables within the examined entities by assuming that the unobserved effect is random and uncorrelated with each independent variable. The model assumes the two key assumptions of the FE model plus the assumption that the unobserved effect is uncorrelated with each independent variable.

In other words, the fixed effect estimator can demonstrate the variations within the region(s), over time whilst the random effect estimator can demonstrate the variation within and across the regions, across the time period. To clarify, since the global variable consists of only one group (global) and thus, only the within variation can be estimated, for consistency, it would make sense to consider only the results from the fixed effect estimator when assessing and comparing impact magnitudes. However, to have a more comprehensive overview of the effects across the impoverished regions, the random effect estimator becomes interesting. Using the random effect estimator, we are able to obtain results which give a good compromise between estimating for group-level variations and simultaneously estimating for variations across all the regions: with the assumption that the deviations of the regions estimates from the global estimate reflect the distribution of a random variable.

The results obtained from the fixed and random regression models are presented in Tables 3, 4, 5, 6 and 7. However, due to the above stated reasons, as well as using the Lagrange multiplier and Hausman tests. Notwithstanding that the estimations of the coefficients across all models are statistically significant, across majority of the tests, the fixed effect estimator is found to be the most efficient estimator.

4 Results

The impacts of education and employment on economic development is a well-researched topic, with evidences that conclude that the cognitive skills of the population within a country are strongly related to individual earnings, income distribution and thus, economic growth Hanushek et al. (2008), Klasen and Lamanna (2009), Hanushek and Woimann (2007). As a result, we investigate the impacts of the traditional use of solid fuels on education and employment. Regarding health outcomes, studies have investigated only the impacts of solid fuel use on indoor pollution and thus, attributable diseases and mortality. As such, existing literature on the subject matter is exiguous.

The impacts of solid fuel use on the variables of interest is explained using the following parameters: the coefficient, the *R*-squared and the *p*-values. The resulting marginal change in the dependent variable, due to a unit change in the independent variable is illustrated by the coefficient parameter. The degree to which the variation can be explained by the estimator model is reported by the *R*-squared parameter. The *p*-values examine the hypotheses that the obtained coefficient values are not zero (0). Therefore, to reject the null hypothesis stating that there is no relationship between the dependent and independent variables, the *p*-values have to be less than 5% (0.05). In doing so, we confirm that the coefficient values are indeed different from zero (0). The impacts of solid fuel use on primary education enrolment is presented in Table 3.

Across both the global and regional models, a significant effect is observed. To determine which models are statistically significant, we firstly consider the *p*-values. We observe that across all models, the null hypotheses of no relationship can be strongly rejected. Thus, for primary education enrolment, all models should be considered.

Starting with the fixed effect models, there is a statistically significant negative effect observed between the use of solid fuels and primary school enrolment. The observed results can be interpreted as such: For the global model, a unit increase in solid fuel use has a statistically significant probability of resulting in a decrease of 0.54 units in primary school enrolment. For the regional model, this decreases further by over 0.03 units. When considering the random effect model, the results become interesting. An almost 30% difference is observed between the global and regional models.

Looking at the *p*-values, the results would suggest that the probability of the coefficient being true is higher in the global estimator model. However, considering the *R*-squared and adjusted *R*-squared values, we observe that the values from the regional estimator indicate the model is a better bit in comparison to the global estimator model. However, considering that other parameter values for the regional (both fixed effect and random effect) models, appear fairly consistent, we accept the hypothesis that the effect of solid fuel use is more substantial within these regions.

Taken together, these obtained results indicate that with increased use of solid fuel, the possibility of children obtaining primary education is significantly impeded.

Moving onto secondary education enrolment, similar to the primary education models, firstly, we confirm that the *p*-values across all models in Table 4 are statistically significant.

Generally speaking, as shown in Table 4, the impact of solid fuel use on secondary school enrolment is observed to be more severe. Across all models, a more than doubling effect in comparison to primary education enrolment is observed in the coefficients. Starting with the fixed effect model, the impact is observed to be higher in the global model. Since across all parameters, *p*-values, *R*-squared and adjusted *R*-squared, the global model shows a stronger estimator fit. We can accept the notion that for fixed effects, there is a higher global impact on secondary school enrolment. For the random effect models, however, the impact is observed to be higher in the regional estimator.

Nonetheless, all results seem consistent, it can be deduced that solid fuel use has a strongly negative impact on secondary school enrolment.

The results obtained for the effects on employment were interesting. Across all models, positive effects on employment were observed. For both the fixed effect and random effect models, the impact can be seen to be higher in the regional estimators. Since all the *p*-values are statistically significant, it appears we can accept the results from all models. However, when the *R*-squared and adjusted *R*-squared values are considered, the values obtained for the regional models demonstrate a much better fit in comparison to the global models. As such, although all statistically considerable, the results from the regional models can be viewed as more significant.

For labour force participation, Table 6, the obtained results are also positive. However, in comparison to results on employment, higher effects are observed. Similar to Table 5, results from the regional estimator models show greater effects compared to the global estimator models.

Finally, with regards to life expectancy, a significant negative effect is observed across all models. In the fixed effects models, a higher impact is observed in the global estimator whilst in the random effect models, the higher impact is seen in the regional estimator.

Nonetheless, across all models, it can be deduced that the use of solid fuel significantly, negatively, impacts on life expectancy. As shown in Table 7, on average, across the globe, a unit increase in solid fuel use would probably result in a 0.30 unit decrease in life expectancy, decreasing further when we consider the energy impoverished regions.

Table 3 Effects of household use of solid fuels on primary education

Effect specification	Primary education			
	Fixed effects (FE)		Random effects (RE)	
	Global	Regional	Global	Regional
Coefficient	−0.5400	−0.5725	−0.2877	−0.5811
Std. Error	0.1190	0.2092	0.0625	0.1496
t-value	−4.5380	−2.7368	−4.6026	−3.8844
p-value	1.63E−05	0.0091	1.19E−09	0.0003
R-squared	0.1751	0.1545	0.1796	0.2600
adj *R*-squared	0.1156	0.0926	0.1625	0.2425
f-stats	20.5935	7.4902	21.1841	15.0883

Table 4 Effects of household use of solid fuels on secondary education

Effect specification	Secondary education			
	Fixed effects (FE)		Random effects (RE)	
	Global	Regional	Global	Regional
Coefficient	−1.1694	−1.1429	−0.9336	−1.0888
Std. Error	0.0900	0.1590	0.0765	0.0956
t-value	−13.0350	−7.1872	−12.2020	−11.3920
p-value	<2.20E−16	8.98E−09	<2.20E−16	1.43E−14
R-squared	0.6366	0.5575	0.5911	0.7511
adj *R*-squared	0.6104	0.5251	0.5871	0.7454
f-stats	169.9190	51.6558	148.8810	129.7880

Table 5 Effects of household use of solid fuels on employment

Effect specification	Employment			
	Fixed effects (FE)		Random effects (RE)	
	Global	Regional	Global	Regional
Coefficient	0.0975	0.3491	0.1000	0.3407
Std. Error	0.0511	0.0361	0.0444	0.0367
t-value	1.9094	9.6647	2.2455	9.2876
p-value	5.92E−02	3.95E−12	2.69E−02	7.74E−12
R-squared	0.0362	0.6950	0.0467	0.6673
adj *R*-squared	−0.0333	0.6726	0.0374	0.6596
f-stats	3.6456	93.4073	5.0425	86.2598

Table 6 Effects of household use of solid fuels on labour force participation

Effect specification	Labour force participation			
	Fixed effects (FE)		Random effects (RE)	
	Global	Regional	Global	Regional
Coefficient	0.2037	0.3983	0.1864	0.3922
Std. Error	0.0390	0.0376	0.0359	0.0378
t-value	5.2205	10.5970	5.2000	10.3679
p-value	1.02E−06	2.60E−13	1.02E−06	2.s5E−13
R-squared	0.2193	0.7326	0.2079	0.7143
adj *R*-squared	0.1630	0.7130	0.2002	0.7076
f-stats	27.2536	112.3000	27.0397	107.4940

Table 7 Effects of household use of solid fuels on life expectancy

Effect specification	Life expectancy			
	Fixed effects (FE)		Random effects (RE)	
	Global	Regional	Global	Regional
Coefficient	−0.4087	−0.3803	−0.3033	−0.4141
Std. Error	0.0470	0.0827	0.0353	0.0630
t-value	−8.6876	−4.5959	−8.6027	−6.5733
p-value	9.09E−14	4.07E−05	9.29E−14	5.36E−08
R-squared	0.4376	0.3400	0.4181	0.5012
adj *R*-squared	0.3970	0.2917	0.4125	0.4896
f-stats	75.4746	21.1223	74.0056	43.2087

5 Discussion and Conclusion

The effects of solid fuel use on primary and secondary education enrolment, employment; labour force participation and life expectancy across the globe, as well as, across the three energy most impoverished regions—East Asia, South Asia and sub-Saharan Africa—have been analysed in this paper.

Undoubtedly, other factors such as income poverty, cultural influences, amongst others, impact on the indicators analysed in this study. However, the results obtained from the models represent the direct impacts of solid fuel use on the aforementioned indicators. By using fixed effects models, potential endogeneity bias due to omitted variable bias, is avoided.

At varying extents, the obtained results have demonstrated that solid fuel use has a substantial effect across all examined variables. Taken together, solid fuel use is seen to have a negative effect on the examined aspects of social development with the exception of labour force participation and employment.

The obtained results from employment and labour force participation are particularly interesting. The positive correlation obtained in these models would suggest that solid fuel use supports employment and labour force participation. Although these positive correlations could appear auspicious and perhaps are in the short term, they can be questioned in the long term.

Many solid fuel-reliant countries[2] tend to have many large or small-scale mining enterprises Drechsler (2001), and increasing demand in solid fuels might indeed create employment opportunities. However, these jobs tend to require uneducated and/or unskilled workers. The effects of the lack of educated and literate workers on economic development has been extensively analysed with studies reporting a negative impact on economic development Brown and Lauder (2000), Berger and Fisher (2013). As such, it can be anticipated that in the long-run, increased solid fuel demand, leading to increased uneducated and/or unskilled workers, could indeed represent a bottleneck for sustainable development within countries. In substantive terms, secondary school enrolment had the strongest correlation (in terms of magnitude) with solid fuel use.

For the interpretation of the education indicators, we found that in the solid fuel-reliant areas (regional models), there are over 30% less enrolment in primary school and almost 20% less enrolment in secondary school. These results agree with studies International Energy Agency (2015) which associate solid fuel use with loss of productive hours (for schooling or employment) for children and women in these regions. In addition to this, the translating high levels of illiteracy, shortage of educated professionals in the labour force could, as mentioned above, represent further impediments to sustainable development. Although the possibility to attribute causality to these observations is restricted, these results can be seen as a good starting point for understanding the role of solid fuel use plays in education enrolment in the energy impoverished regions.

For life expectancy, the significant negative effect observed in our analyses supports previous studies which report a high correlation and causal effect between indoor pollution caused by solid fuel use and respiratory diseases, lungs cancer as well as child mortality rates.

The issue of inaccessibility to clean cooking fuels requires more awareness. However, from a policy-making perspective, for effective solutions to be designed, an improved comprehension of the issue and the underlying factors is crucial. A starting point could be for agencies and governments within solid fuel-reliant regions to establish national databases which provide detailed information on the energy profile of the population. To exemplify, in most countries in Africa,[2] there is a severe lack of data regarding the energy composition of households. Data illustrating what types of stoves households use, the specific type of fuel households utilise for cooking, cost components, amongst other information, are lacking. In addition, data on potential

[2]For example, Malawi, Mozambique, Zambia, amongst others.

fuel sources, stoves types, cost analyses and estimates of the ability and willingness to pay, as a function of income, could equip policy makers with the essential information needed to design more efficacious policies.

Finally, policies to address energy poverty would need to go hand in hand with broader policies aimed at alleviating income poverty and promoting economic development.

References

Berger, N., & Fisher, P. (2013). *A well-educated workforce is key to state prosperity*. Economic Policy Institute. Technical Report: Economic Analysis and research network. https://www.epi.org/publication/states-education-productivity-growth-foundations/

Brown, P., & Lauder, H. (2000). Education, globalization and economic development. In Leske+Budrich (Ed.), *Jahrbuch Bildung und Arbeit 1999/2000. VS Verlag fur Sozialwissenschaften*, Wiesbaden. Chapter Education. https://link.springer.com/book/ https://doi.org/10.1007/97s-3-322-974s7-7

Drechsler, B. (2001). *Small-scale mining and sustainable development within the SADC region*. Technical Report 84. International institute of environment and development. http://commdev.org/files/1798_file_asm_southern_africa.pdf

Fouquet, R., & Pearson, P. J. G. (2012). Past and prospective energy transitions: Insights from history. *Energy Policy, 50*, 1–7. http://www.sciencedirect.com/science/article/pii/S0301421512006s05, https://doi.org/10.1016/j.enpol.2012.08.014.

Garba, I., & Bellingham, R. (2018). The Impact of Lack of Clean Cooking Fuels on Sustainable Development in Developing Countries: ASME 2018 12th International Conference on Energy Sustainability, ASME, Lake Buena Vista (pp. 1–8). http://proceedings.asmedigitalcollection.asme.org/proceeding.aspx?articleid=2706327

Hanushek, E., & Woimann, L. (2007). The role of education quality in economic growth: The role of school improvement in economic development. https://papers.ssrn.com/sol3/papers.cfm?abstract_id=960379, https://doi.org/10.2139/ssrn.960379

Hanushek, E. A., Woessmann, L., Jamison, E. A., & Jamison, D. T. (2008). Education and economic growth. https://search.proquest.com/docview/1237s1s111?accountid=14116

International Energy Agency. (2015). Energy for cooking in developing countries. World Energy Outlook 2006 (pp. 419–445). https://www.iea.org/publications/freepublications/publication/cooking.pdf, https://doi.org/10.1787/weo-2006-16-en

Klasen, S., & Lamanna, F. (2009). The impact of gender inequality in education and employment on economic growth: New evidence for a panel of countries. *Feminist Economics, 15*, 91–132. https://doi.org/10.1080/13545700902893106

Lin, B., & Ouyang, X. (2014). Energy demand in china: Comparison of characteristics between the US and china in rapid urbanization stage. *Energy Conversion and Management, 79*, 128–139. http://www.sciencedirect.com/science/article/pii/S0196s90413007929

Rühl, C., Appleby, P., Fennema, J., Naumov, A., & Schaffer, M. (2012). Economic development and the demand for energy: A historical perspective on the next 20 years. *Energy Policy, 50*, 109–116. http://www.sciencedirect.com/science/article/pii/S0301421512006313, https://doi.org/10.1016/j.enpol.2012.07.039.

United Nations. (2017). *Work of the statistical commission pertaining to the 2030: Agenda for sustainable development*. https://www.un.org/sustainabledevelopment/developmentagenda/

World Bank Group. (2014). *Clean and improved cooking in Sub-Saharan Africa*. Technical Report 98664. World Bank Group. Washington DC. http://documents.worldbank.org/curated/en/16424146s17s757464/pdf/9s664REVISED-WP-P146621-PUBLIC-Box3931s5B.pdf

World Bank Group. (2016). *World development indicators*. Technical Report. World Bank Group (WB). Washington DC. https://data.worldbank.org/indicator

World Bank. (2017). *World development indicators*. https://data.worldbank.org/datacatalog/worlddevelopment-indicators

Evolution of Optimum Landfill Site for Disposal of Solid Waste Through Geoinformatics: A Case Study of Ahmedabad City

Zalak Bhavsar and Gaurav Jain

Abstract

Solid waste management is one of the essential responsibilities of local government agencies, which is a primary goal to maintain the personal satisfaction of individuals and ensuring acceptable standards of well-being and sanitation. It is estimated that the Ahmedabad Municipal Corporation generates nearly 2200 tons of solid waste daily. Most of this solid waste will find its way to the Piranha landfill, which is to the southeast of the city of Ahmedabad. The city has seen the un-point of population growth over the last few years, and the inclusion of many new areas that are possible, there is an increasing demand that the current framework has to be sufficient to meet. The alternative landfill site thus uses geomatics methods based on physical, environmental, and socio-economic parameters and the multivariate index approach. Thematic maps, i.e., Land use/land dispersal, surface waters, groundwater potential, drainage, infrastructure, were prepared using Indian Remote Sensing satellite data. Census records and the Ahmedabad Municipal Corporation provided socio-economic data and other relevant environmental data. Multivariate index analysis was performed using the analytical hierarchy process, taking into account the seventeen parameters. These parameters were weighted using a pair-wise comparison technique to calculate the matrix of the weight-dependent Satty's matrix. The landfill site suitability index is processed using a weighted index strategy. All exclusion criteria were then masked, and the rest were classified into four compatibility classes. The appropriate site was identified as a consequence and proposed for development as a sanitary landfill. This paper demonstrates the value of following the right geomatic and analytical hierarchy method in order not only to identify ideal landfill sites for the removal of solid waste but also to decide the best routes to bring the waste from the city to the landfill site in Ahmedabad (India).

Keywords

Solid waste management • Ahmedabad • Geographic information system • Geoinformatics • Multi-criteria decision matrix

1 Introduction

Solid wastes involve all the wastes arising out of human and animal exercises that are commonly solid and are discarded as inconsequential or unfortunate. Solid waste management may be portrayed as "the order related to the control of generation, storage, collection, transfer and handling and removal of wastes in a way that is according to the best guidelines of general prosperity, budgetary perspectives, designing, security, style, and other natural idea." Solid waste management is the most major undertaking depended to urban nearby government bodies for keeping up a decent personal satisfaction of individuals and for guaranteeing better gauges of well-being and sanitation.

In urban centers of India, average per capita waste generation ranges from 0.2 kg to 0.6 kg per day. The annual increase in per capita waste generation is estimated at 1.33% (CPHEEO, 2005). With the urban population increasing at 2.7–3.5% per annum, the yearly increase in the overall quantity of solid waste in the cities will be more than 5%. The Energy and Resources Institute (Kil & Shin, 1997). It has been estimated that waste generation will exceed 260 million tons per year by 2047—over multiple times the present level. Urban communities with 100,000 or more

Z. Bhavsar (✉)
L. J. Institute of Engineering and Technology,
L J University, Ahmedabad, Gujarat 382210, India
e-mail: zalak.bhavsar@ljinstitutes.edu.in

G. Jain
Space Application Centre (SAC), Indian Space Research Organization (ISRO), Ahmedabad, Gujarat 380015, India

C. Alalouch et al. (eds.), *Towards Implementation of Sustainability Concepts in Developing Countries*,
Advances in Science, Technology & Innovation, https://doi.org/10.1007/978-3-030-74349-9_23

population contribute 72.5% of the waste created in the nation when contrasted with other 3955 urban centers that produce just 17.5% of the total waste and generally, collection efficiency ranges from 70 to 90% in major metro cities whereas in several smaller cities collection efficiency is below 50%.

The municipal solid wastes have to be managed under the framework of the Municipal Solid Wastes (Management and Handling) Rules, 2000 under Environmental Protection Act, 1986. The responsibility of Solid Waste Management is thus entrusted to Urban Local Government bodies like Municipal Corporation and Municipality. In most of the urban centers, landfill sites remain the principal method of solid waste disposal.

Landfills are the physical facilities utilized for the removal of residual waste in the earth's surface soil. Sanitary landfills allude to an engineered facility for the removal of Municipal Solid Waste structured and worked to limit general well-being and ecological effects. Lamentably, these landfill destinations are regularly seen as simply open dumps at some faraway area.

Disposal of municipal solid waste as open dumps or in an improperly designed landfill site (i.e., low lying areas) without proper inspection and separation result in severe environmental impacts like groundwater contamination by the leachate generation from waste dump; surface water contamination by the runoff from the waste dump; air pollution due release of greenhouse gases, dust, and odor; acidity to the surrounding soil; fires within the waste dump due to generation of methane; epidemics from the stray animals; and bird menace at the waste dumps which affect flight of aircraft.

As sanitary landfill site is an inevitable part of Solid Waste Management (Leverenz et al., 2011), the identification of proper site to final disposal of wastes such that it would cause less damage to the environment by considering constraints (exclusionary criteria), environmental, social and economic factors are very significant.

This study presents a method of determining a suitable site for the construction of Municipal Solid Waste landfills (using Analytical Hierarchy Process for multi-criteria evaluation) and optimal route selection in the Geographic Information System environment.

2 Procedural Investigation

Geographical information system can help in managing a few factors all the while, which should be considered while arranging waste management. The prime target of the examination is to build up an incorporated system for solid waste management. The investigation is done in two stages.

1. Selection of sanitary landfill site using remote sensing data and Analytical hierarchy process for multi-criteria decision analysis
2. Optimal route selection for the existing landfill site in a GIS environment.

The focus of this paper is the first phase, i.e., selection of the site for sanitary landfill using the analytical hierarchy process in a GIS environment.

3 Study Area

Ahmedabad Urban Agglomeration is the seventh-largest Urban Agglomeration in India, as per the Census of India (2011). In the north-central part of Gujarat State, Ahmedabad Urban Agglomeration is spread over an area of 1294.65 km^2, accommodating the population of 56,33,927 (Pollution et al., 2012). The map-1 shows the extent of the study area (Fig. 1) along with the location of the landfill site in Ahmedabad.

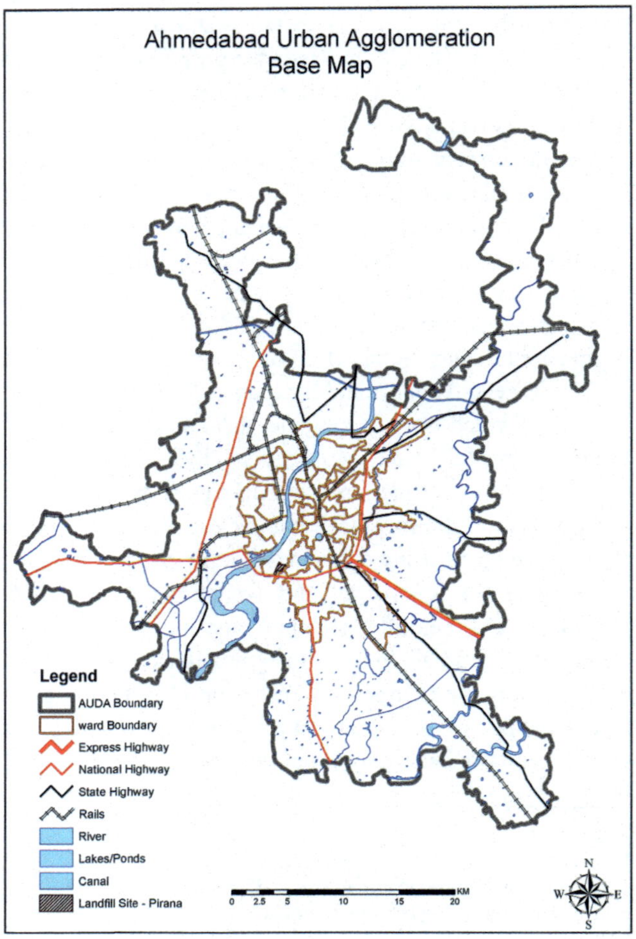

Fig. 1 Study area map

Per capita, municipal solid waste generation ranges between 0.4 and 1.0 kg per day, and average waste generation is 0.69 kg per day in Ahmedabad, which includes domestic, commercial, and institutional waste and also wastes generated from the street sweeping. Daily waste generation by all six zones of Ahmedabad Municipal Corporation is approximately 4500 tons. The newly added areas to Municipal Corporation contribute about 139 tons of waste besides. The solid waste is disposed off at the landfill site at Pirana spread over 84 Acres land area. AUDA has recently developed a landfill site at Fatehwadi on an area of 38 ha, which is yet to start operation. Thus, the Pirana landfill site remains the only available option for solid waste disposal for most of the Ahmedabad.

4 Methodology

Figure 2 provides the conceptual model, demonstrating the adopted method for the use of GIS in Solid Waste Management in this study. Seventeen Parameters were considered for the Analysis.

The Parameters were divided into two categories, firstly those that are constraints binding as per the laws enforced in India, and secondly, those that define the site's suitability about the construction of landfill site, its social, economic, and environmental impacts (Table 1).

As the objective was to classify land (study area) into four suitability classes, the first step was to rank the parameters into four categories. The ranking is based upon a "Compromise Criteria" which categorizes each parameter into four ranks of increasing limitation. The ranks for the nine parameters are shown in Table 2.

4.1 Air Quality Index (AQI) and the Ground Water Quality Index (GWQI) Map

Various air quality and groundwater quality data were obtained from the Gujarat Pollution Control Board. AQI and GQI maps are essential to derive environmental conditions in the study area. The Digital Elevation Model was generated from point data of various environmental parameters. Air quality data comprising a concentration of SPM, RSPM, SO_2, and NO_x were used to generate the AQI map. All concentration values were in mg/l unit. Groundwater quality data comprising of concentration of TDS, Chloride, Na^+, and SO_4^{-2} were used to generate the GQI map. All concentration values were in mg/l unit.

The DEM was reclassified into four categories, as mentioned in Table 3. The AQI and GQI index is computed using the following formula:

$$\mathbf{AQI} = SPM_{rank} * RSPM_{rank} * SO_{2rank} * NO_{Xrank} \quad (1)$$

$$\mathbf{GQI} = TDS_{rank} * Cl_{rank} * Na_{rank} * SO_{4rank} \quad (2)$$

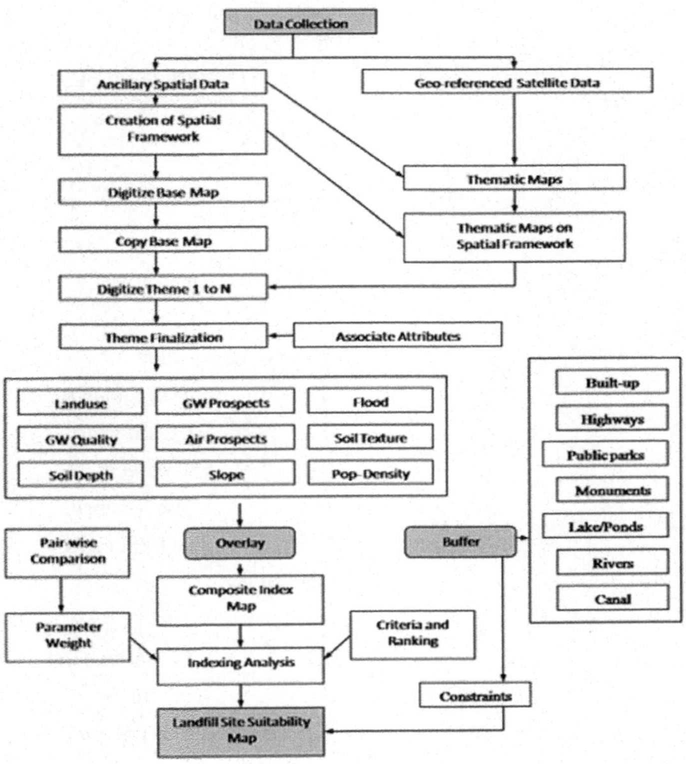

Fig. 2 Methodology

Table 1 Constraint criteria for landfill sitting (distance is in meters)

No.	Criteria	Distance (m)
1	Distance from highways (NH & SH)	200
2	Distance from lake/pond	200
3	Distance from river	100
4	Distance from canal	30
5	Distance from habitation	500
6	Distance from public park	300
7	Distance from historical places and national monuments	500
8	Depth of ground water	>Than 2 m

Source Guideline developed for the management of Urban Development, Govt. India

Table 2 Criteria for landfill suitability

S. No.	Parameter	1st Rank	2nd Rank	3rd Rank	4th Rank
		Min.	Limitation		Max.
	Land use (based on usage of land)	Land without scrub	Scrubland and wasteland	Fallow land	Agriculture, cropland and plantation
2	GW prospectus	Poor	Moderate	Good	Very good
3	Flood (on the basis of occurrences)	No flood hazard	Low flood hazard	High flood hazard	–
4	GWQ index	1–10	10–70	70–150	>150
5	AQ index	1–20	20–55	55–100	>100
6	Soil Texture (based on particle size analysis)	Moderately well-drained, fine loamy soil	Well-drained, fine and loamy soil	Excessively well-drained, coarse loamy soil	–
7	Soil depth (based on soil digging)	Very deep	Deep	Moderately deep	Shallow
8	Slope (rise /run) *100	1–5%	0–1%	5–10%	–
9	Population density (population per sq. km)	34.48–8497.49	8497.49–25,511.18	25,511.18–46,070.26	46,070.26–86,624.86

Note Rank 1 is most suitable

Table 3 AQI and GQI (in mg/l)

S. No.	Parameter	1st Rank	2nd Rank	3rd Rank	4th Rank
		Min.	Limitation		Max.
Groundwater quality					
1	TDS (in mg/l)	<1000	1000–2000	2000–3000	>3000
2	Cl^- (in mg/l)	<200	200–500	550–1000	>1000
3	Na+ (in mg/l)	<200	200–600	600–1000	>1000
4	SO_4^- (in mg/l)	<200	200–450	450–650	>650
	GQI (in mg/l)	1–10	10–70	70–150	>150
Air quality					
1	SPM (in ppm)	<200	200–300	300–400	>400
2	RSPM (in ppm)	<100	100–140	140–180	>180
3	SO_2 (in ppm)	<15	15–25	25–40	>40
4	NO_x (in ppm)	<25	25–30	30–35	>35

Note All values are in mg/l unit

5 Analysis and Results

As described in the methodology, each of the nine suitability parameters was divided into four suitability classes, and ranks were assigned as per Table 2. The weights of importance for suitability criteria were assigned based on the Pair-wise Comparison method. The importance matrix for the nine parameters was generated and the importance numbers are as per Satty's Pair-wise comparison method. The other elements of the 9 × 9 matrix are determined by the reciprocality of the matrix. Based on Satty's scale. The final weight of each parameter is given in Table 4.

All nine layers were integrated into a GIS environment, and the landfill suitability index (LSI) was computed as a cumulative index of nine parameters. The Landfill Suitability Index (LSI) so obtained was reclassified into four categories based upon mean and standard deviation, as described below in Table 5.

6 Conclusion

The landfill site suitability map and constraint map for the study area have been shown. The areas available as per suitability for a landfill site after removing the constraint areas have been given in Table 6.

The spatial analysis for the suitability for the landfill site development has been done using the ArcGIS software. It can be observed that only 4.53% of the total area is classified under class-1 suitability, and these could be the priority areas for development. Class-2 suitability areas, which are about 53.44% of the total area, could also be used for development. Class-3 suitability, which is about 41.56% of the total area, would have a high order of limitation and be the third priority. Class-4 suitability, which is about 0.47% of the total area and has a maximum limitation, could be taken up as the last priority. The study demonstrated that a GIS-based approach to landfill site selection offers the ability to manage conflicting demands on land more logically and acceptably, thereby satisfying social, economic, and environmental requirements. In the second stage of the project, the Central zone of Ahmedabad was chosen for examining the transportation of solid waste from assortment focuses on the removal destinations. The area of solid waste containers was resolved to utilize a palm-top mounted GPS beneficiary. Qpad, the software indigenously developed by Space Applications Center: Indian Space Research Organization, Ahmedabad, was utilized for recording GPS perceptions. The road network obtained from high-goals satellite information of Cartosat-2 was utilized to find an optimum course from the containers to landfill destinations utilizing GIS Network examination. The examination approach followed right now Geomatics and Analytical Hierarchy Process is incredibly valuable for not just distinguishing perfect landfill sites for solid waste removal but also deciding the optimum paths in doing the waste from city to the landfill site. This will help the Urban Local Body to determine the most suitable path for transporting the waste to the landfill site and

Table 4 AQI and GQI parameters baed on pair-wise comparison method using Satty's scale

S. No.	Parameter	Weight
1	Land use	0.2996
2	Ground water prospects	0.2505
3	Flood hazard	0.1364
4	Ground water quality index	0.0973
5	Air quality index	0.0719
6	Soil texture	0.0532
7	Soil depth	0.0444
8	Slope	0.0266
9	Population density	0.0186
		1.0000

Table 5 Landfill site suitability reclassification

Suitability class	Value range	Suitability
Class I	LSI $\leq$ (m − s)	Highly suitable
Class II	(m − s) < LSI $\leq$ m	Moderately suitable
Class III	m < LSI $\leq$ (m + s)	Less suitable
Class IV	LSI > (m + s)	Not suitable

Note m—mean and s—standard deviation

Table 6 Landfill site suitability areas (in sq. mtr.)

S. No.	Class	Area (sq. km)	Percentage of total available area
1	I	24.98	4.53
2	II	294.92	53.44
3	III	229.39	41.56
4	IV	2.6	0.47
Total		551.89	100.00

this study will impart the thorough knowledge of how GIS can be efficiently used to determine the route for transportation of solid waste being generated.

References

CPHEEO. (2005). *Status of water supply, sanitaion and solid waste management in urban areas.*

Kil, D. H., & Shin, F. B. (1997). Automatic road-distress identification using a combination of hierarchical classifiers and expert systems-object processing (pp. 414–417).

Leverenz, H. L., Tchobanoglous, G., & Asano, T. (2011). Direct potable reuse: A future imperative. *Journal of Water Reuse and Desalination, 1*(1), 2–10.

Pollution, C. et al.: No Title. (2012).

The Spatial Intercorrelation of Global Carbon Dioxide Emissions: An Empirical Verification of China's Impacts by Spatial Durbin Panel Data Model

Zhiguang Song

Abstract

As the "Going Out" strategy has been introduced after WTO accession and post-crisis era, besides the economic effects, outward foreign direct investment (OFDI) from China has been boomed, along with its boosted foreign trade during industrial upgrading. However, these efforts have been found to cause significant global environmental issues, which have been rarely mentioned and discussed systematically, especially from the perspective of spatial intercorrelation between environment and cross-border investment, international trade, institutional factors, and rising environmental impacts from China. On that basis, this study is conducted by the spatial econometric models in a macroeconomic panel of 146 countries and regions in 2003–2016 to empirically delve into the spatial intercorrelation and third-country effects of carbon dioxide (CO_2) emissions per capita among the observed areas, under China's impacts. The empirical results first confirm the existence of spatial autocorrelation for CO_2 emissions per capita by Moran's I statistic, suggesting that an interregional prevention system is supposed to be implemented to sufficiently reduce global CO_2 emissions. Furthermore, as suggested by the spatial Durbin model, local government should show a close economic relationship with neighboring countries to attract spatial spillover effects of China's OFDI and then develop a multinational economic partnership that enables the high level of technology and human capital to move freely, as an attempt to lower the level of CO_2 emissions in a whole range. Moreover, political and resource and energy purposes in the foreign investment and trade activities of Chinese government and multinationals should be progressively limited, and economic cooperation with more developed markets in a higher degree of environmental regulations and standards involved in a coordinated multilateral agreement should be conducted for sustainability and global emission reduction

Keywords

China • Carbon dioxide emissions • Spatial econometric models • Spatial intercorrelation

1 Introduction

Since the implementation of "Going Out" strategy after WTO accession and policy guidance on the coordinated development of inward and outward investment in the 2010s, China's OFDI has skyrocketed. The flow of China's foreign investment firstly exceeded its amount of inward foreign direct investment (IFDI) in 2015, as suggested by the "Statistical Bulletin of China's Outward Foreign Direct Investment." Moreover, newly increased overseas OFDI by Chinese domestic investors surged to $1456.7 billion in 2015 from only $28.5 billion in 2003, which was ranked second worldwide, thereby making it a true investor country (Fig. 1).

Besides, for the motivations of market and resource seeking, China's OFDI considered with economic benefit and significant social and environmental impacts, especially in countries with relatively poor governance system and environmental regulation, when remarkable amount of China's rising OFDI falls in energy and resources sectors, and most of which are located in Asia, Africa, and Latin America. For instance, as China's largest public product for the international community thus far, the "One Belt & One Road" initiative (BRI, 2013) is aiming to strengthen China's connectivity with the world (i.e., fields of commodity trade, infrastructure construction, intergovernmental communication, and cultural ties). Among the mentioned, sectors like mining and construction are thought to be likely to cause social and environmental concerns by their natural property.

Z. Song (✉)
Graduate School of Global Environmental Studies, Sophia University, Tokyo, Japan

C. Alalouch et al. (eds.), *Towards Implementation of Sustainability Concepts in Developing Countries*, Advances in Science, Technology & Innovation, https://doi.org/10.1007/978-3-030-74349-9_24

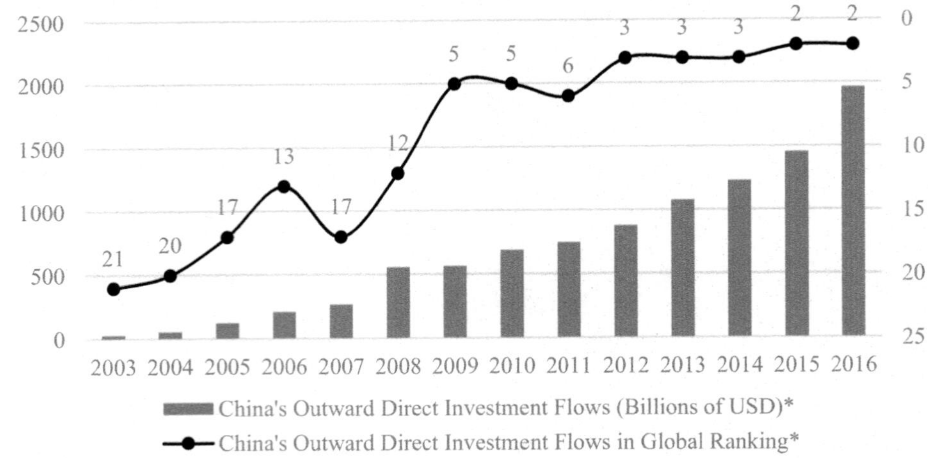

Fig. 1 China's outward capital flows and global ranking over the years. *Source* "Statistical Bulletin of China's Outward Foreign Direct Investment (2016)." *Note* China's OFDI values and its global ranking in left and right axis, respectively

Foreign trade, compared with the development of overseas investment, is considered a vital sector to China's economy for a longer period, especially since accession to the WTO in 2001. According to the "United Nations (UN) Comtrade Database," China's international trade boomed noticeably, the overall value of exports and imports increased from $851 billion in 2003 to $3686 billion in 2016, leading as the world's No.1 trading country since 2013, being the biggest bilateral trading partner for most countries (Fig. 2).

Here, as suggested by the "United Nations Conference on Trade and Development (UNCTAD)," trade competition between China and industrialized countries (i.e., USA, Japan, and Germany) tends to rise, as a result of China's ongoing industrial transformation that shifting its comparative advantage from labor-intensive to capital and technology-intensive products (i.e., communication technology, equipment manufacturing). However, China's complementarity with resource-rich economies (i.e., Russia, Australia) is increasing, suggesting that China's demand for primary products is rising under industrial upgrading. Moreover, China's competition with emerging industrializing economies (i.e., India and Indonesia) is declining, reflecting that China's industrial upgrading and specialization in higher value-added products have been advancing. Moreover, the transfer of China's low-end assembly and manufacturing to the emerging industrializing economies is expected to further improve, especially in labor-intensive industries (Kwan, 2013).

2 Literature Reviews

Growing studies on the relationships between investment facilitation and trade liberalization and environment suggest a potentially high level of interaction, as primarily presented in two aspects. The one focuses on the impacts of foreign direct investment (FDI) on environment; the other refers to the embodied emissions in international trade (Ding et al., 2017).

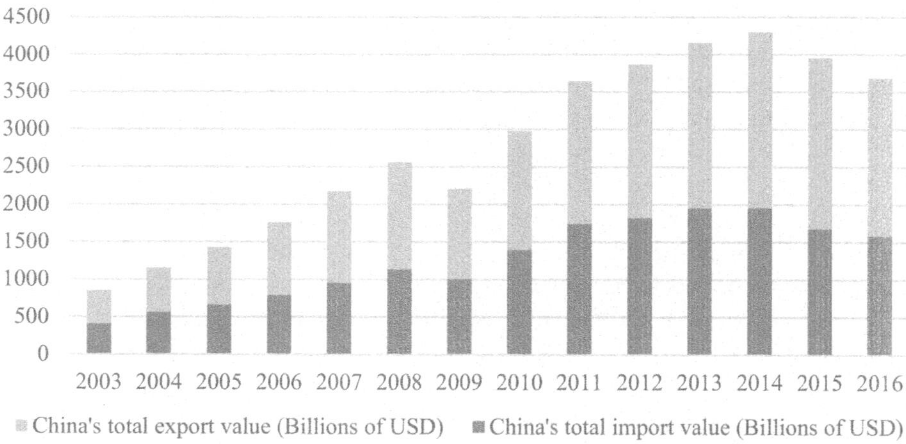

Fig. 2 China's total export and import value over years. *Source* UN Comtrade Database

2.1 Impacts of Foreign Direct Investment on Environment

Grossman and Krueger (1991) initially classified the impacting mechanism of FDI on environment into three channels, including "scale effect, composition effect, and technique effect" (p. 3), which has been referenced in numerous studies, and arising discussions on FDI and environment. Subsequently, Grimes and Kentor (2003) analyzed the panel data of 66 less developed countries, reporting that "foreign investment is more concentrated in those industries that requires more energy." Then, "transnational corporates may relocate highly polluting industries to countries under fewer environmental controls" (p. 1). However, Perkins and Neumayer (2012) delved into a set of dynamic panel data in 77 developed and developing countries, reporting that "less CO_2-efficient countries and countries with higher institutional quality experience stronger FDI-weighted CO_2-efficiency spillovers" (p. 2).

Moreover, for China's recent FDI and environment issues, Hao and Liu (2015) employed a two-equation model to delve into the relationships of FDI, foreign trade, and CO_2 emissions. As indicated from the estimation results, the "increase in per capita FDI helps to inhibit the growth of China's per capita CO_2 emissions," by dividing the "total impact of FDI on emission into the direct and indirect" (p. 3). Likewise, for regional differences in China, Zhang and Zhou (2016) studied the "impact of FDI on China's CO_2 emissions at the national and regional levels using provincial panel data from 1995 to 2010," by adopting the "Stochastic Impacts by Regression on Population, Affluence, and Technology (STIRPAT) model." It suggests that "FDI contributes to CO_2 emission reductions in China," revealing a pollution halo hypothesis. However, "FDI's impact on CO_2 emissions decreases from the western region to the eastern and central regions" (p. 1). In contrast, Lian et al. (2016) explored the "impact of environmental regulations on the location of China's manufacturing," and strong evidence of the pollution haven hypothesis was found, by empirical results showing that "provinces with easy environmental regulation efforts are more likely to attract polluting enterprises and become pollution haven" (p. 1).

2.2 Embodied Emissions in International Trade

In the meantime, since the issue of international trade and environment was exposed at the United Nations Conference on the Human Environment (1972), international attention to the relationships between trade and environment has been rising, along with emerging concern of the embodied emissions in global trade, which has raised the controversies of accounting and responsibility system in greenhouse gas (GHG) emissions, termed as the "Polluter pays principle" and "Consumer pays principle."

Among the mentioned in China's trade activities, Zhao et al. (2016a) analyzed the "driving factors of carbon emissions embodied in China–US trade over the period of 1995–2009," and it suggests that "the increment in carbon emissions embodied in US exports to China was mostly contributed by 'total demands abroad'" (p. 1). Likewise, Zhao et al. (2016b) employed the "bilateral trade input–output (BTIO) approach" to discuss the "CO_2 emissions embodied in China–Japan trade during 1995–2009." As suggested from the results, "CO_2 emissions embodied in China's exports increased by about 100% from 1995 to 2009, whereas those embodied in China's imports increased by about 500% during the same period" (p. 1). More generally, a quantitative research conducted by Liu et al. (2016) suggested that "emissions embodied in Chinese exports, which are larger than the annual emissions of Japan or Germany, are primarily the result of China's coal-based energy mix and the very high emissions intensity" (p. 1). Particularly, based on the "year-to-year (import) non-competitive input–output tables," Ning et al. (2016) revealed that "China's embodied carbon emission from domestic production was higher than that from domestic consumption by 14.8–26.1% because of the balance of embodied emissions in trade during 1997–2010." Meanwhile, "China's embodied carbon emission within export grew from 0.70Gt in 1997 to 1.85Gt in 2010, accounting for 50% of China's embodied carbon emissions in export resulted from trade among China, US, Japan, and EU" (p. 1).

2.3 Spatial Econometric Analysis

From the spatial perspective in China's foreign investment, Ma and Liu (2016) built a spatial econometric model with a "panel data of 42 host countries along 'the Belt and Road' between 2003 and 2014" to empirically test the third-country effect of China's OFDI. The study revealed that "there is significant third-country effect in China's OFDI to countries along 'the Belt and Road,'" and the "third-country effect is mainly represented as crowding-out effect." Besides, the paper also suggested that "China's OFDI is characterized by a guide and a motive, namely with resource-seeking motivation and trade-oriented features" (p. 1). Likewise, based on the spatial Durbin model, Liu (2019) used China's OFDI data to in the BRI countries (2006–2015) to systematically examine the institutional and economic impacts on China's OFDI. The findings suggest that China's foreign capital is relatively prioritized in resource-endowed countries, followed by potential market size and low labor-seeking characteristics; it is also more inclined to the host countries with lower level of political institution and higher level of economic institution. On the other hand, in terms of the relevance to the environmental pollution, Shi et al. (2017) used the "spatial vector autoregressive model (SpVAR)" and "PM2.5 data of 1998–2010 from the satellite monitoring data." As indicated from the result, the "increasing FDI significantly deteriorates the city's Haze phenomenon in the early stage, presenting pollution heaven effect" (p. 1), with various impacts on haze pollution by FDI in different regions. Lately, Wang et al. (2019) employed the "spatial econometric model and the differential generalized method of moments," to "investigate the effects of China's carbon emissions during 2004–2016." They reported that "China's carbon emissions show significant spatial correlation characteristics and interregional diffusion," demonstrating that "regional coordinated cooperative governance is key to carbon emission mitigation" (p. 1).

Overall, considering that China's expanding OFDI with the transitions in industrial and trade structure can bring about globally environmental problems. It is noteworthy that emerging topics like CO_2 emissions attributed to China's foreign investment, as well as the trade relation with China (when China as a carbon consumer by its imports in global market); how the institutional factors will contribute to China's participation in global economic cooperation (and world environment indirectly); what is critical to global emission reduction and sustainable development and others have been put into research subject, as China is gradually becoming one of the world's largest investor and largest trading nation. Furthermore, since most existing studies focused on the relationships of investment facilitation or trade liberalization and environment in a specific area, general findings have been rarely achieved on global environmental effects of both cross-border investment and international trade simultaneously other than economic benefits on the view of spatial intercorrelation. Due to a typical spatial interaction on global CO_2 emissions, based on the property of economic liberalization and environmental externality (Fig. 3), this study implies the spatial econometric models with macroeconomic panel data of 146 countries and regions in 2003–2016 to empirically study the spatial intercorrelation of CO_2 emissions per capita among the observed areas, which is affected by China's rising impacts including its OFDI, imports and other institutional factors, on the view of sustainability and emission reduction.

The plan of following sections is: (1) Spatial weight matrix based on the geographical distance will be constructed to test the Moran's I index of CO_2 emissions per capita, to investigate whether there exists spatial autocorrelation of CO_2 emissions among the 146 countries & regions. (2) Lagrange multiplier (LM) test in the non-spatial model will be conducted to judge the spatial autocorrelation of CO_2 emissions per capita again and provide information for model selection. (3) Based on the Wald and likelihood ratio (LR) test, spatial Durbin model with spatial and time fixed effects is finally used to explain the spatial intercorrelation and third-country effects of CO_2 emissions per capita among the observed areas. (4) Conclusions and implications will be listed.

3 Model Establishment and Empirical Test

According to the Tobler's first law of geography proposed by the American geographer Waldo Tobler in 1969 at the International Geographical Union (IGU): "Everything is related to everything, but near things are more related than distant things." This is to say, spatial effect complies with the principle of distance attenuation that contradicts with the assumption of "observing samples are independent of each other" in the conventional econometric analysis, and if the spatial factor is overlooked, the estimation results will be deviated from reality.

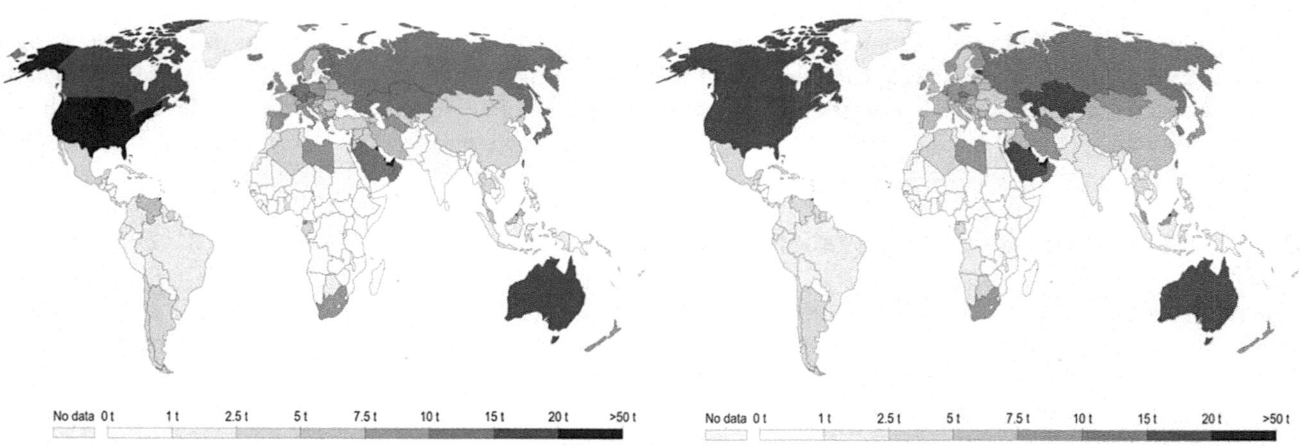

Fig. 3 Spatial distribution of CO_2 emissions per capita in 2003 (left) and 2016 (right). *Source* Our World in Data (OWID) based on CDIAC; Global Carbon Project; Gapminder & UN. *Note* CO_2 emissions per capita measured in tonnes per year

The concept of spatial econometrics as a rising subject was first proposed by Paelinck and Klaassen (1979). On that basis, Anselin (1988) systematically summarized it from estimation to application by a cross section or panel data (Anselin, 1988). Subsequently, the spatial econometric method has aroused rising attention, especially after the new economic geography represented by Krugman (1991) that applied to various studies like the economics of cities and regional disparities, which provides an integrated and micro-founded approach to spatial economics (Krugman, 1991, as cited in Venables, 2008). The spatial econometrics now can explain well in the aggregation effects, spatial spillover effects, or third-country effects of environmental pollution, as well as global economic cooperation (i.e., cross-border investment and international trade). Accordingly, spatial economic method is becoming an emerging trend in the study fields of spatial interaction of international economic activities and environmental issues.

3.1 Spatial Econometric Models

The theory of spatial econometrics suggests three basic spatial interactions to the spatial variables. (1) Endogenous spatial interaction, related to the "spatial lagged model (SLM)," which indicates that spatial relationships originate from dependent variables between different geographic units. (2) Exogenous spatial interaction, which is pointed to the "spatial Durbin model (SDM)," i.e., spatial effects derived from both dependent and explanatory variables in different regions. (3) Spatial interaction in the error terms, equal to the "spatial (autoregressive) error model (SEM)," suggesting that the spatial relationships are derived from unobservable factors (i.e., specific geographic location) in geographic units (Wang et al., 2019, p. 9).

On that basis, the specific model settings of SLM, SDM, and SEM are, respectively, expressed as Eqs. (1)–(3) below.

$$y_{ij} = \alpha + \rho \sum_{j=1}^{n} w_{ij}, y_{ij} + (\beta_1 X_{i1} + \cdots + \beta_k X_{ik}) + \varepsilon_{ij} \quad (1)$$

$$\begin{aligned} y_{ij} = \alpha + \lambda_1 \sum_{j=1}^{n} w_{ij}, y_{ij} + (\beta_1 X_{i1} + \cdots + \beta_k X_{ik}) \\ + \lambda_2 \sum_{j=1}^{n} w'_{ij} + (\beta_1 X_{i1} + \cdots + \beta_k X_{ik}) + \varepsilon_{ij} \end{aligned} \quad (2)$$

$$y_{ij} = \alpha + \beta_1 X_{i1} + \cdots + \beta_k X_{ik} + \mu_{ij} \quad \mu_{ij} = \theta \sum_{j=1}^{n} w_{ij}\mu_{ij} + \varepsilon_{ij} \quad (3)$$

where i and j denote the area and time effect separately; W represents the "spatial weight matrix reflecting the spatial relationship among the units;" y refers to the explained variable; X indicates a set of explanatory variables with coefficients β_i. Moreover, ρ and λ_1 in Eqs. (1) and (2) are the "spatial lag coefficients reflecting the space-dependent relation" between the explained variables, while θ in Eq. (3) is the "spatial error coefficient reflecting the spatial relation existing in the random disturbance item," and λ_2 in Eq. (2) is the "spatial lag coefficient" from explanatory variables. Moreover, u denotes the individual effect; and ε is a random disturbance term (Wang et al., 2019, p. 9).

3.2 Variables and Data

China's overseas investment and foreign trade, as well as institutional distance (i.e., economic development and political proximity with China) will act as core explanatory variables in this study, along with the control explanatory variables that represented by the "level of gross domestic product (GDP) per capita," "industrial output," and "individuals via the Internet of population" among the observed areas, which are, respectively, considered the scale, structure, and technique effect toward the local CO_2 emissions.

Specifically, "CO_2 Emissions Per capita" (CO_2P) of 146 countries and regions in 2003–2016 comes from the "Our World in Data," in which the calculation is based on the "Carbon Dioxide Information Analysis Centre," "Global Carbon Project," "Gapminder," and "UN population" estimates. Moreover, "China's Outward Foreign Direct Investment Stock" (OFDI) to a specific country and region, is taken from the "Statistical Bulletin of China's Outward Foreign Direct Investment," and "China's Import Value" (IV) from a specific area is referenced from the "UN Comtrade," both applied to express the third-country effects of global emissions, from the spatial effects of China's OFDI and emissions embodied in trade imported by China. Likewise, "Gap of Industrial Output" (IG) and "Gap of Worldwide Governance Indicators" (WGIB) between China and a specific country and region in absolute value are collected from the open data and world development indicators of the "World Bank," respectively, as adopted to capture how industrialization level and institutional factors will impact global emissions from the spatial perspective. On the other hand, "GDP Per capita" (GDP) in current prices and "Industrial Output (% of GDP)" (IGG) referenced from the "World Bank," as well as "Individuals Using the Internet (% of Population)" (IUI) from the "Our World in Data" in local countries and regions, respectively, are adopted to capture the scale, structure, and technique effect of CO_2 emissions per capita in this study (Table 1). Moreover, to minimize the impact of heteroscedasticity and make the relationship among variables linear, the variables will be transformed into logarithms.

Table 1 Variable definitions and descriptive statistics

Variables	Abbreviation	Unit	Description	Mean	Std. Dev.	Minimum	Maximum
Dependent variable	CO_2P	Ton	CO_2 emissions per capita	4.881	5.641	0.019	36.001
Core explanatory variables	OFDI	Billion USD	China's Outward Foreign Direct Investment Stock	2.559	28.859	0.001	780.745
	IV	Billion USD	China's import value	7.045	20.935	0.001	194.568
	IG	Trillion USD	Gap of Industrial Output (absolute value)	2.640	1.393	0.049	4.529
	WGIB	-	Gap of Worldwide Governance Indicators (absolute value)	0.811	0.677	0.001	2.533
Control explanatory variables	GDP	10 Thousand USD	GDP per capita	1.332	1.860	0.011	11.882
	IGG	%	Industrial output (% of GDP)	27.910	12.917	3.720	85.659
	IUI	%	Individuals using the Internet (% of population)	33.383	28.932	0.024	98.137

Note Std. Dev.—standard deviation

3.3 Spatial Weight Matrix

Spatial weight matrix W embodied in the spatial econometric models will directly affect estimation that measures the magnitude of interdependence between spatial units, based on different standards like binary, distance, or inverse distance and economic distance weighting. For the research object in this study (CO_2 emissions), an inverse distance weighting will be established, as expressed in Eq. (4), where d denotes geographical distance (capital city) between two spatial units i and j. In other words, lower geographical distance will comply with a higher spatial autocorrelation; higher geographical distance is associated with a lower spatial autocorrelation (Paas & Schlitte, 2006).

$$W_{ij} = \begin{cases} 1/d_{ij} & i \neq j \\ 0 & i \neq j \end{cases} \tag{4}$$

Moreover, regarding the spatial weight matrix of a panel data is NT × NT, where N denotes the number of cross-sectional samples, T represents the sample years, the non-diagonal elements of matrix are overall 0, while the elements on diagonal is the cross-sectional weight matrix of N × N in a specific year. Thus, the spatial weight matrix in this study will be standardized before the estimation, i.e., each element in the matrix will be divided by the sum of elements of row, so the sum of elements in each row is 1 after standardization.

3.4 Spatial Autocorrelation Test

Spatial econometricians have primarily proposed the Moran's I and Geary tests to demonstrate the existence of spatial autocorrelation or spatial dependence between variables at different regions, and the global Moran's I index applied in this paper is calculated as:

$$\text{Moran's I} = \frac{\sum_{i=1}^{n}\sum_{j=1}^{n}\omega_{ij}(x_i-\overline{x})(x_j-\overline{x})}{\sum_{i=1}^{n}\sum_{j=1}^{n}\omega_{ij}\sum_{i=1}^{n}(x_i-\overline{x})^2} = \frac{\sum_{i=1}^{n}\sum_{j=1}^{n}\omega_{ij}(x_i-\overline{x})(x_j-\overline{x})}{S^2\sum_{i=1}^{n}\sum_{j=1}^{n}\omega_{ij}} \tag{5}$$

where $S^2 = \frac{1}{n}\sum_{i=1}^{n}(x_i-\overline{x})^2$, $\overline{x} = \frac{1}{n}\sum_{i=1}^{n}x_i$, x_i represents observation variable of i region, S^2 is sample variance, n means total number of regions, and ω_{ij} denotes spatial weight matrix.

Here, the index ranges in (−1, 1), if Moran's I > 0, indicating a positive spatial autocorrelation in observation variables between geographic units; when Moran's I ≤ 0, it suggests a weak or random spatial autocorrelation. Namely, absolute higher Moran's I is according with a closer spatial relationship, meaning that high value of observation variable can be adjacent to a high value, while low value can be adjacent to a low value between geographic units.

The Moran's I statistics of corresponding variables in this study calculated by Stata 15 is given in Table 2, and the null

Table 2 Estimates of Moran's I statistic

	CO_2P		OFDI	IV	IG	WGIB
	Moran's I	*Z* score	Moran's I	Moran's I	Moran's I	Moran's I
2003	0.435***	7.738	0.024***	0.555***	−0.168***	0.417***
2004	0.409***	7.284	0.030***	0.593***	−0.105*	0.409***
2005	0.393***	7.018	0.022***	0.618***	−0.009	0.417***
2006	0.382***	6.844	0.018***	0.603***	0.112**	0.427***
2007	0.345***	6.197	0.014***	0.591***	0.238***	0.406***
2008	0.344***	6.178	0.013***	0.543***	0.297***	0.383***
2009	0.342***	6.156	0.009***	0.533***	0.254***	0.399***
2010	0.330***	5.956	0.008**	0.528***	0.244***	0.423***
2011	0.328***	5.917	0.006*	0.491***	0.251***	0.422***
2012	0.337***	6.070	0.004	0.466***	0.251***	0.423***
2013	0.363***	6.529	0.002	0.426***	0.254***	0.435***
2014	0.331***	5.982	−0.001	0.432***	0.254***	0.428***
2015	0.328***	5.935	0.002	0.463***	0.225***	0.418***
2016	0.353***	6.337	0.002	0.466***	0.219***	0.401***

Note ***, **, * Denote significance at the 1%, 5%, and 10% levels, respectively. Both *Z* score and *p*-value are used to evaluate the significance of the Moran's I index value, and only *Z* score of the CO_2P is listed in this table

hypothesis (There is no spatial autocorrelation) is rejected in all period for CO_2P, IV, and WGIB with *p*-value of 0.01 or lower, which means that the spatial distribution for the level of local CO_2 emissions, China's imports, and political proximity with China among the 146 countries and regions achieves a certain clustered pattern, while OFDI in 2011–2016 and IG in 2003–2005 display a dispersed pattern.

Furthermore, to clarify the heterogeneity between the 146 countries and regions at a specific year, local Moran's I scatter plot outlined by Anselin et al. (1996) is adopted to classify spatial autocorrelation into four categories, which is consisted by an observation variable on the *x*-axis with its spatially lagged variable on the *y*-axis (Anselin et al., 1996). Thus, there will be four types of spatial relationships in four quadrants, i.e., "High-High (HH), Low–High (LH), Low-Low (LL), and High-Low (HL)," respectively, in the first, second, third, and fourth quadrant, i.e., countries and regions located in the first and third quadrants which exhibits a positive spatial autocorrelation. In contrast, it exists a negative spatial autocorrelation when countries and regions are located in the second and fourth quadrants.

To be specific, local Moran's I of CO_2P of observed areas in this study are presented in Fig. 4, and most of areas are especially located in the third quadrants, which is exhibiting a positive spatial autocorrelation and "Low-Low" agglomeration phenomenon. It is suggested that most countries and regions with low level of CO_2 emissions tend to cluster with countries and regions of low emission level.

4 Empirical Results and Interpretations

As suggested from Anselin (1988), when spatially lagged explained variables are covered in the SLM and SDM as explanatory variables, the endogenous explanatory variables are generated, and then, the ordinary least squares (OLS) estimation is biased and inconsistent. Besides, the SEM does not meet the assumption that the random error term is independent and identically distributed, making the OLS estimation invalid as well. Accordingly, the spatial econometric models can no longer be estimated with the OLS method, but with the maximum likelihood estimation (Anselin, 1988). On that basis, this study will first implement the non-spatial panel model based on the OLS estimation with the LM test to prove the existence of spatial autocorrelation. Subsequently, the spatial Durbin model based on the Wald and LR tests will be implemented by the maximum likelihood estimation.

4.1 Estimations of Non-spatial Panel Data Model

In Table 3, this study first performs the Hausman test to demonstrate that the fixed effects model has a unique advantage at 1% statistical significance, when adopting the non-spatial panel model in OLS estimation by Stata 15.

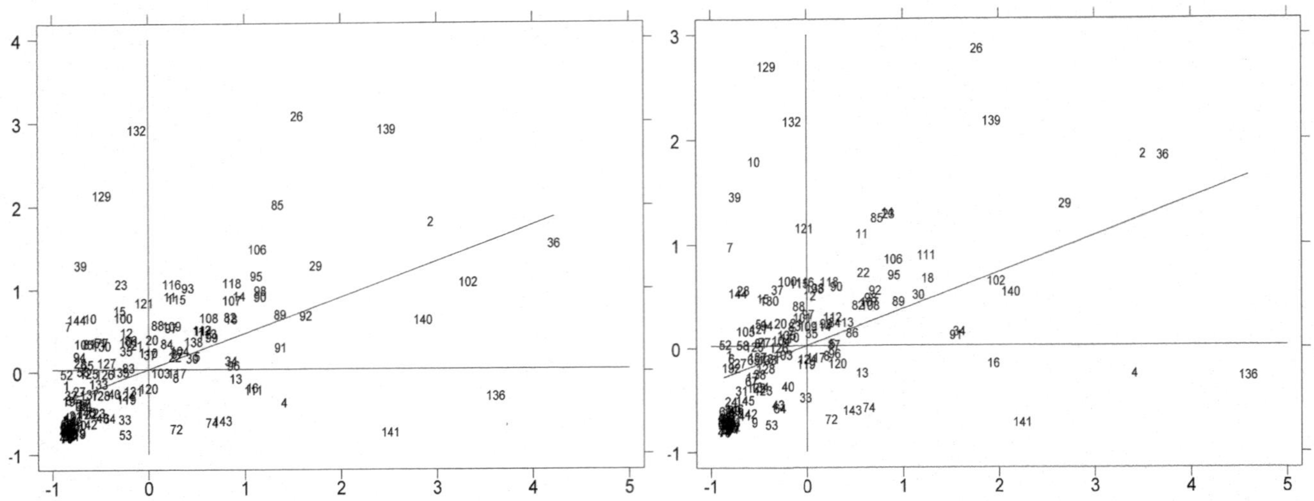

Fig. 4 Local Moran's I scatter plot of CO_2 emissions per capita in 2003 (left) and 2016 (right). *Note* The *x*-axis suggests observation variable, and the *y*-axis is its spatially lagged variable. The number means every specific 146 country and region as observing areas in this paper

Table 3 Estimations of non-spatial panel data model

Sample: 146 countries and regions (2003–2016)				
Dependent variable: $\log CO_2P$				
Observation: 2044				
	No fixed effects	Spatial fixed effects	Time fixed effects	Spatial and time fixed effects
logOFDI	−0.453	0.001	−0.003	−0.002
logIV	−0.001	0.005	−0.003	−0.002
logIG	0.004***	−0.942***	−0.972***	−0.964***
logWGIB	−0.948***	−0.629***	−0.958***	−0.737***
logGDP	−0.854***	0.813***	0.948***	0.805***
logIGG	0.955***	0.146***	0.133***	0.140***
logIUI	0.140***	0.103***	0.105***	0.108***
intercept	0.100			
R^2	0.552	0.567	0.470	0.482
Hausman test	1019.92***			
LMLAG	232.374***	215.974***	189.762***	169.073***
R-LMLAG	0.343	0.247	0.046	0.017
LMERR	413.203***	395.358***	306.088***	280.132***
R-LMERR	181.172***	179.631***	116.373***	111.076***

Note ***, **, * denote significance at the 1%, 5%, and 10% levels, respectively. R^2 = coefficient of determination

Moreover, it can be further split into "spatial (regional) fixed effects" and "time fixed effects." The former refers to the time-invariant geographical characteristics in different regions, and the latter is the time-variant factors (i.e., governmental policies and technological changes).

The LM test in standard and robust form in "No fixed effects" model by MATLAB 2017 (Table 3), proposed by Burridge (1980) and Auselin (1988), is performed to verify the significance of spatial autocorrelation and provide information for model selection between the "spatial lag model" and "spatial error model" (Anselin & Florax, 1995). The null hypothesis is, respectively, no spatial lag effects and no spatial error effects. As a result, the standard LM test for "spatial lag model (LMLAG)" and "spatial error model (LMERR)," along with the "robust LM test for spatial error model (R-LMERR)," are overall rejected significantly, while

Table 4 Estimations of spatial panel data in SDM

Sample: 146 countries and regions (2003–2016)				
Dependent variable: $\log CO_2P$				
Observation: 2044				
	No fixed effects	Spatial fixed effects	Time fixed effects	Spatial and time fixed effects
logOFDI	0.002	0.003	0.001	0.001
logIV	−0.013***	−0.011**	−0.013***	−0.012***
logIG	−0.911***	−0.899***	−0.927***	−0.916***
logWGIB	−1.223***	−0.970***	−1.209***	−0.954***
logGDP	1.100***	0.970***	1.077***	0.943***
logIGG	0.143***	0.150***	0.140***	0.146***
logIUI	0.100***	0.101***	0.103***	0.105***
W * logOFDI	−0.005	−0.005	−0.012*	−0.012**
W * logIV	0.033***	0.032***	0.030***	0.028***
W * logIG	0.107	0.090	−0.025	−0.050
W * logWGIB	1.547***	1.354***	1.544***	1.369***
W * logGDP	−0.785***	−0.740***	−0.800***	−0.778***
W * logIGG	−0.074***	−0.077***	−0.078***	−0.080***
W * logIUI	−0.050***	−0.046***	−0.031***	−0.025**
W * dep. var	0.436***	0.421***	0.366***	0.350***
intercept	0.444			
R^2	0.6465	0.674	0.651	0.680
LogL	−5424.778	−5339.813	−5399.300	−5308.582
Wald test		172.698***	133.311***	129.984***
LR test		45.697***	49.683***	49.632***

Note ***, **, * denote significance at the 1%, 5%, and 10% levels, respectively

the "robust LM test for spatial lag model (R-LMLAG)" is accepted, revealing that the spatial error model is relatively significant.

4.2 Estimations of Spatial Panel Data Model (SDM)

As suggested from meritorious results in Anselin (1988), LeSage and Pace (2009), and Lee et al. (2010), when LM test rejects the non-spatial effects hypothesis and the spatial lag or error model is appropriate, the applicability of the spatial Durbin model should be considered, which is determined by the Wald and LR tests. In fact, the SDM only limits estimation efficiency, whereas the SLM or SEM overlook significant spatial intercorrelation effect of explanatory variables (Anselin, 1988; Lee et al., 2010; LeSage & Pace, 2009). For this reason, the null hypothesis of Wald and LR tests is that the "SDM can be, respectively, simplified into the SLM or SEM." On the whole, the "spatial Durbin model with spatial and time fixed effects" is more appropriate as proved by MATLAB 2017 in this study based on the Wald and LR tests, along with the R^2 (coefficient of determination) and LogL (Log-likelihood) in Table 4.

To be more specific, spatial coefficient of CO_2P (W * dep. var.) for the estimated results is positive and statistically significant (0.350) in the SDM with spatial and time fixed effects, revealing an obvious agglomeration effect of CO_2 emissions per capita in the 146 countries and regions.

Moreover, IV, IG, and WGIB express a negative local coefficient of −0.012, −0.916, and −0.954 at 1% significance level toward the CO_2P, respectively, in the region. For instance, it could be interpreted that 1% increase in China's imports from a specific country and region will cause 0.012% decrease in local CO_2 emissions per capita. Therefore, it generally suggests that local country and region with growing export to China, greater gap of the industrial level and lower political proximity with China tend to ease the pressure of environment in the region. Moreover, GDP, IGG, and IUI represented as the scale, structure, and technique effect of economic development hold statistically positive local coefficient respectively, implying that they will all directly promote the local CO_2 emissions level.

Meantime, OFDI, GDP, IGG, and IUI exhibit a negative spatial coefficient of −0.012, −0.778, −0.080, and −0.025 with p-value of 0.05 or lower, showing a spatial crowding-out effect in CO_2 emissions toward the adjacent regions. For instance, 1% increase of China's investment in surrounding regions will cause 0.012% decreasing of local CO_2 emissions per capita. On the other hand, IV and WGIB have a significantly positive spatial coefficient of 0.028 and 1.369, respectively, which are associated with environmental deterioration toward the adjacent areas.

5 Conclusions and Implications

Overall, this study attempts to delve into the third-country effects of global CO_2 emissions and the spatial intercorrelation between investment facilitation and trade liberalization and environment, during which China's increasing impacts have been especially mentioned, by applying the spatial econometric models with macroeconomic panel data of 146 countries and regions in 2003–2016.

Before the empirical estimations, this study first conducted the Moran's I test and local Moran's I scatter plot of CO_2 emissions per capita among the 146 countries and regions, under the spatial weight matrix of inverse geographic distance standard, which ultimately suggests a strong spatial autocorrelation at overall level. Under such circumstance, any unilateral behavior can generate limited emission reduction in the region, and rather an interregional prevention system for air pollution should be implemented to effectively curb the diffusion of CO_2 emissions (i.e., interregional and multinational emission trading system). To be more specific, for estimated results of spatial Durbin model with spatial and time fixed effects:

(1) IV, IG, and WGIB show significantly negative local coefficients toward the CO_2 emissions per capita in the region, empirically demonstrating that countries and regions with increasing manufacturing exports to China, greater gap in industrial development level (less developed industrial countries), and Worldwide Governance Indicators (lower political proximity) with China tend to be directly correlated with a lower emission level. At this stage, slight significant carbon leakage embodying in global trade exports to China is found in this study. On the other hand, since China is the top global industrial producer, it is implied that big industrial countries tend to emit higher CO_2 emissions. Moreover, political proximity and institutional barriers are still crucial determinants in such process. Thus, for sustainable development and global emission reduction, Chinese government should continuously facilitate the industrial upgrading and development with low-carbon level. Moreover, China's multinationals should take the environmental issue into the consideration from the first in economic cooperation with less developed countries, and political claims with other non-economic and environmental purposes in this process should also be gradually limited.

(2) It is noteworthy that OFDI exhibits a negative spatial coefficient significantly, exerting a spatial crowding-out effect on CO_2 emissions of the adjacent regions, which can be clarified as the existence of technique effect in China's OFDI that helps reduce CO_2 emissions to the entire region. Thus, for local government, close economic relationships should be developed with neighboring countries to attract the spatial technique effects of China's OFDI, and a complete legal system should be created to introduce China's businesses and investment into low-carbon industries. In contrast, both IV and WGIB show significantly positive spatial coefficients, proving the existence of positive third-country effects in CO_2 emissions for surrounding regions. In other words, local government should emphasize the transfer of polluting industries from neighboring countries based on a well-established legal system of environmental protection and development of advanced industries. Moreover, Chinese government and multinationals should seek economic cooperation partners exhibiting higher energy efficiency and conduct economic cooperation with more developed markets in higher degree of environmental standards.

(3) Moreover, GDP, IGG, and IUI, respectively, represented as scale, structure, and technique effect to the emission level in this study, overall show positively local coefficients and negatively spatial coefficients, demonstrating that local country and region should exploit third-country effects proactively from the economic and technique development of adjacent regions. Furthermore, developing a multinational economic partnership agreement that allows green technology and human capital to move freely is empirically demonstrated to lower the level of CO_2 emissions more broadly.

Acknowledgements This paper did not receive any specific grant from funding agencies in the public, commercial, or non-profit sectors.

References

Anselin, L. (1988). Lagrange multiplier test diagnostics for spatial dependence and spatial heterogeneity. *Geographical Analysis, 20* (1), 1–17. https://doi.org/10.1111/j.1538-4632.1988.tb00159.x

Anselin, L., Bera, A. K., Florax, R., & Yoon, M. J. (1996). Simple diagnostic tests for spatial dependence. *Regional Science and Urban Economics, 26*(1), 77–104. https://doi.org/10.1016/0166-0462(95)02111-6

Anselin, L., & Florax, R. J. G. M. (1995). Small sample properties of tests for spatial dependence in regression models: Some further results. In L. Anselin & R. J. G. M. Florax (Eds.), *New directions in spatial econometrics. Advances in spatial science* (pp. 21–74) Springer. https://doi.org/10.1007/978-3-642-79877-1_2

Ding, T., Ning, T. D., & Zhang, Y. (2017). The contribution of China's outward foreign direct investment (OFDI) to the reduction of global CO_2 emissions. *Sustainability, MDPI, Open Access Journal, 9*(5), 1–15. https://doi.org/10.3390/su9050741

Grimes, P., & Kentor, J. (2003). Exporting the greenhouse: Foreign capital penetration and CO_2 emissions 1980–1996. *Journal of World-Systems Research, 9*(2), 261–275. https://doi.org/10.5195/jwsr.2003.244

Grossman, G. M., & Krueger, A. B. (1991). *Environmental impacts of a North American Free Trade Agreement*. NBER Working Papers 3914, National Bureau of Economic Research, Inc. https://doi.org/10.3386/w3914

Hao, Y., & Liu, Y. (2015). Has the development of FDI and foreign trade contributed to China's CO_2 emissions? An empirical study with provincial panel data. *Natural Hazards, 76*, 1079–1091. https://doi.org/10.1007/s11069-014-1534-4

Krugman, P. (1991). *Geography and trade*. The MIT Press.

Kwan, C. H. (2013, June 5). *Trade structure of China becoming more sophisticated: Changing complementary and competitive relationships with other countries*. Research Institute of Economy, Trade and Industry. https://www.rieti.go.jp/en/china/13060502.html

Lee, L. F., Liu, X. D., & Lin, X. (2010). Specification and estimation of social interaction models with network structures. *The Econometrics Journal, 13*(2), 145–176. https://doi.org/10.1111/j.1368-423X.2010.00310.x

LeSage, J. P., & Pace, K. R. (2009). *Introduction to spatial econometrics* (1st ed.). Chapman and Hall/CRC.

Lian, T. H., Ma, T. Y., Cao, J., & Wu, Y. (2016). The effects of environmental regulation on the industrial location of China's manufacturing. *Journal of the International Society for the Prevention and Mitigation of Natural Hazards, 80*, 1381–1403. https://doi.org/10.1007/s11069-015-2008-z

Liu, J. (2019). The spatial proximity effect of host country characteristics on China's OFDI—An empirical data analysis based on "One Belt and One Road" countries. *Economic Survey, 36*(1), 56–63.

Liu, Z., Davis, S. J., Feng, K., Klaus, H., Liang, S., Chen, B., Liu, J. R., Yan, J. Y., Guan, D. B., & Anadon, L. D. (2016). Targeted opportunities to address the climate-trade dilemma in China. *Nature Climate Change, 6*, 201–208. https://doi.org/10.1038/nclimate2800

Ma, S. Z., & Liu, M. H. (2016). Research on the third-country effects of China's outward foreign direct investment in countries along "One Belt and One Road": Based on spatial econometric method. *Journal of International Trade, 07*, 72–83. https://doi.org/10.13510/j.cnki.jit.2016.07.007-en

Ning, Y. D., Ding, T., & Li, X. (2016). Character of China embodied carbon emissions within the trade. *Smart Science, 3*(2), 108–116. https://doi.org/10.1080/23080477.2015.11665644

Paas, T., & Schlitte, F. (2006). *Regional income inequality and convergence processes in the EU-25*. HWWA Discussion Papers 355, Hamburg Institute of International Economics (HWWA). https://www.hwwi.org/typo3_upload/groups/4/Nutzer/Klimapolitik/Alle/Dissertationsprojekte/DP_355.pdf

Paelinck, J. H. P., & Klaassen, L. H. (1979). *Spatial econometrics*. Gower Press.

Perkins, R., & Neumayer, E. (2012). Do recipient country characteristics affect international spillovers of CO_2 efficiency via trade and foreign direct investment. *Climatic Change, 112*, 469–491. https://doi.org/10.1007/s10584-011-0204-8

Shi, Z. K., Shao, J., & Wang, M. C. (2017). Impacts of FDI on the haze pollution in temporal-spatial dimension: Based on the SpVAR model. *Journal of International Trade, 09*, 107–117. https://doi.org/10.13510/j.cnki.jit.2017.09.010-en

Venables, A. J. (2008). *New economic geography* (Palgrave Macmillan, Ed.), The New Palgrave Dictionary of Economics. Palgrave Macmillan, London. (Original work published 2016).

Wang, Y. F., Malik, A., Xu, L. X., & Liao, M. (2019). Carbon emission effects of the coordinated development of two-way foreign direct investment in China. *Sustainability, 11*(8), 2428. https://doi.org/10.3390/su11082428

Zhang, C. G., & Zhou, X. X. (2016). Does foreign direct investment lead to lower CO_2 emissions? Evidence from a regional analysis in China. *Renewable and Sustainable Energy Reviews, 58*, 943–951. https://doi.org/10.1016/j.rser.2015.12.226

Zhao, Y. H., Wang, S., Zhang, Z. H., Liu, Y., & Ahmad, A. (2016a). Driving factors of carbon emissions embodied in China–US trade: A structural decomposition analysis. *Journal of Cleaner Production, 131*, 678–689. https://doi.org/10.1016/j.jclepro.2016.04.114

Zhao, Y. H., Wang, S., Zhang, Z. H., & Yang, J. Q. (2016b). Input-output analysis of carbon emissions embodied in China–Japan trade. *Applied Economics, 48*(16), 1515–1529. https://doi.org/10.1080/00036846.2015.1102845

Strategic Shift from Transmissive to Transformative Higher Education for Sustainable Development

Mohamed Jama Madar

Abstract

Education for sustainable development (ESD) is a global initiative toward transforming education for sustainability. The integration of SDGs into the education is an essential approach that ensures strategic alignment of higher education (HE) with SDGs. Document reviews were employed to explore and discuss the difference between transmissive and transformative education. In this trajectory, it is expected that the concept of transformative education is likely to become more common to meet the emerging social, economic, and environmental issues. Nonetheless, practical challenges remain in Somaliland HE sector, ranging from quality of education and relevance. The roadmap toward addressing transformative education for sustainability is not included in Somaliland's education philosophy; mainly, ESD has not been presented in all policy documents. However, this paper proposed a generic framework that spotlights the integration of education and SDGs. Meanwhile, many countries are now prioritizing structural transformation in their HEIs that are tailored to the national, regional development programs, and beyond. Consistent with the Rio+ 20 outcomes, this paper analyzed the concept of the "sustainable university" and identified that it is divided into three interrelated and complementary categories: social, environmental, and economic-oriented universities in pursuit of actualizing sustainable development (SD). This paper recommends the need for significant reforms in the education system in terms of strategies, structures, and functions. The HEIs in Somaliland need to upgrade their capacities by changing the ways they are operating and shifting into a position that helps them contribute to localization and implementation of the SDGs at the grassroots levels.

Keywords

ESD • Transformative education • Transmissive education • SDGs • HEIs

1 Introduction

Following the global declaration and political commitment to the SDGs and during the Earth Summit in 1992, education for sustainable development (ESD) has become a significant concern within higher education institutions (HEIs). The scheme of ESD came into the global agenda in 1990, where HEIs leaders from 320 universities in 47 countries pledged the Talloires Declaration for Environmental Sustainability (Stewart, 2010). These declarations and other subsequent summits obligated HEIs to respond to the pivotal sustainability dimensions. The awakening calls for sustainability and the challenges of the global dynamics necessitated HEIs to shift and become more transformative.

Furthermore, HEIs were offered to adopt the Higher Education Sustainability Act and the United Nations Decade of Education for Sustainable Development (DESD) in 2005–2015 (Beynaghi et al., 2014). The UN DESD (2005–2015) provides a significant opportunity and responsibility to HEIs worldwide to address sustainability at grassroots and global levels (Corcoran & Koshy, 2010). Besides, ESD is an educational goal that aims to provide the present generations with the intellectual capacity, skills, competencies, and principles that are crucial for societal sustainability (Svanström et al., 2008).

On the other hand, the notion of sustainability is both challenging and intersecting since it combines economic, environmental, and social aspects. In this regard, it is indispensable to propose and develop comprehensive directions for HEIs in Somaliland to deal with the sustainability challenges. On top of that, developing a global framework for ESD as a policy benchmark is crucial for

M. J. Madar (✉)
Hikma Center for Research and Training, Hargeisa, Somaliland

C. Alalouch et al. (eds.), *Towards Implementation of Sustainability Concepts in Developing Countries*, Advances in Science, Technology & Innovation, https://doi.org/10.1007/978-3-030-74349-9_25

all-inclusive solutions to SDGs' current challenges (Svanström et al., 2008). Then, there is to transform universities in ways that contribute to SDGs' targets and find durable solutions to the sustainability challenges for the present and future generations to come (Sachs, 2012). Furthermore, UNESCO's declaration of DESD advised all educational sectors of society to develop and deliver the programs of ESD by integrating sustainability elements into their provisions (Wals, 2014).

The ESD initiatives enabled HEIs to position and restructure their education portfolios in the direction of more transformative programs tailored to the SDGs. In this regard, HE must respond to the social dynamics within the features of sustainability powered by the increasing demand for life-changing knowledge, skills, and attitudes (Ballantyne & Packer, 2005). However, a fruitful strategic change in HE requires transformational leadership as well as social capital among the key actors and stakeholders, including private businesses, government, and civic institutions (Holley, 2009). Since the targets of SDGs are supposed to meet with the needs of the present and the future generation, any educational program outcome shall be an indicator of the sustainability dimensions (Robert et al., 2005). Additionally, education institutions ranging from primary to university levels ought to adopt a more transformative educational design rather than ensuring only a transmissive role.

Given that, HEIs should demonstrate a transformative role in fostering social problems and finding durable solutions for sustainability challenges (Beynaghi et al., 2014). In Somaliland context, all endeavors of SDGs are tailored to the National Education Plan to some extent but are not planned and coordinated in a contextual approach within this trajectory. Nonetheless, the responsibility to lead and ensure whether SDGs are being implemented goes to government institutions in collaboration with companies, professional bodies, education sector, and research institutions (Madar et al., 2018).

The concept of SD has been gaining global attention over the past decades, proving to be the path toward an inclusive social and economic transformation (Milutinović & Nikolić, 2014). One particular factor is that the quality of human capital has proven to be an essential condition for a successful transition toward sustainable development (Milutinovic & Zivkovic, 2014). Therefore, building human capacity is not more than establishing an education system that promotes transformational change in the human life cycle through a twenty-first century curriculum framework. Finally, this review paper will spotlight HE's status in Somaliland and propose critical recommendations on the strategic turning points.

2 Education Toward Sustainable Development

When it comes to Somaliland context, the process of drafting ESD is now a significant concern as there are no such plans set for guiding HEIs toward sustainability endeavors. The current Somaliland National Development Goals and the more extensive Somali states lack the sustainability portfolios with their national priorities. The lack of ESD as a national strategy, in fact, burdens HEIs' sustainability initiatives as numerous challenges ahead. HEIs in the Somaliland context need reform in the areas related to SDGs and structure their educational programs based on the fundamental principles of ESD. A significant pace in establishing structural commitment toward HE initiatives is based on the declarations on sustainability in HE (Grindsted & Holm, 2012). The declarations of HE for SDGs are not conceptually addressed in SNDGs as well as the current Education Sector Plan (ESP) of Somaliland. Therefore, the operationalization of SDGs targets within the SNDGs remains unmeasurable, let alone to analyze the impact.

Among the problems that the Somaliland Education Program is facing are the issue of the educational goals mismatch with the direction of the country's economic development. In the case of Somaliland, education calls for the expansion of the country's supply of SDGs provisions, focusing on industrial development and STI (Armah & Baek, 2015). Despite the number of development initiatives within SNDGs, yet increasingly all universities in Somaliland are engaged in transmissive education without any accreditation and standards. While there is an indispensable need for HEIs to adopt a transformative approach, yet there are sparking issues about relevance and quality of education. As a result, many universities in Somaliland, like any other underdeveloped world, are yet lagging behind the syndicates in assisting societies to become more sustainable with economic prosperity and social well-being. For centuries, universities have been at the forefront in creating and breaking the patterns through educating the future generation entrepreneurs, policymakers, leaders, and others (Etzkowitz & Zhou, 2017). However, universities in Somaliland have remained small in both capacity and competency related to SDGs, let alone act as a transformative icon.

Many educationists and researchers strongly suggested that curriculum innovation is the essential vehicle for the implementation of ESD in universities (Niu et al., 2010). While education institutions are the drivers and factories that produce the knowledge and skills required to boost the economy, nonetheless, they are not the only entities obligated to undertake this role. Similarly, corporations, the small enterprise, and other professional bodies should also integrate SD into their social responsibilities. Private sectors similarly play a shaping role in formulating the SDG agenda, which potentially offers improved social, economic, and environmental conditions (Scheyvens et al., 2016). Since all levels of education are an integral part of social transformation, this paper, however, emphasizes the instrumental role of HEIs toward achieving SDGs. For the contemporary university generations, learners are being prepared to be job creators and not the job seekers in this regard. Likewise, many universities focus on preparing graduates to enter the labor market and emerge with skills to support the SDGs and share their knowledge and innovative ideas with the existing ecosystems. Increasingly, universities and other professional bodies have been integrating SD values and practices into their core activities, including their institutional management and operational systems (Johnston, 2007). However, the debate thus far has focused primarily on the rationale and reasoning for why Somaliland needs broad SD adoption? However, the international discussions have failed to specify the various frameworks that HEIs can implement for them to respond to the SD dimensions. Nevertheless, this paper emphasizes integrating HESD into strategic decisions at the national, regional, and global levels, while Somaliland's current education system does not equip graduates with the desired skills and knowledge required in the marketplace.

Also, the aim of ESD is a universal aspiration, where everyone has the opportunity to benefit from quality education and learn the values, behaviors, and social skills required for a sustainable future and a positive societal transformation (Hansmann, 2010). In contrast, the Somaliland education system does not follow the twenty-first century education framework, which is tailored to the SDGs and its inventory of skills. Finally, if the education system is to create a difference in social, economic, and environmental aspects, Somaliland's current education might not be able to fit the marketplace demands, let alone be transformative and contribute to the national goals.

2.1 Transforming HEIs for Sustainable Development

Concerning the Ministerial Round Table communique 2017, integrating education into SD is paramount, and it is a global trend that shows the turning points of HEIs (Barth et al., 2015). The transformative role of HEIs in all development sectors is all about knowledge creation, transfer, translation, production, and utilization to create innovative solutions for sustainability challenges. The significant role of HEIs in making development more productive and sustainable is now the pressing agenda for all nations to adopt and implement the SDGs ratified by the world of nations in September 2015. In all parts of the world, education should respond to the needs of the present generation, and particularly among the youth who will take charge of the future endeavors (Johnson & Miller, 2000). In this regard, HEIs are required to take due account of the triple constraints of sustainability (environmental, economic, and social). Since all levels of education are significant in social, economic, and environmental, yet there is a need to transform the education system to serve the humanities better. Thus, ESD includes education for poverty reduction, human rights, gender equality, cultural diversity, universal collaboration and partnership, peacebuilding, and many more. However, these targets are the pressing concerns that more developing countries and other underdeveloped countries like Somaliland want to address the challenges posted on the current SDGs.

UNESCO states that the mission of HEIs is to promote human rights, good governance, peace, security, equality, and justice (Orme & Dooris, 2010). These pillars are the key factors that facilitate to achieve the targets of SDGs. Through their crucial and prominent role in society, higher education institutions play an instrumental role in achieving a more sustainable future (Cortese, 2003). Transformative leadership in HE also plays a crucial role in elevating SDGs' importance by incorporating the university's strategy into the 17 goals. While the underlying philosophy of education is bond to the transformation of human behavior, nonetheless education system should meet the social goals to sustainability dimensions, both local and global perspectives.

Furthermore, the current vision of HEIs in Somaliland ought to focus on quality, transformational change, and create values required for creating a sustainable society. For HEIs to undertake a positive societal transformation, the collaboration of other stakeholders is indispensable, including government support, R&D investment of the private sector, and other offers of the professional bodies. HEIs should act as a bridge between the societal desires, the marketplace demand guided by the governmental policies, and strategies.

This paper begins by highlighting a synopsis of transformational change with an effort to capitalize on how such efforts can enable multidisciplinary initiatives in education for sustainability. In this regard, Somaliland's national development goals should focus on the basic needs through domestic education financing and elevating standards of education at all levels. At present, almost all higher learning institutions adopted the slogan of Transforming Higher Education for Sustainability, which is the response to the UN initiative of ESD. Therefore, interlinking the trajectory of education and national goals with a proper strategic framework would ensure the realization of the national development goals. The Nexus between development goals and the education system together should deal with the dynamics of both the ecological, socio-economical, and humanity, which may include spiritual development (Huppè et al., 2013).

Moreover, ESD is clearly stated in target 4.7 of SDG4, which aims to make sure that all learners obtain the knowledge and twenty-first century skills needed to uphold SD and is understood as a significant approach to achieve all the other 16 SDGs (Rieckmann, 2017). As such, HEIs act as the central axis and an active multiplier for achieving the targets of SDGs. The new development plan of Somaliland is based on the global targets within the SDGs, but no attempts were made to clarify HESD and its implications for achieving the target set. Therefore, the following framework highlights the role of HEIs would play toward contributing to the SDGs (Fig. 1).

The following framework links HEIs and the triple helix of SDGs, which presents several disciplines, such as humanity related programs, climate change, environmental education, health sciences, and STI programs, including entrepreneurship, just to mention a few of them. Given the breadth and the interconnectedness of the SD agenda, the SDGs cannot be pursued in isolation. Therefore, contributions of HEIs to achieving SDGs are explicit and paramount. Though, the proposed framework has not been empirically analyzed, it builds upon the prior conceptual frameworks for ESD and could be used as a generic nexus between HEIs and SDGs. Furthermore, the framework demonstrates the bigger picture of national education's interconnectivity and achieving the targets within the SDGs based on national priorities.

There are many frameworks and indices already developed, while there is limited literature on the Somaliland context. For Somaliland to achieve essential targets of SDGs, it requires transformative HE leadership, alongside with private sector, government, and civil society, as the private sector holds much of the advanced technologies and management systems, which would be essential for the success of the SDGs.

As a result of the above discussions, comprehensive views are obligating that HEIs play an irreplaceable and crucial role in ensuring SDGs. The following framework draws the bigger picture of interlinkages between HEIs and SDGs.

The aforementioned integrated framework sets out common SDGs and HEIs roles, including intellectual capital development and leveraging R&D and STI provision into the various aspects of sustainability. HEIs in Somaliland have not yet responded to and initiated relevant policies and strategies to transform universities for sustainability. Although a shift from one stage to another stage should meet preconditions in this regard, absorptive capacities of public and private institutions are not yet in place. Therefore, any plans set, including the current Somaliland development goals, will remain a document on the shelf, and the progress will not also be measurable as there have not been relevant indicators in place. For HEIs to be more transformative rather than transmissive, the framework proposes the following common goals:

1. Localizing and institutionalizing SD principles across HEIs' programs;
2. Introduce a common ESD strategy charting what Somaliland is challenging to achieve and the guiding principles needed to achieve it;
3. Identifying the national education goals of Somaliland and develop a philosophy of education for the SDGs;

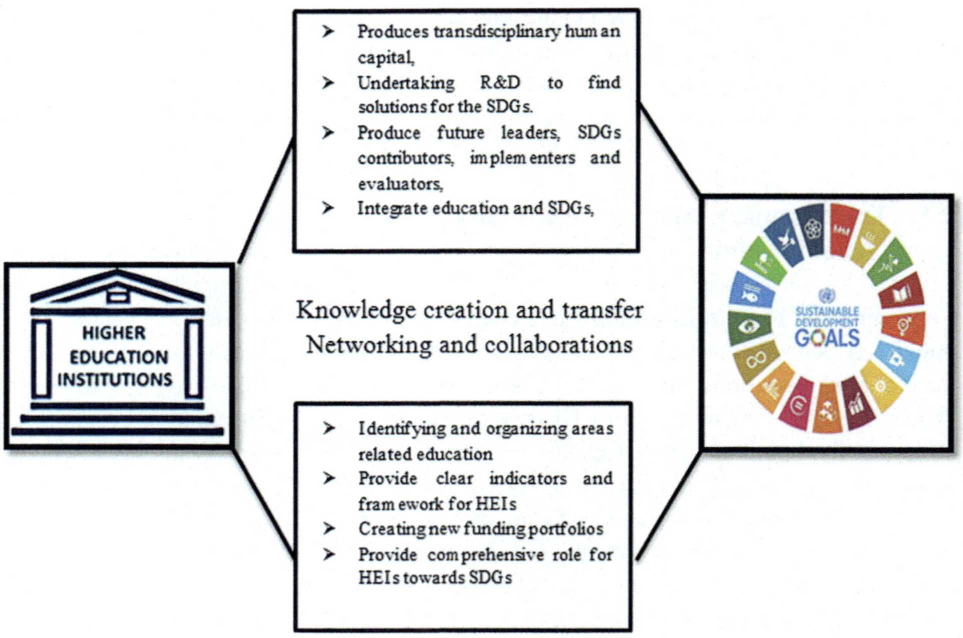

Fig. 1 HESD framework (Madar, 2020)

4. Aligning HEIs with sustainability dimensions and setting both relevant policies and plans for transformational change;
5. Designing suitable instruments/indicators to measure the progress and evaluate the impact.

Furthermore, the framework presented above is the nexus between HEIs' provisions and SDGs. The framework strongly suggests establishing a synergy between HEIs and SD, which has been proved by some empirical evidence in the prior scholarships. As a result of this, the Somaliland education system should rigorously structure their portfolios, so that sustainability issues will be incorporated into their curriculum. Positioning HEIs to take a leading role in the implementation of SD entails designing their leadership structures by adopting the best practices for managing complex organizations.

Similarly, the structural and functional transformation of HE would establish sustainable resources for research and development. The proposed framework also came up with the following concepts; there are undoubted, among others:

1. Shifting from merely poorly structured transmissive to transformative education equips learners with the capacity to contribute to the SDGs;
2. Adapting a vigorous strategy of formulating multidisciplinary provisions; teaching, research, community initiatives, technopreneurship among others;
3. Undertaking community problem-solving activities through research, forums, and public discussions and collaborating with professional bodies to find durable solutions for the sustainability challenges;
4. Establishing institutional ecosystem that can tap into varied expertise around the campus to efficiently and meaningfully share resources;
5. Position transformative leadership in HEIs promotes transformational change and value creation accompanied by accountability and public acceptance, which aim to a long-term transformation of the HEIs and are willing to be responsive to social dynamics.

The UN Declaration on ESD entails that HEIs respond to the sustainability challenges and develop critical competencies in STI and R&D as their core capacities (Reinfried, 2009). For this purpose, a paradigm shift and HEIs turning points are necessary, moving away from merely transmissive to a transformative culture based on the SDGs.

Furthermore, cascading skills for development as a new learning culture is one of the fundamental principles of transforming HESD. Moreover, ESD would enable HE to provide grandaunts with the desired skills to contributing to the SDGs and developing national intellectual capacity as a sustainable asset. In Somaliland context, the need for quality of primary education is paramount, and re-establishing the existing universities toward sustainability is now a national issue. On top of this, good governance, envisioning, and transformative leadership within HEIs are the primary concern in Somaliland context. In this case, the goal is to alleviating education standards, capacitating HEIs for inclusive and holistic development, enabling universities to be able to cope with complex social, economic, and environmental challenges. HESD is a process which considers the requisites of transformative education, including the following strategic points:

1. **Societal orientation**: learning for SD is always also societal learning. Learning takes place in real-life situations that question and changes societal living.
2. **Economic orientation**: the focus of education is the change in skills by obtaining relevant key competencies; this requires a transactional framework that constitutes inventories of competencies.
3. **Environmental orientation**: learning through the context and relating education to environmental issues should be taken into consideration for developing ecological competencies.

For education to be transformative, the Somaliland education system needs to be re-engineered and set goals and strategies that help HEIs become more responsive to social, economic, and environmental issues. Though improving the holistic quality of education at the transmissive stage would be a significant concern, there are potential challenges for transforming the current education system. This review paper is spotlighting the need for education shifting, the current trend of universal education, and how it will go in the future. The three pillars of sustainability, as indicated above, are the most challenging triple constrains that HEIs are expected to contribute through aligning their curriculum with sustainability dimensions.

3 Discussions

In this section, the policy document reviews and prior research works answered the importance of transforming HESD and assured many gaps within ESD, particularly in the universities. ESD is both a multi-sectoral and cross-cutting agenda, and it positions education on the central axis for sustainability aspects. Through their pivotal and significant role in society, HEIs are vital stakeholders in contributing to the targets of SDGs (Leal Filho et al., 2019). As respected thought, HEIs leaders can elevate the importance of SD through scholarly and public discourse.

As knowledge and skills providers, universities should cement a framework for the community in a way that organizes the other stakeholders' efforts and guides them toward contributing to the sustainable future. Conversely, in Somaliland context, universities lack a common platform within the national strategic plan that integrates HEIs' capacities into the development targets (Madar et al., 2016). Also, universities can provide their communities with graduates who have the knowledge and skills necessary to help transform their workplaces and live as responsible global citizens (Berner et al., 2013).

UNESCO (2006) pinpoints four plunges of (ESD):

1. Improving access to quality primary education,
2. Reorienting existing education to address sustainability,
3. Developing public understanding and awareness,
4. Providing training programs for all sectors of private, governmental, and civil society.

These four provisions are becoming fundamental core functions of HEIs, in Somaliland context, education remained to be only transmissive, though the quality and relevance of education programs are questionable. Similarly, Graz's declaration in 2005 concluded and emphasized that university should confront the challenges and opportunities of sustainability through; (1) teaching and learning, (2) research and development, and (3) internal and external social responsibilities.

Though HE creates social, economic, and environmental impact, and again it has the highest multiplier impact in this regard. Table 1 highlights HEIs related SDGs.

The table outlines the interdependence between education and SDGs. The key features of SDG4-Education 2030 and the global commitments expressed in the SDG4 targets, as articulated in the Incheon Declaration, did not explain the multiplier effects of education over the other 17 goals. However, the table could also be used as a guide that helps researchers and policymakers examine the implications of translating these SDGs within, and through, national education development efforts. Nevertheless, the current education system in the Somali context remains poorly transmissive and does not follow the global standards; in this regard, the ESD. For HE to be genuinely transformative, higher education needs to transform itself before transforming other trajectories.

Besides, to achieve the targets of SDGs in the Somali, context is challenging due to the lack of good governance and transformative strategies for HEIs. Targets of SDGs have three pairs of great importance as a prerequisite: (i) good governance and inclusive politics led by transformative and envisioning leadership; (ii) peace and security, and (iii) structural and cultural stability (Berner et al., 2013).

There are encouraging signs that universities around the world are responding to these opportunities and beginning to engage in activities related to SDGs, whether through campus "greening," development of specialized courses on sustainability, or offering collaborative research opportunities. Nevertheless, what is required in Somaliland HE system is to restructure and to reorient their faculties through integrating programs into the SDGs at the national level. Considerable scholarships on ESD highlighted approaches to integrating SDGs into teaching and learning (Felgendreher & Löfgren, 2018).

This paper is arguing that transformational change within universities, particularly in structural and operational organizations, could be a strategic shifting from transmissive to transformative education. Rather than looking at individual universities in Somaliland, this paper's focus is the holistic perspective on the transformation that would enable HEIs to change their direction and more rapidly shift path. Since the 1970s, world universities went through significant evolutions in their roles, inputs, outputs, and their leadership. For instance, first-generation university and the current fifth-generation universities are very different in their provisions, capacities, and outputs and, of course, their leadership approach. What has been marked by universities as first and future generation universities are the extents to which a university produces, translates, and utilizes knowledge to address social, economic, and environmental dimensions of sustainability (Komiyama & Takeuchi, 2006)?

As a result of this, Somaliland universities are far behind to benchmark themselves to the 1970s first-generation universities while the current and the future universities are more oriented in transformation and sustainability. Therefore, in reaction to the community's increasing concerns about sustainability challenges, universities in Somaliland should move toward a sustainable future by offering programs incorporated with SDGs.

4 Conclusions

This paper aims to explore the nexus between HEIs and SDGs through both transmissive and transformative education perspectives. The trend of HE went through different stages, from teaching to a transformational approach and from first-generation universities to emerging universities. The prior research works have shown great interest in ESD, and it has been spotlighted on the interconnectivity of HE and SDGs. Other scholarship works made their conclusions connected to the UNESCO proposition of ESD and the role of universities in achieving SDGs. Bearing in mind the interactive effects between the trends of SD, HE, and SD–HE, especially under the influence of Rio+ 20, shows that

Table 1 Education for sustainable development goals (UNESCO, 2014, 2015)

Relevant SDGs	
Goal 01	Global monitoring report in 2015 concluded that higher levels of education are associated with lower poverty rates. The report calculated a strong correlation between years of schooling and poverty (UNICEF 2015). Likewise, in the Somali context, education enrollment is very low, while poverty rates are very high
Goal 02	Children whose parents are educated are less likely to experience severe poverty than those with no education (UNICEF 2014). Agricultural education increases productivity and sustainability capacities, which in term increases food security. There is a need in Somaliland to attain advanced technological skills to respond to the food demands and tackle down impacts climate changes on food production (UN DESA citing FAO 2014)
Goal 03	Child mortality rates depend on the rate of mothers' education. A schooled mother is more likely to have a skilled birth attendant than a mother with no education. The probability of a child receiving vaccinations would increase with a parent's income and education level (UNESCO, 2014). In Somaliland's situation, reports show that the sincere majority of children with incomplete vaccination have uneducated parents
Goal 04	Available literature all showed that quality of education helps the attainment of career development and improves equity and general health for all. Moreover, high-quality education increases inclusive access to wealth and higher education Despite the universal gains in education enrolment, many children do not access primary education, while many do not acquire basic literacy and numeracy skills Despite considerable progress in primary school enrolment between 2000 and 2014, 9% of primary-school-aged children worldwide were still out of school in 2014
Goal 05	Global trends indicate that girls' higher level of education was associated with delayed marriage, fertility, and childbirth. The level of political participation and decision making is associated with a level of education for the girls. A high level of education for girls was associated with a higher income and employability. It seems that the education level influences gender equality to political participation and decision making, yet this assumption needs evidence to support it
Goals 06	Documents reviewed indicated that educated farmers are more productive and more likely to adopt technology for cultivation, irrigation, and harvesting (UNESCO, 2014). In Somaliland context, farmers rely on traditional farming methods since they lack primary agricultural education, and on top of that, there are no technology transfer mechanisms in place to foster productivity. Nevertheless, poverty and hunger are increasing though Somaliland receives plentiful rains and possesses fertile land that suits different cereals
Goal 07	Research showed that countries with a higher level of education are likely to practice energy-saving mechanisms. Likewise, countries with higher education levels are more likely to employ water-saving habits (UNESCO, 2014). Many countries like Somaliland, water cultivation practices are absent, while areas that practice water conservation are not significant in terms of volume and durability. TVET and other professional institutions are absent in Somaliland that would produce water engineers, who might assist and train farmers on sustainable irrigation systems
Goal 08	Researches showed that there is a strong association between education economic growth. The HDE and GDP are the indicators explained the impact of education on employment, macroeconomic growth. Many reports and empirical researches presented that the increase of schooling years is associated with an increase of GDP (UNICEF 2015 Investment Case for Education)
Goal 09	STI are the current drivers of SDGs as influencing infrastructure development, industrialization, and diversification of economic growth. Specifically, STI is now more valuable than general managerial capacities, while high-tech firms indispensably require primary, intermediate, and high-level technical skills. The twenty-first century has marked the age of information and technology, which enabled many countries inclusive economic growth and service provisions (UN DESA)
Goal 10	Investing in education reduces inequality between individuals. Countries like France, Malaysia, and Brazil, investing education, reduced 10% over 20 years and contributed to the growth of the population qualified to tertiary education (UNICEF 2015 Investment Case for Education). Somaliland yet remained to be very poor in quality of education, and it has been underfunded
Goal 11	Research findings showed that there is an association between tertiary education graduates and their living status. A higher level of education is associated with the possession of advanced life skills, including disaster prevention and preparedness plans and other safety measures (UNESCO, 2014)
Goal 12	Empirical research of farmers in 10 African countries presented that an increase in years of education influence adaption practices to climate change. Government institutions, including the NGOs and other UN agencies, are now more concerned with capacities and other interventions to find solutions to the current sustainability challenges (UNESCO, 2014)
Goal 13	In 2006, 30 OECD countries' assessments showed that an increase in environmental awareness index was associated with education level. Other research on environmental awareness concluded that his level of education was associated with an increase in environmental concerns (UNESCO, 2014)
Goal 14	The prior review of the literature reviewed presented neither empirical evidence for this goal nor a direct relationship with levels of education in this regard. According to the current trend about ESD is showing that learning has multiplier effects over all other sectors in the SDGs. Nonetheless, the indication is that higher levels of education might be associated with a high ecological balance level. As a result, to apprehend this link better necessitates additional research endeavors

(continued)

Table 1 (continued)

Relevant SDGs	
Goal 15	According to the International Social Survey Program (ISSP) in 2010, a higher level of education response was more likely to sign a petition related to environmental issues (UNESCO, 2014)
Goal 16	It has been approved through empirical analysis. The public opinion indicated a relationship between education and political participation. In Somaliland context, due to levels of income and education, political participation seems to be very low, but their other factors influencing the levels of voting, but this needs evidence to support (UNESCO, 2014) Similarly, war and conflict studies show that over 100 low- and middle-income countries have higher rates of education inequalities are more likely to experience conflicts and wars in the following years. For instance, conflicts and political disputes empowered by instability in Somaliland polity compromised the development and state-building endeavors Other studies showed that juvenile crimes are strongly associated with unemployment, poverty, and a decrease in family income education access. Fragile states like Somaliland similar situations might cause armed violence to occur frequently and will continue to happen for generations to come
Goal 17	Universities are not stand-alone, such they are part of the global institutional ecosystem that together co-create values and produce knowledge. In contrast, Somaliland HEIs do not have collaborative efforts in working together for SDGs initiatives and other provisions. The documents reviewed provided no empirical evidence for this goal, though it can be concluded that higher education levels might promote inter-institutional collaboration foster partnerships for SDGs. More research is needed to build an evidence base for this goal

HEIs are taking a new role under the influence of SDGs. According to prior studies and policy documents reviewed, indispensably placed ESD as the soul of the SDGs and has been recognized as a crucial pillar toward achieving the vision of 2030. In this regard, the notion of ESD considers HEIs as an integral part of the strategic shifting from transmissive to transformative education. Education system demands are collective efforts and cooperation among the universities and the other stakeholders, including the government and private business sectors, while civil society organizations would play an integral role if they are appropriately engaged.

The global agenda and the synergy that has been inspired within ESD brought about the current transformational change of many universities across the developed and the developing world, though many challenges are yet ahead. However, in some countries like Somaliland/Somalia, the concept of higher education for sustainable development is not on the national agenda. Although, the government focuses on primary education, which is always an essential element in HDI, yet it needs transformational change to cope with the SDGs' dimensions.

The paper illustrated how to advance the interdisciplinary approach in ESD, in alignment with Somaliland National Development Goals. Thus, education is to be considered a critical agent of development, either as a way of developing human capacity, increasing the skilled workforce for modernization, or personal freedom, developing capability, and empowerment. About learning content such as curricula, the aspects of sustainability and other relevant thematic areas are the present issues that need be addressed and incorporated into Somaliland education system.

Education must, therefore, address critical issues such as climate change, poverty, and sustainable production. ESD promotes the integration of these critical sustainability issues in local and global contexts into the curriculum to prepare learners to understand and respond to the changing world. ESD aims to produce learning outcomes that include core competencies such as critical and systemic thinking, collaborative decision making, and taking responsibility for the present and future generations.

Thus, to address such various and evolving concerns in ESD, the Somali education system should adopt state-of-the-art pedagogy, inspiring teaching and learning in an interactive, student-centered approach that empowers exploratory, outcome-oriented, and transformative learning. Learners should be enabled to think critically, creatively, and systematically develop values and attitudes for a sustainable future. The present traditional single-directional delivery of knowledge without any standards is benchmark which is no longer sufficient to inspire learners to take action as responsible citizens. Building the capacity of academia and policy support at the Somali education system will help drive these changes in transforming universities for sustainability. Empowered youth and local communities interacting with education institutions become key actors in advancing SDGs.

In conclusion, this review paper underscores four clusters of ESD as the basic principles and sets of universal goals including; (i) improving access to quality primary education; (ii) reorienting existing education to address sustainability aspects; (iii) developing public understanding and awareness; and (iv) providing training programs for all sectors of private, governmental, and civil society.

5 Policy Recommendations

1. Establish multidisciplinary programs for research and education;
2. Domestically financing education and research to enhance the roles of higher education institutions;
3. Upgrading and training university staff to collective leadership for transforming higher education for sustainability;
4. Inspiring academic staff for transformation assuring academic freedom to the potential staff at higher education institutions;
5. Establishing SDGs oriented education policies at the regional and global levels;
6. Developing indicators that address and guide the links between higher education institutions and the other sectors of the economy;
7. Integrating SDGs and entrepreneurship into education programs to ensure the needs of the current generation;
8. Adapting new ways teaching and learning strategies ensure the acquisition of twenty-first century skills among the graduates.

6 Future Research Work

Future research directions include:

1. Design and develop a theoretical framework that establishes the multiplier impact education on the global agenda;
2. Outline a framework for transforming higher education into a more SDGs oriented approach;
3. Conceptualize future leadership for higher education and its trajectory in the SDGs;
4. Propose innovative strategies for aligning the curriculum with the future trend of sustainability dimensions.

Acknowledgements First of all, I would like to thank the IsDB for sponsoring my research career building in the academic network. I would also like to thank my peers, who always appreciate my publications in the domain of Higher Education for Sustainable Development, Dr. Idiris Aden Ali, Muse Mohamed Dirie, and Mohamed Sulub.

References

Armah, B., & Baek, S.-J. (2015). Can the SDGs promote structural transformation in Africa? An empirical analysis. *Development, 58*, 473–491.

Ballantyne, R., & Packer, J. (2005). Promoting environmentally sustainable attitudes and behaviour through free-choice learning experiences: What is the state of the game? *Environmental Education Research, 11*, 281–295.

Barth, M., Michelsen, G., Rieckmann, M. & Thomas, I. (2015). *Routledge handbook of higher education for sustainable development*. Routledge.

Berner, A., Lobo, S., & Silva, N. (2013). A strategic and transformative approach to education for sustainable development.

Beynaghi, A., Moztarzadeh, F., Maknoon, R., Waas, T., Mozafari, M., Hugé, J., & Leal Filho, W. (2014). Towards an orientation of higher education in the post Rio+ 20 process: How is the game changing? *Futures, 63*, 49–67.

Corcoran, P. B., & Koshy, K. C. (2010). The Pacific way: Sustainability in higher education in the South Pacific Island nations. *International Journal of Sustainability in Higher Education*.

Cortese, A. D. (2003). The critical role of higher education in creating a sustainable future. *Planning for Higher Education, 31*, 15–22.

Etzkowitz, H., & Zhou, C. (2017). *The triple helix: University–industry–government innovation and entrepreneurship*. Routledge.

Felgendreher, S., & Löfgren, Å. (2018). Higher education for sustainability: Can education affect moral perceptions? *Environmental Education Research, 24*, 479–491.

Grindsted, T., & Holm, T. (2012). Thematic development of declarations on sustainability in higher education. *Environmental economics, 3*.

Hansmann, R. (2010). "Sustainability learning": An introduction to the concept and its motivational aspects. *Sustainability, 2*, 2873–2897.

Holley, K. A. (2009). Interdisciplinary strategies as transformative change in higher education. *Innovative Higher Education, 34*, 331.

Huppè, G., Creech, H., & Buckler, C. (2013). *Education for sustainable development at Manitoba Colleges and Universities*. International Institute for Sustainable Development (IISD).

Johnson, D. B., & Miller, M. T. (2000). Redesigning traditional programs to meet the needs of generation Y. *Journal of College Orientation, Transition, and Retention, 7*.

Johnston, A. (2007). Higher education for sustainable development. *Final Report of International Action Research Project, the Organization for Economic Cooperation and Development (OECD)*.

Komiyama, H., & Takeuchi, K. (2006). *Sustainability science: Building a new discipline*. Springer.

Leal Filho, W., Vargas, V. R., Salvia, A. L., Brandli, L. L., Pallant, E., Klavins, M., Ray, S., Moggi, S., Maruna, M., & Conticelli, E. (2019). The role of higher education institutions in sustainability initiatives at the local level. *Journal of Cleaner Production, 233*, 1004–1015.

Madar, M. J., Subari, M. D., & Baqutayan, S. M. S. (2018). Incorporating science and technology in higher education into strategic planning for sustainable development in Somaliland. *Journal of Science, Technology and Innovation Policy, 3*.

Madar, M. J., Subari, M. D., & Mohamed, S. (2016). A look at Somaliland's science and technology capacity in higher education.

Milutinović, S., & Nikolić, V. (2014). Rethinking higher education for sustainable development in Serbia: An assessment of Copernicus charter principles in current higher education practices. *Journal of Cleaner Production, 62*, 107–113.

Milutinovic, S., & Zivkovic, S. (2014). Planning local sustainable development in Western Balkans. *Management of Environmental Quality: An International Journal, 25*, 19–29.

Niu, D., Jiang, D., & Li, F. (2010). Higher education for sustainable development in China. *International Journal of Sustainability in Higher Education*.

Orme, J., & Dooris, M. (2010). Integrating health and sustainability: The higher education sector as a timely catalyst. *Health Education Research, 25*, 425–437.

Reinfried, S. (2009). Education for sustainable development and the Lucerne Declaration. *International Research in Geographical and Environmental Education, 18*, 229–232.

Rieckmann, M. (2017). *Education for sustainable development goals: Learning objectives*. UNESCO Publishing.

Robert, K. W., Parris, T. M., & Leiserowitz, A. A. (2005). What is sustainable development? Goals, indicators, values, and practice. *Environment: Science and Policy for Sustainable Development, 47*, 8–21.

Sachs, J. D. (2012). From millennium development goals to sustainable development goals. *The Lancet, 379*, 2206–2211.

Scheyvens, R., Banks, G., & Hughes, E. (2016). The private sector and the SDGs: The need to move beyond 'business as usual.' *Sustainable Development, 24*, 371–382.

Stewart, M. (2010). Transforming higher education: A practical plan for integrating sustainability education into the student experience. *Journal of Sustainability Education, 1*, 195–203.

Svanström, M., Lozano-García, F. J., & Rowe, D. (2008). Learning outcomes for sustainable development in higher education. *International Journal of Sustainability in Higher Education.*

Wals, A. E. (2014). Sustainability in higher education in the context of the UN DESD: A review of learning and institutionalization processes. *Journal of Cleaner Production, 62*, 8–15.

Printed by Printforce, the Netherlands